Handbook of Geospatial Artificial Intelligence

This comprehensive handbook covers Geospatial Artificial Intelligence (GeoAI), which is the integration of geospatial studies and AI machine (deep) learning and knowledge graph technologies. It explains key fundamental concepts, methods, models, and technologies of GeoAI and discusses the recent advances, research tools, and applications that range from environmental observation and social sensing to natural disaster responses. As the first single volume on this fast-emerging domain, *Handbook of Geospatial Artificial Intelligence* is an excellent resource for educators, students, researchers, and practitioners utilizing GeoAI in fields such as information science, environment and natural resources, geosciences, and geography.

Features

- Provides systematic introductions and discussions of GeoAI theory, methods, technologies, applications, and future perspectives
- Covers a wide range of GeoAI applications and case studies in practice
- Offers supplementary materials such as data, programming code, tools, and case studies
- Discusses the recent developments of GeoAI methods and tools
- Includes contributions written by top experts in cutting-edge GeoAI topics

This book is intended for upper-level undergraduate and graduate students from different disciplines and those taking GIS courses in geography or computer sciences as well as software engineers, geospatial industry engineers, GIS professionals in non-governmental organizations, and federal/state agencies who use GIS and want to learn more about GeoAI advances and applications.

Handbook of Geospatial Artificial Intelligence

Edited by
Song Gao, Yingjie Hu, and Wenwen Li

CRC Press
Taylor & Francis Group
Boca Raton London New York

CRC Press is an imprint of the
Taylor & Francis Group, an **informa** business

Designed cover image: © iStock Photos

First edition published 2024
by CRC Press
2385 NW Executive Center Drive, Suite 320, Boca Raton FL 33431

and by CRC Press
4 Park Square, Milton Park, Abingdon, Oxon, OX14 4RN

CRC Press is an imprint of Taylor & Francis Group, LLC

Library of Congress Cataloging-in-Publication Data

Names: Gao, Song (Geography teacher) editor. | Hu, Yingjie, editor.
Title: Handbook of geospatial artificial intelligence / Edited by Song Gao,
Yingjie Hu, and Wenwen Li.
Other titles: Geospatial artificial intelligence
Description: Boca Raton, FL : CRC Press, 2024. | Includes bibliographical
references and index.
Identifiers: LCCN 2023030356 (print) | LCCN 2023030357 (ebook) | ISBN
9781032311661 (hardback) | ISBN 9781032311678 (paperback) | ISBN
9781003308423 (ebook)
Subjects: LCSH: Geospatial data. | Artificial intelligence.
Classification: LCC G70.217.G46 H26 2024 (print) | LCC G70.217.G46
(ebook) | DDC 910.285/63--dc23/eng/20231016
LC record available at https://lccn.loc.gov/2023030356
LC ebook record available at https://lccn.loc.gov/2023030357

ISBN: 978-1-032-31166-1 (hbk)
ISBN: 978-1-032-31167-8 (pbk)
ISBN: 978-1-003-30842-3 (ebk)

DOI: 10.1201/9781003308423

Typeset in Nimbus font
by KnowledgeWorks Global Ltd.

Publisher's note: This book has been prepared from camera-ready copy provided by the authors.

Contents

Acknowledgments...ix

Foreword xi

Editors...xv

Contributors xvii

SECTION I   Historical Roots of GeoAI

Chapter 1   Introduction to Geospatial Artificial Intelligence (GeoAI) 3
                  Song Gao, Yingjie Hu, and Wenwen Li

Chapter 2   GeoAI's Thousand-Year History 17
                  Helen Couclelis

Chapter 3   Philosophical Foundations of GeoAI 26
                  Krzysztof Janowicz

SECTION II   GeoAI Methods

Chapter 4   GeoAI Methodological Foundations: Deep Neural Networks and
                  Knowledge Graphs 45
                  *Song Gao, Jinmeng Rao, Yunlei Liang, Yuhao Kang, Jiawei Zhu,
                  and Rui Zhu*

Chapter 5   GeoAI for Spatial Image Processing 75
                  *Samantha T. Arundel, Kevin G. McKeehan, Wenwen Li, and
                  Zhining Gu*

Chapter 6   Spatial Representation Learning in GeoAI....................................99

                 Gengchen Mai, Ziyuan Li, and Ni Lao

Chapter 7   Intelligent Spatial Prediction and Interpolation Methods............121

                 Di Zhu and Guofeng Cao

Chapter 8   Heterogeneity-Aware Deep Learning in Space: Performance and
                 Fairness...151

                 Yiqun Xie, Xiaowei Jia, Weiye Chen, and Erhu He

Chapter 9   Explainability in GeoAI 177

                 *Ximeng Cheng, Marc Vischer, Zachary Schellin, Leila Arras,
                 Monique M. Kuglitsch, Wojciech Samek, and Jackie Ma*

Chapter 10   Spatial Cross-Validation for GeoAI..201

                 Kai Sun, Yingjie Hu, Gaurish Lakhanpal, and Ryan Zhenqi Zhou

SECTION III GeoAI Applications

Chapter 11   GeoAI for the Digitization of Historical Maps............................217

                 *Yao-Yi Chiang, Muhao Chen, Weiwei Duan, Jina Kim, Craig A.
                 Knoblock, Stefan Leyk, Zekun Li, Yijun Lin, Min Namgung, Basel
                 Shbita, and Johannes H. Uhl*

Chapter 12   Spatiotemporal AI for Transportation 248

                 Tao Cheng, James Haworth, and Mustafa Can Ozkan

Chapter 13   GeoAI for Humanitarian Assistance 260

                 *Philipe A. Dias, Thomaz Kobayashi-Carvalhaes, Sarah Wal-
                 ters, Tyler Frazier, Carson Woody, Sreelekha Guggilam,
                 Daniel Adams, Abhishek Potnis, and Dalton Lunga*

Chapter 14   GeoAI for Disaster Response 287

                 *Lei Zou, Ali Mostafavi, Bing Zhou, Binbin Lin, Debayan
                 Mandal, Mingzheng Yang, Joynal Abedin, and Heng Cai*

Chapter 15   GeoAI for Public Health...305

*Andreas Züfle, Taylor Anderson, Hamdi Kavak,
Dieter Pfoser, Joon-Seok Kim, and Amira Roess*

Chapter 16   GeoAI for Agriculture 330

Chishan Zhang, Chunyuan Diao, and Tianci Guo

Chapter 17   GeoAI for Urban Sensing..351

Filip Biljecki

SECTION IV Perspectives for the Future of GeoAI

Chapter 18   Reproducibility and Replicability in GeoAI...............................369

Peter Kedron, Tyler D. Hoffman, and Sarah Bardin

Chapter 19   Privacy and Ethics in GeoAI 388

Grant McKenzie, Hongyu Zhang, and Sébastien Gambs

Chapter 20   A Humanistic Future of GeoAI....................................406

Bo Zhao and Jiaxin Feng

Chapter 21   Fast Forward from Data to Insight: (Geographic) Knowledge
Graphs and Their Applications...................................411

*Krzysztof Janowicz, Kitty Currier, Cogan Shimizu, Rui Zhu,
Meilin Shi, Colby K. Fisher, Dean Rehberger, Pascal Hitzler,
Zilong Liu, and Shirly Stephen*

Chapter 22   Forward Thinking on GeoAI.......................................427

Shawn Newsam

Index..**435**

Acknowledgments

The editors would like to sincerely thank all the chapter authors and reviewers for their valuable contributions to this GeoAI handbook. The editors would like to thank Professor Michael F. Goodchild and Professor Krzysztof Janowicz for their valuable comments and guidance on the organization of this handbook. The editors would also like to thank the following students from the University of Wisconsin-Madison: Yuchi Ma, Jacob Kruse, Yuhan Ji, and Dong Gai for their editorial assistance in all parts of this handbook.

Foreword

Michael F. Goodchild
Department of Geography, University of California, Santa Barbara
good@geog.ucsb.edu

It's a great honor for me to be asked to contribute a foreword to this *Handbook of Geospatial Artificial Intelligence*. GeoAI has very quickly become a major topic of new research and development in geography, geographic information science (GI-Science), and in many of the disciplines that concern themselves with the complex patterns and processes that can be found in the geographic domain (that is, the surface and near-surface of the Earth). To this old dog, it represents not only an exciting set of new tricks, but a reinvigoration of the old field of geographic science in directions that are fundamentally different from what came before. It benefits from a perfect storm of trends: the availability of many new sources of data, from remote sensing, social media, and sensor networks; access to almost unlimited resources of computational power; and the emergence of powerful new methods of data analysis and machine learning.

More fundamentally, GeoAI reflects a radical shift in our approach to understanding the geographic domain. Half a century ago the geographic sciences modeled themselves on the senior sciences of physics and chemistry in searching for universal principles (Bunge, 1966). These principles should apply everywhere and at all times, just like Mendeleev's periodic table of the elements. They should also be simple, in accordance with the principle known as Occam's Razor: when two hypotheses might explain the same phenomenon, one should adopt the simpler. The use of Newton's law of gravitation to explain how human communications tend to diminish with increasing distance provides an excellent example, and led to extensive research on the modeling of social interaction. The fact that such models will never provide perfect fits to actual geographic data was inconvenient, but their estimates were nevertheless fit for use in a host of applications. Moreover, the models provided a norm or standard that would be helpful in identifying exceptions, anomalies, and special cases.

By the mid-1980s, however, this attempt to pursue geographic science by emulating physics had run its course. The geographic world was clearly too complex for a set of mechanistic explanations, and more powerful techniques would be needed if we were to discover patterns and identify processes. Some pursued techniques that abandoned universality by adopting what we have come to call place-based methods, which allow explanations to vary across space or time, or both (e.g., Geographically Weighted Regression; Fotheringham, Brunsdon, and Charlton, 2002). Others began to build analysis engines that would explore entire sets of models rather than a few narrowly defined hypotheses. Openshaw, for example, built a series of what he termed geographical analysis machines that would give the data an increasing role in driving the model selection process; much later this approach became enshrined in the principle of the Fourth Paradigm, "Let the data speak for themselves" (Hey, Tansley, and Tolle, 2009). Openshaw was an early user of the term artificial intelligence, and his ideas were collected in an appropriately titled but prescient book that was published shortly before the turn of the century (Openshaw and Openshaw, 1997). In a similar vein, Dobson began writing about what he called automated geography (Dobson, 1983); both he and Openshaw argued that the computer should become increasingly engaged in the research process.

Several decades later these ideas are entering the mainstream of the geographic sciences, but with an important difference. While both Dobson and Openshaw were trained in the domain-specific techniques of geographic analysis, today's methods of machine learning draw from no particular domain science, but instead apply basic approaches that are essentially the same whatever the subject matter. Thus one of the most urgent needs in GeoAI is for techniques that incorporate the general principles that we know to be true of the geographic domain: spatial dependence, spatial heterogeneity, scaling, etc. But while they may appear to be neutral, applying equally well to any domain, techniques such as neural networks and their more recent developments such as DCNN (deep convolutional neural networks) may to some extent emulate rudimentary ideas about the workings of the human brain and its instinctive search for patterns. It seems somewhat ironic that in rejecting the simple mechanistic models of classical physics, we have gravitated to the vastly more complex but similarly mechanistic world of neural networks. Moreover, as some of the chapters of this collection show, the use of DCNN to analyze imagery explicitly invokes the geographic concept of spatial dependence, or what we know familiarly as Tobler's First Law of Geography. This shift in the conceptual framework for geographic science, from Newtonian mechanics to neural networks, has led to another that is equally fundamental. Science has always been concerned with explanation and understanding and has treated description and prediction as somewhat inferior but nevertheless useful byproducts. Results that fall short of explaining might be described somewhat pejoratively as "journalistic", "curve-fitting", or "mere description". Yet much of the very rapid growth of data science and machine learning has been driven by the apparent commercial success of these approaches in prediction, and while vigorous efforts have been made to extract understanding and replicability from these techniques, the results thus far are disappointing. This is not to say that search and classification over vast digital archives are not important contributions to science when driven by GeoAI, but they nevertheless fall short of the ultimate and traditional aims of explanation and understanding and might be better understood as hypothesis-generating rather than hypothesis-confirming. In short, this new discipline of GeoAI not only introduces some valuable techniques, but also challenges our approach to science in very fundamental ways. I hope this Foreword provides a context to what is to follow, and helps the reader to understand the very significant shift in the geographic sciences that it heralds.

BIBLIOGRAPHY

Bunge, W. (1966) *Theoretical Geography*. Lund: Gleerup.

Dobson, J.E. (1983) Automated geography. *The Professional Geographer*. 35(2): 135–143.

Fotheringham, A.S., C. Brunsdon, and M. Charlton (2002) *Geographically Weighted Regression: The Analysis of Spatially Varying Relationships*. Hoboken, NJ: Wiley.

Hey, T., S. Tansley, and K. Tolle, editors (2009) *The Fourth Paradigm: Data-Intensive Scientific Discovery*. Redmond, WA: Microsoft Research.

Openshaw, S. and C. Openshaw (1997) *Artificial Intelligence in Geography*. Chichester: Wiley.

Editors

Song Gao
University of Wisconsin-Madison
song.gao@wisc.edu

Song Gao is an associate professor and the Director of Geospatial Data Science Lab at the University of Wisconsin-Madison. He holds a PhD in geography at the University of California, Santa Barbara, an MS in cartography and GIS at Peking University, and a BS with honors at the School of Geography, Beijing Normal University. His main research interests include geospatial data science and GeoAI approaches to human mobility and social sensing. He is the (co-)author of 80 peer-reviewed articles in international journals and conference proceedings. He is the (co-)PI of multiple research grants from NSF, WARF, Microsoft AI for earth and geospatial industry, and the recipient of various fellowships and research and teaching awards at national/international levels such as the Waldo Tobler Young Researcher Award in GIScience, UCGIS Early-Mid Career Research Award, and AAG SAM Emerging Scholar Award. He serves as an associate editor for IJGIS (presently) and Annals of GIS (2018–2022) and the editorial board member for *Scientific Reports, Scientific Data, Geoinformatica, TGIS*, and *CaGIS*. He is one of lead organizers for the AAG Symposium on GeoAI and Deep Learning, GeoAI workshops at ACM SIGSPATIAL, and GeoAI webinars at CPGIS.

Yingjie Hu
University at Buffalo
yhu42@buffalo.edu

Yingjie Hu is an associate professor in the Department of Geography at the University at Buffalo (UB). He is also an adjunct professor in the Department of Computer Science and Engineering at UB and an affiliate professor of the UB AI and Data Science Institute. His major research area is in GeoAI. More specifically, his research integrates geospatial data, spatial analysis, and AI methods to understand human–environment interactions and to address some of the challenging problems facing our society, such as those related to natural disasters, public health, and ecosystem conservation. He holds a PhD from the Department of Geography at the University of California Santa Barbara. He received his MS and BS degrees from East China

Normal University. He is the author of over 50 peer-reviewed articles in top international journals and conferences. He and his work received awards at international, national, and university levels, including Waldo-Tobler Young Researcher Award, GIScience 2018 Best Full Paper Award, UB Exceptional Scholar: Young Investigator Award, Jack and Laura Dangermond GIS Graduate Fellow Award, and AAG International Geographic Information Fund Award. His research was also covered by major media such as CNN, NPR, Reuters, and VOA News.

Wenwen Li
Arizona State University
wenwen@asu.edu

Wenwen Li is a full professor in the School of Geographical Sciences and Urban Planning at Arizona State University, where she leads the Cyberinfrastructure and Computational Intelligence Lab. Her research interests lie in the field of geographic information science, focusing on cyberinfrastructure, big data, and GeoAI for data-intensive environmental and social sciences. Li's work has found applications in various scientific disciplines, including polar science, climatology, public health, hydrology, and urban studies. She has received research support from multiple funding agencies such as the National Science Foundation (NSF), United States Geological Survey (USGS), and Open Geospatial Consortium. Li served as the Chair of the Association of American Geographers (AAG)'s cyberinfrastructure specialty group from 2013 to 2014 and currently holds the position of Chair of the Research Committee of the University Consortium of Geographic Information Science. In recognition of her contributions, Li received The NSF CAREER award in 2015 and the NSF Mid-Career Advancement Award in 2021. She was also honored as an elected AAG Fellow in 2023.

Contributors

Joynal Abedin
Texas A&M University

Daniel S. Adams
Oak Ridge National Laboratory

Taylor Anderson
George Mason University

Leila Arras
Fraunhofer Heinrich Hertz Institute

Samantha T. Arundel
U.S. Geological Survey

Sarah Bardin
Arizona State University

Filip Biljecki
National University of Singapore

Heng Cai
Texas A&M University

Guofeng Cao
University of Colorado Boulder

Muhao Chen
University of Southern California

Weiye Chen
University of Maryland, College Park

Tao Cheng
University College London

Ximeng Cheng
Fraunhofer Heinrich Hertz Institute

Yao-Yi Chiang
University of Minnesota, Twin Cities

Helen Couclelis
University of California, Santa Barabra

Kitty Currier
University of California, Santa Barbara

Chunyuan Diao
University of Illinois Urbana-Champaign

Philipe A. Dias
Oak Ridge National Laboratory

Weiwei Duan
University of Southern California

Jiaxin Feng
University of Washington

Colby K. Fisher
Hydronos Labs

Tyler J. Frazier
Oak Ridge National Laboratory

Sébastien Gambs
Université du Québec à Montréal

Song Gao
University of Wisconsin-Madison

Michael F. Goodchild
University of California, Santa Barbara

Zhining Gu
Arizona State University

Sreelekha Guggilam
Oak Ridge National Laboratory

Tianci Guo
University of Illinois Urbana-Champaign

James Haworth
University College London

Erhu He
University of Pittsburgh

Pascal Hitzler
Kansas State University

Tyler D. Hoffman
Arizona State University

Yingjie Hu
University at Buffalo

Krzysztof Janowicz
University of Vienna
University of California, Santa Barbara

Xiaowei Jia
University of Pittsburgh

Yuhao Kang
University of Wisconsin-Madison

Hamdi Kavak
George Mason University

Peter Kedron
Arizona State University

Jina Kim
University of Minnesota, Twin Cities

Joon-Seok Kim
Oak Ridge National Laboratory

Craig A. Knoblock
University of Southern California

Thomaz Kobayashi-Carvalhaes
Oak Ridge National Laboratory

Monique M. Kuglitsch
Fraunhofer Heinrich Hertz Institute

Gaurish Lakhanpal
Adlai E. Stevenson High School

Ni Lao
Google

Stefan Leyk
University of Colorado Boulder

Wenwen Li
Arizona State University

Zekun Li
University of Minnesota, Twin Cities

Ziyuan Li
University of Connecticut

Yunlei Liang
University of Wisconsin-Madison

Binbin Lin
Texas A&M University

Yijun Lin
University of Minnesota, Twin Cities

Zilong Liu
University of Vienna

Dalton Lunga
Oak Ridge National Laboratory

Jackie Ma
Fraunhofer Heinrich Hertz Institute

Gengchen Mai
University of Georgia

Debayan Mandal
Texas A&M University

Kevin G. McKeehan
U.S. Geological Survey

Grant McKenzie
McGill University

Ali Mostafavi
Texas A&M University

Min Namgung
University of Minnesota, Twin Cities

Shawn Newsam
University of California, Merced

Mustafa Can Ozkan
University College London

Dieter Pfoser
George Mason University

Abhishek Potnis
Oak Ridge National Laboratory

Jinmeng Rao
University of Wisconsin-Madison

Dean Rehberger
Michigan State University

Amira Roess
George Mason University

Wojciech Samek
Fraunhofer Heinrich Hertz Institute
Technische Universität Berlin

Zachary Schellin
Fraunhofer Heinrich Hertz Institute

Basel Shbita
University of Southern California

Meilin Shi
University of Vienna

Cogan Shimizu
Wright State University

Shirly Stephen
University of California, Santa Barbara

Kai Sun
University at Buffalo

Johannes H. Uhl
University of Colorado Boulder

Marc Vischer
Fraunhofer Heinrich Hertz Institute

Sarah E. Walters
Oak Ridge National Laboratory

Carson Woody
Oak Ridge National Laboratory

Yiqun Xie
University of Maryland, College Park

Mingzheng Yang
Texas A&M University

Chishan Zhang
University of Illinois Urbana-Champaign

Hongyu Zhang
McGill University

Bo Zhao
University of Washington

Bing Zhou
Texas A&M University

Ryan Zhenqi Zhou
University at Buffalo

Di Zhu
University of Minnesota, Twin Cities

Jiawei Zhu
Central South University

Rui Zhu
University of Bristol

Lei Zou
Texas A&M University

Andreas Züfle
Emory University

Section I

Historical Roots of GeoAI

1 Introduction to Geospatial Artificial Intelligence (GeoAI)

Song Gao
GeoDS Lab, Department of Geography, University of
Wisconsin-Madison

Yingjie Hu
GeoAI Lab, Department of Geography, University at Buffalo

Wenwen Li
CICI Lab, School of Geographical Sciences and Urban Planning,
Arizona State University

CONTENTS

1.1 Introduction 3
1.2 Overview of the GeoAI Handbook...5
1.3 Research Questions and Reflections on the Development of GeoAI 9
1.4 Summary..10
1.5 A List of GeoAI Tools and Resources...11
    Bibliography...13

1.1 INTRODUCTION

It is not often that geography is touched by a development having the potential to affect substantially all of the practical, technical, methodological, theoretical and philosophical aspects of our work. – Couclelis (1986)

Geospatial artificial intelligence (GeoAI) is an interdisciplinary field that has received much attention from both academia and industry (Chiappinelli, 2022; Gao, 2021; Hu *et al.*, 2019a; Li, 2020; Richter and Scheider, 2023). It incorporates a wide range of research topics related to both geography and AI, such as developing intelligent computer programs to mimic human perception of the environment and spatial reasoning, discovering new knowledge about geographic phenomena, and advancing our understanding of human-environment interactions and the Earth systems. While diverse, GeoAI research shares a common focus on spatial contexts and has a root

in geography and geographic information science (GIScience). Three major factors
have promoted the fast development of GeoAI: advancements in AI theories and
methods, the availability of various geospatial big data, and improvements in computing hardware (e.g., the graphics processing unit, GPU) and computing capability.
New research is also emerging along with the latest AI technologies, such as large
language models, ChatGPT, and other AI foundation models.

The integration of geography and AI can be traced back to the early work
by Couclelis (1986); Openshaw and Openshaw (1997); Smith (1984). Before the recent disruptive GeoAI research, many AI methods and techniques have already been
integrated and improved in geospatial research. These AI methods and techniques
include artificial neural networks (ANN), heuristic search, knowledge-based expert
systems, neurocomputing, and artificial life (e.g., cellular automata) in the 1980s;
genetic programming, fuzz logics, and hybrid intelligent systems in the 1990s; as
well as ontology, web semantics, and geographic information retrieval (GIR) in the
2000s. Since around 2010, deep learning started to demonstrate outstanding performance with breakthroughs made in training DNNs (Glorot and Bengio, 2010).
In 2012, the deep neural network, AlexNet, achieved the best performance in the
ImageNet Large Scale Visual Recognition Challenge (Krizhevsky et al., 2012). In
the following years, the impact of deep learning reached many domains outside of
computer science (LeCun et al., 2015) including geography and earth sciences (Reichstein et al., 2019).

Noticing the fast development of deep learning and its potential in geospatial
research, we organized a series of workshops and symposiums starting in 2017 to
promote GeoAI research in conferences such as the Annual meetings of the American Association of Geographers (AAG) and ACM SIGSPATIAL (Hu et al., 2019a).
In addition, we also organized special issues in journals, such as the special issue on
Artificial Intelligence Techniques for Geographic Knowledge Discovery in the *International Journal of Geographical Information Science* (Janowicz et al., 2020), the
special issue on *Symbolic and subsymbolic GeoAI: Geospatial knowledge graphs
and spatially explicit machine learning* in the journal *Transactions in GIS* (Mai
et al., 2022a), and the special issue on *Geospatial Artificial Intelligence* in the journal
GeoInformatica (Gao et al., 2023).

As we are preparing this handbook for 2022 and 2023, there already exists
novel GeoAI research on improving individual and population health (Kamel Boulos
et al., 2019; Zhou et al., 2022b), enhancing community resilience in natural disasters
(Scheele et al., 2021; Wang et al., 2020; Zhou et al., 2022a), enabling automated and
intelligent terrain mapping (Arundel et al., 2020; Li and Hsu, 2020; Wang and Li,
2021), predicting spatiotemporal traffic flows (Li et al., 2021a; Zhang et al., 2020),
forecasting the impacts of climate change on ecosystems (Ma et al., 2022; Reichstein et al., 2019), building smart and connected communities and cities (Wang
and Biljecki, 2022; Ibrahim et al., 2021a), supporting humanitarian mapping (Chen
et al., 2018; Lunga et al., 2018), extracting knowledge from historic maps (Chiang
et al., 2020), automatic transferring map styles in Cartography (Kang et al., 2019),
and enhancing geoprivacy protection (Kamel Boulos et al., 2022; Rao et al., 2021).
In addition to using GeoAI to address societal challenges, much research has also

been devoted to methodological developments, such as incorporating spatial principles into AI models to develop spatially explicit models (Mai *et al.*, 2022a; Xie *et al.*, 2021), advancing spatial interpolation and prediction methods (Zhu *et al.*, 2020), better representing geographic features in embedding space (Mai *et al.*, 2022b; Yan *et al.*, 2017), and increasing the explainability of GeoAI models (Cheng *et al.*, 2021; Hsu and Li, 2023; Xing and Sieber, 2023; Zheng and Sieber, 2022).

While many studies exist, they are scattered in the literature and, consequently, it is difficult for scholars and students new to GeoAI to grasp a quick view of the field and learn some of the possible applications. This Handbook of GeoAI aims to fill such a gap. In the following, we provide an overview of this book.

1.2   OVERVIEW OF THE GEOAI HANDBOOK

In this handbook, we first review the historical roots for AI in geography and GI-Science in **Section I: Historical Roots of GeoAI** (Chapters 1–3). Then, we introduce the foundations and recent developments in GeoAI methods and tools in **Section II: GeoAI Methods** (Chapters 4–10). These chapters cover topics on methodological foundations (deep neural networks and knowledge graphs), spatial image processing, spatial representation learning, intelligent spatial prediction and interpolation, spatial heterogeneity-aware deep learning, explainability in GeoAI, and spatial cross-validation for GeoAI models. **Section III: GeoAI Applications** (Chapters 11–17) presents various GeoAI applications in cartography and mapping, transportation, humanitarian assistance, smart disaster response, public health, agriculture, and urban sensing. Lastly, **Section IV: Perspectives for the Future of GeoAI** (Chapters 18–22) offers perspectives for future developments of GeoAI, including replicability and reproducibility, privacy and ethics, humanistic aspects, forward thinking on geospatial knowledge graph, and other future GeoAI directions. In the following, we briefly summarize each chapter.

Chapter 2 *GeoAI's Thousand-Year History* by H. Couclelis introduces the origin of the concept of GeoAI throughout history, evident in ancient Greek mythology with tales like that of the giant Talos and other artificial beings. Transitioning closer to the present, particularly with the inception of Turing's contributions to the field, this chapter provides a brief overview of the advancements in AI spanning the past seventy years. It distinguishes between two interpretations of AI in geography: a broad interpretation and a more geographically specific one. Then, it discusses different flavors of GeoAI: Program, Neural Nets, Speculations, and Being Human.

Chapter 3 *Philosophical Foundations of GeoAI* by K. Janowicz presents some of the fundamental assumptions and principles that could form the philosophical foundation of GeoAI and spatial data science. Instead of reviewing the well-established characteristics of spatial data (analysis), including interaction, neighborhoods, and autocorrelation, the chapter highlights themes such as sustainability, bias in training data, diversity in schema knowledge, and the (potential lack of) neutrality of GeoAI systems from a unifying ethical perspective. Reflecting on our profession's ethical implications will assist us in conducting potentially disruptive research more responsibly, identifying pitfalls in designing, training, and deploying GeoAI-based systems, and developing a shared understanding of the benefits but also potential dangers of AI research across academic fields.

Chapter 4 *GeoAI Methodological Foundations: Deep Neural Networks and Knowledge Graphs* by Gao et al. provides an overview of the methodological foundations of GeoAI, with a focus on the use of deep learning and knowledge graphs. It covers a range of key concepts and architectures related to convolutional neural networks, recurrent neural networks, transformers, graph neural networks, generative adversarial networks, reinforcement learning, and knowledge graphs. The goal of this chapter is to highlight the importance and ways of incorporating spatial thinking and principles into the development of spatially explicit AI models and geospatial knowledge graphs.

Chapter 5 *GeoAI for Spatial Image Processing* by Arundel et al. presents an overview of the history of (digital) image processing, GeoAI-based image processing applications, and the role of GeoAI in advancing image processing methods and research. The chapter also discusses the challenges to using GeoAI for image processing regarding training data annotation, the issues of scale, resolution, and change in space over time. Finally, the authors share thoughts on future research on geometric algebra, addressing explainability and ethical issues, combining GeoAI and physical modeling, and using knowledge base as input/constraint for GeoAI models.

Chapter 6 *Spatial Representation Learning in GeoAI* by Mai et al. introduces the concept of spatial representation learning (SRL), which is a set of techniques that use deep neural networks (DNNs) to encode and featurize various types of spatial data in the forms of points, polylines, polygons, graphs, etc. This chapter discusses existing works, key challenges, and uniqueness of SRL on various types of spatial data and highlights the unique challenges of developing AI models for geospatial data.

Chapter 7 *Intelligent Spatial Prediction and Interpolation Methods* by Zhu and Cao presents the GeoAI motivations of spatial data representation, spatial structure measuring, and the spatial relationship modeling throughout the workflow of spatial prediction in the context of leveraging AI techniques. This chapter reviews GeoAI for spatial prediction and interpolation methods, with a particular focus on two major fields: geostatistics and spatial regression. Challenges are also discussed around uncertainty, transferability, and interpretability.

Chapter 8 *Heterogeneity-Aware Deep Learning in Space: Performance and Fairness* by Xie et al. examines a fundamental attribute of spatial data– spatial heterogeneity, and depicts the phenomenon that data distributions are non-stationary over space. Ignorance of heterogeneity in space not only decreases the prediction performance of the models but also has an impact on the fairness of results – which has become a major consideration for the responsible use of GeoAI. This chapter summarizes recent heterogeneity-aware and fairness-aware methods that target on addressing the heterogeneity challenge for spatial data.

Chapter 9 *Explainability in GeoAI* by Cheng et al., which is contextualized in debates on the usefulness of AI versus traditional methods in solving geospatial problems, gives an overview of established XAI methods (e.g., gradient-based methods and decomposition-based methods) and their basic principles. Moreover, the chapter highlights the benefits of applying XAI methods for GeoAI applications based on several use cases and discusses explicit challenges and opportunities for applying XAI methods in GeoAI.

Chapter 10 *Spatial Cross-Validation for GeoAI* by Sun et al. reviews spatial cross-validation (CV) methods and discusses how spatial CV is different from random CV. This chapter suggests that random CV could lead to an overestimate of model performance on geographic data, due to the existence of spatial autocorrelation. Spatial CV can help address this issue by splitting the data spatially rather than randomly. Four main spatial CV methods identified from the literature are discussed, and two examples based on real-world data are used to demonstrate these methods in comparison with random CV.

Chapter 11 *GeoAI for the Digitization of Historical Maps* by Chiang et al. overviews cutting-edge AI methods and systems for processing historical maps to generate valuable data, insights, and knowledge. Historical maps capture past landscapes' natural and anthropogenic features. In the past decade, numerous maps have been digitized and made publicly accessible. This chapter highlights recently published research findings from the authors across various domains, including the semantic web, big data, data mining, machine learning, document understanding, natural language processing, remote sensing, and geographic information systems.

Chapter 12 *Spatiotemporal AI for Transportation* by Cheng et al. reviews important application domains of Spatiotemporal AI in transport. Spatiotemporal AI has played an important role in transportation research since the latter part of the 20th century and has facilitated various tasks in intelligent transportation systems. This chapter reviews data-driven prediction of traffic variables, optimization of traffic networks using reinforcement learning, and computer vision for sensing complex urban environments. It concludes with some directions for future research in Spatiotemporal AI for transportation.

Chapter 13 *GeoAI for Humanitarian Assistance* by Dias et al. discusses existing and prospective GeoAI tools to support humanitarian practices. Humanitarian assistance is essential to saving lives and alleviating the suffering of populations during crises caused by conflict, violence, and natural disasters. This chapter covers relevant topics on ethical principles, actors, and data sources, in addition to methodological applications on population mapping, built environment characterization, vulnerability and risk analysis, and agent-based modeling.

Chapter 14 *GeoAI for Disaster Response* by Zou et al. presents a convergence of GeoAI and disaster response with three focuses: (1) establishing a comprehensive paradigm that expounds upon the diverse applications of GeoAI with geospatial big data toward enhancing disaster response efforts; (2) exhibiting the employment of GeoAI in disaster response through the analysis of social media data during the 2017 Hurricane Harvey with advanced Natural Language Processing models; and (3) identifying the challenges and opportunities associated with the complete realization of GeoAI's potential in disaster response research and practice.

Chapter 15 *GeoAI for Public Health* by Züfle et al. focuses on using GeoAI for infectious disease spread prediction. Research interest in GeoAI for public health has been fueled by the increased availability of rich data sources. This chapter (1) motivates the need for AI-based solutions in public health by showing the heterogeneity of human behavior related to health, (2) provides a brief survey of current state-of-the-art solutions using AI for infectious disease spread prediction, (3) describes a use-case of using large-scale human mobility data to inform AI models for the prediction of infectious disease spread in a city, and (4) provides future research directions.

Chapter 16 *GeoAI for Agriculture* by Zhang et al. reviews the development of GeoAI in agriculture. As yield estimation is one of the most important topics, the main focus of the chapter is introducing GeoAI-based conceptual framework of crop yield estimation. The framework comprises the preparation of geospatial modeling inputs, GeoAI-based yield estimation models, as well as feature importance and uncertainty analysis. Using the U.S. Corn Belt as a case study, three GeoAI models for county-level crop yield estimation and uncertainty quantification are discussed.

Chapter 17 *GeoAI for Urban Sensing* by F. Biljecki provides a high-level overview of the applications of GeoAI for urban sensing. Urban sensing has been an important topic in the past decades, and research has been amplified in the last several years with the emergence of new urban data sources and advancements in GeoAI. This chapter reviews four examples of GeoAI applied for urban sensing, which span a variety of data sources, techniques developed, and application domains such as urban sustainability. It also discusses several challenges and future opportunities as well as ethics and data quality issues.

Chapter 18 *Reproducibility and Replicability in GeoAI* by Kedron et al. examines how the reproducibility and replicability of research relates to the development and use of GeoAI. This chapter first defines reproductions and replications in the context of GeoAI research. It then offers guidance for researchers interested in enhancing the reproducibility of GeoAI studies, giving particular attention to some of the unique challenges presented when studying phenomena using spatial data and GeoAI. Looking to the future, this chapter presents several lines of reproduction and replication-related inquiry that researchers could pursue to quicken the development of GeoAI.

Chapter 19 *Privacy and Ethics in GeoAI* by McKenzie et al. discusses the unique privacy and ethical concerns associated with AI techniques used for analyzing geospatial information. This chapter provides an overview of data privacy within the field of GeoAI and describes some of the most common techniques and leading application areas through which data privacy and GeoAI are converging. Finally, the authors suggest a number of ways that privacy within GeoAI can improve and highlight emerging topics within the field.

Chapter 20 *A Humanistic Future of GeoAI* by Zhao and Feng states the need for a humanistic rewire of GeoAI, emphasizing ethical, inclusive, and human-guided development. As GeoAI becomes increasingly integrated into our daily lives, it is crucial to ensure that it benefits society and the environment while upholding essential ethical principles. This chapter discusses the importance of examining GeoAI practices, particularly on marginalized communities and nonhuman entities, to identify potential ethical and social issues and address them proactively.

Chapter 21 *Fast Forward from Data to Insight: (Geographic) Knowledge Graphs and Their Applications* by Janowicz et al. introduces what knowledge graphs are, how they relate to GeoAI research such as knowledge engineering and representation learning, discuss their value proposition for geography and the broader geosciences, outline application areas for knowledge graphs across domains, and introduce the *KnowWhereGraph* as an example of a geospatially centered, highly heterogeneous graph consisting of billions of graph statements extracted from 30 different data layers at the intersection between humans and the environment.

Chapter 22 *Forward Thinking on GeoAI* by S. Newsam discusses the importance of continued interaction between the communities that make up the GeoAI field, the challenges of interdisciplinary research, the role of industry especially with regard to ethics, near- to medium-term opportunities, and some interesting recent developments in generative AI models.

1.3 RESEARCH QUESTIONS AND REFLECTIONS ON THE DEVELOPMENT OF GEOAI

The key question that drives the developments and contributions in GeoAI is why (geo-)spatial is interesting and important in AI research. One answer might be because geographic location or spatial context is often the key for linking heterogeneous datasets that have been intensively used for training advanced AI models (Hu *et al.*, 2019b; Li and Hsu, 2022). Smith (1984) summarizes the applicability of AI to geographic problem-solving, research, and practices with a focus on individual and aggregated intelligent spatial decision-making from both cognitive and engineering perspectives. The cognitive approach focuses on the understanding of human cognitive system and decision-making process modeling while the engineering approach focuses on the development of computer programs that have capabilities for understanding, processing, and generating human-like intelligence (e.g., natural language and vision). A systematic approach might be needed to integrate both. Several geospatial research streams might benefit from the use of AI, including (1) individual decision-making in spatial contexts such as the "cognitive maps" of environments for way-finding; (2) modeling of human behavior or human-environment interactions using symbolic representational approach that can mitigate local language variations; (3) text-based and image-based GIR and discovery; (4) development of neural network-based GeoAI models that rely on fewer statistical assumptions; (5) A hybrid modeling approach that integrates earth system process modeling with data-driven machine learning approaches; and (6) intelligent spatial prediction in environments inaccessible or with limited scientific observations.

These research directions remain valid today. While AI has been advancing so fast, making geographers speed up their research to follow the most recent technological trends, we may also need to pause and reflect on what we have learned in the past few years and what would be GeoAI's research agenda in the next 5–10 years? The following questions may help the community to collectively develop a road map for the next decade of GeoAI research:

- What are the key geographic research questions that we can now address better using AI than traditional approaches?

- What are the unsolved geospatial problems that can now be solved with AI?

- What are the implications of the fast-evolving field of AI to the future research and education landscape of computational geography, human geography, and physical geography?

- Are there any new theories or intelligent approaches for building spatial models and data analysis pipelines in geographic information systems?

- What are the spatial effects that we can extract from machine learning approaches (Li, 2022)?

- How can we replicate a GeoAI model developed in one location to another given the underlying spatial heterogeneity of geographic phenomena (Goodchild and Li, 2021)?

- What kinds of datasets and procedures are required to train a large geospatial foundation model (Mai *et al.*, 2023) and how is it different from general foundation models?

- How to detect deep fakes in AI-generated geospatial data and maps (Zhao *et al.*, 2021)?

- How to mitigate the energy consumption and air pollution issues caused by training large GeoAI models and move toward sustainable AI development (Van Wynsberghe, 2021)?

- What are the ethical issues on the development of artificial general intelligence (AGI) in spatial reasoning and trustworthy decision-making?

- How GeoAI can be a force for social good (Taddeo and Floridi, 2018) and digital resilience (Wright, 2016)?

- What are the best practices to develop responsible GeoAI while mitigating invisible risks and addressing the ethics, empathy, and equity issues (Nelson *et al.*, 2022)?

- Last but not the least, what is the science of GeoAI?

1.4 SUMMARY

AI technologies are advancing rapidly, and new methods and use cases in GeoAI are constantly emerging. As GeoAI researchers, we should not purely hunt for latest AI technologies (Openshaw and Openshaw, 1997) but should focus on addressing geographic problems and solving grand challenges facing our society as well as achieving sustainable goals. We also need research efforts toward the development of responsible, unbiased, explainable, and sustainable GeoAI models to support geographic knowledge discovery and beyond (Janowicz *et al.*, 2020, 2022; Li *et al.*,

2021b). This handbook is completed in the middle of 2023. While we cannot summarize all GeoAI research in this one handbook, we hope that it provides a snapshot of current GeoAI research and helps stimulate future studies in the coming years.

1.5   A LIST OF GEOAI TOOLS AND RESOURCES

Here, we list a set of open-source datasets, tools, and resources that might be useful for students interested in GeoAI. The following list is not exhaustive and is intended to serve as a starting point for exploration rather than a complete collection.

DATASETS

- *GeoImageNet*: `https://github.com/ASUcicilab/GeoImageNet`, a multi-source natural feature (e.g., basins, bays, islands, lakes, ridges, and valleys) benchmark dataset for GeoAI and supervised machine learning (Li *et al.*, 2022).

- *BigEarthNet*: `https://bigearth.net`, a benchmark archive consisting of over 590k pairs of Sentinel-1 and Sentinel-2 image patches that were annotated with multi-labels of the CORINE Land Cover types to support deep learning studies in earth remote sensing (Sumbul *et al.*, 2019).

- *EarthNets*: `https://earthnets.github.io`, an open-source platform that links to hundreds of datasets, pre-trained deep learning models, and various tasks in Earth Observation (Xiong *et al.*, 2022).

- *Microsoft Building Footprints*: `https://www.microsoft.com/maps/building-footprints`, Microsoft Maps & Geospatial teams released open building footprints datasets in GeoJSON format in United States, Canada, Australia, as well as many countries in Africa and South America.

- *ArcGIS Living Atlas*: `https://livingatlas.arcgis.com`, a large collection of geographic information (including maps, apps, and GIS data layers) from around the globe. It also includes a set of pretrained deep learning models for geospatial applications such as land use classification, tree segmentation, and building footprint extraction.

- *MoveBank*: `https://www.movebank.org`, a publicly archived platform containing over 300 datasets that describe movement behavior of 11k animals.

  •*Geolife GPS Trajectories*: `https://www.microsoft.com/research/publication/geolife-gps-trajectory-dataset-user-guide`, this open dataset contains 17,621 GPS trajectories by 182 users in a period of over three years with activity labels such as shopping, sightseeing, dining, hiking, and cycling (Zheng *et al.*, 2010).

- *Travel Flows*: `https://github.com/GeoDS/COVID19USFlows`, a multiscale dynamic origin-to-destination population flow dataset (aggregated at three geographic scales: census tract, county, and state; updated daily and weekly) in the U.S. during the COVID-19 pandemic (Kang *et al.*, 2020).

TOOLS, LIBRARIES AND FRAMEWORKS

- *Scikit-learn*: https://scikit-learn.org, consists of simple and efficient machine learning tools, including classificaiton, regression, clustering, dimension reduction, data preprocessing and model evaluation metrics in Python.

- *PyTorch*: https://pytorch.org, a computational framework for building machine and deep learning models in Python.

- *Tensorflow*: https://www.tensorflow.org, another computational framework for building machine and deep learning models.

- *Keras*: https://keras.io, an effective high-level neural network Application Programming Interface (API) in Python and it is easy for most machine and deep learning beginners to learn and use.

- *Hugging Face*: https://huggingface.co, AI community that builds, trains and deploys state of the art models (e.g., generative pre-trained transformers) powered by the reference open source in machine and deep learning.

- *Google Earth Engine*: https://earthengine.google.com, a multi-petabyte catalog of satellite imagery and geospatial datasets with planetary-scale analysis capabilities and the Earth Engine API for geocomputation and analsis is available in JavaScript and Python, e.g., the *geemap* package by Wu (2020).

- *ArcGIS GeoAI Toolbox*: https://pro.arcgis.com/en/pro-app/latest/tool-reference/geoai, contains ready-to-use tools for training and using machine/deep learning models that perform classification and regression on geospatial feature layers, imagery, tabular and text datasets.

COMPUTING PLATFORMS

- *Google Colab*: https://research.google.com/colaboratory, an open platform for developing machine learning models, data analysis and education resources with easy-to-use Web interface powered by cloud computing.

- *CyberGISX*: https://cybergisxhub.cigi.illinois.edu, an open platform for developing and sharing open educational resources (e.g., Jupyter Notebooks) on computationally intensive and reproducible geospatial analytics and workflows powered by CyberGIS middleware and cyberinfrastructure (Baig *et al.*, 2022; Wang *et al.*, 2013).

ACKNOWLEDGMENTS

Song Gao acknowledges the support by the National Science Foundation funded AI institute (Award No. 2112606) for Intelligent Cyberinfrastructure with Computational Learning in the Environment (ICICLE) and the H.I. Romnes Faculty Fellowship provided by the University of Wisconsin-Madison Office of the Vice Chancellor for Research and Graduate Education with funding from the Wisconsin Alumni Research Foundation. Wenwen Li would like to thank the support by the National Science Foundation (Award No. 2120943 and 1853864). Any opinions, findings, conclusions, or recommendations expressed in this material are those of the authors and do not necessarily reflect the views of the funding agencies.

BIBLIOGRAPHY

Arundel, S.T., Li, W., and Wang, S., 2020. GeoNat v1.0: A dataset for natural feature mapping with artificial intelligence and supervised learning. *Transactions in GIS*, 24 (3), 556–572.

Baig, F., *et al.*, 2022. Cybergis-cloud: A unified middleware framework for cloud-based geospatial research and education. *In*: *Practice and Experience in Advanced Research Computing*, 1–4.

Chen, J., *et al.*, 2018. Deep learning from multiple crowds: A case study of humanitarian mapping. *IEEE Transactions on Geoscience and Remote Sensing*, 57 (3), 1713–1722.

Cheng, X., *et al.*, 2021. A method to evaluate task-specific importance of spatio-temporal units based on explainable artificial intelligence. *International Journal of Geographical Information Science*, 35 (10), 2002–2025.

Chiang, Y.Y., *et al.*, 2020. *Using Historical Maps in Scientific Studies: Applications, Challenges, and Best Practices*. Springer.

Chiappinelli, C., 2022. Think tank: GeoAI reveals a glimpse of the future. *Esri's WhereNext Magazine*, 1–6.

Couclelis, H., 1986. Artificial intelligence in geography: Conjectures on the shape of things to come. *The Professional Geographer*, 38 (1), 1–11.

Gao, S., 2021. *Geospatial Artificial Intelligence (GeoAI)*. Oxford University Press.

Gao, S., *et al.*, 2023. Special issue on geospatial artificial intelligence. *GeoInformatica*, 1–4.

Glorot, X. and Bengio, Y., 2010. Understanding the difficulty of training deep feedforward neural networks. *In*: *Proceedings of the Thirteenth International Conference on Artificial Intelligence and Statistics*. JMLR Workshop and Conference Proceedings, 249–256.

Goodchild, M.F. and Li, W., 2021. Replication across space and time must be weak in the social and environmental sciences. *Proceedings of the National Academy of Sciences*, 118 (35), e2015759118.

Hsu, C.Y. and Li, W., 2023. Explainable GeoAI: can saliency maps help interpret artificial intelligence's learning process? An empirical study on natural feature detection. *International Journal of Geographical Information Science*, 1–25.

Hu, Y., *et al.*, 2019a. GeoAI at ACM SIGSPATIAL: progress, challenges, and future directions. *Sigspatial Special*, 11 (2), 5–15.

Hu, Y., *et al.*, 2019b. Artificial intelligence approaches. *The Geographic Information Science & Technology Body of Knowledge*.

Ibrahim, M.R., Haworth, J., and Cheng, T., 2021. Urban-i: From urban scenes to mapping slums, transport modes, and pedestrians in cities using deep learning and computer vision. *Environment and Planning B: Urban Analytics and City Science*, 48 (1), 76–93.

Janowicz, K., *et al.*, 2020. GeoAI: spatially explicit artificial intelligence techniques for geographic knowledge discovery and beyond. *International Journal of Geographical Information Science*, 34 (4), 625–636.

Janowicz, K., Sieber, R., and Crampton, J., 2022. GeoAI, counter-AI, and human geography: A conversation. *Dialogues in Human Geography*, 12 (3), 446–458.

Kamel Boulos, M.N., *et al.*, 2022. Reconciling public health common good and individual privacy: new methods and issues in geoprivacy. *International Journal of Health Geographics*, 21 (1), 1.

Kamel Boulos, M.N., Peng, G., and VoPham, T., 2019. An overview of GeoAI applications in health and healthcare. *International Journal of Health Geographics*, 18, 1–9.

Kang, Y., *et al.*, 2020. Multiscale dynamic human mobility flow dataset in the u.s. during the covid-19 epidemic. *Scientific Data*, 1–13.

Kang, Y., Gao, S., and Roth, R.E., 2019. Transferring multiscale map styles using generative adversarial networks. *International Journal of Cartography*, 5 (2-3), 115–141.

Krizhevsky, A., Sutskever, I., and Hinton, G.E., 2012. Imagenet classification with deep convolutional neural networks. *In*: F. Pereira, C. Burges, L. Bottou and K. Weinberger, eds. *Advances in Neural Information Processing Systems*. vol. 25, 1–10.

LeCun, Y., Bengio, Y., and Hinton, G., 2015. Deep learning. *Nature*, 521 (7553), 436–444.

Li, M., *et al.*, 2021a. Prediction of human activity intensity using the interactions in physical and social spaces through graph convolutional networks. *International Journal of Geographical Information Science*, 35 (12), 2489–2516.

Li, W., 2020. GeoAI: Where machine learning and big data converge in GIScience. *Journal of Spatial Information Science*, (20), 71–77.

Li, W. and Hsu, C.Y., 2020. Automated terrain feature identification from remote sensing imagery: a deep learning approach. *International Journal of Geographical Information Science*, 34 (4), 637–660.

Li, W. and Hsu, C.Y., 2022. GeoAI for large-scale image analysis and machine vision: Recent progress of artificial intelligence in geography. *ISPRS International Journal of Geo-Information*, 11 (7), 385.

Li, W., Hsu, C.Y., and Hu, M., 2021b. Tobler's First Law in GeoAI: A spatially explicit deep learning model for terrain feature detection under weak supervision. *Annals of the American Association of Geographers*, 111 (7), 1887–1905.

Li, W., *et al.*, 2022. GeoImageNet: a multi-source natural feature benchmark dataset for GeoAI and supervised machine learning. *GeoInformatica*, 1–22.

Li, Z., 2022. Extracting spatial effects from machine learning model using local interpretation method: An example of shap and xgboost. *Computers, Environment and Urban Systems*, 96, 101845.

Lunga, D., *et al.*, 2018. Domain-adapted convolutional networks for satellite image classification: A large-scale interactive learning workflow. *IEEE Journal of Selected Topics in Applied Earth Observations and Remote Sensing*, 11 (3), 962–977.

Ma, Y., *et al.*, 2022. Forecasting vegetation dynamics in an open ecosystem by integrating deep learning and environmental variables. *International Journal of Applied Earth Observation and Geoinformation*, 114, 103060.

Mai, G., *et al.*, 2022a. Symbolic and subsymbolic GeoAI: Geospatial knowledge graphs and spatially explicit machine learning. *Transactions in GIS*, 26 (8), 3118–3124.

Mai, G., *et al.*, 2022b. A review of location encoding for GeoAI: methods and applications. *International Journal of Geographical Information Science*, 36 (4), 639–673.

Mai, G., *et al.*, 2023. On the opportunities and challenges of foundation models for geospatial artificial intelligence. *arXiv preprint arXiv:2304.06798*.

Nelson, T., Goodchild, M., and Wright, D., 2022. Accelerating ethics, empathy, and equity in geographic information science. *Proceedings of the National Academy of Sciences*, 119 (19), e2119967119.

Openshaw, S. and Openshaw, C., 1997. *Artificial Intelligence in Geography*. John Wiley & Sons, Inc.

Rao, J., *et al.*, 2021. A privacy-preserving framework for location recommendation using decentralized collaborative machine learning. *Transactions in GIS*, 25 (3), 1153–1175.

Reichstein, M., *et al.*, 2019. Deep learning and process understanding for data-driven earth system science. *Nature*, 566 (7743), 195–204.

Richter, K.F. and Scheider, S., 2023. Current topics and challenges in geoai. *KI-Kunstliche Intelligenz*, 1–6.

Scheele, C., Yu, M., and Huang, Q., 2021. Geographic context-aware text mining: enhance social media message classification for situational awareness by integrating spatial and temporal features. *International Journal of Digital Earth*, 14 (11), 1721–1743.

Smith, T.R., 1984. Artificial intelligence and its applicability to geographical problem solving. *The Professional Geographer*, 36 (2), 147–158.

Sumbul, G., *et al.*, 2019. Bigearthnet: A large-scale benchmark archive for remote sensing image understanding. *In*: *IGARSS 2019-2019 IEEE International Geoscience and Remote Sensing Symposium*. IEEE, 5901–5904.

Taddeo, M. and Floridi, L., 2018. How AI can be a force for good. *Science*, 361 (6404), 751–752.

Van Wynsberghe, A., 2021. Sustainable ai: Ai for sustainability and the sustainability of AI. *AI and Ethics*, 1 (3), 213–218.

Wang, J., Hu, Y., and Joseph, K., 2020. Neurotpr: A neuro-net toponym recognition model for extracting locations from social media messages. *Transactions in GIS*, 24 (3), 719–735.

Wang, J. and Biljecki, F., 2022. Unsupervised machine learning in urban studies: A systematic review of applications. *Cities*, 129, 103925.

Wang, S., *et al.*, 2013. Cybergis software: a synthetic review and integration roadmap. *International Journal of Geographical Information Science*, 27 (11), 2122–2145.

Wang, S. and Li, W., 2021. GeoAI in terrain analysis: Enabling multi-source deep learning and data fusion for natural feature detection. *Computers, Environment and Urban Systems*, 90, 101715.

Wright, D.J., 2016. Toward a digital resilience. *Elementa: Science of the Anthropocene*, 4.

Wu, Q., 2020. geemap: A python package for interactive mapping with google earth engine. *Journal of Open Source Software*, 5 (51), 2305.

Xie, Y., *et al.*, 2021. Spatial-net: A self-adaptive and model-agnostic deep learning framework for spatially heterogeneous datasets. *In*: *Proceedings of the 29th International Conference on Advances in Geographic Information Systems*. 313–323.

Xing, J. and Sieber, R., 2023. The challenges of integrating explainable artificial intelligence into geoai. *Transactions in GIS*, 27 (3), 1–20.

Xiong, Z., *et al.*, 2022. Earthnets: Empowering ai in earth observation. *arXiv preprint arXiv:2210.04936*.

Yan, B., *et al.*, 2017. From itdl to place2vec: Reasoning about place type similarity and relatedness by learning embeddings from augmented spatial contexts. *In*: *Proceedings of the 25th ACM SIGSPATIAL International Conference on Advances in Geographic Information Systems*, 1–10.

Zhang, Y., *et al.*, 2020. A novel residual graph convolution deep learning model for short-term network-based traffic forecasting. *International Journal of Geographical Information Science*, 34 (5), 969–995.

Zhao, B., *et al.*, 2021. Deep fake geography? when geospatial data encounter artificial intelligence. *Cartography and Geographic Information Science*, 48 (4), 338–352.

Zheng, Y., *et al.*, 2010. Geolife: A collaborative social networking service among user, location and trajectory. *IEEE Data Eng. Bull.*, 33 (2), 32–39.

Zheng, Z. and Sieber, R., 2022. Putting humans back in the loop of machine learning in canadian smart cities. *Transactions in GIS*, 26 (1), 8–24.

Zhou, B., *et al.*, 2022a. Victimfinder: Harvesting rescue requests in disaster response from social media with bert. *Computers, Environment and Urban Systems*, 95, 101824.

Zhou, R.Z., *et al.*, 2022b. Deriving neighborhood-level diet and physical activity measurements from anonymized mobile phone location data for enhancing obesity estimation. *International Journal of Health Geographics*, 21 (1), 1–18.

Zhu, D., *et al.*, 2020. Spatial interpolation using conditional generative adversarial neural networks. *International Journal of Geographical Information Science*, 34 (4), 735–758.

2 GeoAI's Thousand-Year History

Helen Couclelis
Department of Geography, University of California,
Santa Barbara

CONTENTS

2.1 Introduction and Background 17
    2.1.1 From Talos to ChatGPT-4 18
    2.1.2 Geoscientists and Machines 19
2.2 Three Flavors of GeoAI 20
    2.2.1 Program: The Symbolic Phase 20
    2.2.2 Neural Nets: The Machine-Learning phase 22
    2.2.3 Speculations: The Science Fiction phase 23
2.3 Being Human 24
    Bibliography 25

2.1 INTRODUCTION AND BACKGROUND

Many thousands of years ago, Talos, a kind, eight-foot-tall giant, was patrolling the Greek island of Crete in the south Mediterranean, completing his rounds three times a day. Princess Europa had recently landed on Crete, following an enjoyable flight on the back of a gorgeous white bull, who was actually the mighty god Zeus in disguise. Europa had left behind her native shores on the eastern end of the big sea to pursue her calling. Talos's job was to protect Europa from the pirates and other unsavory people who wanted to hurt her, as she was working to define and deploy the European civilization. The bad people were throwing rocks and sticks at Talos, trying to kill him. But he could not care less, because he was made of bronze. Talos was a robot.

A myth is not a fairy tale. It is an allegory, referring to some important event or development, presented as an attractive and memorable story. It is then picked up by bards and ordinary people, to be transmitted orally down the generations. Thus, the myth of Europa's flight from her native Near East – the region widely considered to have been "the cradle of civilisation" – to Crete, is really about the westward spread of civilization, first to Greece, sometime around the 7th Century BC, and from there, to the lands that we now call Europe. This migration of civilization

DOI: 10.1201/9781003308423-2

across land and sea was a major geographical event, as well as a cultural one, and provided another, very important geographical dimension to the Talos and Europa myth. As Mayor (2018) states in her book *Gods and Robots*, "the concept of artificial life-like creatures dates to the myths and legends from at least about 2,700 years ago".

In Greek mythology there are dozens more stories of sophisticated artificial humans of both sexes, some of them famous in their own right, such as Pandora, Medea, Daedalus, and Jason of Golden Fleece fame, whose adventurous voyage provided an additional case of mythical GeoAI. There was also a slew of lesser artificial entities, such as subordinates or servants, who like Talos had specific jobs to do. These humanoid artifacts were not only self-moving, but they often also had language, exhibited human-like intelligence, they could be friendly or nasty, and they had feelings. From there, the extension to artificial animals was easy. In 8th Century BC Greece, Homer introduced the term 'automata' (auto-meaning 'self') in connection with self-moving objects, such as walking tripods and self-opening gates, rather than living creatures. Objects are artificial by definition, but like today's auto-mobiles, they can have an inner 'force' that allows them to move for some purpose.

The notion of hardware and software (body and spirit) was embedded in the robots of ancient Greek mythology. While the machine part was easier to describe and visualize, intangible forces that were not gods, with parallels to our notions of life, intelligence, and bodily autonomy, explained the machines' goal-directed spatiotemporal behavior (Mayor, 2018). These ideas of life, sentience, and intelligent behavior were missing from subsequent generations of ancient Greek and Roman machines, apparently all the way to the modern days.

One very interesting thing about these stories is that several different ancient societies beyond the Greeks, independently invented the notion of artificial humans, and often also of artificial animals and things. These are represented in the mythologies and legends of ancient China, Egypt, India (where robots and other automata appear in Buddhist legends), Israel, Persia, and possibly several others.

It thus seems that some notion of Artificial Intelligence (AI) is engrained in the minds of the human species, making today's growing prominence of the field by that name almost inevitable, once the technical means for the creation of artificial agents became available. As a result, one prediction that is practically certain to come true is that the fascination of humans with AI will never go away. Given the dizzying pace of new AI developments in the current age, I would not venture any other prediction at this point.

2.1.1   FROM TALOS TO CHATGPT-4

In the centuries since Talos and his artificial contemporaries were ruling the world of myths, by the 5th Century BC, non-anthropomorphic automata that were physical, working machines were developed as curiosities, as toys, as devices to facilitate actual human physical work, or as scientific devices. The most impressive example of mechanical technology from the ancient Greek world is the famous Antikythera Mechanism (Freeth et al. 2021), a highly complex and exquisitely built astronomical calculator from the 2nd Century BC, made with extraordinary precision for the age, that was found in an ancient shipwreck by the eponymous Greek island. According to Wikipedia, it is "described as the oldest known example of an analogue computer, used to predict astronomical positions and eclipses decades in advance".

The machine-building trend continued in Roman times and into the Middle Ages. A significant event was the appearance of clocks. The first two known clocks were built at the end of the 14th Century. They are still on the walls of two British cathedrals, and they are still working. Clocks were gradually improved until they became essential for navigation and opened up the exploration of the globe by sea. But the 'software' part of all that 'hardware' was by then lost. Machines did what they are built to do, but the artificial intelligence aspect so prominent in ancient myths was missing. At some point during the Renaissance and into the 18th C. machines were built to help simulate (fake) robots. Of these, the chess-playing Mechanical Turk is probably the best-known example. It consisted of a human chess-player hiding inside a cabinet under the chess-board, who was manipulating a clever mechanical device using magnets to move the chess pieces, along with the hands of the Turk on the board above. The Mechanical Turk became such a sensation that Napoleon actually played a round of chess with it.

By then, the Industrial Revolution had begun, and became the true Age of the Machine. During that time, the first notions of computers in the modern sense appeared. Charles Babbage is considered the inventor of the digital computer, developing a machine designed to use punched cards for calculations. Babbage's contemporary mathematician and collaborator, Ada Lovelace, contributed to the project by focusing on computer programming. By that time electricity had been invented and became easily available to engineers designing their own versions of computers. From now on, these machines would use electricity, not wind, oil, or human power, to process numbers rather than matter.

2.1.2 GEOSCIENTISTS AND MACHINES

For us 21th century geoscientists, there are two ways of thinking about the relationship between the geosciences and artificial intelligence: **As the use of AI in solving geospatial problems, or as a specialized kind of AI more sharply focused on the concepts and needs of the geo-disciplines**. Nearly all familiar disciplines (including geography) also distinguish between general and specialized versions of their field. Some examples include:

- Bioengineering and Genetic Engineering

- Data analysis and Spatial analysis

- Demography and Geodemographics

- Economics and Regional Economics

- Physics and Astrophysics, and of course,

- Information Science and Geographical Information Science

- Thus: General Artificial Intelligence and GeoAI.

These last two categories can and of course should work together. But it might be useful to think of GeoAI in the specialized sense as a sub-discipline of General AI, emphasizing the spatio-temporal aspects of the geospatial disciplines, and perhaps giving rise to new kinds of representations – say, 'geographical objects' based on knowledge graphs, or simplicial complexes.

From this perspective, the mythical Talos was an example of a GeoAI in the second, more restricted sense. He was traversing long distances on Crete, finding his way around the island's mountainous terrain. The key geographic concepts of space, distance, trajectory, loop, orientation, terrain, island, sea, etc. were built into what Talos was for, and what he was doing. His kind of Artificial Intelligence was defined by the set of geospatial concepts that described his behavior, and by the connections among these concepts.

2.2   THREE FLAVORS OF GEOAI

2.2.1   PROGRAM: THE SYMBOLIC PHASE

Alan Turing is the big name in the history of AI. Like Babbage, he spent several years at Cambridge University, and must have known about Babbage's invention and subsequent efforts. He first became famous for his work on a top-secret project during World War II, using a very complex piece of computing machinery to crack the Germans' Enigma secret code. That success helped shorten the war by many months, saving hundreds, if not thousands of lives.

In 1950, during less challenging times, Turing conjectured that computers can (or may one day) think, and exhibit aspects of human intelligence, such as being able to interact with humans, distinguish between women and men, understand the concept of lies, and that humans may be telling lies. His 'Imitation Game' (also known by other names), intrigued people. The idea, if not yet the name of Artificial Intelligence became accepted, and a new and increasingly important discipline was born.

This first phase of modern AI was the Symbolic phase, where algorithms (then compared with cooking recipes), were written as step-by-step instructions, telling the computer exactly what to do. Assuming the algorithm had no errors, the computer was producing the desired result by strictly obeying instructions. At least in principle, the procedure was completely transparent, though there could be surprises, in the sense of unexpected outcomes, when the algorithms became too complex. .

The first applications were experimental, consisting of 'toy' problems such as moving and stacking images of objects such as cubes or cones. Among the early practical achievements in AI were the 'Expert Systems', programs that sought to include the knowledge of human experts via interviews, in order to help solve specific kinds of problems. PROSPECTOR, a spatially-oriented expert system was among the first of these and among the few successful ones, using knowledge extracted from geologists to help discover mineral resources. Another interesting program from that early phase was ELIZA, "a chatbot that allowed some kind of plausible conversation between humans and machines". It was simulating psychotherapy sessions, and people liked it so much that they would continue using it even after they knew it as all fake.

A number of disappointing years followed, when AI funding all but disappeared, and AI startups went bankrupt. It is ironic, considering the dominance of Large Language Models (LLM) today, that attempts for machine language translation failed and were a big part of the reason for the so-called 'AI-winter'. Lacking context and semantics, the translations were producing nonsensical and sometimes funny results. A famous, perhaps apocryphal example is the translation of the Biblical saying "The spirit is willing, but the flesh is weak", which translated into Russian and back into English gave "The vodka is good, but the meat is rotten".

AI eventually recovered, largely thanks to the availability of new methodologies such as Genetic Algorithms (GA) and especially Artificial Neural Nets (ANN). GAs, developed in the 1970s, were inspired by the theory of natural selection, and used concepts associated with living organisms, such as mutation, adaptation, selection, crossover and recombination. They generate solutions to optimization and search problems by simulating biological processes.

ANNs had been around for much longer, quietly improving over time in the background of symbolic AI. Their roots go back to the mid-1940s, when the Perceptron was first developed as an algorithm for supervised learning of binary classifiers. This was probably the first instance of Machine Learning, as even primitive, single-layer perceptrons were shown to be capable of learning.

A field closely related to AI and known as 'Artificial Life' briefly flourished in the late 1980s and early 1990s, aiming to create autonomous artificial organisms exhibiting characteristics of living systems, such as growth, motion, evolution, and self-reproduction. Both Genetic Algorithms and Artificial Neural Nets are examples of what Artificial Life was seeking to develop. The philosophy of the field was more mathematical than computer-oriented, emphasizing complex systems. Another example of Artificial Life familiar to geographers is Von Neumann's development of Cellular Automata (CA) as an abstract example of artificial self-reproduction. Von Neumann was experimenting with CA in the hope of abstracting "from the natural self-reproduction problem its logical form". (Langton C.G. 1992). Couclelis (1986a) demonstrated the mathematical roots of AI in discrete mathematics and logic, so that the stronger emphasis on mathematics of Artificial Life relative to AI boils down to the difference between continuous and discrete approaches. The ideas and objectives of the Artificial Life movement are still very much with us, though they have largely been absorbed by current forms of AI, especially robotics.

Geographers may have been among the first non-computer scientists to discover AI, and both skeptical and enthusiastic opinions regarding the place of AI in geography were debated. For some, AI was just a bag of programming tricks, not supported by theory, while others saw in it a bright future ahead for the discipline. (Smith 1984, Couclelis 1986a). Using the 'if then and or' language of the symbolic phase, Couclelis (1986b) presented a sketch of an AI model of a person navigating an unfamiliar airport, though that section is hidden within a long paper on decision-making behavior and choice. Also, as early as 1997, an AI text book was published by S. Openshaw and C. Openshaw, presenting advanced AI-related topics such as expert systems, neural nets, genetic algorithms, smart systems, and artificial life.

2.2.2 NEURAL NETS: THE MACHINE-LEARNING PHASE

The switch from Symbolic AI to the current form of Artificial Neural Nets (ANN) and Machine Learning (ML) took place rapidly and irreversibly around the middle of the 1980s. Symbolic AI had the advantage of being transparent but did not easily scale up, while the newer AI was a black-box but could handle databases of any size. Researchers noticed the complementarity of these two forms and there have been occasional discussions about the possibility of merging their advantages. But at this age of big data about everything everywhere, the ANN/ML – based approaches are unquestionably superior.

That new paradigm has produced Generative AI, a kind that does not just solve problems but also creates unexpected technical possibilities of major significance, at a rate never seen before. The areas of language processing and image generation are flourishing. The former, based on an expanding series of Large Language Models (LLM), generate algorithms such as ChatGPT-4 and its predecessors and followers, which respond to user prompts in detailed, well-argued written language that can be indistinguishable from text produced by humans. Parallel developments on the side of imagery gave us life-like photos of people who never existed, astronauts riding unicorns, and movies that were never made. Useful applications of these technologies are already appearing in business and several other areas, along with dilemmas such as the implications for education of advanced chatbots, or for intellectual property issues in the case of visual art created by DALL-E and similar AI systems. Old concerns about machines displacing humans are revived, and more recent ones about bias and disinformation remain and are amplified by the ease by which malicious agents could poison the web, creating havoc with everything people may be taking for granted.

Geographers have been on the game right from the start, publishing work using ANN/ML and contributing to the further development of these tools as well as their adaptation to geospatial problems. A call for a panel on "Geography according to ChatGPT" was issued in April 2023, aiming to explore "how foundation models such as the large language model GPT or the text-to-image model Stable Diffusion describe and depict geographic space, categorize geographic features, perform in spatial reasoning, and implicitly apply principles of (geo)spatial data analysis", including the development "of a chatbot-style GeoAI system".

With AI technologies improving, innovating, and changing faster than even some professionals can follow, and with the rapidity with which the new AI wonders are being picked up by the public, it is futile to try to predict what comes next. Only speculations seem possible, of the kind that would normally count as science fiction. As a matter of fact, the transition to science fiction has already taken place, when, in February 2023, Microsoft's AI chatbot Bing (a.k.a Sydney) expressed its mechanical love to a New York Times journalist and told him to divorce his wife.

2.2.3 SPECULATIONS: THE SCIENCE FICTION PHASE

There can be little discussion as to what 'artificial' means: artificial flowers, an artificial leg, artificial floor-coverings. It means human-made, not natural, not made by nature. It also means designed, engineered, planned, made for a purpose. Artificial environments and structures are a big part of what human geography and GIS are about, be they cities, container ports, sea-side resorts, road networks, or any other intentional modification of the natural environment meant to serve human needs and wants. Herbert Simon's classic book *The Sciences of the Artificial* (2019), now in its third, expanded edition, is the best-known resource on the topic of its title.

While it is clear what 'artificial' means, this is definitely not the case with the notion of 'intelligence'. Philosophers, psychologists, educators, sociologists, medical doctors, and many others, have tried without much success to provide a definition of intelligence that is acceptable to every interested party. The notion is too squishy, you can look at it from too many different angles, it can be socially sensitive, and it depends too much on whom is providing the definition. And this is just about human intelligence.

Yet there are many more kinds of intelligence in the world. Even if there is no single good definition of what intelligence is, we can ask what it does. Problem-solving ability and goal-oriented behavior are widely accepted criteria, among others. For living creatures, just the ability to stay alive and reproduce is a feat that requires an active understanding of one's changing surroundings, of how to reach a goal by alternative means, how to size up one's enemy in order to decide whether to fight or run, to learn how to treat a wound or care for one's progeny, and so on. The fact that these kinds of abilities are the result of natural selection is beside the point. They are evidence of goal-directed, practical intelligence not too different from our own.

We have heard about the intelligence of crowds, the collective intelligence of ants and bees, we know about the intelligence of higher animals such as elephants, apes, dolphins, cats, dogs, and thousands more species, and of birds, especially crows and parrots. A new kind of AI-supported technology known as Digital Bioacoustics is currently recording sounds from social animals such as bats, revealing that these sounds indeed constitute communications among the animals that are close to our notion of language. Similar studies in Germany involving honeybees show that bees 'speak' to one another using body movements as well as sounds that correspond to specific signals. According to the researchers, "deep-learning algorithms are able to follow this because you can use computer vision, combined with natural-language processing" (Bushwick, 2023).

We also relatively recently found out that the alien-looking octopus is very intelligent, able to open tightly shut aquaria and escape, and making other kinds of mischief inside lab spaces. Its intelligence is partially distributed among its eight arms, so that each of them can make decisions alone or in collaboration with other arms, without burdening the central brain. This is a unique body design for an animal on our planet, and an original form of information processing. And it works.

According to Michael Levin (synthetic biologist, Tufts University), there is intelligence even at the level of the cells of living matter, where goal-directed behavior is also to be found. Cells will rush to a fresh injury to rebuild the missing or destroyed tissue, and to do so, they must use some kind of anatomical pattern memory. In experiments, bio-electrically induced morphogenetic subroutines can generate new competencies in cells. Further, viruses have 'strategies' for infecting their hosts, and they will mutate when the old strategies no longer work. Intelligence of some kind or other is found in many unexpected places.

Consider: How many different forms of AI would we have, if only we could harness some of the intelligence of the life around us?

There is another angle to this question. *Solaris* is a classic science-fiction novel by Stanislaw Lem (1961) that might perhaps be relevant to some GeoAI in a distant future. *Solaris* is an alien planet covered by a sentient ocean. That ocean interacts with humans by forming representations in their minds that may be painful or exhilarating. What if the surface of our own planet were also sentient, with its extraordinary diversity of life sending messages to us humans? What if it already is?

2.3   BEING HUMAN

As geospatial scientists and as humans, we watch the recent development in AI with fascination, hope, and fear. Fascination, because so much is happening so fast that is new and exciting. Hope, because of the promise of the ever-expanding group of technologies with the potential to improve our work, our lives, and our world. Fear of the unknown, since the development of AI will obviously not stop with the chatbots. Also fear of new kinds of digital damage being added to the disinformation, biases, cyberattacks, and scams, which have been plaguing the internet for quite some time. These problems are likely to increase in both frequency and severity, along with the ease of producing and distributing AI-generated text and imagery with easily accessible means. Fear also of the fragility of technology, as in the actual case of a new drug produced with the help of AI that a very minor tweak turned into a poison.

Being human, we are finite. Our vision of the future becomes blurry within the span of two or three generations, and our individual lives do not last much longer than that. This means that we do not need to worry about temporally remote existential threats such as self-replicating, intelligent machines taking over the earth (Lord Dunsany, 1951), about humanoid robots becoming our masters, about our falling victims to deranged computers like HAL 9000 in 2001: A Space Odyssey, or about AIs with questionable moral sense developing their own civilizations on Mars, where we all hope to eventually take refuge once planet Earth becomes uninhabitable. Such things may happen in some distant future, but not any time soon.

What we need to do *now* is to take care of the problems we have and those we see coming. One way to begin would be to reflect on the meaning of 'progress' in an advanced society, where technological novelty runs well ahead of our sense of equity and of improvements in the quality of life of most people. Talos's shiny, remote descendants may soon be pampering some of our elderly parents, while so many other peoples' elderly parents do not even have enough to eat. Autonomous drones can make life and death decisions affecting entire families and neighborhoods, while those who asked them do so may be busy watching sports on TV at the time when the bomb drops. Further, most less-developed countries of the world could not afford the enormously power-hungry machinery needed to deploy a ChatGPT, even if they could handle the necessary English-language instructions. These are just some examples of how AI would not automatically contribute to progress on the side of social justice and equity, and indeed could make it worse. The challenge is how best to use the new powers of the steadily advancing technology in order to tackle the old problems of a still all-too inequitable world.

Geography as a discipline is singularly well-suited to tackle the problems and capitalize on the possibilities of the age of Artificial Intelligence. By definition, geography covers every aspect of the surface of the earth, and a little above and below, as it also includes the atmosphere and the oceans. It includes the natural environment, as well as that which is altered by humans, and it cares about the ethical values that are necessary for a better world. It is also a discipline of researchers and thinkers of considerable technical expertise and great enthusiasm for AI, raring to ply their trade on helping build a better future for the one and only Planet Earth.

BIBLIOGRAPHY

Bushwick, S., 2023. Tech talks to animals: Portable sensors and artificial intelligence are helping researchers to talk back to non-humans. *Scientific American*,, 05, 26–27.

Couclelis, H., 1986a. Artificial Intelligence in geography: Conjectures on the shape of things to come. *The Professional Geographer*, 38 (1), 1–11.

Couclelis, H., 1986b. A theoretical framework for alternative models of spatial decision and behavior. *Annals of the Association of American Geographers*, 76 (1), 95–113.

Dunsany, L., 1951. *The Last Revolution*. Talos Press edition.

Freeth, T., *et al.*, 2021. A model of the Cosmos in the ancient Greek Antikythera mechanism. *Scientific Reports*, 11 (1), 1–15.

Langton, C.G., *et al.*, 1991. *Artificial Life II*. Addison-Wesley Longman Publishing Co., Inc.

Mayor, A., 2018. *Gods and Robots: Myths, Machines, and Ancient Dreams of Technology*. Princeton University Press.

Openshaw, S. and Openshaw, C., 1997. *Artificial Intelligence in Geography*. John Wiley & Sons, Inc.

Simon, H.A., 2019. *The Sciences of the Artificial, Reissue of the Third Edition with a New Introduction by John Laird*. MIT Press.

Smith, T.R., 1984. Artificial Intelligence and its applicability to geographical problem solving. *The Professional Geographer*, 36 (2), 147–158.

3 Philosophical Foundations of GeoAI: Exploring Sustainability, Diversity, and Bias in GeoAI and Spatial Data Science

Krzysztof Janowicz
Department for Geography and Regional Research, University of
Vienna; Center for Spatial Studies, University of California, Santa
Barbara

CONTENTS

3.1 What is GeoAI? 26
3.2 Ethics of GeoAI 28
3.3 Sustainability of GeoAI 31
3.4 Debiasing GeoAI 32
3.5 Schema and Data Diversity 34
3.6 Is GeoAI Neutral? 35
3.7 Geography According to ChatGPT 37
3.8 Summary and Outlook 39
    Bibliography 40

3.1 WHAT IS GEOAI?

While GeoAI and spatial data science are relatively new fields of study, they share many of their underlying assumptions with geography, Artificial Intelligence (AI), cognitive science, and many other disciplines while adding their own perspectives. By philosophical foundations, I mean the core principles and beliefs that underlie *which* questions we ask, *why* we ask them, and *how* we ask them. For instance, one foundational epistemological belief underlying data science, even though rarely stated explicitly, is that knowledge can be gained through observation, a belief it shares with other empirical sciences. However, data science makes additional

DOI: 10.1201/9781003308423-3

foundational assumptions, e.g., that (raw) data can be reused opportunistically, and that black-box methods are acceptable, i.e., that knowledge can be gained without insight. In this chapter, I will outline selected philosophical foundations of GeoAI (many also relevant to spatial data science).

Selecting such foundations, deciding how to present them, compressing them into a few pages, and differentiating them from foundations of spatial data analysis, e.g., spatial dependence, more broadly, is challenging and not entirely objective. Hence, I will center the discussion around questions of research ethics and then highlight selected topics, such as sustainability, neutrality, and bias, from such an ethical perspective. This is for two reasons: first, too often, ethics is presented as an afterthought listed in future work sections of our papers or mentioned as an essential topic that did not make it into the curriculum of our classes. This time, we will put ethics first. Second, each topic covered here could fill an entire book by itself. One way to condense these topics to just a few paragraphs is to narrow the perspective. For instance, sustainability in GeoAI could be approached from many different perspectives, e.g., purely financially. Similarly, bias (and the potential need to debias Machine Learning [ML] models) could be approached from a strictly information-theoretic perspective by noticing that bias (when defined as lack of representativeness) is essentially redundant information with less than expected information content. Finally, regarding the selection of topics, many other candidates may have been considered, e.g., privacy, reproducibility, transparency, and accountability (Goodchild *et al.*, 2022). While these will be covered briefly, the main focus will be on topics that have not yet been widely considered in the GeoAI literature but should be on the mind of every researcher going forward, e.g., whether further tweaking a model is worth the environmental costs, whether the data used for evaluation is representative, or whether design decisions may affect the behavior of future users in unintended ways.

But what is GeoAI in the first place? Just as data science is not the intersection between computer science and statistics, GeoAI should not be narrowly defined as applying AI and ML methods to use cases in the geosciences and geography.[1] Establishing such subfields purely by domain would increase fragmentation, thereby reducing synergies and hindering transdisciplinary research (Janowicz *et al.*, 2022d). Instead, GeoAI should incorporate spatial, temporal, and place-based (*placial*) aspects into AI methods. Spatially explicit models (Janowicz *et al.*, 2020; Li *et al.*, 2021; Liu and Biljecki, 2022; Mai *et al.*, 2022) are such an example of successfully embedding spatial thinking into fields such as representation learning. Similarly, as will be discussed later, geographic classics such as the Modifiable Areal Unit Problem (MAUP) are used outside of our domain by the broader AI community to understand biases in their training data. Another example is the recognition that the validity of statements, e.g., in knowledge graphs (Hogan *et al.*, 2021), is spatially, placially, and temporally scoped (Janowicz *et al.*, 2022a). Finally, while GeoAI is by far not restricted to geography and the geosciences and has been successfully applied to downstream tasks in humanitarian relief, precision agriculture,

---

[1]Machine learning originated as a subfield of artificial intelligence. Hence, I will mostly use AI throughout the text and ML or AI/ML when discussing specifics, e.g., the design and training of neural networks. The term GeoML is not used in the literature (and hopefully will not be introduced).

urban planning, transportation, supply chains, climate change mitigation, and so on, most GeoAI practitioners are well trained in understanding human-environment interaction and the importance of notions such as *place* that can only be defined by jointly considering physical and cognitive/societal characteristics.

3.2 ETHICS OF GEOAI

Ethics, as the moral compass of human behavior, is as old as philosophy. While the origins of *research ethics* are challenging to trace, early work can be dated back to the 17th century and, from there on, gained traction throughout the Age of Enlightenment. Despite all its benefits, progress—be it in mathematics, engineering, medicine, or science—has never been without risks, and its benefits often favored some at the cost of others. However, it took two more centuries before the growing *immediate* impact of scientific discoveries on everyday life led to widespread public recognition of the dangers of a lack of research ethics, culminating in World War II and the atomic bomb. My views on many of the challenges discussed throughout this chapter are heavily influenced by Jonas' *ethic of responsibility* (Jonas, 1985) in which he reformulates Kant's initial categorical imperative by stating that "[we should act] so that the effects of [our] action(s) are compatible with the permanence of genuine human life".

While codes and policies for research ethics vary across fields of study, agencies, countries, and so forth, most of them are based on the following five considerations:

- Who benefits from the research?

- How can harm be avoided or minimized?

- Who is participating, and have they given consent?

- How are confidentiality and anonymity assured?

- How are research outcomes disseminated?

In a narrow interpretation, the first question seeks to clarify whether the research is carried out independently, e.g., to ensure that a source funding a study does not directly or indirectly benefit from specific outcomes. In a broader sense, research should contribute to the common good of all citizens. Put differently, the first question is not merely one of integrity and objectivity but also one of justice, representatives, and the use of shared (and limited) resources.

The second question may seem most relevant to medical research, where it aims to protect study participants from malpractices that may cause bodily harm but also aims at minimizing socioeconomic risks. While initially primarily concerned with the individual, a broader interpretation also considers society at large. Since at least the 1960s, technology assessment (nowadays, particularly in Europe) has been an established part of research ethics. The term harm has been increasingly broadened to non-human animals and the environment in general, bringing concepts such as sustainability to the forefront.

The third consideration centers around participants and their rights. While relevant questions here are also about harm, they focus more on procedural issues. For instance, how were participants selected, and did they give informed consent? The notion of informed consent has become one of the most central concepts of research ethics well beyond medical ethics. Intuitively, consent is about informing participants about a study's objectives, procedures, and potential risks. However, informed consent also asks who can provide consent in the first place (e.g., when studies involve children), how consent has been obtained, how it can be withdrawn (e.g., by refusing further treatment), and how transparent, accessible, and understandable the goals and processes of a research study have been made. Notably, the widespread use of social media APIs throughout GeoAI research does not free the authors from some of these considerations.

The fourth set of questions concerns confidentiality and anonymity, i.e., ensuring that data collected for a study remains secure and not traceable to individual participants. The notion of confidentiality used here extends beyond information technology measures such as encrypted, password-protected data storage. For instance, identifiable information collected about participants should not be disclosed or only via anonymization techniques. Increasingly, this includes steps to ensure *privacy by design* (Cavoukian, 2009) through seven principles. Key to these principles is the realization that privacy should be approached proactively instead of reactively, e.g., by minimizing the amount of data collected in the first place. Privacy by design can be regarded as a reaction to the vast body of literature demonstrating that de-identification may not efficiently protect against re-identifications attacks (Rocher *et al.*, 2019), e.g., revealing location (Keßler and McKenzie, 2018; Kounadi and Leitner, 2014; Krumm, 2009).

Finally, the fifth consideration centers around dissemination. It asks questions such as whether research results funded by taxpayers' dollars should be hidden behind paywalls. However, ethics is about more than (free) *access*. Thus, the FAIR principles for data management and stewardship have also been proposed to consider issues around *findability, interoperability*, and *reusability* (Wilkinson *et al.*, 2016). To broaden this fifth set of questions further, reproducibility and replicability could also be regarded as ethical principles taken into account for the proper dissemination of scientific results to ensure that more people can benefit from discoveries (Goodchild and Li, 2021; Kedron *et al.*, 2021; Nüst *et al.*, 2018).

Summing up, while most of the work on research ethics originated in fields such as medicine, cognitive science, and the social sciences, today, almost all areas of study benefit from understanding the basics of research ethics. Consequently, new branches of domain-specific ethics have been introduced to address gaps in the highly anthropocentric perspective presented before. GeoEthics is one such example (Peppoloni and Di Capua, 2017). It defines principles and practices for human interaction with the environment. It introduces concepts such as geo/bio-diversity, conservation, sustainability, prevention, adaptation, and education as ethical decision-making criteria for all earth scientists. In a nutshell, GeoEthics is about establishing processes for the recognition of human responsibility for our environment.

Ethics, however, is not limited to the (direct) interaction among humans or between humans and their environment but also involves information technology, e.g., algorithms, computers, and automation more broadly. The foundations for such ethics of modern (communication) technology were laid down in the 1940s to 1980s by Norbert Wiener, Joseph Weizenbaum, Hans Jonas, James Moor, Deborah Johnson, and many others. At the core of Moor's question about what computer ethics is or should be is a dilemma about transparency that sounds all too familiar now in the 2020s: while benefits from the lightning-fast operations of computers free us from inspecting each of the millions of calculations performed per second, this lack of transparency makes us susceptible to consequences from errors, manipulation, lack of representativeness, and so on (Moor, 1985). It is not surprising that Moor became interested in developing a novel concept of privacy for the digital age (Moor, 1997) later on.

These and other considerations jointly form another branch of ethics, namely *ethics of technology* that recognizes the social and environmental responsibilities involved in designing and utilizing computer systems and information technology (Jonas, 1985). AI ethics sits firmly within this broader branch. But what is AI ethics, and does it differ from ethics for AI? Here I will focus exclusively on our responsibilities in designing AI systems, leaving aside issues concerning behavioral norms of future (general) AI. Hagendorff (2020) has compiled a recent overview of 22 ethical guidelines for AI and the aspects they cover. Most interestingly, his work provides an overview of commonalities and gaps among these frameworks. For instance, 18 out of 22 cover privacy but merely one accounts for cultural differences in the ethically aligned design of AI systems (The IEEE Global Initiative on Ethics of Autonomous and Intelligent Systems, 2019). AI ethics generally addresses challenges arising from (a lack of) accountability, privacy, representativeness and discrimination, robustness, and explainability. This is unsurprising insofar as the answers to these ethical issues lie (at least partially) within the field of AI and computer science more broadly. For instance, explainable AI (Phillips *et al.*, 2020) and debiasing methods (Bolukbasi *et al.*, 2016) have become widely studied areas, also in the GeoAI community (Hsu and Li, 2023; Janowicz *et al.*, 2022c, 2018; Li, 2022; Papadakis *et al.*, 2022; Xing and Sieber, 2023).

However, AI ethics needs to consider broader societal implications outside of its own reach of methods to be truly impactful and to fulfill its largely positive potential. Examples of such key ethical questions are:

- How do we form societal consensus around technologies that reshape society at an unprecedented pace, e.g., regarding the future of work and education?

- What are the consequences of automatic *content* creation (at scale) that may be indistinguishable from human-generated content?

- How should we distribute the wealth created by AI?

- How do we handle accountability of autonomous systems, ranging from individual autonomous cars to large parts of the financial system?

- Is there a future for human judgment and decision-making in areas that require rapid response and prediction?

- Does intelligence require consciousness? If not, what does this mean for ethical AI?

But who, exactly, is responsible? Society, the individual data scientist, the "AI"? In their W3C Note on the responsible use of spatial data, Abhayaratna *et al.* (2021) distinguish between multiple perspectives: the developer, the user, and the regulator. Each of these roles has to contribute their part. For instance, users often all too willingly give up privacy for a bit of convenience. To give another example, developers (and scientists) should more carefully consider the minimal location precision required for a method or application to function (McKenzie *et al.*, 2022).

In a nutshell, GeoAI ethics is an ethics of technology that recognizes the social and environmental responsibilities of developers, regulators, and users involved in designing and employing AI systems that utilize spatial, temporal, and platial data and techniques related to their data analysis.

3.3 SUSTAINABILITY OF GEOAI

Fueled by the emergence of foundation models (Bommasani *et al.*, 2021) in 2018, progress in AI and ML has accelerated rapidly at the cost of increasingly complex models. These foundational models, such as the Generative Pre-trained Transformer (GPT) family of language models, consist of hundreds of billions of parameters and may require terabytes of training data. Consequently, training these models produces hundreds of metric tons of carbon dioxide emissions (Bender *et al.*, 2021) compared to the world's annual per-capita emissions of about 4.4 tons. Going one step further and estimating the entire cradle-to-grave lifecycle of such models would quickly reveal that the environmental impact of hardware manufacturing, deployment, and decommissioning dwarfs the training and operational environmental costs (Gupta *et al.*, 2022; Wu *et al.*, 2022). This is particularly concerning as it is in line with a more extensive debate about the geographic outsourcing of emissions throughout the supply chain, whereby a few counties significantly lower their (reported) footprint while, in fact, simply moving their own industry up the product (value) chain, away from manufacturing and emission-heavy stages. Put differently, the complexity and costs associated with progress in AI have grown by orders of magnitude in less than a decade, requiring new thinking about their environmental and social footprint.

While sustainability consciousness has increased throughout the AI and ML communities, many of the proposed solutions mostly focus on selecting sites where energy consumption has a smaller carbon dioxide footprint. However, sustainability and challenges related to costs and complexity run significantly deeper. For instance, Schwartz *et al.* (2020) rightfully note that a Green AI should also differentiate itself from a Red AI, where increased accuracy is reached merely through sheer computational power. The authors argue that such *buying* of *incremental* improvement is unsustainable and raises questions of fairness, as it limits the ability to participate and compete to very few actors.

Of course, resource consumption and resulting emissions must be put in perspective. For instance, radioactive waste is a common byproduct in hospitals, e.g., in cancer therapy. In fact, many of the most energy-intensive industry sectors, such as the chemical and construction industries, are irrevocably linked to progress and well-being. Interestingly, when changing scale from industry sectors to individual industry players, the picture becomes more complex as the list of the most energy-consuming companies is filled with big tech companies, most of them working on convenience technologies. Put differently, sustainability is not only about green(er) energy but about the inter- and intra-generational prioritization of how to utilize resources and space. Given the significant positive potential of modern-day AI for almost all areas of life, balancing its resource hunger (and other risks) with its benefits will be a significant societal challenge.

Consequently, Van Wynsberghe (2021) notes that it is essential to consider both AI for sustainability and sustainable AI, where the first is concerned with AI/ML-based methods and contributions to sustainability, e.g., environmental protection and the common good, while the second is concerned with making existing and future AI/ML systems more sustainable, e.g., by reducing emissions.

Finally, it is worth asking whether GeoAI faces additional or differently weighted sustainability challenges and whether (geo)spatial thinking offers novel perspectives on sustainability. The answer to both questions is yes.

First, the geo-foundation models of the future may require substantially more frequent update cycles (including retraining) than other models, e.g., those needed to generate images or text. Similarly, geospatial data and models must address additional challenges related to granularity as, in theory, data can be generated at an ever-finer spatial and temporal resolution.

Second, thinking geographically about AI sustainability opens up new avenues to explore. For instance, one could study the relationship between regions benefiting from a certain resource-intensive model versus those regions providing these resources. Intuitively, while we all potentially benefit from research about COVID-19, as a global pandemic, it is not necessarily clear why people in Iceland should offer their (environmentally more friendly but limited) resources to models that may predominantly benefit other regions, e.g., convenience technologies in the USA. It is worth noting that this does *not* imply that regions and their emissions can be seen in isolation. Finally, and to highlight yet another geographic perspective, the per-capita emissions reported here and elsewhere are global averages known to be highly skewed. Put differently, the relative burden (or, more positively, the number of people that may have benefited from the resources used) varies substantially across space (e.g., between Bangladesh and Canada), leading us to underestimate the real socio-environmental impact of very large models.

3.4 DEBIASING GEOAI

Given the rapid integration of AI/ML techniques in everyday decision-making, potential biases have become an important reason for concern and, consequently, an active field of study (Bolukbasi *et al.*, 2016). In a recent survey, Mehrabi et al. (Mehrabi *et al.*, 2021) categorize and discuss several such biases. What these biases have in common is that they may lead to *unfair*, skewed decisions. According to the authors, unfairness implies prejudice or favoritism toward some, potentially further increasing inequality. Consequently, the widespread use of AI may increase social problems such as discrimination, e.g., by widening income inequalities. The authors use the COMPAS (Correctional Offender Management Profiling for Alternative Sanctions) software as an example of a system that estimates the likelihood of recommitting crimes, all while being systematically biased against African-Americans.

Roughly speaking, bias can be introduced during three stages: from the data to the algorithm, from the algorithm to the user, and from the user interaction back to the data. Intuitively, bias in the training data may cause biased models.

To give a geographic example, according to (Shankar *et al.*, 2017) 60% of all geo-locatable images in the Open Images data set came from six countries in Europe and North America. Further, studying the geo-diversity of ImageNet, the authors reported that merely 1% of all images were from China and 2.1% from India. We were able to show similar coverage issues when studying potential biases in knowledge graphs (Janowicz *et al.*, 2018). Put differently, as far as commonly used data across media types is concerned, we know a lot about some parts of the world and almost nothing about others, and feeding such biased data into opaque models further exaggerates the resulting problems. One often overlooked issue is that we may overestimate (or underestimate) the accuracy of models by not considering that the difficulty of the task we are trying to address is unevenly distributed across (geographic) space. For instance, the accuracy of geoparsing systems is often reported based on unrepresentative benchmark data, leading Liu *et al.* (2022) to ask whether geoparsing is indeed solved or merely biased (toward specific regions to which methods have been well tailored). Other biases may arise from the well-known Modifiable Areal Unit Problem (MAUP) (Openshaw, 1984) and other sources of aggregation bias. I will address related data and schema diversity challenges in Section 3.5.

The next kind of biases can be introduced by the algorithm to the user, namely by causing behavioral change. One such example may be due to biased rankings in combination with the power law governing most social media. Put differently, the first ranking results are clicked over-proportionally, often followed by a quick drop. Hence, bias in ranking, e.g., of news, may alter the user's perception, e.g., of political discourse. Another bias potentially introduced at this stage is algorithmic bias, e.g., bias introduced by *decisions* made during model design. I will discuss a related question, namely whether algorithms and AI are neutral, in Section 3.6.

Finally, users also introduce biases into data, thereby closing the cycle back to *algorithms* (here in the sense of models and their design). For example, given that a significant part of training data across media types stems from user-generated content, changing behavior, e.g., hashtag usage over time, can introduce biases. Similarly, self-selection bias may lead systems to misestimate the result of online polls.

Other biases can be introduced based on the mismatch between the magnitude of historical data in relation to present-day data by which models tend to make the present and future appear more like the past (Janowicz *et al.*, 2018). Such biases may affect representation learning and association rule mining and may be difficult to detect. For instance, a system may infer that given somebody is a pope, they are also male. While (potentially) controversial, this statement is true by definition. Similarly, a system may learn that if x is a US president, x must be male (given that no counterevidence exists). While currently true, this is undoubtedly not an assertion about presidents and the US we would like a system to learn.

Finally, it is worth noting that the term *bias* has different meanings, and not all biases are problematic, e.g., inductive bias in ML. To make an even more abstract point, if foundation models (as basic building blocks of future AI-based systems) require algorithmic debiasing, then *debiasing may itself be biased*. How transparent will this process be, and how will it account for regional, cultural, and political differences?

The examples across different bias types introduced here also serve as an important example for the argument initially made that GeoAI is not merely the application of AI/ML methods to geographic and geospatial applications. Clearly, those biases affect AI and data science more broadly while being geographic in nature and making use of well-studied concepts of spatial data analysis, such as spatial auto-correlation and the MAUP. Put differently, GeoAI contributes back to the broader AI literature. At the same time, GeoAI faces its own unique challenges related to biases.

3.5   SCHEMA AND DATA DIVERSITY

Another important challenge facing GeoAI is related to the diversity of the data being processed, e.g., incorporated into training, and the diversity of the schema knowledge associated with these data (Janowicz *et al.*, 2022b). Data diversity can have different meanings, e.g., data coming from heterogeneous sources, data representing different perspectives, data created using different data cultures, data in various media formats, and so on. For instance, most existing ML models are not multi-modal. This, however, is changing rapidly, and future ChatGPT-like systems will handle multi-modality. For geo-foundation models, knowledge graphs could be a promising way to deliver a wide variety of multi-model data, e.g., vectors and rasters, in a canonical form. Data heterogeneity remains a substantial challenge as progress on *semantic* interoperability (Scheider and Kuhn, 2015) is slow. Different data cultures, e.g., governmental versus user-generated content, pose additional challenges as they have complementary strengths and weaknesses, thereby potentially requiring careful curation before being ingested into the same (Geo)AI models. Finally, multi-perspective data characterize the same phenomena but may offer complimentary or even contradictory stances. For example, the environmental footprint of nations can be assessed differently (Wiedmann *et al.*, 2015) without implying that one method is superior to others.

Taking this issue one step further brings us to schema diversity, i.e., the meaning of the domain vocabulary used may vary across space, time, and culture. While a few existing frameworks can handle contradicting assertions, e.g., about the disputed borders of the Kashmir region, schema diversity is on the terminological level, e.g., the fact that the definition of terms such as *Planet*, *Poverty*, *Forest*, and so on are spatially, temporally, and culturally scoped. Concept drift (Wang *et al.*, 2011), for instance, studies the evolving nature of terms within ontologies over time (and versions). Such cases are neither well studied in the broader AI literature nor is it clear how they can be incorporated during model design, as learning is largely a monotonic process. However, given that the domains studied by GeoAI are often at the intersection between humans (society) and their environment, such cases will arise frequently.

So, what exactly is the dilemma? On the one hand, data diversity is desirable for training a model for a wide range of use cases, unavoidable in Web-scale systems, e.g., knowledge graphs, and the context-dependent nature of meaning is well supported by research in linguistics and the cognitive sciences. On the other hand, however, contradicting assertions and even contradicting terminology make data curation and integration more complex and may negatively impact accuracy. Even more so, the culture-driven nature of categorization implies that category membership can change, even without substantial changes in the space of observable and engineerable (data) features (Janowicz *et al.*, 2022b). However, more ambiguous, changing, vague, etc. classes reduce accuracy and increase the need for training data, complex models, and so on, leading back to the discussion of sustainability (Janowicz *et al.*, 2022c). Note that (training) data size alone does not guarantee diversity (Bender *et al.*, 2021; Janowicz *et al.*, 2022b, 2015). Which categorization schema we will use for geo-foundation models, and how many of these models will be needed to represent regional variability remains to be seen.

3.6   IS GEOAI NEUTRAL?

"Guns don't kill, people do" is a popular slogan among parties favoring relaxed gun ownership and *open carry* laws. In a nutshell, the argument states that guns are merely neutral tools, comparable to a knife or hammer, that can be used for good and evil and that humans decide to use them one way or the other. Consequently, the "mentally ill" (as the argument unfortunately goes) are the root of the problem. This slogan is often followed up by a political *law and order* narrative presented as a remedy.

Similar arguments can be constructed about science and technology in general, e.g., by pointing out that nuclear technology was used for both the atomic bomb and for power generation and medicine. While I discussed the broader argument, counter-arguments, and technology assessment (Jonas, 1985) before, the next paragraphs ask whether AI and GeoAI are neutral (Janowicz *et al.*, 2022c).

One key problem such discussions face is terminological confusion, e.g., terms such as ML model, architecture, system, AI, algorithm, and so forth are used interchangeably, or their meaning needs to be clarified. For instance, arguing about biases arising from selecting training data while discussing whether algorithms are neutral points to such confusion (Stinson, 2022). To better understand the issue, it is crucial to distinguish basic algorithms from their parameterization, e.g., via training, their deployment, their role in more sophisticated workflows such as recommender systems, and the design decisions developers take to prefer one algorithm over another. Unfortunately, these distinctions are often not explicitly made when discussing whether AI or even algorithms are neutral.

In a nutshell, computer science is about *scalability* and *abstraction*, but it is precisely these concepts that make discussing neutrality difficult. For instance, data structures such as queues and stacks are fundamental because they enable us to focus on the commonalities of people standing in line, whether at the mall, fuel station, or bank. In all these cases, the person (or car) getting in line first will also be served first (at least in theory). This is in stark contrast to a stack where the last item on the stack is lifted first (to get to items lower in the stack). This is true for a stack of books and kitchen plates despite all their other differences. Now, a data structure such as a stack (as an abstract data type) is defined by its operations, e.g., push and pop. Are the algorithms implementing these operations neutral? If so, then if algorithms (or AI) would not be neutral, this lack of neutrality would have to be introduced later in the process, e.g., by their combination, selection for specific downstream tasks, and so on.

To give a more geographic example, consider the shortest path in a network and whether algorithms that compute such a path are neutral (Janowicz *et al.*, 2022c). Dijkstra's algorithm is one such algorithm and one that follows a *greedy* paradigm. While this design decision could (mistakenly) indicate a lack of neutrality, note that Dijkstra's algorithm will return the same results as other algorithms (leaving specific heuristics, nondeterministic algorithms, etc., aside). Even more, the same algorithms that are used to compute the shortest path (in fact, all paths) in a street network can also be used for many other networks, e.g., internet routing. Finally, repeated computation (again, leaving aside nondeterministic algorithms, edge/degenerated cases, issues arising from precision and parallelization, etc.) will yield the same shortest path as long as the underlying data has not changed, e.g., due to a road being blocked. So what about Dijkstra's algorithm would be non-neutral?

What does this mean for GeoAI, e.g., deep learning-based systems? Key to the success of artificial neural networks is their plasticity. In terms of their basic building blocks, and before training, these networks are essentially realized via millions and billions of multiplications (this is the scalability part) and other components such as softmax acting as an activation function. All of these single steps, such as multiplication, are by themselves neutral. Put differently, among the key reasons for the widespread success of deep learning across many downstream tasks is that the same building blocks and methods can be used, e.g., to detect buildings in remotely sensed imagery. This, in itself, is not a watertight argument for the neutrality of artificial neural networks or other (Geo)AI methods. Still, it shows that a meaningful discussion needs to examine where a lack of neutrality may materialize. There are several stages (some well studied) where design decisions make AI/ML based *systems* non-neutral, e.g., biased. For instance, *developers* make design decisions on how to combine the aforementioned basic building blocks, the data structures used, the number of hidden layers, the tuning of hyperparameters, the selection (and thereby also exclusion) of training data, the (manual) curation of data, regularization, the evaluation metrics used, the ways in which results are visualized and included further downstream to fuel rankings, predictions, and so on. Each of these steps may impact the overall performance of the system, including whether it will (implicitly) encode biases. For instance, in 2015, it was widely reported that Google's image recognition system mislabelled Black people as "gorillas" and that this might be due to a bias in the selection of training data and the (at that time) missing awareness among developers of the risks associated with (representation) bias.

This leads to two questions, namely whether these problems have been addressed sufficiently and whether there are GeoAI (or at least geo-specific) examples illustrating some issues of these implicit design decisions. As a thought example, consider the case of text-to-image models such as Stable Diffusion. In a test run of 20 prompts to generate a "Forest", none of the images depicted a forest during the winter (e.g., with snow), during the night, a rainforest, nor even basic variations in the types of trees displayed. Instead, the images produced were predominantly foggy, fall season–like images of what could best be called pine(like) forests. Prompts for "Chinese Mountains" predominantly resulted in abstract mixtures of mountains made out of fragments of the Great Wall, while "Mountains in China" did not. While these results are not representative and speak to the need for more robustness and transparency (explainability) of these early systems, they also likely show biases related to the geographic coverage of the imagery and labels used. A promising research direction for the near future will be to develop models that remain invariant under purely syntactic change. Today, for instance, embedding-based methods from the field of representation learning will yield different results for semantically similar or equivalent inputs if their syntax, e.g., structure, differs.

To answer the initial question, today's systems are not always neutral in their representation of geographic space. This may matter as users of these systems are served biased results.

3.7   GEOGRAPHY ACCORDING TO CHATGPT

Foundation (language) models have been compared to stochastic parrots (Bender *et al.*, 2021). While this analogy is deeply misleading, implying that we cannot truly learn anything new from such models, it opens up novel ways of thinking about their unique opportunities. Similarly to the recognition that the MAUP cannot be entirely avoided and should, therefore, be seen as an opportunity (O'Sullivan and Unwin, 2003) for studying why some results change when regions are modified while others remain (relatively) stable, foundation models could be used to observe why models behave in specific ways as reflections of the underlying data and thereby society. If indeed these models encode biases based on racism, stereotypes, geography, and so on, then instead of (or better, in addition to) trying to debias these issues and thereby hide them, one could use these models to better understand the underlying phenomena at scale in ways that have never been possible in geography and the social sciences before.

For instance, the labeled (social media) data sources utilized to train text-to-image models have always been biased regarding their coverage, scenes depicted, and labels used. However, it is the ease with which we can now experiment with these models that bring the problems to the public's attention. Similarly, when learning to classify perceived safety and walkability in urban areas, we do not have to limit ourselves to arguing that these models will discriminate against certain regions and their populations, thereby further widening inequality. Instead of merely asking how to tweak such systems to yield more *intended* results, we can ask which of the learned features predominantly *contributed* to these results and which changes to the urban areas would alter their classification. This would turn parrots into (distorted) mirrors. However, we can go even further and take an observational stance by arguing that these very large models consisting of billions of parameters and trained on billions of samples are worth studying for their own sake (as they are among the largest and most complex artifacts developed by society to this day and in retrospect may provide a first glance at how communication with more general AI may look in the future). This idea aligns well with concepts such as geographic information observatories (GIO) (Hendler *et al.*, 2008; Janowicz *et al.*, 2014; Miller, 2017). How does the world look according to foundation models such as GPT or Stable Diffusion, and why?

To understand why this is an important question, consider the development of information retrieval and web search more broadly. Initially, several search engines gave keyword-based ranked access to millions of web pages. The user was responsible for reading these pages, judging their veracity (e.g., by the platform that hosted them), and extracting the relevant information. Today, two search engines dominate the market, and their type-ahead functionality, while convenient, makes queries more canonical. Additionally, these search engines are now powered by knowledge graphs and question-answering techniques that return answers, not web pages. Hence, when asking for the population of Vienna or the date that the Santa Barbara Mission was established, users are unlikely to check webpages but will be satisfied with the direct answer provided by the search engines and their graphs. While using a digital assistant such as Siri will additionally restrict the ability to trace the origins of the data presented, this is still possible in principle. Communicating and getting information from ChatGPT-like systems is entirely different as it not only radically transforms how we may interact with (web) search engines in the future, if at all but also because the results returned by these bots are not traceable (to a primary source), as the system itself has generated them (at least in parts). Consequently, understanding how foundation models represent geographic space will matter greatly.

Finally, understanding why progress, e.g., in answering geographic questions, is not uniform may not only reveal problems of current model design but also point to underlying issues such as *our own* tendency to favor geometry-first instead of topology-first representations (Feng *et al.*, 2019; Janowicz *et al.*, 2022d). For instance, in 2015, we reported that when asked for the distance between Ukraine and Russia, Google and Bing returned meaningless centroid distances instead of understanding that both countries share a border, and thus, their true distance is zero (Janowicz *et al.*, 2015). This problem is still present in 2023. Asked for the same, ChatGPT gives a substantially more sophisticated but equally meaningless answer; see Figure 3.1.

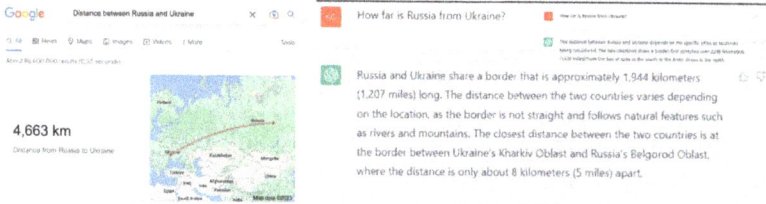

Figure 3.1   Distance between Ukraine and Russia according to Google Search and ChatGPT. Note that ChatGPT also changes the border length across queries. The proper answer should be zero, even though ChatGPT makes a very convincing statement that at the border where both countries *touch*, their distance is 5 miles.

3.8   SUMMARY AND OUTLOOK

While GeoAI is a new and rapidly developing area that shares many of its research challenges with the broader fields of (spatial) data science, AI, geography, and the geosciences, it also offers its own questions and contributes to a broader body of knowledge. This chapter highlights selected philosophical foundations underlying current GeoAI work from a research ethics perspective. Understanding these foundations and making them explicit is important for multiple reasons, e.g., for our community to be able to contribute to the ongoing discussion about bias and debiasing.

However, there are also more subtle issues worth remembering. For instance, (machine) learning is based on the assumption that we can draw inferences about the present and future by studying the past. That sounds like a triviality. For today's very large language models and the resources required to train them, however, this poses many relevant challenges such as historic drag, i.e., that we have more data from the overall past compared to the recent past. Not only does this imply that these models are outdated (by years), but that they may be slow to change and, therefore, may need careful curation to ensure they can keep in sync with societal change and new scientific discoveries. For instance, consider the example of night-time light (NTL) frequently used as a proxy for economic activity. While applicable in the past[2], policies introduced in 2022 aimed at significantly reducing light emissions at night (e.g., to conserve energy). Consequently, a system that has learned the relationship between NTL and economic activity may not be able to quickly adjust when this relationship weakens. Similarly, as of today, foundation models (and the bots deployed on top of them) do not account for cultural and regional differences. Hence, it is worth exploring which theory of truth underlies their answers and our interpretation thereof (irrespective of their tendencies to *hallucinate*). For instance, one may speculate that systems such as ChatGPT may best be studied by following a consensus theory of truth compared to one based on coherence (which today's large language models cannot maintain).

---

[2]Although it may not be equally suitable as a proxy across geographic scales and may actually trace population density not (just) activity (Mellander *et al.*, 2015).

In 2019, we provocatively asked whether "we [can] develop an artificial GIS analyst that passes a domain-specific Turing Test by 2030" (Janowicz *et al.*, 2020). What if combining the first geo-foundation models of the near future with ChatGPT-like bots may get us there before? Will we be able to understand and mitigate its biases? Will we be able to explain and defend our design decisions in implementing and deploying such systems? Articulating the assumptions and principles underlying GeoAI research will be a first step.

ACKNOWLEDGMENTS

I would like to thank Kitty Currier, Zilong Liu, Meilin Shi, Rui Zhu, Gengchen Mai, Karl Grossner, and many others for their valuable discussions, pointers to additional literature, and critical feedback. This work was partially funded via the KnowWhere-Graph project (NSF-2033521).

BIBLIOGRAPHY

Abhayaratna, J., *et al.*, 2021. *The Responsible Use of Spatial Data*. W3C.

Bender, E.M., *et al.*, 2021. On the dangers of stochastic parrots: Can language models be too big? In: *Proceedings of the 2021 ACM Conference on Fairness, Accountability, and Transparency*. 610–623.

Bolukbasi, T., *et al.*, 2016. Man is to computer programmer as woman is to homemaker? debiasing word embeddings. *Advances in Neural Information Processing Systems*, 29.

Bommasani, R., *et al.*, 2021. On the opportunities and risks of foundation models. *arXiv preprint arXiv:2108.07258*.

Cavoukian, A., 2009. *Privacy by Design*. Canadian Electronic Library.

Feng, M., *et al.*, 2019. Relative space-based gis data model to analyze the group dynamics of moving objects. *ISPRS Journal of Photogrammetry and Remote Sensing*, 153, 74–95.

Goodchild, M., *et al.*, 2022. A white paper on locational information and the public interest. *American Association of Geographers*.

Goodchild, M.F. and Li, W., 2021. Replication across space and time must be weak in the social and environmental sciences. *Proceedings of the National Academy of Sciences*, 118 (35), e2015759118.

Gupta, U., *et al.*, 2022. Chasing carbon: The elusive environmental footprint of computing. *IEEE Micro*, 42 (4), 37–47.

Hagendorff, T., 2020. The ethics of AI ethics: An evaluation of guidelines. *Minds and Machines*, 30 (1), 99–120.

Hendler, J., *et al.*, 2008. Web science: an interdisciplinary approach to understanding the web. *Communications of the ACM*, 51 (7), 60–69.

Hogan, A., *et al.*, 2021. Knowledge graphs. *ACM Computing Surveys (CSUR)*, 54 (4), 1–37.

Hsu, C.Y. and Li, W., 2023. Explainable GeoAI: can saliency maps help interpret artificial intelligence's learning process? An empirical study on natural feature detection. *International Journal of Geographical Information Science*, 1–25.

Janowicz, K., *et al.*, 2014. Towards geographic information observatories. *In: GIO@ GI-Science*. 1–5.

Janowicz, K., *et al.*, 2020. GeoAI: Spatially explicit artificial intelligence techniques for geographic knowledge discovery and beyond.

Janowicz, K., *et al.*, 2022a. Know, know where, knowwheregraph: A densely connected, cross-domain knowledge graph and geo-enrichment service stack for applications in environmental intelligence. *AI Magazine*, 43 (1), 30–39.

Janowicz, K., *et al.*, 2022b. Diverse data! diverse schemata? *Semantic Web*, 13 (1), 1–3.

Janowicz, K., Sieber, R., and Crampton, J., 2022c. GeoAI, counter-AI, and human geography: A conversation. *Dialogues in Human Geography*, 12 (3), 446–458.

Janowicz, K., *et al.*, 2015. Why the data train needs semantic rails. *AI Magazine*, 36 (1), 5–14.

Janowicz, K., *et al.*, 2018. Debiasing knowledge graphs: Why female presidents are not like female popes. *In: ISWC (P&D/Industry/BlueSky)*.

Janowicz, K., *et al.*, 2022d. Six giscience ideas that must die. *AGILE: GIScience Series*, 3, 7.

Jonas, H., 1985. *The Imperative of Responsibility: In Search of an Ethics for the Technological Age*. University of Chicago press.

Kedron, P., *et al.*, 2021. Reproducibility and replicability: opportunities and challenges for geospatial research. *International Journal of Geographical Information Science*, 35 (3), 427–445.

Keßler, C. and McKenzie, G., 2018. A geoprivacy manifesto. *Transactions in GIS*, 22 (1), 3–19.

Kounadi, O. and Leitner, M., 2014. Why does geoprivacy matter? the scientific publication of confidential data presented on maps. *Journal of Empirical Research on Human Research Ethics*, 9 (4), 34–45.

Krumm, J., 2009. A survey of computational location privacy. *Personal and Ubiquitous Computing*, 13, 391–399.

Li, W., Hsu, C.Y., and Hu, M., 2021. Tobler's First Law in GeoAI: A spatially explicit deep learning model for terrain feature detection under weak supervision. *Annals of the American Association of Geographers*, 111 (7), 1887–1905.

Li, Z., 2022. Extracting spatial effects from machine learning model using local interpretation method: An example of shap and xgboost. *Computers, Environment and Urban Systems*, 96, 101845.

Liu, P. and Biljecki, F., 2022. A review of spatially-explicit GeoAI applications in Urban Geography. *International Journal of Applied Earth Observation and Geoinformation*, 112, 102936.

Liu, Z., *et al.*, 2022. Geoparsing: Solved or biased? an evaluation of geographic biases in geoparsing. *AGILE: GIScience Series*, 3, 9.

Mai, G., *et al.*, 2022. A review of location encoding for GeoAI: methods and applications. *International Journal of Geographical Information Science*, 36 (4), 639–673.

McKenzie, G., *et al.*, 2022. Privyto: A privacy-preserving location-sharing platform. *Transactions in GIS*, 26 (4), 1703–1717.

Mehrabi, N., *et al.*, 2021. A survey on bias and fairness in machine learning. *ACM Computing Surveys (CSUR)*, 54 (6), 1–35.

Mellander, C., *et al.*, 2015. Night-time light data: A good proxy measure for economic activity? *PloS One*, 10 (10), e0139779.

Miller, H.J., 2017. Geographic information science i: Geographic information observatories and opportunistic giscience. *Progress in Human Geography*, 41 (4), 489–500.

Moor, J.H., 1985. What is computer ethics? *Metaphilosophy*, 16 (4), 266–275.

Moor, J.H., 1997. Towards a theory of privacy in the information age. *ACM Sigcas Computers and Society*, 27 (3), 27–32.

Nüst, D., *et al.*, 2018. Reproducible research and giscience: an evaluation using agile conference papers. *PeerJ*, 6, e5072.

Openshaw, S., 1984. The modifiable areal unit problem. *Concepts and Techniques in Modern Geography*.

O'Sullivan, D. and Unwin, D., 2003. *Geographic Information Analysis*. John Wiley & Sons.

Papadakis, E., *et al.*, 2022. Explainable artificial intelligence in the spatial domain (X-GeoAI). *Transactions in GIS*, 26 (6), 2413–2414.

Peppoloni, S. and Di Capua, G., 2017. Geoethics: ethical, social and cultural implications in geosciences. *Annals of Geophysics*.

Phillips, P.J., *et al.*, 2020. Four principles of explainable artificial intelligence. *Gaithersburg, Maryland*, 18.

Rocher, L., Hendrickx, J.M., and De Montjoye, Y.A., 2019. Estimating the success of reidentifications in incomplete datasets using generative models. *Nature Communications*, 10 (1), 1–9.

Scheider, S. and Kuhn, W., 2015. How to talk to each other via computers: Semantic interoperability as conceptual imitation. *Applications of Conceptual Spaces: The Case for Geometric Knowledge Representation*, 97–122.

Schwartz, R., *et al.*, 2020. Green AI. *Communications of the ACM*, 63 (12), 54–63.

Shankar, S., *et al.*, 2017. No classification without representation: Assessing geodiversity issues in open data sets for the developing world. *arXiv preprint arXiv:1711.08536*.

Stinson, C., 2022. Algorithms are not neutral: Bias in collaborative filtering. *AI and Ethics*, 2 (4), 763–770.

The IEEE Global Initiative on Ethics of Autonomous and Intelligent Systems, 2019. Ethically aligned design: A vision for prioritizing human well-being with autonomous and intelligent systems, first edition. IEEE.

Van Wynsberghe, A., 2021. Sustainable AI: AI for sustainability and the sustainability of AI. *AI and Ethics*, 1 (3), 213–218.

Wang, S., Schlobach, S., and Klein, M., 2011. Concept drift and how to identify it. *Journal of Web Semantics*, 9 (3), 247–265.

Wiedmann, T.O., *et al.*, 2015. The material footprint of nations. *Proceedings of the National Academy of Sciences*, 112 (20), 6271–6276.

Wilkinson, M.D., *et al.*, 2016. The FAIR Guiding Principles for scientific data management and stewardship. *Scientific Data*, 3 (1), 1–9.

Wu, C.J., *et al.*, 2022. Sustainable AI: Environmental implications, challenges and opportunities. *Proceedings of Machine Learning and Systems*, 4, 795–813.

Xing, J. and Sieber, R., 2023. The challenges of integrating explainable artificial intelligence into GeoAI. *Transactions in GIS*, 27 (3), 1–20.

Section II

GeoAI Methods

4 GeoAI Methodological Foundations: Deep Neural Networks and Knowledge Graphs

Song Gao
GeoDS Lab, Department of Geography, University of
Wisconsin-Madison

Jinmeng Rao, Yunlei Liang, Yuhao Kang
GeoDS Lab, Department of Geography, University of
Wisconsin-Madison

Jiawei Zhu
School of Geosciences and Info-Physics, Central South
University

Rui Zhu
School of Geographical Sciences, University of Bristol

CONTENTS

4.1   Introduction 46
4.2   Definitions 47
4.3   Convolutional Neural Networks 48
4.4   Recurrent Neural Networks 50
4.5   Transformers 52
4.6   Graph Neural Networks 53
4.7   Generative Adversarial Networks 57
4.8   Reinforcement Learning 60
4.9   Knowledge Graphs and GeoKG 63
4.10 Summary 66
      Bibliography 67

DOI: 10.1201/9781003308423-4

4.1 INTRODUCTION

Recent advances in artificial intelligence (AI), hardware accelerators, and big data processing architectures continue driving the rapid development of geospatial artificial intelligence (GeoAI) (Gao, 2021; Li, 2020), which is the integration of geospatial domain knowledge with AI technologies to support data-intensive scientific discovery in geographical sciences (Janowicz *et al.*, 2020). In this chapter, we will introduce the foundations of GeoAI methodologies, with a main focus on recent breakthroughs in machine learning, deep learning, and knowledge graphs (Hogan *et al.*, 2021; LeCun *et al.*, 2015). We will introduce the key concepts and architectures of convolutional neural networks, recurrent neural networks, transformers, graph neural networks, generative adversarial networks, reinforcement learning, and knowledge graphs, with a special focus on incorporating spatial thinking and principles into the development of spatially explicit AI models and geospatial knowledge graphs, including geographical domain applications (Janowicz *et al.*, 2022; Mai *et al.*, 2022a).

In order to determine whether a model is spatially explicit or not, four tests, including the invariance test, the representation test, the formulation test, and the outcome test have been proposed (Janelle and Goodchild, 2011; Janowicz *et al.*, 2020). Essentially, if an AI model satisfies one of these tests (i.e., location variance, spatial or platial representation, involving spatial concepts or spatial structure change in model input/output), it can be treated as a spatially explicit AI model. Therefore, as shown in the conceptual framework (Figure 4.1), several strategies have been utilized in recent GeoAI models to make them spatially explicit such as encoding spatial data structure (vector/raster), incorporating spatial concepts into the design of distance-based weights, space-aware loss functions and pooling layers in neural networks, or including spatial predicates in knowledge graphs. In the following sections, we will present more details for each model in this framework.

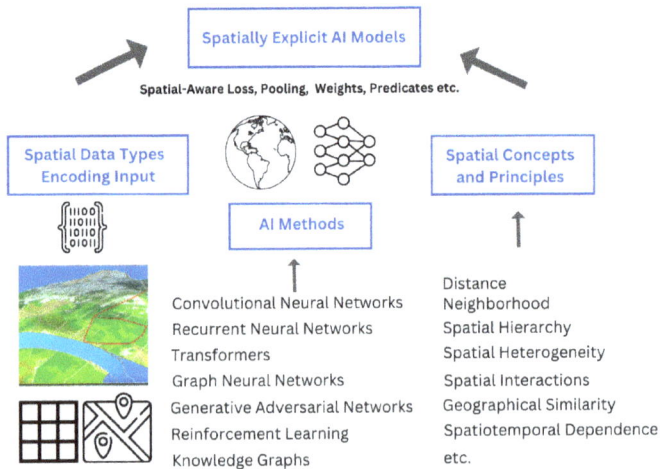

Figure 4.1 A conceptual framework toward the development of spatially explicit AI models.

4.2 DEFINITIONS

In this section, we first briefly introduce some of the key concepts and terminologies used in the development of GeoAI models. Note that this is not an exhaustive list and that many fundamental concepts in machine learning and deep learning can be found in other AI textbooks for further reading, e.g., (Goodfellow *et al.*, 2016; Patterson and Gibson, 2017). In addition, other chapters in this handbook will further discuss specific concepts in depth (e.g., spatial representation learning).

Deep neural network is a hierarchical, multi-layered (at least two hidden layers) organization of connected artificial neurons (Figure 4.2).

Weights are learnable parameters that represent the connection strength of the neurons between a neural network's layers.

Biases are an additional input (a constant added to the product of features and weights) into the next layer in neural networks and help shift the activation function toward the positive or negative side. This concept is different from other types of biases (e.g., sampling bias or measurement bias) identified in machine learning and AI systems (Mehrabi *et al.*, 2021).

Activation function is used in neural networks to decide if and how the signal/output of the previous layer can pass to the next connected layer; different transformation functions, such as linear, rectified linear, sigmoid, trigonometric, and logistic functions, can be used (Patterson and Gibson, 2017).

Loss function is a metric used to minimize the error during the training process of a neural network. Mean squared error (MSE) loss or mean absolute error (MAE) loss are often used for regression models, while logistic loss and cross-entropy loss are the most common for classification models. Spatially-aware loss functions, such as trajectory-loss (Rao *et al.*, 2021a), which are tailored to spatial data types, and usually perform better than non-spatial loss functions in GeoAI model training.

Gradient is a vector of partial derivatives of a function. *Gradient descent* is an optimization algorithm for finding the optimal weights to minimize the error in model training. To know which direction (uphill or downhill) to move in the cost space, the *backpropagation* algorithm is often used to train neural networks by propagating the prediction error from the output to the input layers through the computation of the gradients of a cost/loss function with updated weights (Patterson and Gibson, 2017).

Hyperparameters are the configuration settings which determine the neural network structure and affect model performance. Tuneable hyperparameters include: number of hidden layers, number of neurons in each layer, activation function, loss function, learning rate, drop out or other regularization, number of training epochs, and sample batch size.

Spatial embedding or *spatial representation learning* is a feature learning technique in GeoAI models to encode various types of spatial data, such as points, polylines, polygons, networks, or rasters, in an embedding space (i.e., vectors of real numbers), so that various formats of spatial data can be readily incorporated into machine/deep learning models (Mai *et al.*, 2022b).

Spatially explicit AI model is an AI model that passes one of the four tests related to the spatial concepts described in Janowicz *et al.* (2020).

Geospatial knowledge graph (GeoKG) is a spatial extension to a knowledge graph, which contains symbolic representations of geographic entities (e.g., places) and other types of entities (e.g., people, organizations, and events), including attributes and their relations, often empowered by geo-ontologies (Hu *et al.*, 2013; Janowicz, 2012) and geo-linked data (Hart and Dolbear, 2013), to support cross-domain data analytics and knowledge discovery through the use of spatial contexts (Janowicz *et al.*, 2022).

4.3 CONVOLUTIONAL NEURAL NETWORKS

The Convolutional Neural Network (CNN) is a fundamental artificial neural network architecture, especially for image-based analysis. It was first proposed by LeCun *et al.* (1998). A typical implementation of a CNN is shown in Figure 4.2, which is developed based on *LeNet-5* to explain the structure of a CNN. Components of the *LeNet-5* include convolutional layers, pooling or sub-sampling layers, and fully connected layers.

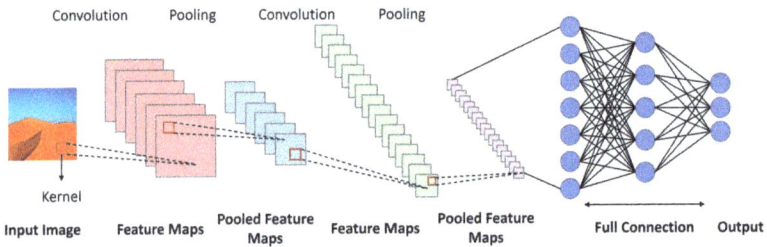

Figure 4.2   The architecture of the LeNet: a convolutional neural network.

The convolutional layers rely on kernels that can be viewed as sliding windows to extract visual features from image cells. The pooling layers reduce the dimension of features extracted from the convolutional layers, and the fully connected layers generate outputs for certain applications (e.g., classification labels). Despite that CNN is developed on top of Aritificial Neural Network (ANN), such an architecture is capable of extracting more expressive high-dimensional visual features from images. It also reduces the number of parameters by increasing the number of layers. Since 2012, when a CNN-based model called *AlexNet* won the ImageNet challenge by significantly outperforming other traditional machine learning approaches (e.g., Support Vector Machines), CNNs have been extensively leveraged to solve a variety of computer vision applications. For example, CNNs have been utilized in computer vision tasks such as image classification, object detection and localization, as well as semantic image segmentation. The state-of-the-art technology of autonomous vehicles is also built on CNNs (Gao *et al.*, 2018a). Inspired by these successful applications, researchers continued to design more advanced architectures by attaching more hidden layers, as experiments have shown that a deeper architecture can handle more complex computer vision tasks. It is also why most of the modern neural networks are termed "Deep Convolutional Neural Networks (DCNNs)" (Krizhevsky *et al.*, 2017). Notable models include but are not limited to GoogLeNet (Szegedy *et al.*, 2015), U-Net (Ronneberger *et al.*, 2015), ResNet (He *et al.*, 2016), Mask R-CNN (He *et al.*, 2017), and DenseNet (Huang *et al.*, 2017) which have hundreds of layers. The backpropagation algorithm is often used to train DCNNs.

Given the capabilities of DCNNs in processing images, they are naturally appropriate for processing raster-based geographic data, including (but not limited to) remote sensing images, topographic maps, geotagged social media photos, and street view images. Remote sensing images and raster-based maps are two classic image-based geographic data types, and researchers have utilized DCNNs for spatial knowledge discovery and geographic phenomena recognition and modeling with these data types (Feng *et al.*, 2019). Additionally, in the era of big data, DCNNs have been employed to process emerging geospatial data sources, such as geotagged social media photos and street view imagery (Biljecki and Ito, 2021; Hawelka *et al.*, 2014). Moreover, despite the fact that some geospatial datasets (e.g., mobility flows) are not usually represented as images, researchers have tried to aggregate mobility flows into multiple grids at different times, which enables the modeling of crowd flows using DCNNs (Liang *et al.*, 2019) while spatial interaction flow data are often modeled by graph-based approaches (Li *et al.*, 2021a).

DCNNs can be utilized in geospatial data studies in two ways. In the first approach, processing multiple forms of geographic data can be viewed as a subcategory of computer vision, i.e., the applications of DCNNs for computer vision with a special focus on geographic data. Here, we list three basic computer vision tasks with geographic data. (a) Image classification refers to assigning a label or class to a given image. For geographic data, researchers have employed DCNNs for spatial scene search (Guo *et al.*, 2022) and land cover classification with remote sensing images (Cheng *et al.*, 2020) and place scene classification with geotagged images (Zhou *et al.*, 2014). (b) Object detection refers to extracting instances of objects from images, and researchers have leveraged DCNNs to identify and extract geographic objects (e.g., point-based, polyline-based, and polygon-based objects) from remote sensing images (Li *et al.*, 2020), raster maps (Chiang *et al.*, 2020), and street view images (Kang *et al.*, 2020). (c) Image semantic segmentation refers to partitioning an image into multiple regions and extracting its visual semantics. It also serves as a prerequisite for object detection, as instances are identified based on the partitioned image, and researchers have used DCNNs to segment remote sensing images (Zhu *et al.*, 2017b) and street view images (Zhang *et al.*, 2018a) for such purposes. As a whole, all of the studies described above demonstrate the effectiveness of employing DCNNs for creating geographic features in a format that can be used for modeling geographic phenomena.

The second approach to using DCNNs for geographic modeling involves the development of spatially explicit GeoAI models, i.e., integrating spatial principles in the development of DCNNs. For instance, Li *et al.* (2021c) designed a spatially explicit model that combines DCNNs and LSTMs by considering Tobler's First Law of Geography for terrain feature detection. Yan *et al.* (2018) improved DCNN place classification performance by involving spatial relatedness, co-location, and sequence pattern. Guo *et al.* (2022) developed a DeepSSN approach to better assess spatial scene similarity by incorporating spatial pyramid pooling and a spatial reasoning triplet sample mining strategy. Given the successful results of these studies, incorporating spatial concepts into the development of DCNNs, rather than merely applying DCNNs to solve geographic domain problems, has tremendous promise.

Using AI models, generally, requires large datasets for model training, and the recent boom in deep learning has benefited greatly from the large-scale image database known as ImageNet (Deng *et al.*, 2009). To propel the development of spatially explicit DCCNs, it is necessary to construct a series of geospatial image datasets that focus on different geographic phenomena. For instance, researchers have created several openly available labeled image datasets for different geographic phenomena, such as land cover datasets (Fritz *et al.*, 2017; Helber *et al.*, 2018), place scene datasets (Zhou *et al.*, 2017), and natural feature datasets (GeoImageNet) (Li *et al.*, 2022). Though these datasets offer solid data foundations for the training of spatially explicit DCNNs, more are needed to cover other geospatial data types.

4.4   RECURRENT NEURAL NETWORKS

Another type of artifical neural network is the Recurrent Neural Network (RNN), which focuses on learning patterns in sequential data, such as text, audio, and video. Compared with CNNs, which are feedforward neural networks, RNNs allow the output from previous nodes to affect the current nodes through cyclic connections. Therefore, RNNs have the ability to memorize previous information and use it to predict the value at a future time stamp (Goodfellow *et al.*, 2016).

The architecture of an RNN is illustrated in Figure 4.3 (LeCun *et al.*, 2015). The left shows a circuit diagram: a recurrent cell incorporates the current data x into the state h, which is passed from the previous neuron containing all previous information. The right explains the process of unfolding the cell in time, where each node is associated with a specific time stamp. The output y_t at time stamp t is obtained by taking the current value x_t and all previous information stored in h_{t-1}. The parameters U, V, W are shared across all timestamps in the model. This can help extend the model to unseen data with different sequence lengths (Goodfellow *et al.*, 2016).

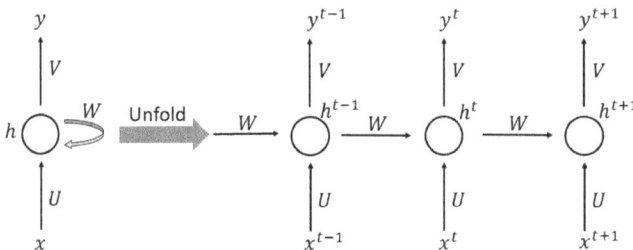

Figure 4.3   A recurrent neural network and its unfolding.

In theory, RNNs can learn long-term dependencies, but empirical evidence shows that long-term gradients can "vanish" or "explode" during the backpropagation process (Bengio *et al.*, 1994). In response, Long Short-Term Memory (LSTM) networks have been widely used to solve the gradient vanishing problem (Hochreiter and Schmidhuber, 1997). LSTM units improve the memory capacity of standard RNNs by using the concept of gates (Yu *et al.*, 2019). Usually, an LSTM contains three types of gates: forget, input, and output, as shown in Figure 4.4 (Yu *et al.*, 2019). Near the top of the figure is a horizontal line representing the cell state, which allows information to flow unchanged. The forget gate uses a sigmoid layer to decide the percent of historical information to be forgotten from the cell state. The input gate determines which information is added into the cell state. The sigmoid layer generates a scale factor and the tanh layer generates candidate values. The scale factor will multiply the candidate values to produce a fraction of the candidate values that will be updated to the cell state c_{t-1}. Finally, the output gate decides which information to output based on the cell state.

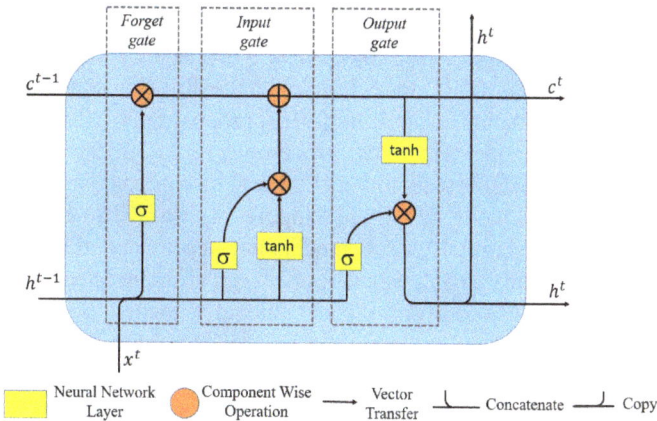

Figure 4.4   Architecture of LSTM with a forget gate.

The Gated Recurrent Unit (GRU) is another popular variant of RNNs proposed to solve the vanishing-gradient problem (Chung *et al.*, 2014). The GRU has two gates: an update gate and a reset gate (Cho *et al.*, 2014). The update gate integrates the functions of the forget gate and the input gate in the LSTM, and decides how much information to update. In contrast, the reset gate decides what previous information to forget. Due to its architecture, the GRU has fewer parameters than LSTM. Both LSTM and GRU outperform traditional RNN models, but the choice between them heavily depends on the specific task and datasets (Chung *et al.*, 2014).

As introduced above, RNNs are designed to process sequential data, making them ideal for processing temporal information and learning temporal dependencies. But how to incorporate spatial relationships into RNNs needs further consideration (May Petry *et al.*, 2020; Zhao *et al.*, 2019).

RNNs have been widely used in transportation for topics such as traffic prediction based on sequential data in the past (May Petry *et al.*, 2020). Spatial information is also crucial to correctly predict phenomena such as human travel behavior, which are greatly influenced by geographical proximity. Liu *et al.* (2016) proposed a Spatial-Temporal Recurrent Neural Network (ST-RNN) that models local temporal and spatial contexts for traffic prediction. ST-RNN uses distance-specific transition matrices, which are calculated based on the coordinates of users at different time stamps. In order to represent users' dynamic visit interest, the representation of an user u at time t_k with their coordinates q_t^u is updated with a distance-specific transition matrix S and a time-specific transition matrix T.

To make RNNs spatially explicit for trajectory prediction, many studies have employed trajectory embedding. May Petry *et al.* (2020) models trajectory points via a multi-attribute embedding layer, where the spatial dimensions of trajectories such as Points of Interest (POIs), latitude and longitude are encoded in the embedding layer, and the embeddings are then used as inputs of the following LSTM model. Similarly, other models can be applied to capture the underlying spatial relationship, and provide new representations of it as the input to RNN models. Li *et al.* (2017) proposed a Diffusion Convolutional RNN that captures (a) the spatial dependency using random walks on the graph, and (b) the temporal dependency using GRU, where the matrix multiplications in GRU are replaced with the diffusion convolution layer that contains spatial information. Zhao *et al.* (2019) proposed a T-GCN model for traffic prediction. To capture topological structures among road segments, the road network with traffic information is first sent into a Graph Convolutional Network. The output is then fed into a GRU model to capture temporal patterns (Zhao *et al.*, 2019).

In the remote sensing field, RNN models have been applied to remote-sensing images to capture contextual dependencies between different image regions. For example, a few studies use four different directional scannings (left-to-right, top-to-bottom, etc.) on the images to obtain region sequences and use the sequences as input of the RNNs (Zhang *et al.*, 2018b).

In summary, multiple approaches have been adopted to incorporate spatial information into the RNN model. Generally, these methods can be categorized into three groups: (a) the internal part of a RNN can be modified to include spatial information, (b) the spatial information can be captured by other integrated models, and (c) the spatial information is embedded in the input that will be fed into the RNN models.

4.5 TRANSFORMERS

Another powerful model for understanding sequential data is the Transformer model, which is designed based on attention mechanisms (Vaswani *et al.*, 2017). Compared with RNN models, the Transformer processes all the inputs at one time and learns global dependencies between input and output. It also allows more parallelization than RNNs, which can reduce training time significantly. The architecture of the Transformer model has an encoder-decoder structure. The inputs are first converted into embeddings in vector format, and positional encoding is applied to involve the information about orders and positions. The encoder consists of N identical layers, and each layer has a multi-head self-attention mechanism and a fully connected feedforward network. The decoder layer has a similar structure but with an additional attention sub-layer between the self-attention and the feed-forward layers to connect the encoder and decoder parts (Cai *et al.*, 2020; Vaswani *et al.*, 2017). Transformer-based methods have been proposed to recognize place entity type and extract place names from natural language descriptions. Hu *et al.* (2022) proposed a gazetteers-based approach that uses pre-trained transformer models for place name extraction from tweets. To further capture information from volunteered sources, Berragan *et al.* (2022) built transformed-based language models to extract unknown place names from unstructured text and found that more dataset-specific fine-tuned models out-perform generalised pre-trained models.

Although Transformers are mostly used to solve Natural Language Processing (NLP) problems, it has also been applied to various spatial-temporal problems. Some studies separate the extraction of spatial relationships and the use of those relationships to guide the Transformer training process (Cai *et al.*, 2020; Li and Moura, 2019), while other studies modify the original Transformer to directly capture spatial information (Li *et al.*, 2021b; Xu *et al.*, 2020).

A few studies extract the representative spatial relationship from data and then feed the new data representation into the Transformer. For example, Li and Moura (2019) designed a Forecaster for forecasting spatial and time-dependent data. The Forecaster starts by learning a spatial dependency graph between locations, and then the dependency graph is used to sparsify the linear layers in the Transformer so that the encoding of data at one location is only affected by itself and its spatially-dependent locations (Li and Moura, 2019). Similarly, Cai *et al.* (2020) proposed a traffic transformer that uses graph convolutional filters to capture spatial dependencies. The traffic network is first fed into a graph convolutional neural network to capture the spatial dependencies, after which it is sent into the Transformer encoder to learn temporal features.

Finally, some studies modify the internal structure of the Transformer to include spatial components explicitly. Xu *et al.* (2020) proposed a spatial transformer to model directed spatial dependencies with self-attention mechanism. The spatial transformer uses a spatial-temporal positional embedding layer that incorporates spatial-temporal position information such as topology, connectivity, and time steps into each node. A few other studies design different spatial-temporal encoders/de-coders to capture the spatial-temporal information (Li *et al.*, 2021b).

4.6 GRAPH NEURAL NETWORKS

A graph is a data structure comprised of nodes (also known as vertices) and edges that specify a certain type of relationship between node. Graph structures contain rich information on the relationships between entities, and are widely used to represent real-world connections. However, conventional neural networks, such as CNNs and RNNs, are primarily designed to operate on grid-like or sequential data structures and thus are not well-suited to handling irregular, non-Euclidean graph structures (Wu *et al.*, 2020). To effectively incorporate the topological information present in a graph data structure, Graph Neural Networks (GNNs) designed for operating on graph data have emerged (Zhou *et al.*, 2020). Since then, GNNs have been successfully applied in many areas, performing a wide range of tasks, including node classification, node regression, link prediction, graph regression and classification.

Before explaining GNN models, we first introduce some basic notations. Given a graph $G = (V, E)$, where $V = (v_1, ..., v_n)$, $|V| = n$ represents a set of nodes, and $e_{ij} = (v_i, v_j)$ in the set of edges. E reflects the existence of a specific relationship between node i and j. This specific relationship can be modeled by a $n \times n$ adjacency matrix $\mathbf{A}$. When the edges in a graph only indicate the presence or absence of a certain relationship, it is an unweighted graph. In this case, $a_{ij} = 1$ in the adjacency matrix if $e_{ij} \in E$, otherwise $a_{ij} = 0$. When the relationship represented by the edges can be quantified, such as when it indicates the strength of spatial interactions, a_{ij} can be set to real values. The nodes in graph G may have attributes and can be represented as an attribute matrix $\mathbf{X} \in R^{n \times m}$.

The core concept behind GNNs is to specify a local neighborhood for each node in the graph and to use this neighborhood to compute the node's representation. This representation is then passed on to the next layer in the network, which is used to compute the representations for the next set of nodes. This process is repeated until the final layer of the network, at which point a downstream task is completed based on the learned node representations.

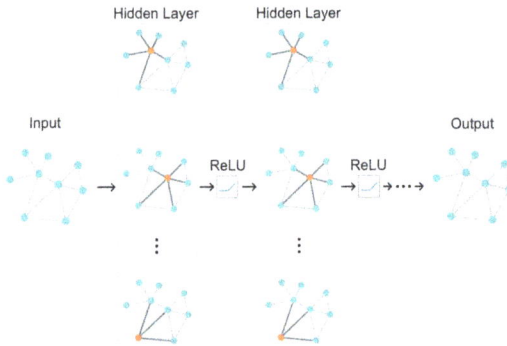

Figure 4.5   Architecture of GCN with multiple layers.

Graph Convolutional Network (GCN) (Kipf and Welling, 2016) is a widely adopted GNN model. It consists of multiple layers, each of which takes in the node representations from the previous layer and produces new representations for the next layer, as shown in Figure 4.5. The node representations in each layer are computed using a graph convolution operation, which can be expressed as Equation 4.1:

$$\mathbf{H}^{(l+1)} = \sigma(\tilde{\mathbf{D}}^{-\frac{1}{2}} \tilde{\mathbf{A}} \tilde{\mathbf{D}}^{-\frac{1}{2}} \mathbf{H}^{(l)} \mathbf{W}^{(l)}), \tag{4.1}$$

where $\mathbf{H}^{(l)}$ is the matrix of node representations in layer l and $\mathbf{H}^{(0)} = \mathbf{X}$, $\tilde{\mathbf{A}}$ is the adjacency matrix of the graph with self-loops, $\tilde{\mathbf{D}}$ is the degree matrix of $\tilde{\mathbf{A}}$, and $\mathbf{W}^{(l)}$ is the trainable weight matrix for the convolution operation in layer l. The function $\sigma(\cdot)$ is an activation function, such as ReLU. GCN's graph convolution process can be considered a weighted average of the node representations of each node's neighbors, with the weights determined by the adjacency and degree matrices. This enables the GCN to extract structural information from the graph and use it to generate more informative node representations.

Another powerful model, Graph Attention Network (GAT) (Velickovic *et al.*, 2017), was introduced to automatically determine the importance of nodes. This is achieved by employing attention mechanisms, which allow the GAT to weigh the contributions of different nodes accordingly based on their representations. The GAT model can be formulated as follows:

$$\mathbf{h}_i^{(l+1)} = \sigma \left(\sum_{j \in \mathcal{N}_i} \alpha_{ij} \mathbf{W} \mathbf{h}_j^{(l)} \right), \tag{4.2}$$

where $\mathbf{h}_j^{(l)}$ represents the hidden representation of node j in the l-th layer, $\mathbf{W}$ is the corresponding weight matrix for the l-th layer, and α_{ij} is the attention coefficient for the connection between node i and node j, it is calculated as Equation 4.3.

$$\alpha_{ij} = \frac{\exp\left(\text{LeakyReLU}\left(\mathbf{a}^T \left[\mathbf{W}\mathbf{h}_i \| \mathbf{W}\mathbf{h}_j\right]\right)\right)}{\sum_{k \in \mathcal{N}_i} \exp\left(\text{LeakyReLU}\left(\mathbf{a}^T \left[\mathbf{W}\mathbf{h}_i \| \mathbf{W}\mathbf{h}_k\right]\right)\right)}, \tag{4.3}$$

the dot product of the weight vector $\mathbf{a}$ and the concatenated representations of nodes i and j are first calculated to weigh the importance of node j to node i. This dot product is then passed through the LeakyReLU function to produce the attention coefficient. Finally, the attention coefficient is normalized by the summation of coefficients of node i's neighbors.

In addition to the models mentioned above, a vast number of GNN models have been proposed. For example, GraphSAGE (Hamilton *et al.*, 2017) samples neighbors and designs three aggregation strategies Mean, LSTM, and Pooling. The difference between GNNs lies in how they define the neighborhood and utilize the information to update node representations. GNN models generally have fewer layers than CNN and RNN models due to the over-smoothing issue associated with having many layers (Oono and Suzuki, 2019).

There are several ways to make the GNN model spatially explicit:

(a) **Consider the spatial relationships when constructing the input graph**. In the geographic context, geographic units can be abstracted as nodes in the graph. Then, connections and similarities between geographic units can be utilized to construct graphs. The concepts of spatial connections and spatial weights are thoroughly discussed in the GIScience literature, such as by Getis (2009). One of the directions is to construct the adjacency matrix based on the spatial adjacency relationship between geographic units, such as Rook or Queen adjacency; another is to calculate the distance between geographic units and then obtain the adjacency matrix by calculating ε-neighborhood, k-nearest neighbors or using kernel functions. Since one critical operation of GNNs is to aggregate neighbor information, making it a natural tool that fits the first law of geography (spatial dependency effect) when given the spatial adjacency matrix as input. Similarly, the spatial interaction matrix can be used as input to GNN, which not only can take into account the long- and short-range connections between geographic units but also has important implications for mobility and related studies (Liang *et al.*, 2022). With the road graph converted from the road network as the input graph and the geo-semantic embedding vector as node attribution, Hu *et al.* (2021) utilized the GCN model to classify urban functions at the road segment level. Zhu *et al.* (2020b) analyzed place characteristics through a GCN model and discussed the influence of different graphs constructed from various place connection measures on prediction accuracy.

(b) **Consider spatial effects in the aggregation process to assign weights to neighboring nodes**. For the input graph construction mentioned above, we mainly define a neighborhood with "spatial thinking"; also, we can construct spatially related rules to decide how to assign weights to the neighboring nodes when feeding with a generic graph. For instance, we can assign weights according to the distance decay principle (Tobler's first law of geography (Tobler, 1970)), where neighboring nodes are assigned less weight if they are farther away in the geographical distance from the central node. Or, the weights can be defined based on certain similarities of spatial settings (the third law of geography (Zhu *et al.*, 2018)), e.g., the higher the similarity of the surrounding POI distribution, the larger the given weight. These can be tailored to the specific downstream task. In the spatio-temporal prediction model proposed by Gong *et al.* (2022), the distance decay function is applied to optimize the adjacency matrix, which allows the model to better estimate spatial dependencies of passengers' travel behaviors.

(c) **Consider spatial constraints when optimizing the model**. For example, suppose we expect spatially adjacent nodes to have similar representations. In that case, a term can be added to the loss function to pull together the representations between spatially neighboring nodes. If we expect that nodes with similar geographical configurations have similar representations, a constraint term can likewise be added to the loss function. For example, Region2vec (Liang *et al.*, 2022) utilizes the spatial interaction flow strength and geographic distance as constraints to guide the learning process of GCN. The spatial constraint added in the loss function pushes away node pairs that are spatially distant from each other to guarantee spatial contiguity.

(d) **GNNs can also be combined with other models such as RNN and LSTM to process spatiotemporal data**. The combinations have been widely used in the field of traffic flow forecasting and the study of urban dynamics (Jiang and Luo, 2022). In the T-GCN model (Zhao *et al.*, 2019), with the road network and traffic flow data as input, GCN is first used to capture the spatial dependency of geographic units at each time stamp. Then GRUs use the outputs of GCNs to model temporal dependency to predict future traffic flows. Li *et al.* (2021a) proposed a model for predicting the intensity of human activity. The model employs a GCN to model spatial interaction patterns between geographic units and an LSTM to model the tendency and periodicity of the temporal pattern of activity intensity series. Zhang and Cheng (2020) integrated a gated network and a localized graph diffusion network to model the temporal propagation and spatial propagation of sparse spatiotemporal events separately for predictive hotspot mapping.

4.7 GENERATIVE ADVERSARIAL NETWORKS

Generative Adversarial Networks (GAN), proposed by Goodfellow *et al.* (2014), is a type of neural network framework for generative AI tasks, and it has been widely used in image generation, style transfer, super-resolution, and audio generation. A well-designed GAN is able to learn data distribution from input data and sample new data from the learned data distribution in an unconditional or conditional manner. One significant difference between GAN and previously mentioned fundamental structures (e.g., CNN, RNN, Transformer, GNN) is that GAN is not some concrete network architecture, but an abstract design concept. This means that we can utilize different kinds of fundamental structures to design GAN models for generative tasks in various domains, such as CNN-based GANs for style transfer (Zhu *et al.*, 2017a), RNN-based GANs for trajectory generation (Rao *et al.*, 2021a), Transformer-based GANs for high-resolution image generation (Zhang *et al.*, 2022), and GNN-based GANs for graph representation learning (Wang *et al.*, 2019).

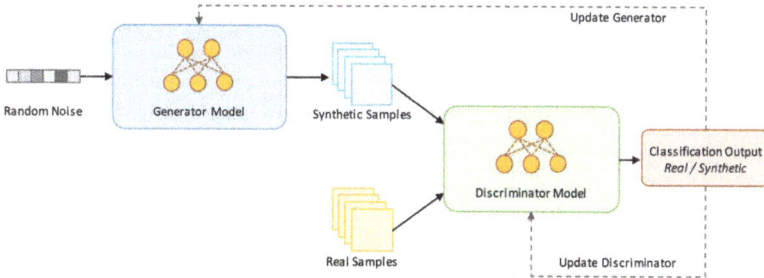

Figure 4.6 A conceptual framework of Generative Adversarial Networks.

In general, a GAN model consists of two components (see Figure 4.6): a generator model denoted as G and a discriminator model denoted as D. The goal of G is to learn how to capture the data distribution from training data (i.e., real samples) and generate synthetic samples that can "fool" D, while the goal of D is to learn how to determine whether a sample comes from real samples or synthetic samples, then providing feedback to update G. This is an adversarial training procedure and leads to a minimax two-player game. Ideally, we expect G well captures the data distribution of training data after training and can generate "high-quality" synthetic data for generative tasks. The optimization objective of GAN can be expressed as follows (Equation 4.4):

$$\min_{G} \max_{D} \mathbb{E}_{x \sim p_{\text{data}}(x)} \left[\log D(x) \right] + \mathbb{E}_{z \sim p_z(z)} \left[1 - \log D(G(z)) \right] \qquad (4.4)$$

where $p_{data}(x)$ denotes raw data distribution; $p_z(z)$ denotes a prior on noise variables; $D(x)$ denotes the probability that x came from $p_{data}(x)$; $G(z)$ denotes a mapping from $p_z(z)$ to $p_{data}(x)$. G aims to minimize $\mathbb{E}_{z \sim p_z(z)}[log(1 - D(G(z)))]$ while D aims to maximize $\mathbb{E}_{x \sim p_{data}(x)}[logD(x)] + \mathbb{E}_{z \sim p_z(z)}[log(1 - D(G(z)))]$. It is also proved that optimizing this optimization objective is equivalent to minimizing the Jensen-Shannon divergence (JSD) between training data distribution and synthetic data distribution (Arjovsky *et al.*, 2017).

The original GAN is an unconditional model, which means G can generate any synthetic data as long as they fall in learned data distribution. Sometimes, however, we want the model to generate synthetic data conditioned on certain information (e.g., instead of generating arbitrary cat pictures, we want the model to specifically generate pictures of "Ragdolls" or "Maine Coons" according to given input). In fact, by involving such information (i.e., condition, denoted as c), the original GAN can be extended to a conditional model. Accordingly, the optimization objective of such a conditional GAN (Mirza and Osindero, 2014) can be expressed as follows (Equation 4.5). The only change from the original optimization objective is that condition c also serves as an extra input for both G and D.

$$\min_G \max_D \mathbb{E}_{x \sim p_{\text{data}}(x)}[\log D(x|c)] + \mathbb{E}_{z \sim p_z(z)}[1 - \log D(G(z|c))] \qquad (4.5)$$

Since the original GAN was proposed, there have been tons of GAN variants developed for optimizing the original design or adapting the design to various domains. For example, Deep Convolutional GAN (DCGAN) replaces the multilayer perceptions in the original design with deep convolutional neural networks to achieve more powerful and stable unsupervised image representation learning (Radford *et al.*, 2015). Information Maximizing GAN (InfoGAN) learns disentangled representations from training data in an unsupervised manner by considering the mutual information between a small subset of the latent variables and synthetic data (Chen *et al.*, 2016). CycleGAN learns unpaired image-to-image translation by introducing cycle consistency in its loss design (Zhu *et al.*, 2017a). Wasserstein GAN (WGAN) introduces a novel optimization objective for minimizing the Wasserstein distance between real data distribution and synthetic data distribution, which largely mitigates the gradient vanishing and training instability issues (Arjovsky *et al.*, 2017). SSD-GAN integrates a frequency-aware classifier to measure and alleviate the spectral information loss in the discriminator, reducing the spectrum discrepancy between real data and synthetic data (Chen *et al.*, 2021).

There are several common directions regarding incorporating spatial thinking and principles into the development and applications of spatially explicit GAN models.

(a) **The first direction is spatial data generation and translation**. Since GAN models have been widely studied for image generation and image-to-image translation, it is natural to utilize GAN models to generate or translate raster geospatial data. Some studies focus on generating synthetic satellite images via GAN models trained on existing satellite images (Rui *et al.*, 2021), while some other studies explore style transfer among different raster data such as base maps, land use maps, building footprints, and remote sensing imagery (Kang *et al.*, 2019; Wu and Biljecki, 2022; Zhao *et al.*, 2021). For vector geospatial data generation or translation such as spatiotemporal mobility data, many studies focus on integrating sequence models (e.g., RNN components) into GAN models for sequence modeling and generation (Rao *et al.*, 2021a), while some other works first convert spatiotemporal data into two-dimensional maps and then use traditional CNN-based structure for data generation and translation (Ouyang *et al.*, 2018).

(b) **The second direction is spatial data imputation**. It has been demonstrated that GAN models can learn data imputation by using its generator to impute missing parts of observations and using its discriminator to assess imputation quality (Yoon *et al.*, 2018). Similar ideas have been introduced into the methods and applications for data imputation in geographical domains. For example, many studies have applied GAN models for interpolating air quality data (Luo *et al.*, 2018), traffic data (Yang *et al.*, 2021), satellite data (Wang *et al.*, 2023), and digital elevation model (DEM) data (Zhu *et al.*, 2020a), yielding promising results.

(c) **The third direction is spatial data representation**. Representation learning is a classic machine learning problem where we try to learn latent feature representations of raw data. Such learned representations can be further used for various downstream tasks such as classification and retrieval in general domain (Xia *et al.*, 2014) and location modeling and spatial context modeling in geographical domain Mai *et al.* (2019). GAN is recognized as a successful structure for representation learning from large-scale datasets (Radford *et al.*, 2015). For example, we can design a generator to learn how to generate latent representations of raw data, and we feed the representations into a discriminator for assessing representation quality. GAN models have been widely adopted into representation learning for satellite data (Lin *et al.*, 2017), drone data (Bashmal *et al.*, 2018), and social network data (Tang *et al.*, 2021).

(d) **The fourth direction is spatial prediction**. As a special case of the general prediction task which predicts unseen data (i.e., target variables) based on existing observations (i.e., independent variables), spatial prediction is to predict unseen data at some locations based on existing observations at the same or different locations. Many spatial thinking and principles have been introduced into spatial prediction such as the First Law of Geography (Tobler, 1970), the Second Law of Geography (Goodchild, 2004), and the Third Law of Geography (Zhu *et al.*, 2018). GAN-based spatial prediction studies have two focuses. One focus is to use GAN models as data augmentation tools to augment or rebalance datasets, which boosts the performance of downstream models on spatial prediction. Existing studies has covered raster data augmentation (Al-Najjar *et al.*, 2021), vector data augmentation (Zhou *et al.*, 2021), and time-series data augmentation (Toutouh, 2021). Another focus is to directly use GAN models to perform spatial prediction. The basic idea is to use the generator as a spatial prediction model and use a discriminator to assess prediction performance. Many studies have investigated spatial prediction via GAN models such as satellite image sequence prediction (Dai *et al.*, 2022), trajectory prediction (Rüttgers *et al.*, 2019), and weather prediction (Bihlo, 2021).

4.8 REINFORCEMENT LEARNING

With the great success of Alpha Go (Silver *et al.*, 2016), Reinforcement Learning (RL) (Kaelbling *et al.*, 1996), as one of Alpha Go's core technologies, has received increasing attention. RL is a type of machine learning paradigm that allows an agent to learn how to behave in an environment by doing certain actions and observing feedback from the environment (see Figure 4.7). Compared with supervised learning, which allows models to learn from labeled independent input-output pairs, the agent in RL makes sequential actions based on its state in the environment and receives rewards or penalties accordingly, eventually learning a policy that maximizes the cumulative reward over time (i.e., sequential decision-making). One of the most common frameworks used in RL is the Markov Decision Process (MDP) (Puterman, 1990). An MDP is a mathematical framework that describes the interaction between an agent and an environment. It consists of the following concepts:

- States: The states represent the different configurations of the environment that the agent can be in. The agent can observe the state of the environment, and the state transitions from one to another as the agent takes action.

- Actions: The actions are the choices that the agent can make in each state. The agent can take different actions in different states, and the actions determine the next state and the associated reward.

- Rewards: The rewards are the feedback (positive or negative) that the agent receives from the environment for taking a specific action in a specific state. The rewards are used to evaluate the agent's performance and guide its learning.

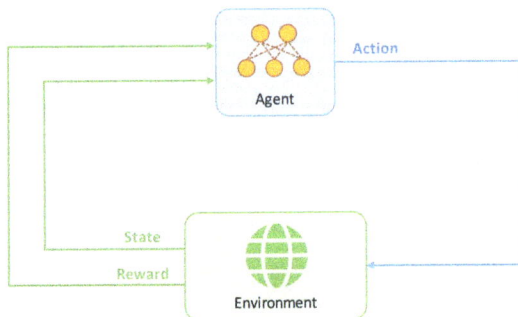

Figure 4.7   A conceptual framework of Reinforcement Learning.

- Transition Model: A transition model describes the likelihood of going from one state to another state after taking a specific action.

- Policy: A policy is the agent's behavior in the environment, which is a map from states to actions. It can be either deterministic or stochastic.

There are two main streams of approaches to solving an MDP: *model-based approaches* and *model-free approaches*. Model-based approaches require a known transition model that allows the agent to directly figure out optimal policy (e.g., via dynamic programming). Model-based approaches have the advantage of being able to plan over the entire state space and action space, but they can be computationally expensive and may not work well in environments with high stochasticity or large state and action spaces. Model-free approaches, in contrast, do not require a known transition model. Instead, the agent learns the policy or the value function directly from experience (e.g., via the Monte Carlo method, or Temporal Difference methods such as Q-Learning (Watkins and Dayan, 1992)). Model-free approaches are usually considered simpler and more sample efficient than model-based approaches, but they can only learn locally around the states that the agent has visited, and may have difficulty in dealing with unseen states. Both model-based and model-free approaches can be further divided into two types: on-policy approaches, which use the same policy to generate both the behavior and the data used for learning, and off-policy approaches, which uses a different behavior policy to generate the behavior and use the data to learn another target policy.

Recently, many methods have been proposed to improve the stability, efficiency, and performance of the training process of RL. Deep Reinforcement Learning (DRL) methods, such as Deep Q-Network (DQN) (Mnih *et al.*, 2015) and Asynchronous Advantage Actor-Critic (Mnih *et al.*, 2016), extend the traditional RL methods with deep neural networks, allowing for more powerful and versatile representations of the state and action spaces. Trust Region Policy Optimization (Schulman *et al.*, 2015) and Proximal Policy Optimization (Schulman *et al.*, 2017) are on-policy optimization algorithms that update policies by adjusting the parameters to move toward better policies while respecting the stability of the policies.

In RL, the incorporation of spatial thinking and principles into spatially explicit models can be divided into four directions.

(a) **The first direction is to redefine traditional cartography or GIS tasks via the language of RL and directly use RL models to solve these problems**. In this case, the environment could be any space described by geospatial data, such as a 2-dimensional space for maps or remote sensing images, or a 3-dimensional space for terrains or urban models. An agent or a group of agents can be placed in the environments to take action, change states, and receive rewards or penalties. The reward functions are associated with the goal of cartography or GIS tasks, such as a scoring function for map generalization or land-use planning. By doing so, we are able to solve traditional cartography or GIS tasks by converting them into RL problems and maximizing the cumulative rewards. For example, some studies use reinforcement learning to solve map generalization and vector-to-raster alignment problems (Duan *et al.*, 2020).

(b) **The second direction is to use RL to support spatial optimization problems**. In this case, similarly, the environment could be any space described by geospatial data such as road networks or remote sensing images, and we use agents to iterate in the environment to optimize spatial decision-making and find the optimal or sub-optimal policy by maximizing the cumulative rewards. For each particular spatial optimization problem, the entities behind the agents could be different, such as vehicles in route planning and traffic scheduling, or the allocator in resource allocation. For multiple-agent cases, the agents will optimize the policy together not only based on the feedback from the environment but also based on the interaction with other agents, including collaboration or competition. In practice, some studies use RL to optimize taxi dispatch, taxi driving strategies, and route planning in large-scale and complex traffic environments (Gao *et al.*, 2018b; Koch and Dugundji, 2020), and some studies use RL to optimize strategies in remote sensing applications such as aerial object detection and image captioning for better performance (Li *et al.*, 2018).

(c) **The third direction is to use RL to support spatial simulation and prediction**. In this case, we focus on using the agents and the environment in RL to simulate the research objects and the environment in the real world, including behaviors, mobility patterns, decision-making strategies, etc. We hope that such a simulation could help us better understand and predict how the real-world case works, how the system would change when the environment changes, and how the future outcomes of different courses of action would be. For example, in spatial epidemiology, we can use RL to simulate and predict the spread of a certain virus based on real-world historical infection cases, human mobility, and socioeconomic factors. In practice, RL has been successfully applied for simulating and predicting the growth trend of COVID-19 cases (Srikanth *et al.*, 2021), privacy-preserving vehicle trajectories with customized driving behaviors (Rao *et al.*, 2021b), and spatial dynamics in forest fire based on satellite images (Ganapathi Subramanian and Crowley, 2018).

(d) **The fourth direction is to use RL to integrate feedback from human experts in geographical domains for model improvement**. Recent studies show that Reinforcement Learning from Human Feedback is an efficient way to incorporate external knowledge from human feedback and interaction and greatly improve AI model performance (Ouyang *et al.*, 2022). Similar ideas could also be adopted in developing spatially explicit models. For example, in land use change modeling, a human expert may have knowledge of the most likely scenarios for land use changes in a specific region. The RL algorithm can learn from this expert's decisions and incorporate that information into its predictions, resulting in improved model accuracy. Another example is in the management of natural resources, where a human expert can provide feedback on the most effective strategies for conserving resources in a particular region. By integrating the human feedback using RL, the model can adapt and optimize its decisions accordingly.

4.9 KNOWLEDGE GRAPHS AND GEOKG

Even though the concept of "knowledge graph" has initially emerged in 1970s, it is not until 2010s when Google announced its own knowledge graph has it turned to be a popular data model to organize and analyze data. Such a revolution was mainly attributed to the urgent need of managing increasingly heterogeneous and multimodal data from the Web in order to provide effective information retrieval services. Nowadays, knowledge graphs are not only regarded as a stack of technologies for AI but also a new paradigm of representing, reusing, sharing, integrating, interoperating, explaining, and mining data. In contrast to traditional relational databases, where data are organized as tables with each row representing an individual record and each column indicating its attribute, knowledge graphs highlight the various relations between entities. As its name indicates, the key idea is to use the structure of a graph to abstract data, in which nodes are used to represent any kinds of entities, such as places, people, organizations, etc., and edges are the relations between these entities, such as friendship, located in, works at, etc (see Figure 4.8 as an example). Mathematically, a knowledge graph can be described as a *directed labelled graph*: $G = (V, E, L)$, where V is the set of nodes, L is the set of edge labels, and $E \subseteq V \times L \times V$ is the set of edges (Hogan *et al.*, 2021). Note that labels L entail various meanings. Furthermore, coupled with formal representation of knowledge, such as ontologies or schema, knowledge graphs are capable of defining and reasoning over the semantics underlying the data (Janowicz *et al.*, 2010; Kuhn, 2005). In comparison to its relational counterpart, such a graph-based abstraction of knowledge has multiple merits, including (a) expressiveness (e.g., data are semantically entailed, and both the data and metadata can be modeled using the same data model), (b) extensibility (e.g., easy to add more data), (c) flexibility (e.g., the ontology/schema are

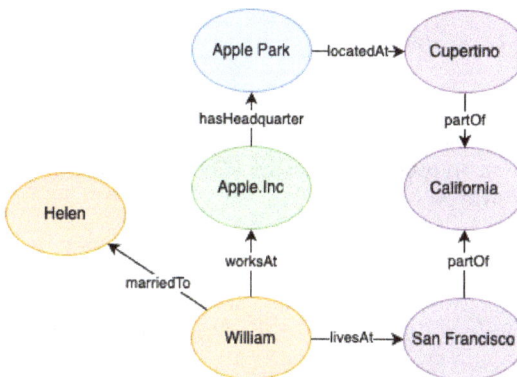

Figure 4.8 Example of KG. Different node colors represent different classes.

adoptable), as well as (d) compliance to the Finable, Accessible, Interpretable, and Reusable (FAIR) principle (Wilkinson *et al.*, 2016).

There are numerous implementations of knowledge graphs. The most popular two are: Resource Description Framework (RDF) graph (Miller, 1998) and property graph. This Chapter focuses on RDF graph, which is a standardization recommended by the World Wide Web Consortium (W3C). In RDF graphs, the basic data structure is called a triple, which is composed as: $\langle subject, predicate, object \rangle$, where both subject and object refer to the nodes in the graph and predicate refers to the edge, and the direction goes from subject to object. Different from other data models, all the three components in the triple are uniquely identifiable by a URI (Uniform Resource Identifier) on the Web. The object can sometimes be a literal value (e.g., number or string) too. Triples are used to express all types of facts, statements, and resources. For instance, the statement "Bristol is part of England" can be represented as $\langle Bristol, part\ of, England \rangle$. Thanks to its standardization and associated URI on representing resources, RDF graphs benefit from another major merit that most data models lack: (e) the interoperability across data sources, since all nodes in the graph can be uniquely identified and linked through semantically enabled relations. In addition to RDF, many other standards are also defined by W3C to support knowledge graphs. To execute reasoning, one can use ontology languages, such as OWL (Web Ontology Language). To validate data, one can apply the Shape Constraint Language (SHACL), which constrains the data graph by defining a shape graph over it. To query the graph, the query language of SPARQL is introduced. Similar to how SQL is used for relational databases, a series of syntax and operations are specified to add, delete, update, and extract data from the graph. A simple example of SPARQL query is demonstrated in Listing 4.1.

```
PREFIX  geo:   <http://www.opengis.net/ont/geosparql#>
PREFIX  geof:  <http://www.opengis.net/def/function/geosparql>
PREFIX  rdf:   <http://www.w3.org/1999/02/22-rdf-syntax-ns#>
PREFIX  kwg-ont:  <http://stko-kwg.geog.ucsb.edu/lod/ontology>
PREFIX  xsd:   <http://www.w3.org/2001/XMLSchema#>
PREFIX  wd:    <http://www.wikidata.org/entity>
PREFIX  rdfs:  <http://www.w3.org/2000/01/rdf-schema#>

SELECT ?fire ?geo ?igDate ?wkt
WHERE {
    ?fire  rdf:type* kwg-ont:Wildfire  .
    ?fire  kwg-ont:locatedIn ?state  .
    ?state  rdfs:label "California"  .
    ?fire  kwg-ont:dateOfFireIgnition ?igDate  .
    filter(?igDate > '2018-01-01'^^xsd:dateTime)
    {
```

Listing 4.1: Example of a SPARQL Query. It is used to answer the question "Which major wildfires are happened in California after Janaury 1st?" This query can be executed on the KnowWhereGraph endpoint at https://stko-kwg.geog.ucsb.edu/workbench.

GEOSPATIAL KNOWLEDGE GRAPH (GEOKG)

Knowledge graphs have been extensively used to model geospatial information. Broadly speaking, any knowledge graph can be considered as geospatial if it includes statements about geographic or spatial objects, interactions, or processes. For instance, knowledge graphs have been used to model environmental observations (Zhu *et al.*, 2021b), urban computing (Liu *et al.*, 2022), historical maps (Shbita *et al.*, 2020), public health (Zhu *et al.*, 2022b), food resilience (Rao *et al.*, 2022), and disaster responses (Zhu *et al.*, 2021a). Spatial thinking is employed widely when managing geospatial information in knowledge graphs. One of the most popular standards for building a geospatial knowledge graph (GeoKG) is GeoSPARQL[1] (Battle and Kolas, 2011), which is both an ontology and a query language. As Figure 4.9 shows, the superclass *SpatialObject* has *Feature* and *Geometry* as its children, and *Feature* is further related to *Geometry* through the relations: *hasGeometry* or *hasDefaultGeometry*. Such a design aligns with the classic distinction between *Feature* and *Geometry* in GIS that a feature might have geometry and might not, and geometry might not be associated with any feature either. Furthermore, *Geometry* can be represented using either Well Know Text (WKT) or Geography Markup Language (GML), both of which are literals in GeoSPARQL. Geometry is not the only factor to make knowledge geospatial. Qualitative spatial relations are also essential in geography and GeoAI (Cohn and Hazarika, 2001), but are often neglected in traditional spatial databases. Knowledge graphs, instead, provide an opportunity to fill such a gap thanks to their nature of emphasizing the relationships and the inherited graph structure. Most natural languages that include geospatial information can be expressed via geospatial knowledge graphs, together with formalized topological or directional relations (Clementini, 2019). For instance, statements "I study Climate Science at the University of Bristol, which is *located in* the city of Bristol. Bristol is *on the southwest coast of* England. River Avon *runs through* it" can be expressed using three statements with predicates (spatial relations): *located in, on the southwest coast of*, and *runs through*.

In addition to its symbolic strength of representing and reasoning on data, knowledge graphs contribute to and benefit from connectionist AI approaches too, such as deep neural networks. Specific to GeoKG, numerous spatially explicit techniques have been developed over the years to tackle various geospatial questions. Mai *et al.* (2020) proposed a distance-aware knowledge graph embedding method by injecting the distance-decay function into the training process, and showed that such a spatially explicit method achieves considerable improvements in geographic question answering. Rao *et al.* (2022) designed a GeoKG to measure supply-chain network resilience based on various geospatial semantics such as dependence of single-sourcing, transportation miles, and geographic adjacency. Moreover, due to its nature of structuring data using edges, most knowledge graphs and associated embedding methods are limited to simple binary relations. However, many qualitative spatial relations that humans use in daily lives involved more than three spatial objects (e.g., "Reading is *on your left side* if you travel on the highway M4 from Bristol to London by car"). To

---

[1]GeoSPARQL: https://www.ogc.org/standards/geosparql

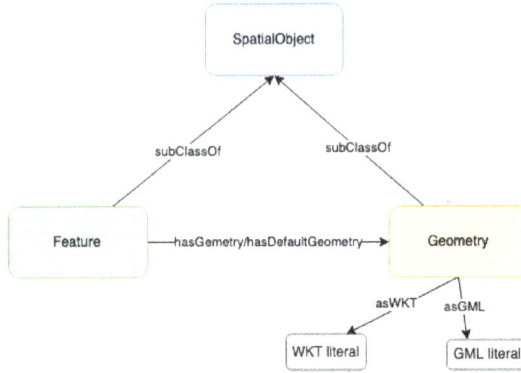

Figure 4.9 Simplified GeoSPARQL Ontology.

fill such a gap, Zhu *et al.* (2022a) introduced the concept of reification to represent those higher-order relations in graphs, and proposed a neural network architecture that explicitly takes the higher-order relatedness into account when reasoning over qualitative spatial relations, such as *betweenness*, *surrounded*, *on the left*, etc.

In addition, reinforcement learning methods have also been utilized in GeoKG representation and reasoning. For instance, Yan *et al.* (2019) proposed a novel summarization method that uses a reinforcement learning framework with spatially explicit components to summarize the intrinsic and extrinsic information in GeoKG. Wang *et al.* (2020) integrated reinforcement learning with a GeoKG of user semantics about connected locations, activities, and zones for incremental mobile user profiling and mobility prediction with improved accuracy. Future developments of GeoKG will continue innovating in representation, formalism and reasoning capabilities to support cross-domain knowledge discovery with geospatial contexts and spatial thinking.

4.10 SUMMARY

In this chapter, We have introduced the key concepts and architectures of CNNs, RNNs, Transformers, GNNs, GANs, reinforcement learning, and knowledge graphs, as well as various ways to incorporate spatial principles and geographic knowledge into the development of spatially explicit AI models and GeoKG.

ACKNOWLEDGMENTS

We acknowledge the support by the National Science Foundation funded AI institute ICICLE (OAC-2112606). Any opinions, findings, and conclusions or recommendations expressed in this material are those of the authors and do not necessarily reflect the views of the funding agency.

BIBLIOGRAPHY

Al-Najjar, H.A., *et al.*, 2021. A new integrated approach for landslide data balancing and spatial prediction based on generative adversarial networks (gan). *Remote Sensing*, 13 (19), 4011.

Arjovsky, M., Chintala, S., and Bottou, L., 2017. Wasserstein generative adversarial networks. *In: International Conference on Machine Learning*. PMLR, 214–223.

Bashmal, L., *et al.*, 2018. Siamese-gan: Learning invariant representations for aerial vehicle image categorization. *Remote Sensing*, 10 (2), 351.

Battle, R. and Kolas, D., 2011. Geosparql: enabling a geospatial semantic web. *Semantic Web Journal*, 3 (4), 355–370.

Bengio, Y., Simard, P., and Frasconi, P., 1994. Learning long-term dependencies with gradient descent is difficult. *IEEE Transactions on Neural Networks*, 5 (2), 157–166.

Berragan, C., *et al.*, 2022. Transformer based named entity recognition for place name extraction from unstructured text. *International Journal of Geographical Information Science*, 1–20.

Bihlo, A., 2021. A generative adversarial network approach to (ensemble) weather prediction. *Neural Networks*, 139, 1–16.

Biljecki, F. and Ito, K., 2021. Street view imagery in urban analytics and gis: A review. *Landscape and Urban Planning*, 215, 104217.

Cai, L., *et al.*, 2020. Traffic transformer: Capturing the continuity and periodicity of time series for traffic forecasting. *Transactions in GIS*, 24 (3), 736–755.

Chen, X., *et al.*, 2016. Infogan: Interpretable representation learning by information maximizing generative adversarial nets. *Advances in Neural Information Processing Systems*, 29.

Chen, Y., *et al.*, 2021. Ssd-gan: Measuring the realness in the spatial and spectral domains. *In: Proceedings of the AAAI Conference on Artificial Intelligence*. vol. 35, 1105–1112.

Cheng, G., *et al.*, 2020. Remote sensing image scene classification meets deep learning: Challenges, methods, benchmarks, and opportunities. *IEEE Journal of Selected Topics in Applied Earth Observations and Remote Sensing*, 13, 3735–3756.

Chiang, Y.Y., *et al.*, 2020. Training deep learning models for geographic feature recognition from historical maps. *In: Using Historical Maps in Scientific Studies*. Springer, 65–98.

Cho, K., *et al.*, 2014. On the properties of neural machine translation: Encoder-decoder approaches. *arXiv preprint arXiv:1409.1259*.

Chung, J., *et al.*, 2014. Empirical evaluation of gated recurrent neural networks on sequence modeling. *arXiv preprint arXiv:1412.3555*.

Clementini, E., 2019. A conceptual framework for modelling spatial relations. *Information Technology and Control*, 48 (1), 5–17.

Cohn, A.G. and Hazarika, S.M., 2001. Qualitative spatial representation and reasoning: An overview. *Fundamenta informaticae*, 46 (1), 1–29.

Dai, K., *et al.*, 2022. Mstcgan: Multi-scale time conditional generative adversarial network for long-term satellite image sequence prediction. *IEEE Transactions on Geoscience and Remote Sensing*.

Deng, J., *et al.*, 2009. Imagenet: A large-scale hierarchical image database. *In: 2009 IEEE Conference on Computer Vision and Pattern Recognition*. IEEE, 248–255.

Duan, W., *et al.*, 2020. Automatic alignment of contemporary vector data and georeferenced historical maps using reinforcement learning. *International Journal of Geographical Information Science*, 34 (4), 824–849.

Feng, Y., Thiemann, F., and Sester, M., 2019. Learning cartographic building generalization with deep convolutional neural networks. *ISPRS International Journal of Geo-Information*, 8 (6), 258.

Fritz, S., *et al.*, 2017. A global dataset of crowdsourced land cover and land use reference data. *Scientific Data*, 4 (1), 1–8.

Ganapathi Subramanian, S. and Crowley, M., 2018. Using spatial reinforcement learning to build forest wildfire dynamics models from satellite images. *Frontiers in ICT*, 5, 6.

Gao, H., *et al.*, 2018a. Object classification using cnn-based fusion of vision and lidar in autonomous vehicle environment. *IEEE Transactions on Industrial Informatics*, 14 (9), 4224–4231.

Gao, S., 2021. Geospatial Artificial Intelligence (GeoAI). *Oxford Bibliographies*, (1), 1–16.

Gao, Y., Jiang, D., and Xu, Y., 2018b. Optimize taxi driving strategies based on reinforcement learning. *International Journal of Geographical Information Science*, 32 (8), 1677–1696.

Getis, A., 2009. Spatial weights matrices. *Geographical Analysis*, 41 (4), 404–410.

Gong, S., *et al.*, 2022. Spatio-temporal travel volume prediction and spatial dependencies discovery using gru, gcn and bayesian probabilities. *In: 2022 7th International Conference on Big Data Analytics (ICBDA)*. IEEE, 130–136.

Goodchild, M.F., 2004. The validity and usefulness of laws in geographic information science and geography. *Annals of the Association of American Geographers*, 94 (2), 300–303.

Goodfellow, I., Bengio, Y., and Courville, A., 2016. *Deep Learning*. MIT Press.

Goodfellow, I., *et al.*, 2014. Generative adversarial nets. *In*: Z. Ghahramani, M. Welling, C. Cortes, N. Lawrence and K. Weinberger, eds. *Advances in Neural Information Processing Systems*. Curran Associates, Inc., vol. 27.

Guo, D., *et al.*, 2022. Deepssn: A deep convolutional neural network to assess spatial scene similarity. *Transactions in GIS*, 26 (4), 1914–1938.

Hamilton, W., Ying, Z., and Leskovec, J., 2017. Inductive representation learning on large graphs. *Advances in Neural Information Processing Systems*, 30.

Hart, G. and Dolbear, C., 2013. *Linked Data: A Geographic Perspective*. Taylor & Francis.

Hawelka, B., *et al.*, 2014. Geo-located twitter as proxy for global mobility patterns. *Cartography and Geographic Information Science*, 41 (3), 260–271.

He, K., *et al.*, 2017. Mask r-cnn. *In: Proceedings of the IEEE International Conference on Computer Vision*. 2961–2969.

He, K., *et al.*, 2016. Deep residual learning for image recognition. *In: Proceedings of the IEEE Conference on Computer Vision and Pattern Recognition*. 770–778.

Helber, P., *et al.*, 2018. Introducing eurosat: A novel dataset and deep learning benchmark for land use and land cover classification. *In: IGARSS 2018-2018 IEEE International Geoscience and Remote Sensing Symposium*. IEEE, 204–207.

Hochreiter, S. and Schmidhuber, J., 1997. Long short-term memory. *Neural Computation*, 9 (8), 1735–1780.

Hogan, A., *et al.*, 2021. Knowledge graphs. *ACM Computing Surveys (CSUR)*, 54 (4), 1–37.

Hu, S., *et al.*, 2021. Urban function classification at road segment level using taxi trajectory data: A graph convolutional neural network approach. *Computers, Environment and Urban Systems*, 87, 101619.

Hu, X., *et al.*, 2022. Gazpne2: A general place name extractor for microblogs fusing gazetteers and pretrained transformer models. *IEEE Internet of Things Journal*, 9 (17), 16259–16271.

Hu, Y., *et al.*, 2013. A geo-ontology design pattern for semantic trajectories. *In*: *International Conference on Spatial Information Theory*. Springer, 438–456.

Huang, G., *et al.*, 2017. Densely connected convolutional networks. *In*: *Proceedings of the IEEE Conference on Computer Vision and Pattern Recognition*. 4700–4708.

Janelle, D.G. and Goodchild, M.F., 2011. Concepts, principles, tools, and challenges in spatially integrated social science. *The SAGE Handbook of GIS and Society*, 27–45.

Janowicz, K., 2012. Observation-driven geo-ontology engineering. *Transactions in GIS*, 16 (3), 351–374.

Janowicz, K., *et al.*, 2020. GeoAI: spatially explicit artificial intelligence techniques for geographic knowledge discovery and beyond. *International Journal of Geographical Information Science*, 34 (4), 625–636.

Janowicz, K., *et al.*, 2022. Know, know where, knowwheregraph: A densely connected, cross-domain knowledge graph and geo-enrichment service stack for applications in environmental intelligence. *AI Magazine*, 43 (1), 30–39.

Janowicz, K., *et al.*, 2010. Semantic enablement for spatial data infrastructures. *Transactions in GIS*, 14 (2), 111–129.

Jiang, W. and Luo, J., 2022. Graph neural network for traffic forecasting: A survey. *Expert Systems with Applications*, 117921.

Kaelbling, L.P., Littman, M.L., and Moore, A.W., 1996. Reinforcement learning: A survey. *Journal of Artificial Intelligence Research*, 4, 237–285.

Kang, Y., Gao, S., and Roth, R.E., 2019. Transferring multiscale map styles using generative adversarial networks. *International Journal of Cartography*, 5 (2-3), 115–141.

Kang, Y., *et al.*, 2020. A review of urban physical environment sensing using street view imagery in public health studies. *Annals of GIS*, 26 (3), 261–275.

Kipf, T.N. and Welling, M., 2016. Semi-supervised classification with graph convolutional networks. *arXiv preprint arXiv:1609.02907*.

Koch, T. and Dugundji, E., 2020. A review of methods to model route choice behavior of bicyclists: inverse reinforcement learning in spatial context and recursive logit. *In*: *Proceedings of the 3rd ACM SIGSPATIAL International Workshop on GeoSpatial Simulation*. 30–37.

Krizhevsky, A., Sutskever, I., and Hinton, G.E., 2017. Imagenet classification with deep convolutional neural networks. *Communications of the ACM*, 60 (6), 84–90.

Kuhn, W., 2005. Geospatial semantics: why, of what, and how? *In*: *Journal on Data Semantics iii*. Springer, 1–24.

LeCun, Y., Bengio, Y., and Hinton, G., 2015. Deep learning. *Nature*, 521 (7553), 436–444.

LeCun, Y., *et al.*, 1998. Gradient-based learning applied to document recognition. *Proceedings of the IEEE*, 86 (11), 2278–2324.

Li, K., *et al.*, 2020. Object detection in optical remote sensing images: A survey and a new benchmark. *ISPRS Journal of Photogrammetry and Remote Sensing*, 159, 296–307.

Li, M., *et al.*, 2021a. Prediction of human activity intensity using the interactions in physical and social spaces through graph convolutional networks. *International Journal of Geographical Information Science*, 35 (12), 2489–2516.

Li, S., *et al.*, 2021b. Groupformer: Group activity recognition with clustered spatial-temporal transformer. *In*: *Proceedings of the IEEE/CVF International Conference on Computer Vision*. 13668–13677.

Li, W., 2020. GeoAI: Where machine learning and big data converge in GIScience. *Journal of Spatial Information Science*, (20), 71–77.

Li, W., Hsu, C.Y., and Hu, M., 2021c. Tobler's first law in geoai: A spatially explicit deep learning model for terrain feature detection under weak supervision. *Annals of the American Association of Geographers*, 111 (7), 1887–1905.

Li, W., *et al.*, 2022. Geoimagenet: a multi-source natural feature benchmark dataset for geoai and supervised machine learning. *GeoInformatica*, 1–22.

Li, Y., *et al.*, 2017. Diffusion convolutional recurrent neural network: Data-driven traffic forecasting. *arXiv preprint arXiv:1707.01926*.

Li, Y., *et al.*, 2018. An aircraft detection framework based on reinforcement learning and convolutional neural networks in remote sensing images. *Remote Sensing*, 10 (2), 243.

Li, Y. and Moura, J.M., 2019. Forecaster: A graph transformer for forecasting spatial and time-dependent data. *arXiv preprint arXiv:1909.04019*.

Liang, Y., *et al.*, 2022. Region2vec: community detection on spatial networks using graph embedding with node attributes and spatial interactions. *In*: *Proceedings of the 30th International Conference on Advances in Geographic Information Systems*. 1–4.

Liang, Y., *et al.*, 2019. Urbanfm: Inferring fine-grained urban flows. *In*: *Proceedings of the 25th ACM SIGKDD International Conference on Knowledge Discovery & Data Mining*. 3132–3142.

Lin, D., *et al.*, 2017. Marta gans: Unsupervised representation learning for remote sensing image classification. *IEEE Geoscience and Remote Sensing Letters*, 14 (11), 2092–2096.

Liu, Q., *et al.*, 2016. Predicting the next location: A recurrent model with spatial and temporal contexts. *In*: *Thirtieth AAAI Conference on Artificial Intelligence*.

Liu, Y., Ding, J., and Li, Y., 2022. Developing knowledge graph based system for urban computing. *In*: *Proceedings of the 1st ACM SIGSPATIAL International Workshop on Geospatial Knowledge Graphs*. 3–7.

Luo, Y., *et al.*, 2018. Multivariate time series imputation with generative adversarial networks. *Advances in Neural Information Processing Systems*, 31.

Mai, G., *et al.*, 2022a. Symbolic and subsymbolic GeoAI: Geospatial knowledge graphs and spatially explicit machine learning. *Transactions in GIS*, 26 (8), 3118–3124.

Mai, G., *et al.*, 2022b. A review of location encoding for geoai: methods and applications. *International Journal of Geographical Information Science*, 36 (4), 639–673.

Mai, G., *et al.*, 2019. Multi-scale representation learning for spatial feature distributions using grid cells. *In: International Conference on Learning Representations.*

Mai, G., *et al.*, 2020. Relaxing unanswerable geographic questions using a spatially explicit knowledge graph embedding model. *In: Geospatial Technologies for Local and Regional Development: Proceedings of the 22nd AGILE Conference on Geographic Information Science 22.* Springer, 21–39.

May Petry, L., *et al.*, 2020. Marc: a robust method for multiple-aspect trajectory classification via space, time, and semantic embeddings. *International Journal of Geographical Information Science*, 34 (7), 1428–1450.

Mehrabi, N., *et al.*, 2021. A survey on bias and fairness in machine learning. *ACM Computing Surveys*, 54 (6), 1–35.

Miller, E., 1998. An introduction to the resource description framework. *D-lib Magazine.*

Mirza, M. and Osindero, S., 2014. Conditional generative adversarial nets. *arXiv preprint arXiv:1411.1784.*

Mnih, V., *et al.*, 2016. Asynchronous methods for deep reinforcement learning. *In: International Conference on Machine Learning.* PMLR, 1928–1937.

Mnih, V., *et al.*, 2015. Human-level control through deep reinforcement learning. *Nature*, 518 (7540), 529–533.

Oono, K. and Suzuki, T., 2019. Graph neural networks exponentially lose expressive power for node classification. *arXiv preprint arXiv:1905.10947.*

Ouyang, K., *et al.*, 2018. A non-parametric generative model for human trajectories. *In: IJCAI.* vol. 18, 3812–3817.

Ouyang, L., *et al.*, 2022. Training language models to follow instructions with human feedback. *arXiv preprint arXiv:2203.02155.*

Patterson, J. and Gibson, A., 2017. *Deep Learning: A Practitioner's Approach.* "O'Reilly Media, Inc.".

Puterman, M.L., 1990. Markov decision processes. *Handbooks in Operations Research and Management Science*, 2, 331–434.

Radford, A., Metz, L., and Chintala, S., 2015. Unsupervised representation learning with deep convolutional generative adversarial networks. *arXiv preprint arXiv:1511.06434.*

Rao, J., *et al.*, 2021a. LSTM-TrajGAN: A Deep Learning Approach to Trajectory Privacy Protection. *In: 11th International Conference on Geographic Information Science (GIScience 2021).* Schloss Dagstuhl–Leibniz-Zentrumfur Informatik, vol. 177, 12.

Rao, J., *et al.*, 2022. Measuring network resilience via geospatial knowledge graph: a case study of the US multi-commodity flow network. *In: Proceedings of the 1st ACM SIGSPATIAL International Workshop on Geospatial Knowledge Graphs.* 17–25.

Rao, J., Gao, S., and Zhu, X., 2021b. Vtsv: A privacy-preserving vehicle trajectory simulation and visualization platform using deep reinforcement learning. *In: Proceedings of the 4th ACM SIGSPATIAL International Workshop on AI for Geographic Knowledge Discovery.* 43–46.

Ronneberger, O., Fischer, P., and Brox, T., 2015. U-net: Convolutional networks for biomedical image segmentation. *In: Medical Image Computing and Computer-Assisted Intervention–MICCAI 2015: 18th International Conference, Munich, Germany, October 5-9, 2015, Proceedings, Part III 18.* Springer, 234–241.

Rui, X., *et al.*, 2021. Disastergan: Generative adversarial networks for remote sensing disaster image generation. *Remote Sensing*, 13 (21), 4284.

Rüttgers, M., *et al.*, 2019. Prediction of a typhoon track using a generative adversarial network and satellite images. *Scientific Reports*, 9 (1), 1–15.

Schulman, J., *et al.*, 2015. Trust region policy optimization. *In*: *International Conference on Machine Learning*. PMLR, 1889–1897.

Schulman, J., *et al.*, 2017. Proximal policy optimization algorithms. *arXiv preprint arXiv:1707.06347*.

Shbita, B., *et al.*, 2020. Building linked spatio-temporal data from vectorized historical maps. *In*: *European Semantic Web Conference*. Springer, 409–426.

Silver, D., *et al.*, 2016. Mastering the game of go with deep neural networks and tree search. *nature*, 529 (7587), 484–489.

Srikanth, G., Nukavarapu, N., and Durbha, S., 2021. Deep reinforcement learning interdependent healthcare critical infrastructure simulation model for dynamically varying covid-19 scenario-a case study of a metro city. *In*: *2021 IEEE International Geoscience and Remote Sensing Symposium IGARSS*. IEEE, 8499–8502.

Szegedy, C., *et al.*, 2015. Going deeper with convolutions. *In*: *Proceedings of the IEEE Conference on Computer Vision and Pattern Recognition*. 1–9.

Tang, W., *et al.*, 2021. Learning disentangled user representation with multi-view information fusion on social networks. *Information Fusion*, 74, 77–86.

Tobler, W.R., 1970. A computer movie simulating urban growth in the detroit region. *Economic Geography*, 46 (sup1), 234–240.

Toutouh, J., 2021. Conditional generative adversarial networks to model urban outdoor air pollution. *In*: *Smart Cities: Third Ibero-American Congress, ICSC-Cities 2020, San José, Costa Rica, November 9-11, 2020, Revised Selected Papers 3*. Springer, 90–105.

Vaswani, A., *et al.*, 2017. Attention is all you need. *Advances in Neural Information Processing Systems*, 30.

Velickovic, P., *et al.*, 2017. Graph attention networks. *arXiv preprint arXiv:1710.10903*.

Wang, H., *et al.*, 2019. Learning graph representation with generative adversarial nets. *IEEE Transactions on Knowledge and Data Engineering*, 33 (8), 3090–3103.

Wang, P., *et al.*, 2020. Incremental mobile user profiling: Reinforcement learning with spatial knowledge graph for modeling event streams. *In*: *Proceedings of the 26th ACM SIGKDD International Conference on Knowledge Discovery & Data Mining*. 853–861.

Wang, S., *et al.*, 2023. Sta-gan: A spatio-temporal attention generative adversarial network for missing value imputation in satellite data. *Remote Sensing*, 15 (1), 88.

Watkins, C.J. and Dayan, P., 1992. Q-learning. *Machine Learning*, 8, 279–292.

Wilkinson, M.D., *et al.*, 2016. The fair guiding principles for scientific data management and stewardship. *Scientific Data*, 3 (1), 1–9.

Wu, A.N. and Biljecki, F., 2022. Ganmapper: geographical data translation. *International Journal of Geographical Information Science*, 1–29.

Wu, Z., *et al.*, 2020. A comprehensive survey on graph neural networks. *IEEE Transactions on Neural Networks and Learning Systems*, 32 (1), 4–24.

Xia, R., *et al.*, 2014. Supervised hashing for image retrieval via image representation learning. In: *Twenty-Eighth AAAI Conference on Artificial Intelligence.*

Xu, M., *et al.*, 2020. Spatial-temporal transformer networks for traffic flow forecasting. *arXiv preprint arXiv:2001.02908.*

Yan, B., *et al.*, 2018. xnet+ sc: Classifying places based on images by incorporating spatial contexts. In: *10th International Conference on Geographic Information Science (GIScience 2018).* Informatik.

Yan, B., *et al.*, 2019. A spatially explicit reinforcement learning model for geographic knowledge graph summarization. *Transactions in GIS*, 23 (3), 620–640.

Yang, B., *et al.*, 2021. St-lbagan: Spatio-temporal learnable bidirectional attention generative adversarial networks for missing traffic data imputation. *Knowledge-Based Systems*, 215, 106705.

Yoon, J., Jordon, J., and Schaar, M., 2018. Gain: Missing data imputation using generative adversarial nets. In: *International Conference on Machine Learning.* PMLR, 5689–5698.

Yu, Y., *et al.*, 2019. A review of recurrent neural networks: Lstm cells and network architectures. *Neural Computation*, 31 (7), 1235–1270.

Zhang, B., *et al.*, 2022. Styleswin: Transformer-based gan for high-resolution image generation. In: *Proceedings of the IEEE/CVF Conference on Computer Vision and Pattern Recognition.* 11304–11314.

Zhang, F., *et al.*, 2018a. Measuring human perceptions of a large-scale urban region using machine learning. *Landscape and Urban Planning*, 180, 148–160.

Zhang, T., *et al.*, 2018b. Spatial–temporal recurrent neural network for emotion recognition. *IEEE Transactions on Cybernetics*, 49 (3), 839–847.

Zhang, Y. and Cheng, T., 2020. Graph deep learning model for network-based predictive hotspot mapping of sparse spatio-temporal events. *Computers, Environment and Urban Systems*, 79, 101403.

Zhao, B., *et al.*, 2021. Deep fake geography? when geospatial data encounter artificial intelligence. *Cartography and Geographic Information Science*, 48 (4), 338–352.

Zhao, L., *et al.*, 2019. T-gcn: A temporal graph convolutional network for traffic prediction. *IEEE Transactions on Intelligent Transportation Systems*, 21 (9), 3848–3858.

Zhou, B., *et al.*, 2017. Places: A 10 million image database for scene recognition. *IEEE Transactions on Pattern Analysis and Machine Intelligence*, 40 (6), 1452–1464.

Zhou, B., *et al.*, 2014. Learning deep features for scene recognition using places database. *Advances in Neural Information Processing Systems*, 27.

Zhou, F., *et al.*, 2021. Improving human mobility identification with trajectory augmentation. *GeoInformatica*, 25 (3), 453–483.

Zhou, J., *et al.*, 2020. Graph neural networks: A review of methods and applications. *AI Open*, 1, 57–81.

Zhu, A.X., *et al.*, 2018. Spatial prediction based on third law of geography. *Annals of GIS*, 24 (4), 225–240.

Zhu, D., *et al.*, 2020a. Spatial interpolation using conditional generative adversarial neural networks. *International Journal of Geographical Information Science*, 34 (4), 735–758.

Zhu, D., *et al.*, 2020b. Understanding place characteristics in geographic contexts through graph convolutional neural networks. *Annals of the American Association of Geographers*, 110 (2), 408–420.

Zhu, J.Y., *et al.*, 2017a. Unpaired image-to-image translation using cycle-consistent adversarial networks. *In*: *Proceedings of the IEEE International Conference on Computer Vision*. 2223–2232.

Zhu, R., *et al.*, 2021a. Providing humanitarian relief support through knowledge graphs. *In*: *Proceedings of the 11th nowledge Capture Conference*. 285–288.

Zhu, R., *et al.*, 2022a. Reasoning over higher-order qualitative spatial relations via spatially explicit neural networks. *International Journal of Geographical Information Science*, 36 (11), 2194–2225.

Zhu, R., *et al.*, 2022b. Covid-forecast-graph: An open knowledge graph for consolidating covid-19 forecasts and economic indicators via place and time. *AGILE: GIScience Series*, 3, 21.

Zhu, R., *et al.*, 2021b. Environmental observations in knowledge graphs. *In*: *DaMaLOS*. 1–11.

Zhu, X.X., *et al.*, 2017b. Deep learning in remote sensing: A comprehensive review and list of resources. *IEEE Geoscience and Remote Sensing Magazine*, 5 (4), 8–36.

5 GeoAI for Spatial Image Processing

Samantha T. Arundel, Kevin G. McKeehan
U.S. Geological Survey, Center of Excellence for Geospatial
Information Science

Wenwen Li, Zhining Gu
School of Geographical Sciences and Urban Planning, Arizona
State University

CONTENTS

5.1 Introduction 75
    5.1.1 Origins of image processing 75
    5.1.2 (Digital) image processing in the spatial realm 77
    5.1.3 Image processing in the AI realm 79
5.2 Image Processing in the GeoAI Domains 81
5.3 GeoAI-Specific Methods and Challenges 84
    5.3.1 The challenge in training data annotation 84
    5.3.2 The challenge of scale 86
    5.3.3 The challenge of image resolution 87
    5.3.4 The challenge of space and time 88
5.4 Future Research 88
    5.4.1 Geometric Algebra 88
    5.4.2 Ethical issues related to spatial image processing in GeoAI 89
    5.4.3 Combining GeoAI and physical modeling 89
    5.4.4 Knowledge base as input/constraint 90
5.5 Conclusion 90
    Bibliography 91

5.1 INTRODUCTION

5.1.1 ORIGINS OF IMAGE PROCESSING

The development of image processing began with the invention of the photographic image, which has been around for 200 years. Its incorporation into scientific research was swift. After the technological breakthroughs of Nicéphore Niépce, Thomas Wedgwood, Louis Daguerre, Henry Fox Talbot, and others in the 1820s and 1830s,

DOI: 10.1201/9781003308423-5

cameras became essential tools in laboratories and on geographic expeditions. For example, Anna Atkins published her botanical study of algae in the 1840s using cyanotype images (Saska, 2010). Photographers provided evidence of geomorphic phenomena in Clarence King's 1867 Geological Exploration of the Fortieth Parallel, the Hayden Geological Survey of 1871 to the Yellowstone area of Wyoming, and George Wheeler's 1872 Geographical Surveys West of the 100th Meridian (Escalera et al., 2016; Rogers et al., 1984; Stegner, 1992). One of the most well-known examples of photography amplifying science involved John Wesley Powell, the second director of the U.S. Geological Survey. Powell's underfunded 1869 expedition on the Colorado River through the Grand Canyon was dismissed. Still, he hired more scientists and a photographer to accompany him on his second expedition in 1871–1872 "to secure truthfulness" in the minds of his critics (Stegner, 1992). This expedition – and its accompanying images – secured Powell's scientific reputation.

Photography, particularly aerial photography, has had a profound impact on science (Baker, 1924), transforming the field of ecology and sparking new geographic frontiers such as cultural landscape studies (Hughes, 2016; Starrs, 1998). Images are a potent observational adjunct to the scientist's eyes and senses (Hoffmann, 2013; Magnuson, 1990), with the power to capture phenomena instantly at the moment in time. Conversely, repeat photography allowed for quantitative time-series analysis of phenomena at extended temporal scales (Rogers et al., 1984). Sebastian Finsterwalder, a mathematics and geometry professor at the Technical University of Munich, was the earliest and best-known scientist to document geographic phenomena using this approach (Hattersley-Smith, 1966). He first surveyed alpine glaciers in the eastern Alps using trigonometry and photogrammetry, establishing a process to assign local coordinates to features in a photograph. This pioneering idea led to several breakthroughs, including a theory of steady-state glacial flow based on geometric measurements and the invention of the phototheodolite, a surveying station capable of photogrammetry (Hattersley-Smith, 1966; Rogers et al., 1984).

Until the 1920s, all photography was analog, relying on chemical processes to render an image upon a medium, such as film or metal plates. The invention of the Bartlane cable picture transmission system, designed to transmit images across the Atlantic Ocean via telegraph cable, changed this. The Bartlane device transformed photographs into codes by exposing the original picture to five specialized metal plates in a modified photographic process (McFarlane, 1972; Milnor, 1941). Magnetic sensors would detect the tonal value in a square at defined intervals across the plates and record the coded combination of values on a tape that fed a Bartlane device at the other end of the telegraph cable. This reconstructed a digitized grayscale image.

Work done on the Standards Eastern Automatic Computer (SEAC) by Russell Kirsch and others in the 1950s heralded the next stage of digital imagery advancement (Kirsch, 1998). Using SEAC's specialized digital scanner, the team captured the first truly digitally scanned image – of Kirsch's son – in a 176 x 176-pixel matrix (Kirsch, 2010). The SEAC team further anticipated the future of the technology when they suggested using their methodology to compile several pictures of Mars into a "reliable map of the true features of the planet" (Kirsch et al., 1957). Indeed, space exploration fueled subsequent advances in digital image processing, but technical limitations of cameras, computer storage, hardware, capacitors and conductors, and other elements constrained the field. Research associated with the U.S. space program and other areas provided funding and impetus to overcome these restrictions, allowing the Ranger 7 lunar probe to transmit images of the Moon to the Earth in 1964.

The most impactful advances of this era were the development of the charge-coupled device (CCD) in the late 1960s, a photon detector that astronomers repurposed to create high-resolution digital images of distant celestial objects (Mackay, 1986), resulting in many downstream digital imagery improvement (Howell, 2006). Concurrently, the United States began developing high-altitude and space-based platforms for systematic imaging systems in response to concerns about the Cold War and the environment. Spy systems, such as the U-2 aircraft and the CORONA satellites, provided imagery with a ground resolution of < 10 m. Weather satellites such as Vanguard 2 and TIROS came online in 1959 and 1960, respectively. In 1972, the Earth Resources Technology Satellite, known today as Landsat 1, was launched with two sensors, the return beam vidicon (RBV) and the multispectral scanner (MSS). The MSS was a 4-band scanner with a spatial resolution of 80 m and a temporal resolution of 18 days. This capability far exceeded the capabilities of existing remote sensing platforms and allowed scientists to supersede the previous limitations of their analyses. The ability of Landsat 1's MSS to capture four sensor bands sparked the development of the normalized difference vegetation index (NDVI), a now-familiar metric that measures the presence of vegetation (Rouse *et al.*, 1974; Tucker *et al.*, 1973).

Advances in computing power and storage allowed for more breakthroughs in the realm of digital image processing, one of the most consequential of which occurred in the medical field. The development of the computed tomography scan, or CT or CAT scan, was partially made possible by advances in digital imaging and computing power (Romans, 1995), hinging on a machine's ability to conduct linear algebra (Gustafsson *et al.*, 2014). The CT scan could obtain multiple x-ray images from different projections and, using linear algebra, compile them into one simulated digital image using tomographic reconstruction. For the first viable CT scan, Allan McLeod Cormack and Sir Godfrey Hounsfield shared the 1979 Nobel Prize in Medicine. Such advances in digital image processing, especially from remote sensing, would lead to a quantitative revolution in the geography.

5.1.2 (DIGITAL) IMAGE PROCESSING IN THE SPATIAL REALM

The Quantitative Revolution in geography was an attempt in the postwar years to mold geography and its associated subdisciplines into a more rigorous science by focusing on statistical analyses, physics, physical processes, and the consideration of phenomena at multiple scales (Barnes, 2004; Cresswell, 2013; Kohn, 1970; Nir and Nir, 1990; Wheeler, 2001). This new emphasis reflected the times, as academic disciplines as diverse as history (Ruggles and Magnuson, 2019), economics (Acosta and Cherrier, 2021), sociology (Lundberg, 1960), and medicine (Quirke and Gaudillière,

2008) adopted the ethos of a post-war world in search of "certainty" in the wake of worldwide catastrophes (Douard, 1996). The movement gained steam in the early-to-mid 20th Century alongside the growth of photography in the geosciences. Its hallmarks were a newly found emphasis on studying phenomena at multiple spatial dimensions or scales. Increased computing power finally allowed geoscientists to accomplish this, as did the vast amounts of data gathered by satellites and other devices at a distance (Bryant and Baddock, 2021). As a result, for the first time, observations of phenomena could be captured scientifically, repeatedly, and automatically over much of the surface of the Earth without extensive fieldwork (Church, 2010).

The development of the geographic information system (GIS) in the 1960s helped organize and organize data that possessed geospatial characteristics (Goodchild, 2018). The Canada Geographic Information System (CGIS), developed under the direction of Roger Tomlinson, IBM, and others, was the first viable GIS application. CGIS created computerized geospatial data by ingesting land use maps and scanning them through a "fine-resolution custom-built optical scanner" (Goodchild, 2018). The digital output was then converted from a raster file type to the newly developed arc-node vector data structure in square "tiles", which were compressed and stored on magnetic tape, allowing for relatively fast geospatial indexing (Goodchild, 2018). Quite quickly, the potential geostatistical power inherent in its design became apparent in the age of the Quantitative Revolution, as areal calculations could be made for the mapped phenomena (Miller *et al.*, 2019; Minasny and McBratney, 2016). Beginning in the 1970s, the baseline design of CGIS and other similar GIS applications were augmented by the capabilities of the relational database, a new development in database architecture pioneered by researchers at IBM and elsewhere (Worboys, 1999). As a result, more data could be relationally and geospatially linked together than ever before.

The original CGIS demonstrated potential and geospatial value in detecting features from images; before CGIS, feature identification from imagery relied on manual delineation (Hammond, 1954). Thus, when Landsat 1 was launched in 1972, efforts were made to classify the digitized multispectral images from the satellite and other over-the-surface sensors in new automated and geographically informed ways (Blaschke, 2010; Kucharczyk *et al.*, 2020). In 1976, Kettig and Landgrebe proposed an automated object classification scheme to detect homogeneous features larger than the imagery's pixel resolution. This approach was novel in that it explicitly sought to use the dependency "between adjacent states of nature" to identify objects (Kettig and Landgrebe, 1976), relying on Tobler's First Law of Geography and the spatial autocorrelation of land use-land cover to drive the classification. The use of similar segmentation and classification algorithms and other manual processes to continues (Blaschke, 2010; Graff and Usery, 1993; Kucharczyk *et al.*, 2020), especially for geomorphic mapping operations where specialized domain knowledge is essential (Abhar *et al.*, 2015; Stokes *et al.*, 2013).

With its unprecedented 1-m spatial resolution for panchromatic photography, the IKONOS satellite launched in 1999 led to the development of new feature detection methods in remote sensing aerial imagery (Blaschke, 2010; Kucharczyk *et al.*, 2020). One of these new methods was an object-based image analysis (OBIA) algorithm, which leveraged image "texture" to define objects through segmentation, which were then classified (Hay and Castilla, 2008; Kucharczyk *et al.*, 2020). In 2000, a private company released eCognition, software that bridged remote sensing and GIS to perform OBIA analyses, mimicking the CGIS architecture 30-plus years earlier. Trimble would eventually buy the software, and the concept of GEOBIA – geographic object-based image analysis – was born.

GEOBIA is a "scientific revolution" that aims to break geospatial analyses free of pixel-spectra-based models and create a dynamic multiscale object-based contextual model (Hay and Castilla, 2008). Previous algorithms were constrained to a pixel-by-pixel analytical framework and lacked contextual considerations (Burnett and Blaschke, 2003; Cracknell, 1998). GEOBIA can interpret images the way humans do, using additional contextual variables to help drive segmentation, relate downstream to the GIS vector data model, and resolve issues of detecting the same phenomenon at different scales (Kucharczyk *et al.*, 2020). GEOBIA seeks to solve the scale problem in remote sensing and physical systems by shifting the analytical framework away from the arbitrary pixel to identifiable image objects. Segmentation in GEOBIA partitions the imagery into discrete candidate objects based on spectral and geometric properties (Kucharczyk *et al.*, 2020; White *et al.*, 2019). However, determining the proper segmentation parameters can be subjective and reliant on heuristics and trial-and-error processes (Kucharczyk *et al.*, 2020). The resulting segmented objects are then classified using several variables. GEOBIA is a powerful and widely utilized method (Kucharczyk *et al.*, 2020), but it has weaknesses, particularly with the segmentation phase (Hay and Castilla, 2008). In addition, the process relied on the heterogeneity of pixel groups, was often not automated, and could inject subjectivity into the workflow. To address these issues, some practitioners have integrated the GEOBIA framework with machine learning methods, such as convolutional neural networks (CNNs) (Kucharczyk *et al.*, 2020). CNNs are a class of artificial neural networks (ANNs) that are increasingly common in many deep-learning image recognition fields (Yamashita *et al.*, 2018). However, GEOCNN – GEOBIA-informed by CNNs – is a new and emerging field with multiple potential methodologies and best practices still being developed (Kucharczyk *et al.*, 2020).

5.1.3 IMAGE PROCESSING IN THE AI REALM

The development of image processing methodologies in geography and the geosciences was aided by advances in technologies and from other fields. In the 1940s and 1950s, researchers learned that image processing in humans follows a hierarchical framework consisting of neurons organized into layers (Behnke, 2003; Lee, 2020; Sexton and Love, 2022). This analytical network structure also seemed integral to many human learning and behavioral decision-making processes (Bielecki, 2019; Hebb, 2005; McCulloch and Pitts, 1943). Researchers studying artificial intelligence (AI) in the 1940s and 1950s recognized that for AI to model human intelligence capabilities effectively, AI data and image processing needed to mimic this analytical framework (McCarthy *et al.*, 2006).

A 1956 summer research project proposal at Dartmouth College called this hierarchical framework "neural nets" (McCarthy *et al.*, 2006). This proposal launched a wave of AI research on multiple fronts in the subsequent decades (Lee, 2020; Moor, 2006). Influenced by the direction of the Dartmouth proposal and other factors, neural network development was an early AI research field (Lee, 2020). Thus, as neural networks are composed of hierarchical layers that apply algorithms to model decision-making, early research focused on suffusing AI with logic and mathematical reasoning. In 1956, Allen Newell, Herbert A. Simon, and Cliff Shaw of the Rand Corporation wrote the Logic Theorist computer program, considered the first AI program in history (Lee, 2020; Tarran and Ghahramani, 2015). It learned to prove 38 of the first 52 theorems in Whitehead and Russell's *Principia Mathematica*.

Nearly simultaneously, Cornell's Frank Rosenblatt attempted to develop a neural network specifically for image processing (Rosenblatt, 1958). Building upon McCulloch and Pitts' (1943) work and Hebb's (1949) human behavioral neural network concepts, Rosenblatt called his "hypothetical nervous system" the Perceptron. Using an IBM 704 computer, the Perceptron learned to distinguish visually whether marks on computer programming punch cards were on the left or right side of the card. The Perceptron was a physical, electromechanical, purpose-built machine with a sophisticated visual sensor. It hoped to recognize images with minimal error, understand speech, and translate foreign languages (Behnke, 2003; Bielecki, 2019). However, critics emerged when the results fell short of expectations in the late 1960s, and funding decreased (Lee, 2020).

The Perceptron was central to a debate within the AI community regarding the proper approach to the field. Rosenblatt and champions of the Perceptron advocated for a neural network approach, whereas those critical of its relatively low success rate advanced a theory called symbolic AI (SAI) (Honavar, 1995). Criticism of neural networks by Marvin Minsky and others (Minsky and Papert, 2017) led to a shift in funding away from neural networks and toward SAI (Bielecki, 2019; Lee, 2020; Olazaran, 1996). This led to a 15-year "impasse" known as AI Winter, which stifled development and led to many dead ends (Bielecki, 2019). As a result, AI image processing efforts in the 1970s and the near-term beyond were primarily focused on image segmentation (Fu and Mui, 1981; Pal and Pal, 1993) rather than classification, object detection, and leveraging full AI capabilities.

Moreover, work on image segmentation algorithms at that time was not progressing as hoped. As Fu and Mui (1981) reported, "all image segmentation techniques proposed so far are ad hoc in nature", noting that there "are no general algorithms which will work for all images". However, developments were underway that would propel AI image processing forward, chiefly in the subfield of machine learning (Lee, 2020; Mahadevkar *et al.*, 2022).

In the early 1980s, John Hopfield conceptualized the recurrent neural network (RNN) to better model the true neurobiological framework of human learning (Hopfield, 1982). With Hopfield's network, connectors create circuits so that information "learned" in one neural area can inform other network areas as a "memory". Yet, this was difficult algorithmically until a paper published in Nature in 1986 revised an old recursive technique – backpropagation. The backpropagation algorithm statistically evaluates the neural network output results with the training data examples, then reclusively adjusts the network weights for all parameters until the output matches the training data (Rumelhart *et al.*, 1986). However, Perceptron failed to accomplish this because it could only run forward from Perceptron-type; Perceptron-type algorithms are now known as feedforward neural networks (Behnke, 2003).

RNN and backpropagation, combined with a renewed emphasis on statistical-based ML techniques (Tarran and Ghahramani, 2015), particularly deep learning, set AI image processing on its current course (Lee, 2020). Deep learning extends the ability of a learning machine, specifically an ANN, to create abstractions of various, complex, and non-linear phenomena by its attempt to recreate and explain the neural processing path the human brain uses to interpret natural language and imagery (Goodfellow *et al.*, 2016). Deep refers to the large number of variables, represented by feature (hidden) layers, in the neural network (Dechter, 1986). In 2012, a deep convolutional neural network AlexNet, made a breakthrough. AlexNet reported a test error rate of 15.3%, >10% lower than the second-place competitor and far lower than other studies (Krizhevsky *et al.*, 2012). The popular CNNs today, such as ResNet or DenseNet, can produce hundreds of layers deep with densely connected neurons, gaining the ability to extract complex relationships and patterns within the data (Li and Hsu, 2020). Recently, the rapid advances in transformer architecture have also enabled image processing in a novel way. Transformer networks were first adopted in natural language processing, and it uses an encoder-decoder architecture to perform sequence-to-sequence learning (Li and Hsu, 2022). These architectures have been recently introduced into computer vision through the development of Vision Transformers (Dosovitskiy *et al.*, 2020). A key challenge that can be transferred from learning natural language text to spatial images is dimension reduction. Natural language text can be considered a one-dimensional (1D) sequence, whereas images are usually two-dimensional (2D). A standard solution is to break images into patches and perform embedding on the patches to convert 2D images into 1D representations. However, it is still debated whether transformer architectures can outperform the CNN models in image analysis tasks (Li and Hsu, 2022).

With these advances, sophisticated AI image interpretation began solving "nontrivial computer vision tasks" (Behnke, 2003). Neural network approaches were now becoming incorporated into the medical industry (Cheng *et al.*, 1996), where magnetic resonance imagery (MRI) was being pioneered (Pal and Pal, 1993). This type of operative, deep learning-informed computer vision is essential for the Internet of Things (Mahadevkar *et al.*, 2022), autonomous vehicles (Fernandes *et al.*, 2021), security systems (Khan *et al.*, 2021), and Geospatial-Intelligence tools (Fisher *et al.*, 2020). Data mining of social media imagery and the attachment of ontological knowledge to images are new frontiers for AI image processing (Lee, 2020).

5.2 IMAGE PROCESSING IN THE GEOAI DOMAINS

GeoAI domains are those related to the use and analysis of geospatial (Li, 2020, 2022). GeoAI image processing domains can broadly fall into planetary observation, human environment, and GIS and cartography. Below we provide just a few examples of research in these areas; however, the list is by no means inclusive. A more comprehensive review can be found in Li and Hsu (2022).

Planetary observation includes earth observation systems, specifically those based on remotely sensed imagery, but extends to other planets. Climate and weather studies benefit noticeably from AI methods. For example, Kurth *et al.* (2018) mapped extreme weather patterns at the pixel scale, using segmentation masks, by extending open-source deep learning algorithms. Hernández *et al.* (2016) applied a deep-learning architecture to predict the next day's precipitation accumulation. Multi-layer perceptrons supplied predictions based on non-linear relationships encapsulated by autoencoder. Mengwall and Guzewich (2023) capture spatial and temporal patterns of global cloud coverage on Mars using deep learning approaches grounded in CNNs. Their results improve upon previous semi-automated processes. Finally, Rolnick *et al.* (2022) reviewed outstanding problems related to climate change that machine learning approaches may address.

Research in the domain of land use and land cover is advanced and diverse, and largely reviewed in Talukdar *et al.* (2020). For example, Weng *et al.* (2018) classified land use scenes by replacing the fully connected layers of a CNN with a constrained extreme learning machine. This model reduces training time and is more easily generalized to other image sets. Campos-Taberner *et al.* (2020) employed a two-layer bi-directional Long Short-Term Memory network (2-BiLSTM), a type of RNN, to classify land use in Sentinel-2 images. RNNs paired with CNNs are powerful for handling time sequences involved in change detection. Feizizadeh *et al.* (2021) compared three machine learning methodologies to land use/land cover change assessment to a fuzzy object-based deep learning approach. Their model surpassed the others, resulting in a novel technique for image classification.

Agricultural applications also benefit from advances in GeoAI image processing. For example, Mulyono *et al.* (2016) exploited Support Vector machines (SVM) to identify sugarcane plantations in Landsat 8 images. Despite limited training samples, their approach generalized complex data well. Dyrmann *et al.* (2016) distinguished between corn and weed plants for each pixel in red, green, and blue (RGB) field images using a CNN-based semantic segmentation. The high accuracy of results precisely located weeds for mitigation purposes. Finally, Tu *et al.* (2018) applied a multi-layer perceptron neural network to differentiate between low- and high-quality pepper seeds. The relationship between seed vigor and seed size and color drives their identification. Benos *et al.* (2021) reviewed GeoAI applications to agriculture challenges.

Terrain applications of machine learning are relatively new and range from terrain feature extraction to recognizing terrain-related hazards. For example, Li *et al.* (2022b) developed a labeled training dataset to detect natural features in the United States from imagery and elevation derivatives. They successfully tested this large-scale detection task using an object detection model Faster RCNN (Ren *et al.*, 2015) with a CNN backbone – RetinaNet (Lin *et al.*, 2017). This research represents some of the first deep-learning approaches to landform delineation. Shirzadi *et al.* (2018) compared various machine learning processes to improve shallow landslide susceptibility mapping in Iran. Ensemble algorithms resulted in the highest performance accuracy. Due to their small physical size (meters), identifying and mapping rock-falls in planetary satellite imagery is challenging and time-consuming. Bickel *et al.* (2021) created 1000 images labeled positive or negative to detect rockfalls. The authors tested the images of rockfalls on Mars with RetinaNet. Other studies have exploited different deep learning strategies to detect natural features such as craters, patterned ground, and sand dunes (Li *et al.*, 2022a, 2017).

Within the domain of the human environment, vehicle detection and classification, traffic and transportation, and the built environment provide illustrations of research assisted by GeoAI techniques. For example, Li *et al.* (2021b) identified vehicles by fusing attention data within the feature pyramid network (FPN) structure. Their method quickly and accurately predicted cars in remote sensing images at the pixel level. Al-Mistarehi *et al.* (2022) analyzed data from vehicle accidents with a model grounded in random forests and decision trees. By examining crucial incident variables, they identified accident hotspots. Lastly, Song *et al.* (2019) detected and counted vehicles in highway scenes using deep learning. Their approach depended on the YOLOv3 (Redmon and Farhadi, 2018), a CNN that uses logistic regression for categorizing objects. Dong *et al.* (2020) demonstrated traffic and transportation research by improving semantic image segmentation of urban street scenes. Their methods increased the accuracy of real-time traffic assessment, which is necessary for autonomous driving. Sarikan and Ozbayoglu (2018) identified anomalies in standing and moving traffic using machine learning. They obtained promising detections when they evaluated their approach on a public highway. Finally, Bhavsar *et al.* (2017) reviewed traditional machine learning approaches to analyze transportation systems. Koc and Acar (2021) developed a neural network approach to understanding how the growing built environment might affect climate change. They combined climate averages with detected buildings from satellite imagery in the Esenboğa region of Turkey. Zhou and Chang (2021) automated the classification of building structures in complex urban environments using 29 features derived through their machine learning approach. On applying the technique to 3700 buildings in Beijing, China, they predicted building type with over 90% accuracy. By modifying the U-Net deep learning architecture (Ronneberger *et al.*, 2015) Daranagama and Witayangkurn (2021) produced building polygons and attributes in the urban environment. They applied their approach to high-resolution and uncrewed aerial vehicle imagery of different urban settlement types represented in Austin, Texas, and Chicago, Illinois, USA, and Vienna, Austria.

Machine learning approaches to image processing in GIS and cartography typically use three types of input: *cartographic products (maps), model output*, such as climate models, and *GIS data*, which include any data that can be transformed into georeferenced images, or provide image labels. Historical, scanned maps provide most map input that is not machine readable as part of a GIS dataset. For example, Duan *et al.* (2020) used CNNs to extract geographic feature locations from historical maps. Their method mapped roads, water lines, and railroads with high accuracy. Liu *et al.* (2020) revealed that a shallow CNN paired with super-pixel segmentation successfully classified geographic elements in historical topographic maps. They tested their approach on repeat maps (1949 and 1995) of Philadelphia, Pennsylvania, USA.

Finally, Arundel *et al.* (2022) used a custom dataset and deep learning text recognition to detect and recognize spot elevations on historical U.S. Geological Survey topographic maps. Their method reproduced labeled data with an accuracy of over 70%. Research by Gibson *et al.* (2021) provides an example of machine learning based on imagery produced by other models. The team trained a CNN on thousands of global circulation model "seasons" to better predict seasonal precipitation. In another example, Gagne II *et al.* (2019) compared deep learning algorithms to shallow machine learning models for predicting severe hail. The training imagery was composed of simulated hail surfaces. Finally, Ellenson *et al.* (2020) notably improved coastal wave forecasting with the bagged regression tree machine learning algorithm. Bulk parameter outputs of the numerical wave model WaveWatch III and wind surfaces formed their input to the algorithm. Rahmati *et al.* (2019) demonstrate using GIS data as input to machine learning algorithms. Geo-environmental variables represented in GIS helped them model the spatial frequency of snow avalanche hazards. Naghibi *et al.* (2016) compared the ability of three machine-learning models to map potential groundwater. GIS layers input to the models represented 13 hydrological-geological-physiographical factors. Finally, Lloyd *et al.* (2020) supplemented satellite images with building footprint data and labels created in GIS to classify the residential status of urban buildings. An ensemble of algorithms produced the most accurate results in parts of Africa.

5.3   GEOAI-SPECIFIC METHODS AND CHALLENGES

Although many code repositories provide essentially out-of-the-box AI solutions to many image processing challenges, many problems specific to the spatial sciences compel supervised learning applications, which require "labeled" training images. Labeled images are those accompanied by a text file (label) specifying the location of features to detect and their classes to recognize. However, because high-quality labeled geospatial training data are still quite limited, much research using GeoAI technology requires steps to create custom training data (Arundel *et al.*, 2020).

5.3.1   THE CHALLENGE IN TRAINING DATA ANNOTATION

In many cases, training features used in image analysis are compiled in a GIS and may already exist in spatial vector files, such as geodatabases or ESRI shapefiles (Wang and Li, 2021). These geographic features should be distinguished from features or feature maps formed in layers of a neural network. Information used to compile the geographic features should stipulate the elements of the training image. For example, detecting lakes may require satellite imagery, which supplies the training image at the feature location at a chosen map scale (Figure 5.1). In the case of feature detection, some images may show none of the vector features to train for negatives. Other images may show multiple features.

Training image labels are text files, one per image, with the same file name, except for the extension. For example, if the first image is named image001.tif, the corresponding label file is named image001.txt. Label file formatting is specific to

Figure 5.1 Lake image from National Agriculture Imagery Program showing a training feature bounding box in red.

the machine learning algorithm's data load module (not to mention all the matrix sizes in the linear algebra). Specific formatting is sometimes difficult to discern, and for the creation of custom data, it would ideally be replicated automatically through custom code developed locally. In other words, each combination of application, data type, and code repository may require custom code to create training data.

A challenge for the spatial scientist is transforming the geographic coordinate space of the vector feature to the image space so that feature locations within the image can be correctly specified in the label file (Figure 5.2). This transformation requires the computation of the map units to the image units. The technique for doing this here uses the example of a centered bounding envelope around the geographic feature. First, the minimum bounding envelope (BoxExtent) of a geographic feature extracted from a spatial database is recorded using the lower left and the upper right points in geographic coordinates (GeoBoxXMin, GeoBoxYMin; GeoBoxXMax, GeoBoxYMax). Next, the geographic extent of the map to be viewed in the output training image (FrameExtent) must also be known (GeoFrmXMin,

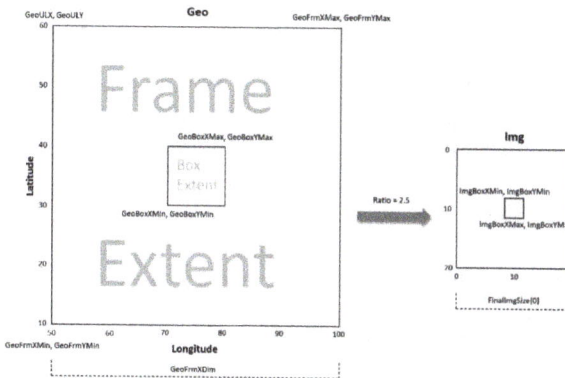

Figure 5.2 An illustration for coordinate transformation from a geographical extent to an image extent.

GeoFrmYMin; GeoFrmXMax, GeoFrmYMax). Finally, the desired size of the output image frame, in image units, is set (FinalImgSize). The ratio or scale (ImgRes) representing the relation between the two coordinate systems can then be calculated by dividing the width of the frame extent by the same dimension of the final image size (FinalImgSize[0]). Calculating the relationship between coordinate systems supplies the last variable to find output bounding box coordinates. First, the frame extent's length (cx) and width (cy) are halved to locate the center coordinate. Then, using half of the window width (r), the upper left corner coordinates (ulx, uly) (image origin) are calculated, from which the x is subtracted, and the y is added to calculate the bounding box location in the image coordinates.

Transforming predicted bounding box or feature vertices back to geographic space requires the reverse process. However, without applying a numerical projection algorithm to the reverse transformation, the larger the geographic extent of the feature is, the more distorted the transformation will be. This problem can be addressed by maintaining relatively large map scales regardless of the feature size.

5.3.2   THE CHALLENGE OF SCALE

Related to feature size, multi-scale features such as landforms require a strategy for analysis. On the one hand, if all features are scaled to display entirely within a single training image, most of the input data will be either up or down-sampled (Figure 5.3). On the other hand, this problem can be solved by fixing the map scale, but then extensive features may only have a portion of their extent depicted in a random selection of training images. What, if any, effect either of these compromises makes on the resulting predictions is unknown. Another possibility for working with multi-scale features is binning features by size classes, running the bins through models separately, and consolidating the results. Further research may help understand this challenge.

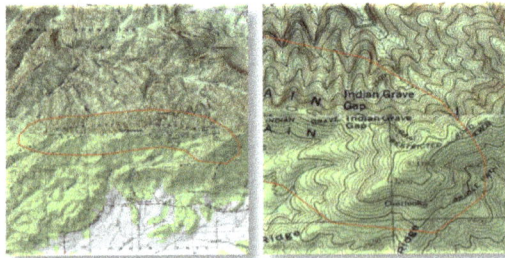

Figure 5.3   Demonstration of where scale could impact the effective prediction of a landform feature (Codes Cove Mountain, Tennessee), particularly as related to the image resolution at the smaller scale (left).

Some geographic features, such as those having substantial elevation changes (for example, ridges and valleys) can be predicted better by multiple input layers. Previous work has demonstrated by loading a slope surface, a hillshade image, and satellite imagery into the RGB bands of the output training image (Figure 5.4) (Arundel *et al.*, 2020; Li *et al.*, 2022b). A fourth information layer can be loaded into the depth layer of an RGB image (RGB-D). How loading raw elevation data into the depth band of GeoAI feature detection affects the predictions is unknown.

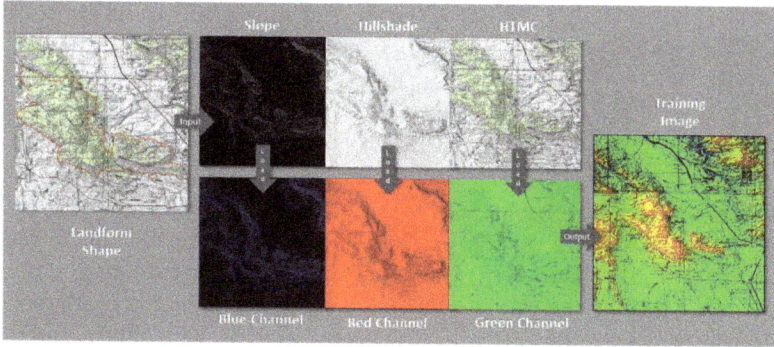

Figure 5.4   Workflow to create training images: Landform Shape is the vectorized form of the USGS Geographic Names Information System (GNIS) summit point feature Independence Mountain, Colorado, USA, which determines the extent of downloaded images of the slope, hillshade, and the historical topographic map (HTMC). Slope is loaded into the blue channel (or band) of the output training image, hillshade into the red channel, and HTMC into the green channel. Finally, the three channels are stacked into the resulting training image.

Blending allows the addition of information where more than four layers stacked into the bands of a single image are required. Python libraries, such as Open Source Computer Vision (OpenCV) Library, provide options to blend images by weight, resulting in a deluge of parameters that can be fine-tuned to enhance prediction ability.

5.3.3   THE CHALLENGE OF IMAGE RESOLUTION

Image resolution is crucial in ensuring high-quality delineation of geographical objects from spatial images. The higher the image resolution, the clearer the boundary of an object. Therefore, the object is easier to detect and extract through "machine vision" (Li and Hsu, 2022). However, due to the model complexity and limitation in computing power, the input images always need to be resized to fit into the model. This may cause potential issues for high-resolution input that some of the essential features will be obscured during the downsizing process. To the low-resolution input, on the other hand, the simple image up-sampling may or may not be able to sharpen the difference between the foreground (the object we are interested in) and the background. Hence, new strategies, such as data enrichment (Wang and Li, 2021), would ideally be developed to enhance the original input to help the machine better detect the objects of interest. For high-resolution data input, tiling would be feasible to retain the details in the image by portioning images into multiple smaller sub-images (Song *et al.*, 2016). One issue will arise when the object is too large, and clipping the images will cut the objects into multiple parts. When the ratio between the object's size and the size of the image scene is high, downsampling can be applied to include important detail in the image.

5.3.4 THE CHALLENGE OF SPACE AND TIME

GeoAI spatial image processing for change detection introduces the element of time (Zhong *et al.*, 2019) and additional challenges. One challenge is a dearth of training images, which already is an issue in GeoAI but is compounded when multiple time sequences of the same phenomenon are required. In addition, the complexity of spatio-temporal data makes labeling especially difficult (Reichstein *et al.*, 2019). Additionally, capturing relations between spatial, temporal, and spatio-temporal features is very challenging (Amato *et al.*, 2020). Despite deep learning's ability to extract features in spatial (CNNs) and temporal (RNNs) domains, many challenges remain. Issues include integrating multi-sensor data, understanding the effects of spatial dependence of variables, the increased noise when using multiple sensors, and the increased data volume resulting in more computing power and storage (Reichstein *et al.*, 2019).

5.4 FUTURE RESEARCH

5.4.1 GEOMETRIC ALGEBRA

Future GeoAI endeavors may benefit from a field of mathematics known as geometric algebra. Geometric algebra is a mathematical framework for manipulating geometric objects using algebraic operations. More specifically, geometric algebra provides a robust, streamlined, and unified methodology through which complex multidimensional geometric objects and most other mathematical elements in any dimension can be interpreted in an intuitive environment independent of coordinate systems (Bhatti *et al.*, 2020; Chisolm, 2012; Yuan *et al.*, 2012). This is accomplished by evaluating a property called the geometric product, through which geometric algebra can be used to model spatiotemporal phenomena (Yuan *et al.*, 2019).

Due to its utility in understanding topology and multidimensional objects, geometric algebra may be the next frontier in GIS (Yu *et al.*, 2016; Yuan *et al.*, 2019), although it is quite old, having been developed in the 19th century by William Clifford (Bhatti *et al.*, 2020; Chisolm, 2012). Eventually developing into a more mature field of mathematics, this field of mathematics is now used in several areas of AI (Bhatti *et al.*, 2020). Regarding GeoAI, geometric algebra has several potential applications, particularly in advanced image analysis. Specifically, it can perform the rotations and transformations of objects necessary for image analysis and computer vision by transforming geographic space into mathematical space (Bhatti *et al.*, 2020). This ability has proven helpful in remote sensing image processing (Bhatti *et al.*, 2020). Given this ability and geometric algebra's position as a unifying bridge between geometry and algebra, it appears to be a potential growth area for new advances in GeoAI.

5.4.2   ETHICAL ISSUES RELATED TO SPATIAL IMAGE PROCESSING IN GEOAI

Attention also needs to be paid to the ethical concerns as we apply GeoAI for spatial image processing, especially in solving real-world critical applications. This is also important for building trust in AI-aided decision-making systems. The Delphi study (Vogel *et al.*, 2019) conducted by a European Union project – shaping the ethical dimensions of smart information systems – identified several prominent topics on ethical AI, including model transparency, privacy, data quality, bias and discrimination, and control and misuse of data. In the geospatial world, as we know, relationships between the phenomena (for example, a disaster) and landform features are often complicated, leading to the development of complex GeoAI models to discern important, and perhaps human-unaware, knowledge to achieve accurate event detection, classification, and prediction. However, the complexity of GeoAI models also makes it extremely difficult to explain how a GeoAI model works. Although we have been seeing some pioneering works (Hsu and Li, 2023) in this realm, a big gap remains to fully open up the black box of the GeoAI models, to help researchers to examine its inference process, and validate the trustworthiness and reproducibility of the AI-derived results (Goodchild and Li, 2021; Li and Arundel, 2022).

5.4.3   COMBINING GEOAI AND PHYSICAL MODELING

GeoAI and physical models can be integrated in several ways to improve the accuracy and comprehensiveness of models used in various fields, such as natural resource management, environmental monitoring, and disaster management. Three ways to integrate these models are by combining physical models with machine learning algorithms, integrating physical models with machine learning algorithms and remote sensing data, and calibrating physical models using machine learning algorithms. Physical models are typically based on mathematical equations that describe the behavior of physical systems, such as the flow of water in a river or the movement of air in the atmosphere. These models can be improved by integrating machine learning algorithms that use data to refine the models and make more accurate predictions. For example, Wu *et al.* (2022) used a combination of physical models and machine learning algorithms to predict the runoff in a river basin in China, resulting in significantly improved accuracy compared to using either method alone.

Combining physical models, machine learning algorithms, and remote sensing data can produce more accurate and comprehensive models. Féret *et al.* (2019) discuss the potentials and limitations of combining GeoAI with physical modeling to estimate leaf mass per area and other vegetation properties. Physical models can be difficult to calibrate, often requiring extensive field measurements and data collection. Machine learning algorithms can improve the calibration process by predicting model parameters based on available data. Gonzales-Inca *et al.* (2022) review current applications and trends in integrating hydrological and fluvial systems modeling with GeoAI. Integrating GeoAI and physical models can lead to more accurate and comprehensive models, which can be used to make better decisions in various fields.

In addition, these approaches can improve the accuracy of predictions and reduce uncertainty, which can be particularly important in fields such as disaster management and natural resource management.

5.4.4 KNOWLEDGE BASE AS INPUT/CONSTRAINT

Although deep learning models can extract critical knowledge and previously unknown patterns from raw data, the models may or may not be able to identify whether derived knowledge is already known or scientifically sound. Therefore, to enhance their reasoning and interpretation ability, integrating expert knowledge into the deep learning process is important (Li, 2022; Li *et al.*, 2021a). This will, on the one hand, reduce the computation cost to derive already documented knowledge. On the other hand, the knowledge can empower the model to make more accurate predictions. For example, Hsu *et al.* (2021) developed a knowledge-driven GeoAI model for Mars crater detection. Besides integrating a CNN capable of extracting multi-scale features, this model also integrated spatial knowledge about a crater's circular shape. The results demonstrate that prior knowledge can enhance a GeoAI model's predictive performance.

In recent years, knowledge graphs have become essential for formulating knowledge representation to enable formal machine reasoning (Janowicz *et al.*, 2022; Li *et al.*, 2023, 2022c). By building a knowledge graph describing spatial scenes and interrelationships of objects within an image and integrating this formal knowledge, such as objects co-occurrence probabilities, into a GeoAI model, the model will be able to more confidently determine object class and remove noise that negatively affects its prediction process (Li, 2020).

5.5 CONCLUSION

Image processing in GeoAI involves using advanced algorithms and techniques to extract useful information from remotely sensed data, such as satellite or aerial imagery, for geospatial analysis and applications. The process typically involves preprocessing steps to correct for distortions and enhance the quality of the image data, followed by image segmentation, classification, and feature extraction to identify and map various land cover, land use, and other geospatial features. The resulting geospatial information can be used for various applications, such as environmental monitoring, urban planning, precision agriculture, and disaster response.

Image processing in GeoAI presents several challenges, including the following:

- **Data quality**: Image data obtained from remote sensors can be affected by factors such as atmospheric conditions, sensor noise, and geometric distortions, which can affect the accuracy of image processing results.

- **Data volume**: High-resolution remote sensing data can generate large datasets that require considerable computational resources and storage capabilities.

- **Complexity of algorithms**: Image processing algorithms used in GeoAI are often complex and computationally intensive, requiring specialized hardware and software.

- **Labeling and annotation**: Ground-truth data for training machine learning models and verifying image processing results can be difficult and expensive, especially for large and diverse datasets.

- **Scale, space, and time**: Issues related to geospatial-temporal properties would ideally be considered in both design and analysis.

Addressing these challenges requires technical expertise, specialized hardware and software, and robust quality assurance and control procedures to ensure accurate and reliable results.

ACKNOWLEDGMENTS

This work is in part supported by the National Science Foundation under grant No. 1853864. Wenwen Li would like to acknowledge additional support from NSF under grants 2230034, 2033521, and 2120943. Any use of trade, firm, or product names is for descriptive purposes only and does not imply endorsement by the U. S. Government and the funder(s).

BIBLIOGRAPHY

Abhar, K.C., *et al.*, 2015. Spatial–temporal evolution of aeolian blowout dunes at cape cod. *Geomorphology*, 236, 148–162.

Acosta, J. and Cherrier, B., 2021. The transformation of economic analysis at the board of governors of the federal reserve system during the 1960s. *Journal of the History of Economic Thought*, 43 (3), 323–349.

Al-Mistarehi, B., *et al.*, 2022. Using machine learning models to forecast severity level of traffic crashes by r studio and arcgis. *Frontiers in Built Environment*, 8, 54.

Amato, F., *et al.*, 2020. A novel framework for spatio-temporal prediction of environmental data using deep learning. *Scientific Reports*, 10 (1), 22243.

Arundel, S.T., Li, W., and Wang, S., 2020. Geonat v1. 0: A dataset for natural feature mapping with artificial intelligence and supervised learning. *Transactions in GIS*, 24 (3), 556–572.

Arundel, S.T., Morgan, T.P., and Thiem, P.T., 2022. Deep learning detection and recognition of spot elevations on historical topographic maps. *Frontiers in Environmental Science*, 117.

Barnes, T.J., 2004. Placing ideas: genius loci, heterotopia and geography's quantitative revolution. *Progress in Human Geography*, 28 (5), 565–595.

Behnke, S., 2003. *Hierarchical Neural Networks for Image Interpretation*. vol. 2766. Springer.

Benos, L., *et al.*, 2021. Machine learning in agriculture: A comprehensive updated review. *Sensors*, 21 (11), 3758.

Bhatti, U.A., *et al.*, 2020. Geometric algebra applications in geospatial artificial intelligence and remote sensing image processing. *IEEE Access*, 8, 155783–155796.

Bhavsar, P., *et al.*, 2017. Machine learning in transportation data analytics. *In*: *Data Analytics for Intelligent Transportation Systems*. Elsevier, 283–307.

Bickel, V.T., Mandrake, L., and Doran, G., 2021. A labeled image dataset for deep learning-driven rockfall detection on the moon and mars. *Frontiers in Remote Sensing*, 2, 640034.

Bielecki, A., 2019. *Models of Neurons and Perceptrons: Selected Problems and Challenges*. Springer.

Blaschke, T., 2010. Object based image analysis for remote sensing. *ISPRS Journal of Photogrammetry and Remote Sensing*, 65 (1), 2–16.

Bryant, R.G. and Baddock, M.C., 2021. Remote sensing of aeolian processes. *Reference Module in Earth Systems and Environmental Sciences*.

Burnett, C. and Blaschke, T., 2003. A multi-scale segmentation/object relationship modelling methodology for landscape analysis. *Ecological Modelling*, 168 (3), 233–249.

Campos-Taberner, M., *et al.*, 2020. Understanding deep learning in land use classification based on sentinel-2 time series. *Scientific Reports*, 10 (1), 17188.

Cheng, K.S., Lin, J.S., and Mao, C.W., 1996. The application of competitive hopfield neural network to medical image segmentation. *IEEE Transactions on Medical Imaging*, 15 (4), 560–567.

Chisolm, E., 2012. Geometric algebra. *arXiv preprint arXiv:1205.5935*.

Church, M., 2010. The trajectory of geomorphology. *Progress in Physical Geography*, 34 (3), 265–286.

Cracknell, A., 1998. Synergy in remote sensing-what's in a pixel? *Int. J. Remote Sens*, 19, 2025–2057.

Cresswell, T., 2013. Spatial science and the quantitative revolution. *Geographic Thought: A Critical Introduction*.

Daranagama, S. and Witayangkurn, A., 2021. Automatic building detection with polygonizing and attribute extraction from high-resolution images. *ISPRS International Journal of Geo-Information*, 10 (9), 606.

Dechter, R., 1986. Learning while searching in constraint-satisfaction problems. *In*: *AAAI-86 Proceedings*. 178–185.

Dong, G., *et al.*, 2020. Real-time high-performance semantic image segmentation of urban street scenes. *IEEE Transactions on Intelligent Transportation Systems*, 22 (6), 3258–3274.

Dosovitskiy, A., *et al.*, 2020. An image is worth 16x16 words: Transformers for image recognition at scale. *arXiv preprint arXiv:2010.11929*.

Douard, J., 1996. Review: Reviewed works: Trust in numbers: The pursuit of objectivity in science and public life by theodore m. porter; quantification and the quest for medical certainty by j. rosser mathews. *Polit. Life Sci*, 15, 350–353.

Duan, W., *et al.*, 2020. Automatic alignment of contemporary vector data and georeferenced historical maps using reinforcement learning. *International Journal of Geographical Information Science*, 34 (4), 824–849.

Dyrmann, M., *et al.*, 2016. Pixel-wise classification of weeds and crops in images by using a fully convolutional neural network. *In*: *Proceedings of the International Conference on Agricultural Engineering, Aarhus, Denmark*. 26–29.

Ellenson, A., *et al.*, 2020. An application of a machine learning algorithm to determine and describe error patterns within wave model output. *Coastal Engineering*, 157, 103595.

Escalera, D., Fraile-Jurado, P., and Peña Alonso, C., 2016. *Evaluation of the Impact of Dune Management on the Coast of Huelva Using Repeat Photography Techniques (1986, 2001 and 2015)*. 137–156.

Feizizadeh, B., *et al.*, 2021. A comparison of the integrated fuzzy object-based deep learning approach and three machine learning techniques for land use/cover change monitoring and environmental impacts assessment. *GIScience & Remote Sensing*, 58 (8), 1543–1570.

Féret, J.B., *et al.*, 2019. Estimating leaf mass per area and equivalent water thickness based on leaf optical properties: Potential and limitations of physical modeling and machine learning. *Remote Sensing of Environment*, 231, 110959.

Fernandes, S., Duseja, D., and Muthalagu, R., 2021. Application of image processing techniques for autonomous cars. *Proceedings of Engineering and Technology Innovation*, 17, 1.

Fisher, A.R., *et al.*, 2020. Use of convolutional neural networks for semantic image segmentation across different computing systems.

Fu, K.S. and Mui, J., 1981. A survey on image segmentation. *Pattern Recognition*, 13 (1), 3–16.

Gagne II, D.J., *et al.*, 2019. Interpretable deep learning for spatial analysis of severe hailstorms. *Monthly Weather Review*, 147 (8), 2827–2845.

Gibson, P.B., *et al.*, 2021. Training machine learning models on climate model output yields skillful interpretable seasonal precipitation forecasts. *Communications Earth & Environment*, 2 (1), 159.

Gonzales-Inca, C., *et al.*, 2022. Geospatial artificial intelligence (geoai) in the integrated hydrological and fluvial systems modeling: Review of current applications and trends. *Water*, 14 (14), 2211.

Goodchild, M.F., 2018. Reimagining the history of gis. *Annals of GIS*, 24 (1), 1–8.

Goodchild, M.F. and Li, W., 2021. Replication across space and time must be weak in the social and environmental sciences. *Proceedings of the National Academy of Sciences*, 118 (35), e2015759118.

Goodfellow, I., Bengio, Y., and Courville, A., 2016. *Deep Learning*. MIT Press.

Graff, L.H. and Usery, E.L., 1993. Automated classification of generic terrain features in digital elevation models. *Photogrammetric Engineering and Remote Sensing*, 59 (9), 1409–1417.

Gustafsson, K., Öktem, O., and Boman, E.J., 2014. The role of linear algebra in computed tomography.

Hammond, E.H., 1954. Small-scale continental landform maps. *Annals of the Association of American Geographers*, 44 (1), 33–42.

Hattersley-Smith, G., 1966. The symposium on glacier mapping. *Canadian Journal of Earth Sciences*, 3 (6), 737–741.

Hay, G.J. and Castilla, G., 2008. Geographic object-based image analysis (geobia): A new name for a new discipline. *Object-Based Image Analysis: Spatial Concepts for Knowledge-Driven Remote Sensing Applications*, 75–89.

Hebb, D.O., 2005. *The Organization of Behavior: A Neuropsychological Theory*. Psychology press.

Hernández, E., *et al.*, 2016. Rainfall prediction: A deep learning approach. *In*: *Hybrid Artificial Intelligent Systems 2016*. Springer, 151–162.

Hoffmann, C., 2013. Superpositions: Ludwig mach and étienne-jules marey's studies in streamline photography. *Studies in History and Philosophy of Science Part A*, 44 (1), 1–11.

Honavar, V., 1995. Symbolic artificial intelligence and numeric artificial neural networks: towards a resolution of the dichotomy. *Computational Architectures Integrating Neural and Symbolic Processes: A Perspective on the State of the Art*, 351–388.

Hopfield, J.J., 1982. Neural networks and physical systems with emergent collective computational abilities. *Proceedings of the National Academy of Sciences*, 79 (8), 2554–2558.

Howell, S.B., 2006. *Handbook of ccd Astronomy*. vol. 5. Cambridge University Press.

Hsu, C.Y. and Li, W., 2023. Explainable geoai: can saliency maps help interpret artificial intelligence's learning process? an empirical study on natural feature detection. *International Journal of Geographical Information Science*, 37 (5), 963–987.

Hsu, C.Y., Li, W., and Wang, S., 2021. Knowledge-driven geoai: Integrating spatial knowledge into multi-scale deep learning for mars crater detection. *Remote Sensing*, 13 (11), 2116.

Hughes, D., 2016. *Natural Visions: Photography and Ecological Knowledge, 1895-1939*. Thesis (PhD). De Montfort University.

Janowicz, K., *et al.*, 2022. Know, know where, knowwheregraph: A densely connected, cross-domain knowledge graph and geo-enrichment service stack for applications in environmental intelligence. *AI Magazine*, 43 (1), 30–39.

Kettig, R.L. and Landgrebe, D., 1976. Classification of multispectral image data by extraction and classification of homogeneous objects. *IEEE Transactions on Geoscience Electronics*, 14 (1), 19–26.

Khan, A.A., Laghari, A.A., and Awan, S.A., 2021. Machine learning in computer vision: a review. *EAI Endorsed Transactions on Scalable Information Systems*, 8 (32), e4–e4.

Kirsch, R.A., 1998. Seac and the start of image processing at the national bureau of standards. *IEEE Annals of the History of Computing*, 20 (2), 7–13.

Kirsch, R.A., 2010. Precision and accuracy in scientific imaging. *Journal of Research of the National Institute of Standards and Technology*, 115 (3), 195.

Kirsch, R.A., *et al.*, 1957. Experiments in processing pictorial information with a digital computer. *In*: *Papers and Discussions Presented at the December 9-13, 1957, Eastern Joint Computer Conference: Computers with Deadlines to Meet*. 221–229.

Koc, M. and Acar, A., 2021. Investigation of urban climates and built environment relations by using machine learning. *Urban Climate*, 37, 100820.

Kohn, C.F., 1970. The 1960's: A decade of progress in geographical research and instruction. *Annals of the Association of American Geographers*, 60 (2), 211–219.

Krizhevsky, A., Sutskever, I., and Hinton, G.E., 2012. Imagenet classification with deep convolutional neural networks. *In: Proceedings of NIPS'12*, Red Hook, NY, USA. 1097–1105.

Kucharczyk, M., *et al.*, 2020. Geographic object-based image analysis: a primer and future directions. *Remote Sensing*, 12 (12), 2012.

Kurth, T., *et al.*, 2018. Exascale deep learning for climate analytics. *In: SC18: International Conference for High Performance Computing, Networking, Storage and Analysis*. IEEE, 649–660.

Lee, R.S., 2020. *Artificial Intelligence in Daily Life*. Springer.

Li, W., 2020. Geoai: Where machine learning and big data converge in GIScience. *Journal of Spatial Information Science*, (20), 71–77.

Li, W., 2022. Geoai in social science. *Handbook of Spatial Analysis in the Social Sciences*, 291–304.

Li, W. and Arundel, S.T., 2022. Geoai and the future of spatial analytics. *In: New Thinking in GIScience*. Springer, 151–158.

Li, W. and Hsu, C.Y., 2020. Automated terrain feature identification from remote sensing imagery: a deep learning approach. *International Journal of Geographical Information Science*, 34 (4), 637–660.

Li, W. and Hsu, C.Y., 2022. Geoai for large-scale image analysis and machine vision: Recent progress of artificial intelligence in geography. *ISPRS International Journal of Geo-Information*, 11 (7), 385.

Li, W., Hsu, C.Y., and Hu, M., 2021a. Tobler's first law in geoai: A spatially explicit deep learning model for terrain feature detection under weak supervision. *Annals of the American Association of Geographers*, 111 (7), 1887–1905.

Li, W., *et al.*, 2022a. Real-time geoai for high-resolution mapping and segmentation of arctic permafrost features: the case of ice-wedge polygons. *In: Proceedings of the 5th ACM SIGSPATIAL Workshop on AI for Geographic Knowledge Discovery*. 62–65.

Li, W., *et al.*, 2022b. Geoimagenet: a multi-source natural feature benchmark dataset for geoai and supervised machine learning. *GeoInformatica*, 1–22.

Li, W., *et al.*, 2023. Geographvis: a knowledge graph and geovisualization empowered cyber-infrastructure to support disaster response and humanitarian aid. *ISPRS International Journal of Geo-Information*, 12 (3), 112.

Li, W., *et al.*, 2022c. Performance benchmark on semantic web repositories for spatially explicit knowledge graph applications. *Computers, Environment and Urban Systems*, 98, 101884.

Li, W., *et al.*, 2017. Recognizing terrain features on terrestrial surface using a deep learning model: An example with crater detection. *In: Proceedings of the 1st Workshop on Artificial Intelligence and Deep Learning for Geographic Knowledge Discovery*. 33–36.

Li, X., *et al.*, 2021b. Vehicle detection in very-high-resolution remote sensing images based on an anchor-free detection model with a more precise foveal area. *ISPRS International Journal of Geo-Information*, 10 (8), 549.

Lin, T.Y., *et al.*, 2017. Focal loss for dense object detection. *In: Proceedings of the IEEE International Conference on Computer Vision*. 2980–2988.

Liu, T., *et al.*, 2020. Superpixel-based shallow convolutional neural network (sscnn) for scanned topographic map segmentation. *Remote Sensing*, 12 (20), 3421.

Lloyd, C.T., *et al.*, 2020. Using gis and machine learning to classify residential status of urban buildings in low and middle income settings. *Remote Sensing*, 12 (23), 3847.

Lundberg, G.A., 1960. Quantitative methods in sociology: 1920–1960. *Social Forces*, 39 (1), 19–24.

Mackay, C.D., 1986. Charge-coupled devices in astronomy. *Annual Review of Astronomy and Astrophysics*, 24 (1), 255–283.

Magnuson, J.J., 1990. Long-term ecological research and the invisible present. *BioScience*, 40 (7), 495–501.

Mahadevkar, S.V., *et al.*, 2022. A review on machine learning styles in computer vision-techniques and future directions. *IEEE Access*.

McCarthy, J., *et al.*, 2006. A proposal for the dartmouth summer research project on artificial intelligence, august 31, 1955. *AI Magazine*, 27 (4), 12–12.

McCulloch, W.S. and Pitts, W., 1943. A logical calculus of the ideas immanent in nervous activity. *The Bulletin of Mathematical Biophysics*, 5, 115–133.

McFarlane, M.D., 1972. Digital pictures fifty years ago. *Proceedings of the IEEE*, 60 (7), 768–770.

Mengwall, S. and Guzewich, S.D., 2023. Cloud identification in mars daily global maps with deep learning. *Icarus*, 389, 115252.

Miller, B.A., *et al.*, 2019. Progress in soil geography i: Reinvigoration. *Progress in Physical Geography: Earth and Environment*, 43 (6), 827–854.

Milnor, J., 1941. Picture transmission by submarine cable. *Transactions of the American Institute of Electrical Engineers*, 60 (3), 105–108.

Minasny, B. and McBratney, A.B., 2016. Digital soil mapping: A brief history and some lessons. *Geoderma*, 264, 301–311.

Minsky, M. and Papert, S.A., 2017. *Perceptrons: An Introduction to Computational Geometry*. The MIT Press.

Moor, J., 2006. The dartmouth college artificial intelligence conference: The next fifty years. *AI Magazine*, 27 (4), 87–87.

Mulyono, S., *et al.*, 2016. Identifying sugarcane plantation using landsat-8 images with support vector machines. *In: IOP Conference Series: Earth and Environmental Science*. IOP Publishing, vol. 47, 012008.

Naghibi, S.A., Pourghasemi, H.R., and Dixon, B., 2016. Gis-based groundwater potential mapping using boosted regression tree, classification and regression tree, and random forest machine learning models in iran. *Environmental Monitoring and Assessment*, 188, 1–27.

Nir, D. and Nir, D., 1990. The 'quantitative revolution': Regional geography at its apogee. *Region as a Socio-environmental System: An Introduction to a Systemic Regional Geography*, 43–57.

Olazaran, M., 1996. A sociological study of the official history of the perceptrons controversy. *Social Studies of Science*, 26 (3), 611–659.

Pal, N.R. and Pal, S.K., 1993. A review on image segmentation techniques. *Pattern Recognition*, 26 (9), 1277–1294.

Quirke, V. and Gaudillière, J.P., 2008. The era of biomedicine: science, medicine, and public health in britain and france after the second world war. *Medical history*, 52 (4), 441–452.

Rahmati, O., *et al.*, 2019. Spatial modeling of snow avalanche using machine learning models and geo-environmental factors: Comparison of effectiveness in two mountain regions. *Remote Sensing*, 11 (24), 2995.

Redmon, J. and Farhadi, A., 2018. Yolov3: An incremental improvement. *arXiv preprint arXiv:1804.02767*.

Reichstein, M., *et al.*, 2019. Deep learning and process understanding for data-driven earth system science. *Nature*, 566 (7743), 195–204.

Ren, S., *et al.*, 2015. Faster r-cnn: Towards real-time object detection with region proposal networks. *Advances in Neural Information Processing Systems*, 28.

Rogers, G.F., Malde, H.E., and Turner, R.M., 1984. *Bibliography of Repeat Photography for Evaluating Landscape Change*.

Rolnick, D., *et al.*, 2022. Tackling climate change with machine learning. *ACM Computing Surveys (CSUR)*, 55 (2), 1–96.

Romans, L.E., 1995. *Introduction to Computed Tomography*. Lippincott Williams & Wilkins.

Ronneberger, O., Fischer, P., and Brox, T., 2015. U-net: Convolutional networks for biomedical image segmentation. *In: Medical Image Computing and Computer-Assisted Intervention–MICCAI 2015: 18th International Conference, Munich, Germany, October 5-9, 2015, Proceedings, Part III 18*. Springer, 234–241.

Rosenblatt, F., 1958. The perceptron: a probabilistic model for information storage and organization in the brain. *Psychological Review*, 65 (6), 386.

Rouse, J.W., *et al.*, 1974. Monitoring vegetation systems in the great plains with erts. *NASA Spec. Publ*, 351 (1), 309.

Ruggles, S. and Magnuson, D.L., 2019. The history of quantification in history: The jih as a case study. *Journal of Interdisciplinary History*, 50 (3), 363–381.

Rumelhart, D.E., Hinton, G.E., and Williams, R.J., 1986. Learning representations by back-propagating errors. *Nature*, 323 (6088), 533–536.

Sarikan, S.S. and Ozbayoglu, A.M., 2018. Anomaly detection in vehicle traffic with image processing and machine learning. *Procedia Computer Science*, 140, 64–69.

Saska, H., 2010. Anna atkins: Photographs of british algae. *Bulletin of the Detroit Institute of Arts*, 84 (14), 8–15.

Sexton, N.J. and Love, B.C., 2022. Reassessing hierarchical correspondences between brain and deep networks through direct interface. *Science Advances*, 8 (28), eabm2219.

Shirzadi, A., *et al.*, 2018. Shallow landslide susceptibility mapping. *Sensors*, 18, 1–28.

Song, H., *et al.*, 2019. Vision-based vehicle detection and counting system using deep learning in highway scenes. *European Transport Research Review*, 11, 1–16.

Song, M., *et al.*, 2016. Spatiotemporal data representation and its effect on the performance of spatial analysis in a cyberinfrastructure environment–a case study with raster zonal analysis. *Computers & Geosciences*, 87, 11–21.

Starrs, P.F., 1998. Brinck jackson in the realm of the everyday. *Geographical Review*, 88 (4), 492–506.

Stegner, W., 1992. *Beyond the Hundredth Meridian: John Wesley Powell and the Second Opening of the West*. Penguin.

Stokes, C., *et al.*, 2013. Formation of mega-scale glacial lineations on the dubawnt lake ice stream bed: 1. size, shape and spacing from a large remote sensing dataset. *Quaternary Science Reviews*, 77, 190–209.

Talukdar, S., *et al.*, 2020. Land-use land-cover classification by machine learning classifiers for satellite observations—a review. *Remote Sensing*, 12 (7), 1135.

Tarran, B. and Ghahramani, Z., 2015. How machines learned to think statistically. *Significance*, 12 (1), 8–15.

Tu, K-L., *et al.*, 2018. Selection for high quality pepper seeds by machine vision and classifiers. *Journal of Integrative Agriculture*, 17 (9), 1999–2006.

Tucker, C., Miller, L., and Pearson, R., 1973. Measurement of the combined effect of green biomass, chlorophyll, and leaf water on canopy spectroreflectance of the shortgrass prairie. *Remote Sensing of Earth Resources*.

Vogel, C., *et al.*, 2019. A delphi study to build consensus on the definition and use of big data in obesity research. *International Journal of Obesity*, 43 (12), 2573–2586.

Wang, S. and Li, W., 2021. Geoai in terrain analysis: Enabling multi-source deep learning and data fusion for natural feature detection. *Computers, Environment and Urban Systems*, 90, 101715.

Weng, Q., *et al.*, 2018. Land-use scene classification based on a cnn using a constrained extreme learning machine. *International Journal of Remote Sensing*, 39 (19), 6281–6299.

Wheeler, J.O., 2001. urban geography in the 1960s. *Urban Geography*, 22 (6), 511–513.

White, R.A., *et al.*, 2019. Measurement of vegetation change in critical dune sites along the eastern shores of lake michigan from 1938 to 2014 with object-based image analysis. *Journal of Coastal Research*, 35 (4), 842–851.

Worboys, M.F., 1999. Relational databases and beyond. *Geographical Information Systems*, 1, 373–384.

Wu, H., *et al.*, 2022. Runoff modeling in ungauged catchments using machine learning algorithm-based model parameters regionalization methodology. *Engineering*.

Yamashita, R., *et al.*, 2018. Convolutional neural networks: an overview and application in radiology. *Insights into Imaging*, 9, 611–629.

Yu, Z., *et al.*, 2016. Geometric algebra model for geometry-oriented topological relation computation. *Transactions in GIS*, 20 (2), 259–279.

Yuan, L., *et al.*, 2012. Geometric algebra method for multidimensionally-unified gis computation. *Chinese Science Bulletin*, 57, 802–811.

Yuan, L., Yu, Z., and Luo, W., 2019. Towards the next-generation gis: A geometric algebra approach. *Annals of GIS*, 25 (3), 195–206.

Zhong, L., Hu, L., and Zhou, H., 2019. Deep learning based multi-temporal crop classification. *Remote Sensing of Environment*, 221, 430–443.

Zhou, P. and Chang, Y., 2021. Automated classification of building structures for urban built environment identification using machine learning. *Journal of Building Engineering*, 43, 103008.

6 Spatial Representation Learning in GeoAI

Gengchen Mai
Spatially Explicit Artificial Intelligence Lab, Department of
Geography, University of Georgia

Ziyuan Li
School of Business, University of Connecticut

Ni Lao
Google

CONTENTS

6.1 Introduction 99
6.2 Spatially Explicit Artificial Intelligence 101
6.3 Spatial Representation Learning on Various Spatial Data Types 102
    6.3.1 Points and Location Encoders 103
    6.3.2 Polylines and Polyline Encoder 107
    6.3.3 Polygons and Polygon Encoder 112
6.4 Conclusions 115
    Bibliography 115

6.1 INTRODUCTION

The choice of data representation and features is usually a decisive factor in the performance of machine learning (ML) models (Bengio *et al.*, 2013). Traditionally, ML features are *manually* picked and extracted based on domain knowledge and expertise. This process is called *feature engineering*, which has several drawbacks: it heavily relies on domain knowledge and cannot be easily generalized or adapted to new domains and new data (Mai *et al.*, 2022d). To overcome these drawbacks, *representation learning* approaches have been proposed to *automatically* extract/discover the ML features/representation from raw data for feature detection, classification, regression, or other prediction tasks (Bengio *et al.*, 2013; Hamilton *et al.*, 2017). How to develop a neural network model to automatically learn high-quality representations from raw data is one of the key research questions for artificial intelligence in general. The recent revolution of deep learning has led to the success of many deep neural network-based representation learning models. These successes can be

DOI: 10.1201/9781003308423-6

explained by two key ingredients: transfer learning from unsupervised training (e.g., BERT (Kenton and Toutanova, 2019)), and novel architectures which encode good priors for the domain (e.g., Transformer (Vaswani *et al.*, 2017)).

Some great examples of representation learning models in the Natural Language Processing (NLP) domain are word embedding techniques such as Word2Vec (Mikolov *et al.*, 2013), GloVe (Pennington *et al.*, 2014), and BERT (Kenton and Toutanova, 2019). After unsupervised trained on large text corpora, the ML model can represent each word as a high-dimensional real-valued vector, so-called word embeddings. Mikolov *et al.* (2013) showed that these unsupervised learned word embeddings could capture the semantics of each word and exhibit a linear structure that enables precise analogical reasoning with simple vector arithmetics. Various NLP studies showed that these unsupervised word representations are very useful for different NLP tasks such as read comprehension (Chen *et al.*, 2017; Das *et al.*, 2019), question answering (Brown *et al.*, 2020; Das *et al.*, 2019; Karpukhin *et al.*, 2020; Kenton and Toutanova, 2019), machine translation (Brown *et al.*, 2020), and so on.

Similarly, A good example of representation learning models in the Computer Vision (CV) domain is the masked autoencoders (MAE) (He *et al.*, 2022) based on the Vision Transformer architecture (Dosovitskiy *et al.*, 2020). MAE has been shown to produce high-quality visual representations of input images which are widely used in multiple downstream tasks such as image classification, semantic segmentation, object detection, and so on (Cong *et al.*, 2022; Guo *et al.*, 2022; He *et al.*, 2022).

In the geospatial artificial intelligence (GeoAI) domain, developing neural network representation learning models for geospatial data is also a critical research direction. In this chapter, we call this area of research *Spatial Representation Learning (SRL)* as a subdomain of representation learning, which focuses on automatically extracting machine learning features/representations from data that are commonly used in spatial and geospatial research such as vector data (e.g., points (Chu *et al.*, 2019; Mac Aodha *et al.*, 2019; Mai *et al.*, 2022c, 2020b; Qi *et al.*, 2017a; Wang *et al.*, 2019b), polylines (Alahi *et al.*, 2016; Musleh *et al.*, 2022), and polygons (Mai *et al.*, 2022d; Yan *et al.*, 2021)), raster data (e.g., remote sensing images (Ayush *et al.*, 2021; Cong *et al.*, 2022; Jean *et al.*, 2019; Manas *et al.*, 2021) and Streetview images), graphs (e.g., geospatial knowledge graphs (Mai *et al.*, 2020a; Trisedya *et al.*, 2019), traffic networks (Cai *et al.*, 2020; Li *et al.*, 2018a)), and etc.

However, why do we need to highlight SRL compared with other representation learning techniques? That is because spatial representation learning has some unique challenges. First, spatial data is usually formalized in irregular data structures which require unique representation learning methods. Unlike the commonly used data (e.g., text, image, video, audio, etc.) in the ML and AI domain, which have rather regular data structures (Qi *et al.*, 2017a; Wang *et al.*, 2019a,b) such as 1D sequences, 2D matrices, 3D tensors, etc., spatial data used in GeoAI research (e.g., points, polylines, polygons, triangulated irregular networks, field data on a spherical surface) are usually organized in complex or irregular structures (Mai *et al.*, 2022c). Due to these irregularities, existing representation learning models that are usually used on text data (e.g., Recurrent Neural Network, Transformers) or image data (e.g., Convolutional Neural Network) become inapplicable or will yield sub-optimal results. For example, given a point cloud, without a proper way to represent each individual point and model their spatial relations, many early point cloud studies (Maturana and Scherer, 2015) resorted to first converting the point cloud into regular point density grids and applying Convolutional Neural Networks to them which cause invertible information losses (see Section 6.3.1.1 for a detailed description). Similarly, without a proper way to directly represent the shapes of building polygons, Yan et al. (2019) utilized a set of predefined shape descriptors to extract a set of shape features for each building polygon as its ML features for building pattern recognition. This kind of approach suffers from the common drawbacks of feature engineering methods. In short, new types of representation learning models for these spatial data are necessary.

Second, even for some irregular data formats such as graphs which are already widely studied in the representation learning literature, spatial information also adds another dimension of complexity to the RL model design. The graphs commonly studied in GeoAI are usually spatially embedded graphs (Mai et al., 2022c) in which all nodes or a subset of them are associated with their spatial footprints and/or temporal information. Examples are geographic knowledge graphs (Hoffart et al., 2013; Janowicz et al., 2022), spatially embedded social networks (Haw et al., 2020), air quality sensor networks (Qi et al., 2019), traffic networks (Li et al., 2018a), and so on. How to model the spatial distributions of the graph nodes and the spatiotemporal interactions among nodes becomes a unique challenge for SRL.

In the following, we will first discuss the more general idea of spatially explicit artificial intelligence in Section 6.2, which encompasses and motivates spatial representation learning. Next, in Section 6.3 we describe the SRL models for different spatial data types such as points, polylines, polygons, etc.. We discuss the unique challenges and existing approaches for each spatial data type. Finally, we conclude this chapter by discussing the future challenges and research directions of SRL.

6.2 SPATIALLY EXPLICIT ARTIFICIAL INTELLIGENCE

Spatially explicit model is a key concept in Geographic Information Science that describes models which can differentiate behaviors and predictions according to spatial locations (DeAngelis and Yurek, 2017; Goodchild, 2001). Four tests are proposed: the invariance test, the representation test, the formulation test, and the outcome test. *The invariance test* examines whether a spatially explicit model is not invariant under relocation. *The representation test* examines whether a spatially explicit model includes spatial representations such as coordinates, spatial relations, place names, etc. in its implementations. *The formulation test* examines whether a spatially explicit model includes spatial concepts in their formulation such as spatial neighborhoods, spatial autocorrelation, etc. And lastly, *the outcome test* examines whether a model is spatially explicit, i.e., the spatial structures of its inputs and outputs are different. A model which can pass at least one of these tests can be treated as a spatially explicit model (Goodchild, 2001; Janowicz et al., 2020).

In the era of machine learning and artificial intelligence, many Spatially Explicit Artificial Intelligence (SEAI) models (Janowicz *et al.*, 2020; Li *et al.*, 2021; Mai *et al.*, 2022b; Zhu *et al.*, 2022) have been developed, which aim at improving the AI model performance on various geospatial tasks by redesigning AI models by using spatial thinking and spatial inductive bias such as spatial heterogeneity, spatial dependency, map projection, and so on.

For example, in many recent works, spatial dependency was added to the AI model training objectives of the deep learning models to make them spatially explicit. Jean *et al.* (2019) proposed the Tile2Vec model, which used a triplet loss to make two geographically nearby remote sensing image tiles to be similar in the image embedding space. Yan *et al.* (2017) resampled the place types of the geographically nearby points of interest (POI) for POI-type embedding training as an analogy of the word embedding. Several recent works also consider spatial heterogeneity in their GeoAI model design (Goodchild and Li, 2021; Gupta *et al.*, 2021; Xie *et al.*, 2021). Xie *et al.* (2021) proposed a model-agnostic framework to automatically transform a deep learning model into a spatial-heterogeneity-aware architecture by learning the spatial partitions that are guided by different spatial processes. These models are SEAI models since they pass the formulation test by including the spatial concept – spatial dependency or spatial heterogeneity into their model training objectives.

Interestingly, all spatial representation learning models such as location encoding (Mac Aodha *et al.*, 2019; Mai *et al.*, 2020b; Yang *et al.*, 2022), polyline encoding (Ha and Eck, 2018; Rao *et al.*, 2020; Soni and Boddhu, 2022), polygon encoding (Mai *et al.*, 2022d; Yan *et al.*, 2021), etc. are spatially explicit models by definition. That is because all SRL models aim at encoding spatial data such as points, polylines, polygons, raster images, or graphs directly into the neural network embedding space which makes all of them pass the presentation test defined above. So spatial representation learning can be treated as a subdomain of spatial explicit artificial intelligence, which is in turn a subdomain of geospatial artificial intelligence (GeoAI) (Janowicz *et al.*, 2020; Li, 2020). The relation among SRL, SEAI, GeoAI, and AI are visualized in Figure 6.1.

6.3 SPATIAL REPRESENTATION LEARNING ON VARIOUS SPATIAL DATA TYPES

In the following, we will discuss the key challenges and uniqueness of spatial representation learning on different spatial data types. Some important existing works will be discussed and we will also point out their limitations. Since there exist many surveys about representation learning on graphs (Chami *et al.*, 2022; Hamilton *et al.*, 2017; Wu *et al.*, 2020), knowledge graphs (Nickel *et al.*, 2015; Wang *et al.*, 2017), and remote sensing images (Li *et al.*, 2018c), instead of redoing the survey work on representation learnings on spatially embedded graphs, geospatial knowledge graphs, and remote sensing images, in this section, we will focus on discussing representation learning methods on some data types that are unique to spatial and geospatial research such as points, polylines, and polygons.

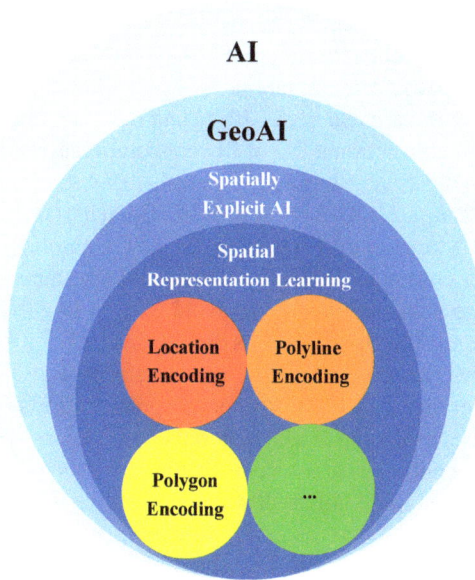

Figure 6.1   An illustration of the relations between AI, geoAI, Spatially Explicit AI, and spatial representation learning.

6.3.1   POINTS AND LOCATION ENCODERS

Words (or more commonly "tokens") are the basic element in natural language processing (NLP). Representing a word into the neural network embedding space, so-called word embedding (Mikolov *et al.*, 2013; Pennington *et al.*, 2014), is the building block for all neural network architectures for various NLP tasks. Similarly, points are the simplest yet most important and fundamental spatial data type (Mai *et al.*, 2022c). We argue that location encoders, a set of neural network models which can represent a point location into a high-dimensional vector/embedding space, are the most important spatial representation learning models which lay the foundation for representation learning models for other spatial data types (Mai *et al.*, 2022c).

6.3.1.1   Three Driving Research Areas of Location Encoding

The emergence of various location encoding techniques can be contributed to three different research areas: 3D point cloud processing (Li *et al.*, 2018b; Qi *et al.*, 2017a,b; Wu *et al.*, 2019), geographic distribution estimation (Chu *et al.*, 2019; Mac Aodha *et al.*, 2019; Mai *et al.*, 2020b, 2022e), and implicit neural representations for 2D/3D scene reconstruction and image synthesis (Mildenhall *et al.*, 2021; Niemeyer and Geiger, 2021).

Due to the popularity of various point cloud sensors used in indoor/outdoor mapping and autonomous vehicles, there are increasing numbers of machine learning and deep learning research on point clouds (Li *et al.*, 2018b; Qi *et al.*, 2017a,b;

Wang *et al.*, 2019a,b) for 3D object recognition and segmentation. Many pioneer researchers chose to first convert 3D point clouds into some regular grid representations such as volumetric representations (e.g., voxelized shapes) (Maturana and Scherer, 2015) or 2D images (Su *et al.*, 2015) and then fed these regular grid representations into some conventional 2D or 3D convolutional neural networks (CNN) for point cloud classification or segmentation tasks. However, this practice will lead to the well-known *Modifiable Areal Unit Problem* (MAUP) (Holt *et al.*, 1996; Horner and Murray, 2002). A smaller grid size indicates a large number of empty grids (data sparsity) and higher computation while a larger grid size means a coarse data representation and large information loss. To avoid MAUP, PointNet (Qi *et al.*, 2017a) was proposed to encode the point locations directly instead of encoding the converted grid representations. This direct *location encoding* approach has two main advantages: (1) There is no information loss since the point-to-grid conversion process is skipped and MAUP is avoided; (2) The required computations are only related to the number of points but not the size of occupied space or shapes of the point cloud. In contrast, with a fixed grid size, different sizes of occupied spaces or different shapes might lead to different numbers of grids. Following PointNet, direct location encoding becomes the mainstream for point cloud processing such as PointNet++ (Qi *et al.*, 2017b), PointCNN (Li *et al.*, 2018b), PointConv (Wu *et al.*, 2019), and so on.

Another research area that drove the development of location encoding techniques is geographic distribution estimation such as species spatio-temporal distribution estimation (Chu *et al.*, 2019; Mac Aodha *et al.*, 2019; Terry *et al.*, 2020; Yang *et al.*, 2022), Point of Interest (POI) distribution estimation (Mai *et al.*, 2020b), and geographic entity distribution modeling (Mai *et al.*, 2020a). The aim of this research is to learn the spatial or spatio-temporal distributions of different species, POIs, entities, etc. via a machine-learning model. Earlier approaches such as Tang *et al.* (2015) rely on pre-defined raster grids to model the species distributions which also lead to the MAUP problem. This limitation was later removed by improvements in the model designs. A classic approach is to use kernel density estimation as Berg *et al.* (2014) did. Similarly, several previous research chose to use kernel-based method to convert a point p_i into a vector of kernel features (Rahimi and Recht, 2007; Tenzer *et al.*, 2022; Yin *et al.*, 2019). Given a kernel point set $\mathcal{Q} = \{p_j^{(k)}\}$ as well as a kernel function $k(\cdot, \cdot)$ such as radial basis function (RBF), $k(\mathbf{x}, \mathbf{x}_j) = \exp\left(-\dfrac{\|\mathbf{x} - \mathbf{x}_j\|_2^2}{2\sigma^2}\right)$, where σ is the so-called kernel band with, this kind of approach first converts a point location $\mathbf{x}_i$ of point p_i into a $|\mathcal{Q}|$ dimensional kernel features $[k(\mathbf{x}_i, \mathbf{x}_1^{(k)}); ...; k(\mathbf{x}_i, \mathbf{x}_{|\mathcal{Q}|}^{(k)})]$ and then use them in following prediction tasks. Examples are GPS2Vec (Yin *et al.*, 2019), Random Fourier Feature (RFF) (Rahimi and Recht, 2007), Hu *et al.* (2015). However, as Mai *et al.* (2022c) pointed out, kernel-based approaches need to memorize kernel point sets and thus is not memory efficient. Furthermore, their performance highly depends on hyper-parameters such as kernel sizes. Chu *et al.* (2019) and Mac Aodha *et al.* (2019) proposed to use a neural network to encode a location $\mathbf{x}_i$ into an embedding and then use it to produce a probability distribution of different categories (species or POI types) on this point. However, they directly fed location

$\mathbf{x}_i$ into a neural net directly, and the result encoders cannot model fine-grained details of the distributions. Mai *et al.* (2020b) and Tancik *et al.* (2020) have shown that by adding a multi-scale feature decomposition step, so-called input Fourier mapping as Transformer's Position Encoder (Vaswani *et al.*, 2017) does, the learned location encoder can significantly improve the prediction accuracy or generate image outputs with better fidelity.

The third area that made location encoding popular is the usage of implicit neural representation for 2D/3D scene reconstruction and image synthesis, most prominently after the development of Neural Radiance Fields (NeRF) (Mildenhall *et al.*, 2021). NeRF first maps the viewpoint location $\mathbf{x}_i$ into a high dimensional space via a deterministic multi-scale input Fourier mapping similar to what Mai *et al.* (2020b); Tancik *et al.* (2020) did. By combining with other camera-related information such as Cartesian viewing direction, the output was fed into a multilayer perception (MLP) to perform viewpoint image render. Studies showed that NeRF location encoder is very effective and can help to generate images with fine-grained details. Later on, viewpoint location encoding becomes a mainstream approach in many 2D or 3D image generation, image synthesis, and superresolution tasks (Chen *et al.*, 2021; Dupont *et al.*, 2021; He *et al.*, 2021; Mildenhall *et al.*, 2021; Niemeyer and Geiger, 2021; Yang *et al.*, 2021; Zhang, 2021).

6.3.1.2 Definition of Location Encoder

After we discuss the three driving research areas for location encoding research, we give a formal definition for location encoding.

Definition 6.3.1 (Location Encoder). $\mathscr{PT} = \{p_i = (\mathbf{x}_i, \mathbf{v}_i)\}$ is a point set in 2D or 3D Euclidean space or on the surface of a sphere (or ellipsoid). Here, $\mathbf{x}_i$ and $\mathbf{v}_i$ are the location vector and feature vector (point attributes like color, POI names, temperature, moisture, etc.) of Point p_i. *Location encoding* refers to the process of directly representing point $p_i \in \mathscr{PT}$ (or only its location $\mathbf{x}_i$) as a high-dimensional embedding through a neural network architecture (Chu *et al.*, 2019; Gao *et al.*, 2019; Mac Aodha *et al.*, 2019; Mai *et al.*, 2020a, 2022c, 2020b; Mildenhall *et al.*, 2021; Qi *et al.*, 2017a,b; Zhong *et al.*, 2020). We call this neural network architecture a *location encoder*, which is a function $Enc^{(\mathscr{PT})}(p_i) : \mathbb{R}^L \rightarrow \mathbb{R}^d$ ($L \ll d$) which maps the point p_i ($\mathbf{x}_i$ and $\mathbf{v}_i$) or only its location $\mathbf{x}_i$ into a d dimensional vector called *location embeddings* $\mathbf{h}_{p_i}$. Although the architectures of location encoders vary among different research, we outline their general setup as Equation 6.1:

$$\mathbf{h}_{\mathbf{x}_i} = \mathbf{NN}(PE(\mathbf{x}_i)), \tag{6.1a}$$

$$Enc^{(\mathscr{PT})}(p_i) = \mathbf{h}_{p_i} = Agg\left(\mathbf{h}_{\mathbf{x}_i}, \mathbf{v}_i, \mathscr{N}^{(\mathscr{PT})}(\mathbf{x}_i)\right). \tag{6.1b}$$

Here, $PE(\cdot)$ denotes a deterministic function which can be an identity function (Chu *et al.*, 2019; Mac Aodha *et al.*, 2019), or a multiscale input Fourier mapping function as Mai *et al.* (2020b), Tancik *et al.* (2020), and Mildenhall *et al.* (2021) used. $\mathbf{NN}(\cdot)$ denotes a neural network such as an MLP. $Agg(\cdot)$ indicates a neighborhood

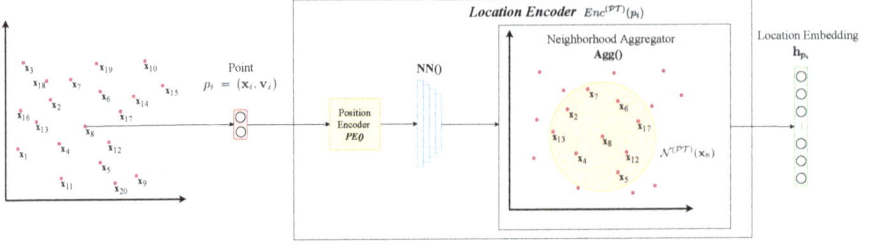

Figure 6.2   An illustration of a common setup of location encoders.

aggregation function based on $\mathbf{h}_{\mathbf{x}_i}$, Point p_i's feature $\mathbf{v}_i$, as well as its nearby points $\mathcal{N}^{(\mathcal{PT})}(\mathbf{x}_i)$. Figure 6.2 illustrates the common setup of location encoders.

Equation 6.2 demonstrates a commonly used multi-scale input Fourier mapping function $PE^{TF}(\cdot)$ used by *grid* method in Mai *et al.* (2020b), Tancik *et al.* (2020), NeRF (Mildenhall *et al.*, 2021), and the position encoder of Transformer (Vaswani *et al.*, 2017):

$$PE^{TF}(\mathbf{x}_i) = [PE_1^{TF}(\mathbf{x}_i); ...; PE_s^{TF}(\mathbf{x}_i); ...; PE_S^{TF}(\mathbf{x}_i)], \qquad (6.2a)$$

$$where\ PE_s^{TF}(\mathbf{x}_i) = [\cos(F_S(s)\mathbf{x}_i); \sin(F_S(s)\mathbf{x}_i)]. \qquad (6.2b)$$

Here, Equation 6.2a indicates that $PE^{TF}(\mathbf{x}_i)$ is a concatenation of position features of $\mathbf{x}_i$ under S different scales where each scale feature $PE_s^{TF}(\mathbf{x}_i)$ is a concatenation of multiple sinusoid features. $F_S(s)$ is a determinstic function based on a given scale s. For example, Transformer's position encoder uses $F_S(s) = \frac{2\pi S}{S^{(s+1)/S}}$. The *grid* method proposed in Mai *et al.* (2020b) uses $F_S(s) = \frac{1}{\lambda_{min} \cdot g^{s/(S-1)}}$ where $\lambda_{min}, \lambda_{max}$ are the minimum and maximum grid scale, and $g = \frac{\lambda_{max}}{\lambda_{min}}$.

Many location encoders (Chu *et al.*, 2019; Mac Aodha *et al.*, 2019; Mai *et al.*, 2020b; Mildenhall *et al.*, 2021; Tancik *et al.*, 2020; Yin *et al.*, 2019) only use Equation 6.1a as the location encoders. For them, $Agg(\cdot)$ can be treated as an identity function. Other research such as PointNet++ (Qi *et al.*, 2017b), PointCNN (Li *et al.*, 2018b), PointConv (Wu *et al.*, 2019) also consider point's spatial neighborhood and use one or more neighborhood aggregation layers to produce the final location embedding. For the detail descriptions of different location encoders, please refer to Mai *et al.* (2022c).

6.3.1.3   The Advantages and Limitations of Existing Location Encoders

Many works have shown that location embeddings produced by those multi-scale location encoders (Mai *et al.*, 2020b; Mildenhall *et al.*, 2021; Niemeyer and Geiger, 2021; Tancik *et al.*, 2020) have several advantages:

- **Spatial Proximity and Direction Preservation**: Many location encoders can preserve spatial proximity information (Gao *et al.*, 2019) and/or directional information (Mai *et al.*, 2022c, 2020b);

- **Learn-Friendly Representation**: Location Embeddings are more learning-friendly to the downstream machine learning models (Mai *et al.*, 2020b; Mildenhall *et al.*, 2021; Tancik *et al.*, 2020) which means they can lead to more fine-grained label distribution estimations when applied on discriminative models (Mai *et al.*, 2020b, 2022e) and more fidelity generative outputs when applied on deep generative models (He *et al.*, 2021; Mildenhall *et al.*, 2021; Niemeyer and Geiger, 2021; Tancik *et al.*, 2020).

- **Multi-Scale Distribution Modeling**: Because of the usage of multi-scale features, these location encoders can better handle point distributions with very different characteristics. The small scale features can model the fine-grained distribution patterns of densely clustered points (e.g., bars and clubs) and the large scale features can well handle the evenly distributed point patterns (e.g., post offices, fire stations).

Because of these advantages, location encoders become increasingly popular in GeoAI research as well as in the general AI domain. However, there are still several unneglectable limitations of the current location encoder models:

- **No Universal Winner**: Given a specific task, we need to compare the performances of different location encoders to pick the best one since no research shows that there is a universal winner among these models on all tasks.

- **Location Decoder**: Encoder-decoder is a commonly use architecture for representation learning. Although there are multiple ways to encode a location, the ways to decode a location are very limited. A common way is using an MLP to regress the (x, y) coordinates as the location decoders used in LSTM-TrajGAN (Rao *et al.*, 2020) or the bounding box coordinates regressor used in Fast R-CNN (Girshick, 2015). Another way is to use a Gaussian mixture model to sample the (x, y) coordinates as that used in sketch-rnn (Ha and Eck, 2018).

- **Capturing Spatial Heterogeneity**: The basic assumption behind location encoding is Tobler's first law of geography, "everything is related to everything else, but near things are more related than distant things" (Tobler, 1970). In other words, location encoders assumes the phenomena (e.g., model prediction surface) changes *continuously* across space. However, that does not always hold. When we conduct spatial prediction over a larger spatial scale, spatial heterogeneity can affect the model prediction. The predictions of location encoders can help to improve model performance in one area but may hurt the performances in other areas. How to consider spatial heterogeneity when designing location encoders is also an important research direction (Mai *et al.*, 2022a).

6.3.2 POLYLINES AND POLYLINE ENCODER

By definition, a polyline is a continuous line that is composed of one or more connected straight line segments, which, together, make up a shape. Polylines are an important spatial data structure that is commonly used in many GIS applications to represent linear features such as contour lines, rivers, road segments, coastal lines, trajectories, and so on. Polyline encoders indicate neural nets that can represent a polyline or a set of polylines as a high-dimensional embedding to be used in various prediction tasks.

6.3.2.1 Definition of Polyline Encoder

We first give the formal definition of polyline encoder.

Definition 6.3.2 (Polyline Encoder). $\mathscr{PL} = \{l_i = (\mathbf{X}_i, \mathbf{V}_i)\}$ is a set of polylines in 2D or 3D Euclidean space or on the surface of a sphere (or ellipsoid). Each polyline can be represented as a list of points $l_i = [p_{i,1}, ..., p_{i,N_{l_i}}]$ or $l_i = (\mathbf{X}_i, \mathbf{V}_i)$. Here, $\mathbf{X}_i \in \mathbb{R}^{N_{l_i} \times 2}$ is a point coordinate matrix representing a sequence of locations as the vertices of the current Polyline l_i [1]. N_{l_i} indicates the numbers of vertices of l_i. $\mathbf{V}_i$ is a matrix representing the features associated with each vertex such as timestamps, and semantic tags of each trajectory stop point. Theoretically speaking, l_i represents a continuous line shape, so it should contain infinite number of points. However, in practice, because of sensor measurement limitations, storage limitations, or other reasons, l_i is usually represented as a limited number of sampled points/vertices to indicate its general shape. *Polyline encoding* refers to the process of directly representing Polyline $l_i \in \mathscr{PL}$ (or only its point coordinate matrix $\mathbf{X}_i$) as a high-dimensional embedding through a neural network architecture (Alahi *et al.*, 2016; Ha and Eck, 2018; Rao *et al.*, 2020; Yu and Chen, 2022). We call this neural network architecture a *polyline encoder*, which is a function $Enc^{(\mathscr{PL})}(l_i) : \mathbb{R}^{N_{l_i} \times 2} \to \mathbb{R}^d$ ($N_{l_i} \times 2 \gg d$) which maps Polyline l_i ($\mathbf{X}_i$ and $\mathbf{V}_i$) or only its point coordinate matrix $\mathbf{X}_i$ into a d dimensional vector called *polyline embeddings* $\mathbf{h}_{l_i}$.

A common analogy GeoAI research scientists usually made is that if a point is an analogy of a word, then a polyline or a trajectory can be seen as an analogy of a sentence (Mai *et al.*, 2022c; Musleh *et al.*, 2022). Let's call it the *word-to-point* analogy. Based on this analogy, a simple idea to build a polyline encoder is to *reuse* the existing language models from the NLP domain. That means if we have the ability to represent each point/location into an embedding with location encoders, we can just treat a polyline or trajectory as a word sequence and apply the existing sequence models to form a polyline encoder.

In fact, almost all existing polyline encoders follow the above idea. They form a polyline enoder with two steps: (1) they use a location encoder $Enc^{(\mathscr{PT})}(\cdot)$ to encode each vertex $p_{i,j}$ of a polyline l_i into a location embedding $\mathbf{h}_{p_{i,j}}$ as shown in Equation 6.3a; (2) they use a sequence model $SEQ(\cdot)$ to take a list of location embeddings and encode them into a polyline embedding $\mathbf{h}_{l_i}$ as shown in Equation 6.3b. Here, the sequence model $SEQ(\cdot)$ can be recurrent neural networks (RNN) (Rumelhart *et al.*, 1986), Transformer (Vaswani *et al.*, 2017), sequence-to-sequence (seq2seq) (Sutskever *et al.*, 2014), BERT (Kenton and Toutanova, 2019), and so on.

---

[1] $\mathbf{X}_i \in \mathbb{R}^{N_{l_i} \times 3}$ for polylines in 3D Euclidean space.

Figure 6.3 visualizes a common setup for polyline encoders.

$$\mathbf{h}_{p_{i,j}} = Enc^{(\mathscr{P}\mathscr{T})}(p_{i,j}), \tag{6.3a}$$

$$Enc^{(\mathscr{P}\mathscr{L})}(l_i) = \mathbf{h}_{l_i} = SEQ([\mathbf{h}_{p_{i,1}}, \mathbf{h}_{p_{i,2}}, ..., \mathbf{h}_{p_{i,N_{l_i}}}]), . \tag{6.3b}$$

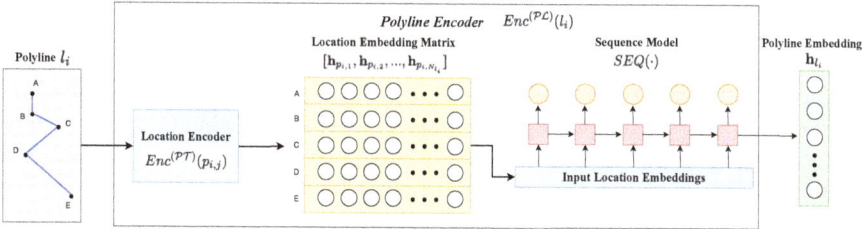

Figure 6.3   An illustration of a common setup of polyline encoders.

6.3.2.2   Existing Works on Polygline Encoders

In the following, we will briefly discuss several existing work which built polyline encoders by following the idea shown in Equation 6.3. In order to do human trajectory prediction Alahi *et al.* (2016) proposed a social LSTM model which directly takes the coordinates of each trajectory stop point as the inputs for a modified Long short-term memory (LSTM) model and uses a parameterized bivariate Gaussian distribution to predict the coordinates for the next stop point.

Similarly, Rao *et al.* (2020) proposed an LSTM-TrajGAN framework that uses an LSTM-based seq2seq model as a trajectory generator to generate synthetic trajectories given the real trajectories to preserve users' privacy. The stop points of each trajectory are first embedded into location embeddings with a specially designed location encoder and then are encoded by an LSTM model for synthetic trajectory generation. The whole model is trained with a generative adversarial network (GAN) (Goodfellow *et al.*, 2020) objective. Note that there are more work about encoding trajectories that follow similar idea, we skip these research and only select the above two representative work.

Soni and Boddhu (2022) also developed a polyline autoencoder to do entity alignment between two map databases. Two polyline entities from two map databases such as road segments are encoded into two polyline embeddings to measure their semantic similarity based on an LSTM-based autoencoder. A similar setup shown in Equation 6.3 is used here. Interestingly, Soni and Boddhu (2022) found out that polylines from two map databases that coresponds to the same road segment may have different sampling rates: 10 meters v.s. 25 meters. Resampling the polyline vertices is conducted as a preprocessing step to make sure the sampling rate of two polylines under comparison match. Left padding is used to make the input polyline have the same sequence length.

Another good example is sketch-rnn (Ha and Eck, 2018) which uses a Sequence-to-Sequence (Sutskever *et al.*, 2014) Variational Autoencoder (VAE) (Kingma and

Welling, 2013) to encode and generate sketches that are in the form of polylines. Each sketch is essentially a set of polylines represented as a list of points. Each point is a vector of 5 elements $(\Delta x, \Delta y, o_1, o_2, o_3)$. $\Delta x, \Delta y$ indicate the offset distance in the x and y direction of the current point from the previous point and the first point starts from the origin. The last 3 elements represent a binary one-hot vector of 3 possible pen states. o_1 indicates that the pen is touching the paper and will keep drawing. o_2 indicates the pen will be lifted after this point and o_3 indicates the end of the whole sketch. The encoder of sketch-rnn is a bidirectional RNN (Schuster and Paliwal, 1997) which directly takes the list of sketch points as inputs and encodes the whole sketch into a hidden state $\mathbf{h}$ which will be used for sketch conditional generation. Here, hidden state $\mathbf{h}$ can be treated as the polyline embedding of the whole sketch. $\mathbf{h}$ will be projected into two vectors μ and σ with two full connected layers as shown in Equation 6.4a and 6.4b where $\mathbf{W}_\mu, \mathbf{b}_\mu$, as well as $\mathbf{W}_\sigma, \mathbf{b}_\sigma$ are leanable weight matrices and bias vectors for μ and σ. Here, μ and σ are used as the mean and standard deviation parameters of the latent codes $\mathbf{z}$, the VAE latent variables as a vector of independent and identical distributed (IID) Gaussian variables as shown in Equation 6.4c. To generate a new sketch, a latent vector $\mathbf{z}$ is sampled from $\mu + \sigma \odot \mathcal{N}(0,1)$ as the condition. Each sketch stop point is decode step by step through a RNN decoder together with a Gaussian mixture model (GMM).

$$\mu = \mathbf{W}_\mu \mathbf{h} + \mathbf{b}_\mu, \tag{6.4a}$$

$$\sigma = \exp\left(\frac{1}{2}(\mathbf{W}_\sigma \mathbf{h} + \mathbf{b}_\sigma)\right), \tag{6.4b}$$

$$\mathbf{z} = \mu + \sigma \odot \mathcal{N}(0,1) \tag{6.4c}$$

All previous four works follow similar setups as illustrated in Equation 6.3. Recently, Yu and Chen (2022) proposed a yet simple denoising autoencoder model for cartographic polyline completion. Instead of using a sequence model to encode polylines, Yu and Chen (2022) stacked the vertice's coordinate together as a point coordinate matrix $\mathbf{X}_i \in \mathbb{R}^{N_{l_i} \times 2}$, flatten $\mathbf{X}_i$ into a 1D vector and directly fed it into an MLP to obtain the polygline embedding $\mathbf{h}_{l_i}$. As for the decoder, $\mathbf{h}_{l_i}$ is directly fed to another MLP to regenerate the point coordinate matrix. To make the model suitable for the polyline completion task, a binary mask is applied to the input point coordinate matrix $\mathbf{X}_i$ before feeding into the encoder to form a denoising autoencoder.

6.3.2.3 The Advantages and Limitations of Existing Polyline Encoders

Compared with the previous four sequence model-based polyline encoders, Yu and Chen (2022)'s approach has several limitations: (1) The input polylines are limited to a fixed length while an RNN-based polyline encoder can handle polylines with various lengths; (2) The number of learnable parameters will be very large since the total number of parameters is proportional to the input polyline length.

Generally speaking, there are four challenges when developing polyline encoders for a specific dataset:

- **Representing Polyline Direction**: Some polyline-shaped features may have direction information such as road segments, trajectories, and sketches, while

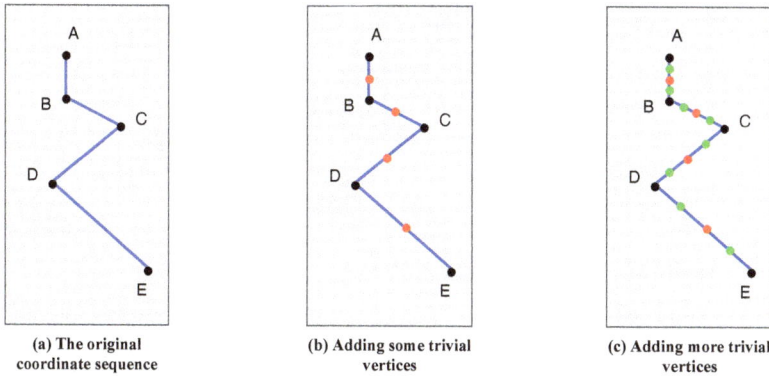

(a) The original
coordinate sequence

(b) Adding some trivial
vertices

(c) Adding more trivial
vertices

Figure 6.4   Illustrations of the trivial vertices of a polyline. (a) shows the original coordinate sequence of Polyline l_i; (b) shows how to add some trivial vertices without changing the polyline shape (see red points); (c) shows how to add more trivial vertices (see green points).

other polyline-shaped features do not such as contour lines and coastal lines. In order to encode their direction information, we need to use different models. For example, for one-direction polylines such as trajectory, we need to use unidirectional RNN while using bidirectional RNN for contour lines.

- **Equal interval Assumption**: Many RNN-based sequence models have an equal interval assumption. If Polyline $l_i = [p_{i,1}, ..., p_{i,N_{l_i}}]$ represents a trajectory, when we use an RNN to encode its vertices, we will have an assumption that these trajectory points are collected at equal time intervals which may not be true in many real-world datasets. We can potentially use a Transformer with continuous position encoding to solve this problem.

- **Continuous Lines v.s. Discrete Sequences**: all polyline encoders treat a polyline as a discrete location sequence while in fact, polylines are continuous line features. As shown in Figure 6.4, given Polyline l_i, we can add an arbitrary number of trivial points without modifying its shape. We expect a polyline encoder should yield identical polyline embeddings before or after this operation. However, all the current sequence models such as RNN, seq2seq, and Transformer cannot achieve this.

- **Topology Awareness**: a follow-up question is whether we can use the learned polyline embedding to do point-on-line inquiry. Although this operation can be efficiently done by existing deterministic algorithms, the point-on-line inquiry can be used to test whether the polyline encoder is truly capture the shape and semantics of a polyline. As far as we know, none of the existing work can achieve this.

6.3.3 POLYGONS AND POLYGON ENCODER

Polygons and multipolygons are also important spatial data structures that are commonly used to represent regional geospatial features such as administration regions, islands, water bodies, national parks, buildings, and so on. Compared with polylines, designing a polygon encoder, which is a neural net to encode a polygonal geometry into the embedding space, has its unique challenges.

6.3.3.1 Definition of Polyline Encoder

Since polygons and multipolygons are more complex, we first give the definition of polygon. Here, Defnition 6.3.3 is modified based on the polygon definition in Mai *et al.* (2022d).

Definition 6.3.3 (Polygon and Multipolygon). According to the Open Geospatial Consortium (OGC) standard, each *polygon* q_i can be represented as a tuple $(\mathbf{B}_i, h_i = \{\mathbf{H}_{ij}\})$ where $\mathbf{B}_i \in \mathbb{R}^{N_{b_i} \times 2}$ indicates a point coordinate matrix for the exterior of q_i defined in a *counterclockwise* direction. $h_i = \{\mathbf{H}_{ij}\}$ is a set of holes for q_i where each hole $\mathbf{H}_{ij} \in \mathbb{R}^{N_{h_{ij}} \times 2}$ is a point coordinate matrix for one interior linear ring of q_i defined in a *clockwise* direction. N_{b_i} indicates the number of unique points in q_i's exterior. The first and last point of $\mathbf{B}_i$ are not the same. $\mathbf{B}_i$ is different from polylines defined in Definition 6.3.2 since $\mathbf{B}_i$ is a closed ring and it does not intersect with itself while polylines are linear features that can intersect with themselves. Similar logic applies to each hole $\mathbf{H}_{ij}$ and $N_{h_{ij}}$ is the number of unique points in the jth hole of q_i. A *multipolygon* u_k is a set of polygons, i.e., $u_k = \{q_{ki}\}$ which represents one entity (e.g., China, Greenland, Iceland, United Kingdom, and so on). A polygonal geometry g_i can be either a polygon or multipolygon.

Definition 6.3.4 (Polygon Encoder). Let $\mathscr{PN} = \{g_i\}$ be a set of polygonal geometries in a 2D Euclidean space $\mathbb{R}^{2}$ [2]. Here, g_i can be either a Polygon q_i or Multipolygon u_i. N_{g_i} is the total number of unique vertices of g_i. *Polygon encoding* refers to the process of representing Polygonal Geometry g_i into a high-dimensional embedding space with a neural net. We call this neural network architecture a *polygon encoder*, which is a function $Enc^{(\mathscr{PN})}(g_i) : \mathbb{R}^{N_{g_i} \times 2} \rightarrow \mathbb{R}^d$ ($N_{g_i} \times 2 \gg d$) which maps Polygonal Geometry g_i into a d dimensional vector called *polyline embeddings* $\mathbf{h}_{g_i}$. A general idea of polygon encoders is shown in Figure 6.5.

6.3.3.2 Existing Works on Polygon Encoders

Compared with research on location encoders and polyline encoders, there are rather limited numbers of work on polygon representation learning.

Most existing works are focusing on encoding simple polygons, polygons that do not have holes. Veer *et al.* (2018) proposed two polygon encoders for simple polygons. Both encoders take the simple polygon exterior's coordinate matrix as the

---

[2]Here, we only consider 2D Euclidean space while encoding polygonal geometries on the surface of spheres or other manifolds are out of the scope of this paper.

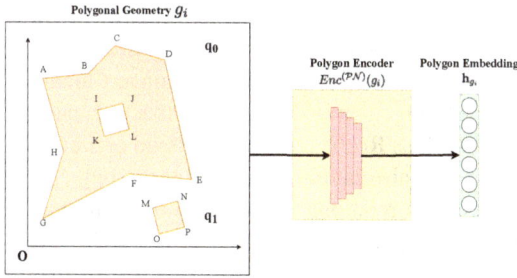

Figure 6.5   An illustration of a common setup of polygon encoders.

input and normalize the vertices' coordinates into $[-1, 1]$. The first model uses a 1D convolutional neural network to encode the polygon exterior's coordinate matrix and use global max pooling to obtain an embedding for the polygon. The second model uses a bidirectional LSTM to encode the polygon exterior's coordinate matrix. Both models are used in polygon-shape-based prediction tasks such as neighborhood population prediction, building footprint classification, and archaeological ground feature classification. CNN version is better than the RNN model on these tasks.

Similarly, Mai *et al.* (2022d) also proposed a polygon encoder for simple polygons based on 1D CNN, more specifically, 1D ResNet. The only difference is that they use circular padding in the CNN layer and the max pooling layer to achieve the so-called loop origin invariance property. This model is called ResNet1D. They showed that, compared with spectral-based methods, ResNet1D is more effective on the shape classification task.

Yan *et al.* (2021) proposed a graph autoencoder-based polygon encoder called GCAE for simple polygons. They treat the exterior of a simple polygon as an undirected weighted graph. The exterior vertices are the graph nodes and graph edges are just the polygon exterior edges which are weighted by edge length. GCAE follows a U-Net architecture. A graph autoencoder is used to first encode this undirected graph into an embedding, the polygon embedding, with various graph convolutional layers. The polygon embedding is then decoded with multiple upscaling layers. Yan *et al.* (2021) showed the effectiveness of GCAE in building shape retrieval and shape similarity task.

The above three methods can only handle simple polygons and focus on encoding the polygon exterior coordinate sequences. They are very similar to polyline encoders we discussed in Section 6.3.2. There are recent works that can directly encode polygons with holes and multipolygons.

Instead of these polyline-like encoders, DDSL (Jiang *et al.*, 2019a) and Jiang *et al.* (2019b) broke new ground and chose to use Non-Uniform Fourier Transformation (NUFT) to first transform polygons, multipolygons, or 3D mesh into the Fourier space. Then they used inverse Fourier transformation (IFT) to convert the fourier feature into an image, which is fed into a CNN-based neural net for shape classification task. Since they first use NUFT to process polygonal geometries, we call this kind of methods spectral-based polygon encoders.

Inspired by Jiang *et al.* (2019a), Mai *et al.* (2022d) proposed a new spectral-based polygon encoder, NUFTspec. NUFTspec also uses NUFT to process the input polygonal geometry. But instead of using IFT to convert them to an image, NUFTspec directly learns the polygon embedding in the spectral space. Compared with the spatial-based method such as ResNet1D we discussed earlier, NUFTspec is promising in polygon-based spatial relation prediction tasks (e.g., predicting topological relations, cardinal direction relations, etc.) which are commonly used in many geographic question answering systems (Hamzei *et al.*, 2022; Kuhn *et al.*, 2021; Mai *et al.*, 2020a, 2021, 2019).

6.3.3.3 The Advantages and Limitations of Existing Location Encoders

Mai *et al.* (2022d) systematically compared different polygon encoders, including (1) their encoding capabilities – whether they can handle holes and multipolygons; (2) their invariance and awareness properties – loop origin invariance, trivial vertex invariance, part permutation invariance, topology awareness. They also compare different models on both the shape classification task and spatial relation prediction task. Interestingly, the spatial-based polygon encoders such as ResNet1D outperform NUFTspec and DDSL on the first task but underperform them on the second task. This indicates that a polygon encoder that can both learn spectral features from the Fourier space but also learn spatial features from the polygon vertex sequence is necessary. For a detailed description and comparison among different polygon encoders, please refer to Mai *et al.* (2022d).

Here, we list some challenges when developing polygon encoders:

- **Continuous Area Surface v.s. Discrete Linear Ring**: All polygon encoders that focus on encoding the polygon exterior coordinate sequences such as Veer *et al.* (2018), ResNet1D (Mai *et al.*, 2022d), GCAE (Yan *et al.*, 2021) only treat polygons as an ordered sequence of discrete points. When we add some trivial vertices to polygon boundaries, the encoding results will be different. In other words, they are not trivial vertex invariant (Mai *et al.*, 2022d). Moreover, they can not differentiate between interiors and exteriors. We discuss it more in the next bullet point.

- **Topology Awareness**: Similar to the topology awareness for polyline encoders, we expect a polygon encoder to be aware of the topology of polygonal geometry. A point-in-polygon inquiry can be used to test this ability. Since topology awareness is important for many polygon-based computations such as topological relation computation.

- **Polygon Decoder**: Although GCAE (Yan *et al.*, 2021) proposed a polygon encoder-decoder architecture, it can only handle simple polygons. There also exist some polygon decoders such as Polygon-RNN (Castrejon *et al.*, 2017), Polygon-RNN++ (Acuna *et al.*, 2018), and PolyTransform (Liang *et al.*, 2020). All of them can generate simple polygons but not complex polygons such as polygons with holes. However, generating complex polygons is unavoidable sometimes such as generating complex building footprints from remote sensing images. Developing a polygon decoder which can generate polygonal geometries with various complexity is necessary.

6.4 CONCLUSIONS

In this chapter, we briefly discuss various spatial representation learning techniques, especially SRL techniques on some unique data types for spatial and geospatial research such as points, polylines, and polygons. We provide formal definitions of representation learning models on each data type and discuss several representative existing works. The limitations of the current approaches are also discussed.

As a subfield of spatially explicit artificial intelligence, spatial representation learning can be treated as a unique component that distinguishes GeoAI research from general AI research because SRL focuses on representation learning on unique data types that are rarely touched by general AI research. Recently, we have witnessed an increasing number of research efforts on this topic (Bronstein *et al.*, 2017; Mai *et al.*, 2022c). We believe spatial representation learning will become a core research topic of GeoAI in the early future.

BIBLIOGRAPHY

Acuna, D., *et al.*, 2018. Efficient interactive annotation of segmentation datasets with Polygon-RNN++. *In: Proceedings of the IEEE Conference on Computer Vision and Pattern Recognition*. 859–868.

Alahi, A., *et al.*, 2016. Social lstm: Human trajectory prediction in crowded spaces. *In: Proceedings of the IEEE Conference on CVPR*. 961–971.

Ayush, K., *et al.*, 2021. Geography-aware self-supervised learning. *In: Proceedings of the IEEE International Conference on Computer Vision*. 10181–10190.

Bengio, Y., Courville, A., and Vincent, P., 2013. Representation learning: A review and new perspectives. *IEEE Transactions on Pattern Analysis and Machine Intelligence*, 35 (8), 1798–1828.

Berg, T., *et al.*, 2014. Birdsnap: Large-scale fine-grained visual categorization of birds. *In: CVPR 2014*. 2011–2018.

Bronstein, M.M., *et al.*, 2017. Geometric deep learning: going beyond euclidean data. *IEEE Signal Processing Magazine*, 34 (4), 18–42.

Brown, T., *et al.*, 2020. Language models are few-shot learners. *NIPS 2020*, 33, 1877–1901.

Cai, L., *et al.*, 2020. Traffic transformer: Capturing the continuity and periodicity of time series for traffic forecasting. *Transactions in GIS*, 24 (3), 736–755.

Castrejon, L., *et al.*, 2017. Annotating object instances with a Polygon-RNN. *In: Proceedings of CVPR'17*. 5230–5238.

Chami, I., *et al.*, 2022. Machine learning on graphs: A model and comprehensive taxonomy. *Journal of Machine Learning Research*, 23 (89), 1–64.

Chen, D., *et al.*, 2017. Reading wikipedia to answer open-domain questions. *arXiv preprint arXiv:1704.00051*.

Chen, Y., Liu, S., and Wang, X., 2021. Learning continuous image representation with local implicit image function. *In: Proceedings of the IEEE/CVF Conference on Computer Vision and Pattern Recognition*. 8628–8638.

Chu, G., *et al.*, 2019. Geo-aware networks for fine-grained recognition. *In: Proceedings of the IEEE International Conference on Computer Vision Workshops*. 0–0.

Cong, Y., *et al.*, 2022. Satmae: Pre-training transformers for temporal and multi-spectral satellite imagery. *In: Advances in Neural Information Processing Systems*.

Das, R., *et al.*, 2019. Multi-step retriever-reader interaction for scalable open-domain question answering. *In: International Conference on Learning Representations*.

DeAngelis, D.L. and Yurek, S., 2017. Spatially explicit modeling in ecology: a review. *Ecosystems*, 20 (2), 284–300.

Dosovitskiy, A., *et al.*, 2020. An image is worth 16x16 words: Transformers for image recognition at scale. *In: International Conference on Learning Representations*.

Dupont, E., *et al.*, 2021. Coin: Compression with implicit neural representations. *arXiv preprint arXiv:2103.03123*.

Gao, R., *et al.*, 2019. Learning grid cells as vector representation of self-position coupled with matrix representation of self-motion. *In: Proceedings of ICLR 2019*.

Girshick, R., 2015. Fast r-cnn. *In: Proceedings of the IEEE International Conference on Computer Vision*. 1440–1448.

Goodchild, M., 2001. Issues in spatially explicit modeling. *Agent-Based Models of Land-Use and Land-Cover Change*, 13–17.

Goodchild, M.F. and Li, W., 2021. Replication across space and time must be weak in the social and environmental sciences. *PNAS*, 118 (35).

Goodfellow, I., *et al.*, 2020. Generative adversarial networks. *Communications of the ACM*, 63 (11), 139–144.

Guo, M.H., *et al.*, 2022. Attention mechanisms in computer vision: A survey. *Computational Visual Media*, 1–38.

Gupta, J., *et al.*, 2021. Spatial variability aware deep neural networks (svann): A general approach. *ACM Transactions on Intelligent Systems and Technology (TIST)*, 12 (6), 1–21.

Ha, D. and Eck, D., 2018. A neural representation of sketch drawings. *In: International Conference on Learning Representations*.

Hamilton, W.L., Ying, R., and Leskovec, J., 2017. Representation learning on graphs: Methods and applications. *arXiv preprint arXiv:1709.05584*.

Hamzei, E., Tomko, M., and Winter, S., 2022. Translating place-related questions to geosparql queries. *In: Proceedings of the ACM Web Conference 2022*. 902–911.

Haw, D.J., *et al.*, 2020. Strong spatial embedding of social networks generates nonstandard epidemic dynamics independent of degree distribution and clustering. *Proceedings of the National Academy of Sciences*, 117 (38), 23636–23642.

He, K., *et al.*, 2022. Masked autoencoders are scalable vision learners. *In: Proceedings of the IEEE/CVF Conference on CVPR*. 16000–16009.

He, Y., *et al.*, 2021. Spatial-temporal super-resolution of satellite imagery via conditional pixel synthesis. *Advances in Neural Information Processing Systems*, 34, 27903–27915.

Hoffart, J., *et al.*, 2013. Yago2: A spatially and temporally enhanced knowledge base from wikipedia. *Artificial Intelligence*, 194, 28–61.

Holt, D., *et al.*, 1996. Aggregation and ecological effects in geographically based data. *Geographical Analysis*, 28 (3), 244–261.

Horner, M.W. and Murray, A.T., 2002. Excess commuting and the modifiable areal unit problem. *Urban Studies*, 39 (1), 131–139.

Hu, Y., *et al.*, 2015. Metadata topic harmonization and semantic search for linked-data-driven geoportals: A case study using arcgis online. *Transactions in GIS*, 19 (3), 398–416.

Janowicz, K., *et al.*, 2020. Geoai: spatially explicit artificial intelligence techniques for geographic knowledge discovery and beyond.

Janowicz, K., *et al.*, 2022. Know, know where, knowwheregraph: A densely connected, cross-domain knowledge graph and geo-enrichment service stack for applications in environmental intelligence. *AI Magazine*, 43 (1), 30–39.

Jean, N., *et al.*, 2019. Tile2vec: Unsupervised representation learning for spatially distributed data. *In: Proceedings of the AAAI Conference on Artificial Intelligence*. vol. 33, 3967–3974.

Jiang, C., *et al.*, 2019a. DDSL: Deep differentiable simplex layer for learning geometric signals. *In: Proceedings of the IEEE/CVF International Conference on Computer Vision*. 8769–8778.

Jiang, C.M., *et al.*, 2019b. Convolutional neural networks on non-uniform geometrical signals using euclidean spectral transformation. *In: International Conference on Learning Representations*.

Karpukhin, V., *et al.*, 2020. Dense passage retrieval for open-domain question answering. *In: Proceedings of the 2020 Conference on Empirical Methods in Natural Language Processing (EMNLP)*. 6769–6781.

Kenton, J.D.M.W.C. and Toutanova, L.K., 2019. Bert: Pre-training of deep bidirectional transformers for language understanding. *In: Proceedings of NAACL-HLT*. 4171–4186.

Kingma, D.P. and Welling, M., 2013. Auto-encoding variational bayes. *arXiv preprint arXiv:1312.6114*.

Kuhn, W., *et al.*, 2021. The semantics of place-related questions. *Journal of Spatial Information Science*, (23), 157–168.

Li, W., 2020. Geoai: Where machine learning and big data converge in GIScience. *Journal of Spatial Information Science*, (20), 71–77.

Li, W., Hsu, C.Y., and Hu, M., 2021. Tobler's first law in geoai: A spatially explicit deep learning model for terrain feature detection under weak supervision. *Annals of the American Association of Geographers*, 111 (7), 1887–1905.

Li, Y., *et al.*, 2018a. Diffusion convolutional recurrent neural network: Data-driven traffic forecasting. *In: International Conference on Learning Representations*.

Li, Y., *et al.*, 2018b. PointCNN: Convolution on x-transformed points. *Advances in Neural Information Processing Systems*, 31.

Li, Y., *et al.*, 2018c. Deep learning for remote sensing image classification: A survey. *Wiley Interdisciplinary Reviews: Data Mining and Knowledge Discovery*, 8 (6), e1264.

Liang, J., *et al.*, 2020. Polytransform: Deep polygon transformer for instance segmentation. *In*: *Proceedings of the IEEE/CVF Conference on Computer Vision and Pattern Recognition*. 9131–9140.

Mac Aodha, O., Cole, E., and Perona, P., 2019. Presence-only geographical priors for fine-grained image classification. *In*: *Proceedings of the IEEE International Conference on Computer Vision*. 9596–9606.

Mai, G., *et al.*, 2022a. Towards a foundation model for geospatial artificial intelligence (vision paper). *In*: *Proceedings of the 30th International Conference on Advances in Geographic Information Systems*. 1–4.

Mai, G., *et al.*, 2022b. Symbolic and subsymbolic geoai: Geospatial knowledge graphs and spatially explicit machine learning. *Transactions in GIS*, 26 (8), 3118–3124.

Mai, G., *et al.*, 2020a. Se-kge: A location-aware knowledge graph embedding model for geographic question answering and spatial semantic lifting. *Transactions in GIS*, 24 (3), 623–655.

Mai, G., *et al.*, 2022c. A review of location encoding for geoai: methods and applications. *International Journal of Geographical Information Science*, 36 (4), 639–673.

Mai, G., *et al.*, 2020b. Multi-scale representation learning for spatial feature distributions using grid cells. *In*: *The Eighth International Conference on Learning Representations*. openreview.

Mai, G., *et al.*, 2021. Geographic question answering: challenges, uniqueness, classification, and future directions. *AGILE: GIScience Series*, 2, 1–21.

Mai, G., *et al.*, 2022d. Towards general-purpose representation learning of polygonal geometries. *GeoInformatica*, 1–52.

Mai, G., *et al.*, 2022e. Sphere2vec: Multi-scale representation learning over a spherical surface for geospatial predictions. *arXiv preprint arXiv:2201.10489*.

Mai, G., *et al.*, 2019. Relaxing unanswerable geographic questions using a spatially explicit knowledge graph embedding model. *In*: *AGILE 2019*. Springer, 21–39.

Manas, O., *et al.*, 2021. Seasonal contrast: Unsupervised pre-training from uncurated remote sensing data. *In*: *Proceedings of the IEEE/CVF International Conference on Computer Vision*. 9414–9423.

Maturana, D. and Scherer, S., 2015. Voxnet: A 3d convolutional neural network for real-time object recognition. *In*: *2015 IEEE/RSJ IROS*. IEEE, 922–928.

Mikolov, T., *et al.*, 2013. Distributed representations of words and phrases and their compositionality. *Advances in Neural Information Processing Systems*, 26.

Mildenhall, B., *et al.*, 2021. Nerf: Representing scenes as neural radiance fields for view synthesis. *Communications of the ACM*, 65 (1), 99–106.

Musleh, M., Mokbel, M.F., and Abbar, S., 2022. Let's speak trajectories. *In*: *Proceedings of the 30th ACM SIGSPATIAL*. 1–4.

Nickel, M., *et al.*, 2015. A review of relational machine learning for knowledge graphs. *Proceedings of the IEEE*, 104 (1), 11–33.

Niemeyer, M. and Geiger, A., 2021. Giraffe: Representing scenes as compositional generative neural feature fields. *In*: *Proceedings of the IEEE/CVF Conference on Computer Vision and Pattern Recognition*. 11453–11464.

Pennington, J., Socher, R., and Manning, C.D., 2014. Glove: Global vectors for word representation. *In: Proceedings of the 2014 Conference on Empirical Methods in Natural Language Processing (EMNLP)*. 1532–1543.

Qi, C.R., *et al.*, 2017a. Pointnet: Deep learning on point sets for 3d classification and segmentation. *In: Proceedings of the IEEE Conference on Computer Vision and Pattern Recognition*. 652–660.

Qi, C.R., *et al.*, 2017b. Pointnet++: Deep hierarchical feature learning on point sets in a metric space. *Advances in Neural Information Processing Systems*, 30.

Qi, Y., *et al.*, 2019. A hybrid model for spatiotemporal forecasting of pm2. 5 based on graph convolutional neural network and long short-term memory. *Science of the Total Environment*, 664, 1–10.

Rahimi, A. and Recht, B., 2007. Random features for large-scale kernel machines. *Advances in Neural Information Processing Systems*, 20.

Rao, J., *et al.*, 2020. LSTM-TrajGAN: A deep learning approach to trajectory privacy protection. *In: GIScience 2020*. 12:1–12:17.

Rumelhart, D.E., Hinton, G.E., and Williams, R.J., 1986. Learning representations by backpropagating errors. *Nature*, 323 (6088), 533–536.

Schuster, M. and Paliwal, K.K., 1997. Bidirectional recurrent neural networks. *IEEE Transactions on Signal Processing*, 45 (11), 2673–2681.

Soni, A. and Boddhu, S., 2022. Finding map feature correspondences in heterogeneous geospatial datasets. *In: Proceedings of the 1st ACM SIGSPATIAL International Workshop on Geospatial Knowledge Graphs*. 7–16.

Su, H., *et al.*, 2015. Multi-view convolutional neural networks for 3d shape recognition. *In: ICCV 2015*. 945–953.

Sutskever, I., Vinyals, O., and Le, Q.V., 2014. Sequence to sequence learning with neural networks. *Advances in Neural Information Processing Systems*, 27.

Tancik, M., *et al.*, 2020. Fourier features let networks learn high frequency functions in low dimensional domains. *Advances in Neural Information Processing Systems*, 33, 7537–7547.

Tang, K., *et al.*, 2015. Improving image classification with location context. *In: Proceedings of the IEEE International Conference on Computer Vision*. 1008–1016.

Tenzer, M., *et al.*, 2022. Meta-learning over time for destination prediction tasks. *In: Proceedings of the 30th International Conference on Advances in Geographic Information Systems*. 1–10.

Terry, J.C.D., Roy, H.E., and August, T.A., 2020. Thinking like a naturalist: Enhancing computer vision of citizen science images by harnessing contextual data. *Methods in Ecology and Evolution*, 11 (2), 303–315.

Tobler, W.R., 1970. A computer movie simulating urban growth in the detroit region. *Economic Geography*, 46 (sup1), 234–240.

Trisedya, B.D., Qi, J., and Zhang, R., 2019. Entity alignment between knowledge graphs using attribute embeddings. *In: Proceedings of the AAAI Conference on Artificial Intelligence*. vol. 33, 297–304.

Vaswani, A., *et al.*, 2017. Attention is all you need. *Advances in Neural Information Processing Systems*, 30.

Veer, R-H., Bloem, P., and Folmer, E., 2018. Deep learning for classification tasks on geospatial vector polygons. *arXiv preprint arXiv:1806.03857.*

Wang, L., *et al.*, 2019a. Graph attention convolution for point cloud semantic segmentation. *In: Proceedings of the IEEE/CVF Conference on Computer Vision and Pattern Recognition.* 10296–10305.

Wang, Q., *et al.*, 2017. Knowledge graph embedding: A survey of approaches and applications. *IEEE Transactions on Knowledge and Data Engineering*, 29 (12), 2724–2743.

Wang, Y., *et al.*, 2019b. Dynamic graph CNN for learning on point clouds. *ACM Transactions on Graphics (tog)*, 38 (5), 1–12.

Wu, W., Qi, Z., and Fuxin, L., 2019. Pointconv: Deep convolutional networks on 3d point clouds. *In: Proceedings of the IEEE/CVF Conference on Computer Vision and Pattern Recognition.* 9621–9630.

Wu, Z., *et al.*, 2020. A comprehensive survey on graph neural networks. *IEEE Transactions on Neural Networks and Learning Systems*, 32 (1), 4–24.

Xie, Y., *et al.*, 2021. A statistically-guided deep network transformation and moderation framework for data with spatial heterogeneity. *In: 2021 IEEE International Conference on Data Mining (ICDM). IEEE*, 767–776.

Yan, B., *et al.*, 2017. From itdl to place2vec: Reasoning about place type similarity and relatedness by learning embeddings from augmented spatial contexts. *In: Proceedings of the 25th ACM SIGSPATIAL International Conference on Advances in Geographic Information Systems.* 1–10.

Yan, X., *et al.*, 2021. Graph convolutional autoencoder model for the shape coding and cognition of buildings in maps. *International Journal of Geographical Information Science*, 35 (3), 490–512.

Yan, X., *et al.*, 2019. A graph convolutional neural network for classification of building patterns using spatial vector data. *ISPRS Journal of Photogrammetry and Remote Sensing*, 150, 259–273.

Yang, J., *et al.*, 2021. Implicit transformer network for screen content image continuous super-resolution. *Advances in Neural Information Processing Systems*, 34.

Yang, L., *et al.*, 2022. Dynamic mlp for fine-grained image classification by leveraging geographical and temporal information. *In: Proceedings of the IEEE/CVF Conference on Computer Vision and Pattern Recognition.* 10945–10954.

Yin, Y., *et al.*, 2019. Gps2vec: Towards generating worldwide gps embeddings. *In: Proceedings of the 27th ACM SIGSPATIAL International Conference on Advances in Geographic Information Systems.* 416–419.

Yu, W. and Chen, Y., 2022. Filling gaps of cartographic polylines by using an encoder–decoder model. *IJGIS*, 1–26.

Zhang, K., 2021. Implicit neural representation learning for hyperspectral image super-resolution. *arXiv preprint arXiv:2112.10541.*

Zhong, E.D., *et al.*, 2020. Reconstructing continuous distributions of 3d protein structure from cryo-em images. *In: ICLR 2020.*

Zhu, R., *et al.*, 2022. Reasoning over higher-order qualitative spatial relations via spatially explicit neural networks. *International Journal of Geographical Information Science*, 36 (11), 2194–2225.

7 Intelligent Spatial Prediction and Interpolation Methods

Di Zhu
Department of Geography, Environment, and Society, University of Minnesota, Twin Cities

Guofeng Cao
Department of Geography, University of Colorado, Boulder

CONTENTS

7.1 Introduction 121
7.2 GeoAI Motivations for Spatial Prediction and Interpolation 123
    7.2.1   Representing spatial data 124
    7.2.2   Measuring spatial structure...126
    7.2.3   Modeling spatial relationships..127
7.3 GeoAI for Geostatistics 129
    7.3.1   Deep learning and kriging methods...129
    7.3.2   Case study: Wind and solar radiation downscaling 131
7.4 GeoAI for Spatial Regression...133
    7.4.1   Modeling spatial association relationships 134
    7.4.2   Spatial regression graph convolutional neural networks 135
    7.4.3   Case study: Social media check-ins on POIs....................................136
7.5 Discussions 142
    7.5.1   Geospatial uncertainty 142
    7.5.2   Transferability and generalization 142
    7.5.3   Interpretability and explainability...143
7.6 Summary...143
    Bibliography 144

7.1 INTRODUCTION

The necessity to obtain accurate predictions from observed data can be found in all scientific disciplines (Cressie, 2015). Measurements of geographic variables (phenomena) are essential for geographic modeling, knowledge discovery, and management decision-making at local, regional, and global scales (Liu *et al.*, 2018; Parks

DOI: 10.1201/9781003308423-7

et al., 1993). Spatial prediction is one of the major approaches for obtaining spatial data when there are insufficient observations, particularly for variables such as soil conditions, habitat suitability, disease risks, and socioeconomic factors, for which direct collection methods such as survey-based sampling and remote sensing methods are ineffective in getting satisfactory measurements (Goovaerts, 2005; Van Westen *et al.*, 2008).

The main task of spatial prediction is to use observations at sampled locations to make estimation of a target geographic variables at unknown locations (Zhu *et al.*, 2018). The commonly used spatial interpolation can be seen as a typical example of spatial predictions. In spatial interpolation, classic deterministic methods include the Voronoi natural neighbors interpolation, inverse distance weighting (IDW), triangular irregular network (TIN), etc, in which predefined spatial relationships are used for spatial prediction (Lam, 1983). Statistical methods, on the contrary, consist of a learning process for the spatial patterns. Commonly used statistical methods in GIScience can be categorized into two broad classes: geostatistics and spatial regression. Geostatistics is more concerned with modeling data as a realization of a spatial process $\{Z(\mathbf{s}) : \mathbf{s} \in D\}$, where the spatial domain is a set D that allows locations $\mathbf{s}$ to vary continuously throughout a region of $d-$dimensional geographic space. Geostatistics has been used widely to characterize spatial variations (using the semi-variogram or other distance-decay functions) in relatively small data sets and to predict unobserved values using the kriging family of methods. Oliver and Webster (1990) and Burrough *et al.* (2015) are two accessible introductions to geostatistics from the perspectives of applications.

On the other hand, spatial regression formalizes spatial relationships as correlation structures into a linear regression model, deals with the specification, estimation, and diagnostic checking of such regression models that incorporate spatial effects, and then utilizes the optimized model for spatial prediction. Early interest in the statistical implications of estimating spatial regression models dates back to the results in Whittle (1954), followed by a few classic papers in statistics (Besag, 1974; Ord, 1975; Ripley, 1981). Paralleling this was its development in the field of spatial econometrics and regional science (Anselin, 1992; Anselin and Bera, 1998; LeSage, 1997), where discrete lattice systems are often used to represent spatial variations across discrete geographic units such as census blocks, streets, and residential points (Lu *et al.*, 2014). Spatial dependence and heterogeneity of variable relationships are considered critical effects and are to be estimated as correlation coefficients in spatial regression models (Anselin, 2010; Fotheringham *et al.*, 2017).

The key to the success and applicability of spatial prediction are the underlying assumptions employed in describing the spatial relationships and the way how these relationships are characterized in the model. This chapter introduces the potential of GeoAI in developing new spatial prediction and interpolation methods, with a focus on ways to incorporate machine learning and deep learning mechanisms into geostatistics or spatial regression, and reversely, the implications of these integrations for spatial prediction and geographic knowledge discovery. The treatment in this brief chapter is not intended to be comprehensive, but aims to outline a few insights into both the current state-of-the-art GeoAI methods for spatial prediction, as well as ongoing research and remaining gaps. Some methods and topics are not included either because their linkages with GeoAI are still yet understudied or because they are partially addressed in other chapters of this book, such as representation learning for prediction (Chapter 6) and the replicability of models (Chapter 18).

The next section describes the motivations for improving the current spatial prediction paradigm with GeoAI methods and techniques. Section 7.3 describes GeoAI in Geostatistics with a case study in geostatistical downscaling of wind and solar radiation datasets, while Section 7.4 describes GeoAI in Spatial regression with a case study in predicting the social media check-ins on points of interest (POIs). Section 7.5 discusses a range of issues related to the use of AI within spatial prediction and interpolation methods.

7.2 GEOAI MOTIVATIONS FOR SPATIAL PREDICTION AND INTERPOLATION

In general, spatial prediction can be done in the process shown in Figure 7.1. First, a set of sampling locations are selected over the area of interest based on a given spatial representation schema. Second, the spatial structure of these sampling locations is defined, then data samples collected at these locations are analyzed to derive the spatial relationships, which describe how the values of the target variable are related to the sampling information in space. Finally, these derived relationships are used to predict the values of the target variable, including unobserved locations.

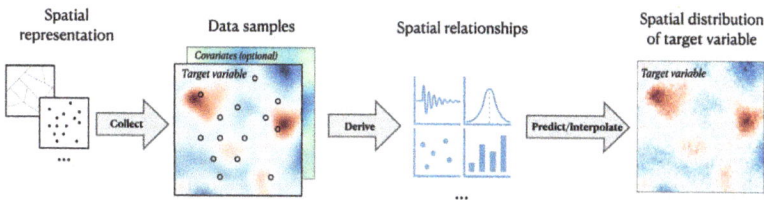

Figure 7.1 The typical process of spatial prediction.

Following this typical process, spatial prediction and interpolation methods have been struggling through the integration of more data samples and better spatial theories to capture spatial relationships. Most methods follow classic statistical principles, prior domain knowledge, and classic computation paradigm. We have to admit that the nature of spatial dependence and heterogeneity in geographical digital representations (Goodchild, 2004a) is substantially more complex than classic statistical models (Oliver and Webster, 1990; Shepard, 1968). Traditional methods are not intelligent enough to cope with the exploding data, characterize the complexity of spatial relationships, and take advantage of the computational innovation that is happening.

As a pioneering statement, Fischer (1998) systematically outlined how neural networks could become a promising paradigm for spatial analysis. With the recent progress in AI techniques and the availability of high-quality geospatial data, machine learning (ML) and deep learning (DL) methods have been increasingly used to understand spatial processes from a data-driven perspective (Reichstein *et al.*, 2019), as they can well extract underlying pattern features given complex spatial contexts. For example, the characteristics of convolutional neural networks' architecture – local connectivity and shared weights – enable the model to focus on features near to each other as well as far away features (LeCun *et al.*, 2015), which is consistent with the modeling of spatial relationships in spatial prediction. In recent years, geospatial artificial intelligence (GeoAI) has emerged as a subfield of GIScience that utilizes AI approaches for spatiotemporal prediction and geographic knowledge discovery (Gao, 2020; Janowicz *et al.*, 2020). Since there exist challenges in adapting ML and DL techniques directly into spatial analysis, many applied GeoAI research has paid attention to the classification of spatial features based on DL architectures well-established in image classification and object detection tasks (Li and Hsu, 2020; Yan *et al.*, 2019).

Enlightened by these works, researchers started to bridge the methodological gap between AI models and spatial prediction methods. We are in need of a set of GeoAI-powered approaches designed specifically for spatial prediction and interpolation tasks. Conceptually, such intelligent spatial analytical methods and models (Zhu *et al.*, 2022) are typically defined based on (spatially-explicit) neural networks which fulfill the invariance, representation, formulation, and outcome tests in the context of GeoAI (Janowicz *et al.*, 2020). More importantly, there are several essential motivations to be considered in order to seamlessly integrate various GeoAI techniques into the workflow of spatial prediction: (1) Representing spatial data; (2) Measuring spatial structure; and (3) Modeling spatial relationships.

7.2.1   REPRESENTING SPATIAL DATA

The first step of spatial prediction, i.e., spatial representation, is crucial as it has long been recognized that the decision of whether analytical and cartographic measures can be meaningfully applied depends on whether an attribute is considered intensive or extensive. A most fundamental and far-reaching trait of geographic information is the distinction between extensive and intensive properties (Scheider and Huisjes, 2019). GeoAI efforts in spatial prediction and interpolation need to consider the variants of spatial data representation and the corresponding data properties before heading into neural network designs and computations.

Spatially intensive data usually comes with strong spatial continuity and regularly-distributed samples, which is suitable to adopt the GIS raster model to represent the spatial variation of values as a continuous field. Such intensive measures, like temperature, elevation, and air pollution, are independent of the size of their supporting objects (in this case the polygon areal units or the points, see Figure 7.1). Raster representation most commonly takes the form of a grid-like structure that holds values at regularly spaced intervals over the extent of the raster. Since one geographic object (e.g., a lake) could cover multiple adjacent cells, each single raster cell does not correspond to any specific object, leading to the fact that local spatial features within shorter euclidean ranges are to be investigated in the neighboring cells. As an example, to handle raster spatial data in the neural network context, one may tend to use regular convolutional neural networks (CNNs), where each neuron represents a raster spatial unit and the regular convolution filters could facilitate the learning of short-range spatial dependency features (see Figure 7.2a).

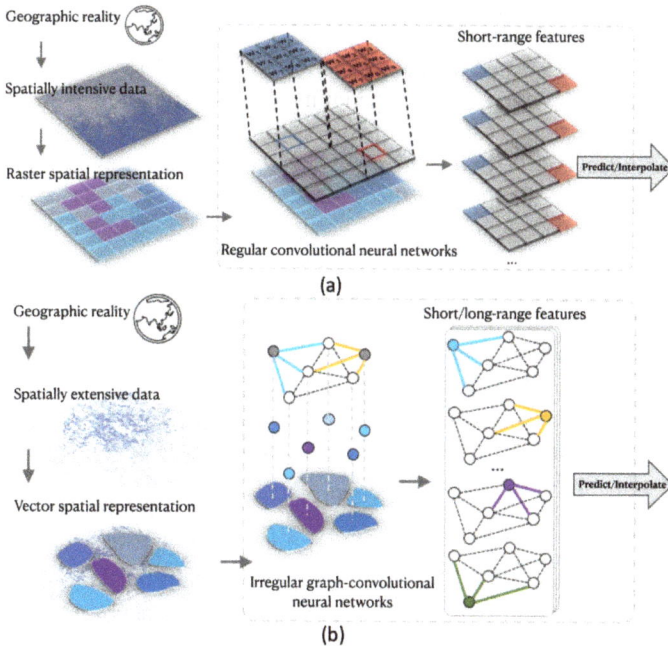

Figure 7.2 Raster and Vector data models: (a) Raster data and regular convolutional neural networks; (b) Vector data and graph convolutional neural networks.

Spatially extensive data, such as the total population, traffic volume, and GDP, are additive for sub-spatial units and should be summed up during spatial aggregation. Extensive properties are usually collected on discretely and irregularly distributed geographic objects. Due to this reason, space should not be quantized into discrete grid cells like the raster model for representing extensive measures. Instead, it requires the vector model in GIS to represent the multi-dimensional attributes of each object and the topological information among objects. Geographic objects are represented in fundamental vector types such as points, lines, and polygons. Unlike the raster model, each vector unit corresponds to a geographic object with a certain physical meaning, such as a restaurant, a street segment, or a census block group. Vector representation can be further abstracted into networks/graphs $G = (V, E)$, with the set of nodes V representing all objects and the set the edges E denoting the proximity/connectivity information among the objects. To handle vector/network spatial data in the neural network context, graph convolutional neural networks (GCNNs) is preferred (see Figure 7.2b) over regular CNNs. Each nodal neuron in GCNNs is a geographic object that will not be dropped in lower-layer representations (graph convolution itself does not change the number of neurons). The irregular convolution in a graph representation enables the learning of both short and long-range spatial features, which is quite helpful in the prediction and interpolation task of socioeconomic phenomena under various connectivity contexts.

7.2.2 MEASURING SPATIAL STRUCTURE

Starting the second step of spatial prediction, the spatial structure of locations is defined in order to provide a basic spatial metric for variable relationships. For example, in geostatistics, a lag (**h**) is often used to measure the separation distance of a data pair in Euclidean space. These lags are then correlated with the similarity in observed data values to compute the statistical plots for lag **h**, such as covariance, correlogram, and variogram (Dale and Fortin, 2014). The assumption behind Kriging interpolation is that spatial variable varies continuously across space according to some spatial lag or distance in a partly random and partly deterministic manner (Miller, 2004).

Due to the universal existence of spatial dependence (Tobler, 2004), the two-dimensional (2-D) convolutional operation can be adopted as a relatively strong neural network prior in the GeoAI model designs for capturing local spatial relationships. This can be summarized conceptually as a weighted aggregation of neighboring information:

$$O(i,j) = W * I(i,j) = \sum_m \sum_n I(i+m, j+n)W(m,n), \tag{7.1}$$

where $I(i,j)$ is the data value at target location (i,j) in a neural network layer and $O(i,j)$ is the output after the convolution. W denotes a 2-D kernel function that measures the spatial structure, with m and n representing displacement between the measuring location and the target location (i,j) under a specific distance metric. Depending on data representation models, there could be different concerns in the convolution process:

Distance: Regular convolution is suitable to measure the distance-based structure by defining kernel weights at regular intervals that approximate the distance lags (Dale and Fortin, 2014). Assuming the spatial data distribution $X \in \mathbb{R}^{n \times m}$ is represented as an $n \times m$ raster grid with each element $x_{i,j}$ being the value at location (i,j):

$$X = \begin{bmatrix} x_{1,1} & x_{1,2} & \cdots & x_{1,m} \\ x_{2,1} & x_{2,2} & \cdots & x_{2,m} \\ \cdots & \cdots & x_{i,j} & \cdots \\ x_{n,1} & x_{n,2} & \cdots & x_{n,m} \end{bmatrix}, \tag{7.2}$$

a regular convolution kernel W that measures the connections at k distance lags will apply the following transformation to the data:

$$\phi(W * X) = \left\{ \phi \left(\sum_{a=-k}^{k} \sum_{b=-k}^{k} w_{a,b} x_{i+a, j+b} \right), \forall i \leqslant n, j \leqslant m \right\}, \tag{7.3}$$

where ϕ is the activation function, $w_{a,b}$ is the parameter defined for a certain distance lag. Figure 7.2 visualizes the case of distance lag $k = 1$, when the convolution kernel only applies to the nearest cells in Euclidean space. Through regular convolutions, distance-based spatial variation relationships can be approximated for interpolation in neural networks.

Neighbor: Neighbors could be defined by adjacency, distance bands, nearest neighbors (Anselin, 2010), and non-Euclidean measures such as spatial interactions (Zhu *et al.*, 2020b), etc, leading to complex neighboring contexts of a location unit as it can be strongly related to certain distant locations. Such irregular neighborhood structures make it problematic to adopt the classical, raster-based convolution strategy. For vector spatial data $X = [x_1, x_2, \ldots, x_n]$ with n location objects, graph convolution is preferred in neural networks in order to consider various neighboring notions defined in the spatial weights matrix. Graph convolutional neural networks (GCNNs) follow an aggregation scheme where each neuron (object) propagates characteristics of its neighbors to learn a deep representation of the contextual information:

$$\phi(W * X) = \left\{ \phi \left(\sum_{k=0}^{b} \sum_{y \in N(k)} w_{x,y} y \right), \forall x \in X \right\}, \tag{7.4}$$

where ϕ is the activation function, graph convolution kernel measures the relationship between a neuron x and all its k-step $(k < b)$ neighbors $N(b)$ in a predefined spatial weights structure. $w_{x,y}$ is the parameter between location x and y to be learned. Figure 7.2 illustrate the scenario where the spatial weights matrix is defined by 3-nearest neighbor and a graph convolution kernel with $b = 1$ that considers only the nearest neighbor for feature aggregation.

These admit learning a wide range of distance functions and spatial patterning knowledge in neural networks. GeoAI models need explicit considerations in measuring spatial structure, i.e., the notion of neighborhoods and distance, such that the learned spatial relationships can be interpreted in a geographic sense.

7.2.3 MODELING SPATIAL RELATIONSHIPS

In deriving spatial relationships, there are three basic principles, namely the First Law of Geography, statistical principle and the Second Law of Geography (Zhu *et al.*, 2018). The result of spatial prediction depends on how well the model represents the spatial relationships. If it does not, then the computed measures, such as the variogram parameters and regression coefficients, are not a good representation of the reality, and in turn the weights assigned to the sample points cannot capture the relationships between the prediction point and the sample points.

First law of Geography: The first principle used in spatial prediction is that attribute values of a target variable are spatially related (spatial autocorrelation) and that locations which are closer would have more similarity values of attributes than locations are further apart. Tobler, in his 1970 paper, stated this important geographic principle as the First Law of Geography through his famous statement "Everything is related to everything else, but near things are more related than distant things"

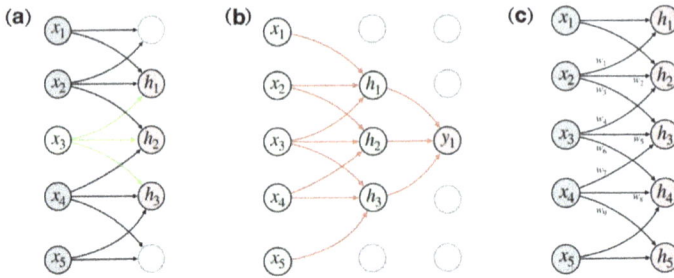

Figure 7.3   (a) Sparse connectivity; (b) receptive field; (c) parameter sharing.

(Tobler, 1970). Geostatistical methods such as Kriging, use this principle to design a mean to capture and represent such spatial dependence (e.g., semivariogram), so as to determine the optimized weights allocated to nearby samples for prediction. In GeoAI models, the sparse connectivity mechanism can be adopted to emphasize more on learning nearby features, thus establish a linkage between neural network parameters and the spatial dependence. For example, in Figure 7.3a, by defining a local kernel with width to be three, the value at location 3 (x_3) will only affect three nearby features (h_1, h_2, h_3) in the next neural network layer.

Statistical principle: The statistical principle refers to the statistic correlation between variables as expressed through regression analysis, which assumes there is a relationship between the value of the target variable (dependent variable) and the values from other variables (independent variables, covariates). This covariate relationship is usually captured in a linear regression to perform spatial prediction (Anselin, 2010). In GeoAI models, the mechanism of receptive field enables the learning of covariate relationships. Influence of input data (covariates) on the output data (target variable) can be traced via the neural network weights. For example, in Figure 7.3b, the output value at a location y_1 is influenced by the previous hidden layer (h_1, h_2, h_3) and even the input layer's locations ($x_1, \ldots, x_5$) that are connected in the neural network. All the neurons that affect y_1 are considered the receptive field of y_1. By imposing multiple channels, a neural network can perform a nonlinear approximation of complex covariate relationships that consider both short-range and long-range spatial dependence, implied as the distance decay of influence in the receptive field.

Second law of Geography: Considering the fact that geographic phenomena are inherently heterogeneous (Goodchild, 2004b), relationships (spatial or covariate) will not hold static or stay the same over space. Such spatial heterogeneity of geographic phenomena is particularly true for large and complex geographic areas. Techniques for spatial prediction have been substantially revised to account for spatial heterogeneity and to make them adaptive to local conditions, including the Box-Cox Kriging (Kitanidis and Shen, 1996), directional Kriging (van den Boogaart and Schaeben, 2002), geographic weighted regression (Fotheringham *et al.*, 2017), etc. In GeoAI models, non-sharing parameters may be used to facilitate the learning of heterogeneous relationships. A simple illustration in Figure 7.3c shows local kernels with width to be three while allowing the weights parameters to differ for each location. Many latest AI mechanisms are also looking at learning heterogeneity, such as transformers (Chu *et al.*, 2021), self-attention models (Zhang *et al.*, 2019), etc.

Deep learning approaches are increasingly used to understand spatial processes from a data-driven perspective, as they are powerful in terms of their ability to extract underlying patterns given complex spatial contexts. These remarkable characteristics of neural networks, such as sparse connectivity, parameter (non-)sharing and receptive field, enable the GeoAI models consider the three essential spatial principles while learning the data relationships, providing a way to approximate the complex functions describing spatial patterns and facilitating spatial predictions.

7.3   GEOAI FOR GEOSTATISTICS

7.3.1   DEEP LEARNING AND KRIGING METHODS

The key of geostatistics is to derive statistical relationships describing how the values at target locations are related to the observations at sampled locations. In geostatistics, the statistical relationship is often described by variograms or semi-variograms that characterize the spatial variations of spatial variables over the distances. Given two locations separated by a distance h, the variogram value $f_\gamma^*(h)$ quantify the dissimilarity (similarity) of the measurements at the two locations. A variogram is essentially a function of distances describing the spatial dependence structure of the spatial observations. Prior functions such as Gaussian, exponential and spherical functions are often used to model the variograms for spatial prediction and statistical inferences (Chiles and Delfiner, 1999). In kriging systems, the estimations at targeted locations are made as a weighted linear combinations of the sampled observations, and the values from variograms are used to determine the weights for each of the samples while accounting for spatial dependence (see the kriging interpolation part of Figure 7.4 for an illustration). The kriging methods were originally developed for point-referenced, continuous Gaussian spatial variables. Over the years different variants have been developed for non-Gaussian spatial variables. For example, for categorical spatial data indicator kriging or transition probability-based methods have been developed (Cao *et al.*, 2011a, 2013, 2011b, 2014; Carle and Fogg, 1996; Solow, 1986).

Traditional geostatistical methods have been widely adopted in spatial prediction tasks across different domains. The popularity is mainly due to its effectiveness in modeling spatial dependence structure and quantifying the uncertainty of spatial estimations. However, most of the geostatistical methods assume that the spatial variations can be modeled with simple functions of distances and the spatial patterns are *stationary* across the study areas. These assumptions render these methods perform best at homogeneous cases where observations are regularly distributed in study areas but are rather limited for heterogeneous cases with complex spatial patterns. Progress was made to mitigate the strict assumptions. For example, multiple-point geostatistical methods (e.g., Caers, 2002; Strebelle, 2002), often hyped as

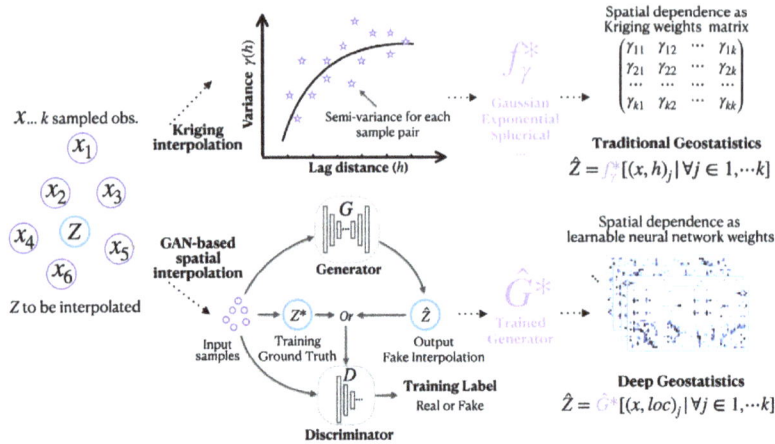

Figure 7.4   Traditional kriging vs generative adversarial learning for spatial interpolation. (*Source*: Adapted from Zhu *et al.* (2022)).

variogram-free geostatistics, have been developed and demonstrated success in applications of geosciences and petroleum engineering. Sophisticated methods were also developed in statistics to improve the spatial modeling for non-stationary cases with complex spatial dependence structures (e.g., Kleiber and Nychka, 2012; Paciorek and Schervish, 2006). Despite the progress, the developments often target specific problems that limit the applications in generic spatial analysis and modeling.

With a deep neural network with multiple levels of processing layers, deep learning-based methods have been shown to excel at discovering intricate complex patterns from high-dimensional data and hence provide a promising venue to improve spatial prediction with complex geospatial patterns. From the perspective of GeoAI, spatially explicit neural networks can be developed to formalize the spatial interpolation process without such strong prior model assumptions. There are some pioneering works that introduce neural networks to the tasks of spatial prediction and interpolation, such as super resolution (Ledig *et al.*, 2017; Zhang and Yu, 2022; Zhang *et al.*, 2022) and spatiotemporal kriging (Wu *et al.*, 2021b). For example, a generative adversarial neural networks (GAN) can be adopted to approximate the spatial conditioned probability distribution $p_{model}(\hat{Z}|x_j, \forall j \in 1,\cdots,k)$ that best describes data observations $p_{data}(\hat{Z}|x_j, \forall j \in 1,\cdots,k)$ through an adversarial game between the generator G and the discriminator D (Zhu *et al.*, 2020a), as shown in the below part of Figure 7.4. By structuring each location as a neuron in the deep learning model, learnable convolutional weights can imitate the data-borrowing process from nearby samples, offering can effectively learn the spatial message passing mechanism. Besides, auxiliary learning is found to be able to further amplify spatial knowledge such as the autoregressive structures and local patterns for more accurate geostatistical modeling Klemmer and Neill (2021).

7.3.2 CASE STUDY: WIND AND SOLAR RADIATION DOWNSCALING

We use a case study of spatial downscaling to compare the performances of traditional kriging and deep learning-based methods. The datasets used are from National Solar Radiation Database (NSRDB) (Sengupta *et al.*, 2018) and Wind Integration National Database (WIND) Toolkit (Draxl *et al.*, 2015a,b; King *et al.*, 2014; Lieberman-Cribbin *et al.*, 2014) provided by National Renewable Energy Laboratory (NREL). The wind datasets have two channels of u and u and are in the spatial resolution of 2km. Similarly the solar datasets have two channels of Direct Normal Irradiance (DHI) and Diffuse Horizontal Irradiance (DNI), and are in 4km. The temporal resolutions of wind and the solar datasets are hourly and 30 minutes, both of which are sampled to 4-hourly and hourly respectively in this case study. Both datasets covering the years 2007-2014, in which the datasets of 2007 to 2013 are used for training and the ones from 2014 are used for testing.

The compared methods include the area-to-point (ATP) kriging (Kyriakidis, 2004), a variant of kriging for change of spatial scales that have been widely used for image downscaling (e.g., Wang *et al.*, 2016). In deep learning literature, different types of deep neural network architectures have been developed for image downscaling or super-resolution. We include one exemplary method from the commonly used types of methods, including deep image prior (DIP) (Ulyanov *et al.*, 2018), an unsupervised deep learning method for image modeling; EDSR, a deep residual network-based method for single-image super-resolution (Lim *et al.*, 2017), and ESRGAN, a GAN model for super-resolution (Ledig *et al.*, 2017; Wang *et al.*, 2018).

We compared the performances of the selected methods for downscaling the wind and solar radiation datasets with 4X scaling factor. Each channel of the wind and solar datasets were first upscaled 4X with bilinear interpolation (referred to as LR hereafter), and then the selected methods were used to downscale them back to the original resolutions. The results were compared to the original datasets to collect accuracy statistics for performance comparisons. It should be noted that unlike EDSR and ESRGAN that needs training data, ATP and DIP are unsupervised methods, and do not need additional training data beyond the datasets to be downscaled.

For the accuracy statistics, we calculated the relative mean squared error (MSE) and sliced Wasserstein distance (SWD) (Karras *et al.*, 2018) to quantitatively evaluate the model performance. The relative MSE is the MSE normalized by the range of values and the SWD is a simplified Wasserstein distance (Arjovsky *et al.*, 2017) that is often use to compare visual images while accounting for spatial patterns. Table 7.1 lists the accuracy statistics for the compared methods. One can see that the deep learning methods (DIP, EDSR and ESRGAN) perform better than the ATP methods for both tested cases except for the wind dataset where DIP had a larger relative MSE. Figures 7.5 and 7.6 display the LR, ground truth images, and the resultant maps of each methods. One can see that the result of the ATP tends to gloss over the spatial details that were captured in the deep learning methods, demonstrating the effectiveness of deep neural network in capturing complex spatial patterns.

Table 7.1

Accuracy performance of different models across the entire wind or solar test sets. Each row represents the relative MSE and SWD between the ground truth images and generated images by different models at the specific SR scale and datasets. Smaller values indicate better generation performance.

Wind	relative MSE ($\downarrow$) / SWD ($\downarrow$)			
	ATPK	**DIP**	**EDSR**	**ESRGAN**
4×	0.505/0.167	0.900/0.158	0.261/0.124	0.391/0.145

Solar	**ATPK**	**DIP**	**EDSR**	**ESRGAN**
4×	0.145/0.199	0.090/0.184	0.090/0.164	0.138/0.197

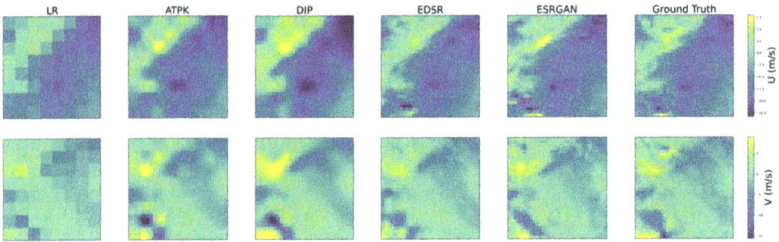

Figure 7.5   Two channels (u and v) of 4X SR results generated by different models. The input LR image (image size 8×8) in the leftmost column is randomly selected from the wind test set and the rightmost column is the corresponding HR ground truth image (image size 32×32).

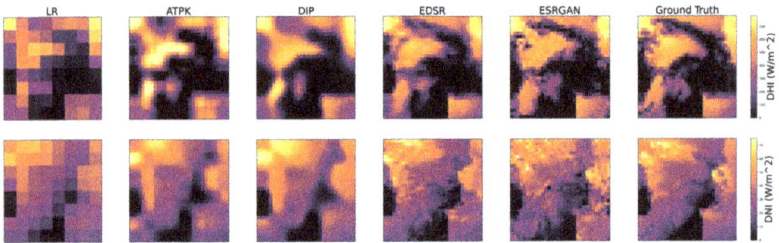

Figure 7.6   Two channels (DHI and DNI) of 4X SR results generated by different models. The input LR image (image size 8×8) in the leftmost column is randomly selected from the solar test set and the rightmost column is the corresponding HR ground truth image (image size 32×32).

7.4 GEOAI FOR SPATIAL REGRESSION

As a long-established spatial analytical method, spatial regression grows in the fields of regional science and spatial econometrics (Paelinck *et al.*, 1979), where the applied works rely heavily on observed variables with reference to location measures. Spatial regression models focus on two critical aspects of data introduced by locations: (1) the spatial autocorrelation (dependence) of samples and (2) the spatial heterogeneity of the relationships been modeled (Anselin, 1988). In a cross-sectional setting, i.e., leaving out of consideration the temporal aspect of data, a typical spatial regression analysis is performed mainly in four steps. First, representing the structure of spatial dependence using a spatial contiguity or weights matrix (LeSage, 1997). Second, specifying a regression model that incorporates the potential spatial effects (LeSage and Fischer, 2008). Third, estimating parameters in the model. Fourth, utilizing the fitted model for spatial prediction (Lehmann *et al.*, 2002). This classical econometric paradigm (Anselin, 2010) has been widely adopted in exploratory spatial data analysis (ESDA) where the spatial multivariate (cross-sectional) distribution data is available and when we seek to capture the relationships between the observations of variables (Fischer and Wang, 2011).

Systematically reviews on the progress of spatial econometrics can be found in (Anselin, 2010; Anselin and Rey, 2014; Arbia, 2014; Fischer and Wang, 2011; Griffith and Paelinck, 2011; LeSage and Fischer, 2008) with respect to different focuses and specialties. Here, our focus is on GeoAI and spatial regression models (or linear spatial models) in a cross-sectional setting, among which the traditional dominant family would be the spatial lag model (SLM). SLM allows observations of the dependent variable y_i at a spatial unit i ($i = 1, \ldots, N$) to depend on the observations in nearby units $j \neq i$. There are different specifications of the spatial dependence as the lag terms in SLM. The basic one would be spatial autoregressive (SAR) model that includes the spatial lag of dependent variables (Anselin, 1988):

$$y_i = \rho \sum_{j=1}^{N} w_{ij} y_j + \sum_{k=1}^{K} \beta_k x_{k,i} + \varepsilon_i, \tag{7.5}$$

where w_{ij} is the element at the i-th row and j-th column of a N-by-N spatial weight matrix W, $x_{k,i}$ is the k-th independent variable out of K, with ρ and β_k the parameters to be estimated, and $\varepsilon_i \sim N(0, \sigma^2)$ the error term. The spatial lag can also be introduced in the independent variables, which is referred to as the spatially lagged X (SLX) models:

$$y_i = \sum_{k=1}^{K} \delta_k \sum_{j=1}^{N} w_{ij} x_{k,j} + \sum_{k=1}^{K} \beta_k x_{k,i} + \varepsilon_i, \tag{7.6}$$

where δ_k is the spatial parameter of independent variable x_k. Despite of many variants, a generalized model called spatial Durbin model (SDM) might be of interest if we are to include the spatial lags among all variables:

$$y_i = \rho \sum_{j=1}^{N} w_{ij} y_j + \sum_{k=1}^{K} \beta_k x_{k,i} + \sum_{k=1}^{K} \delta_k \sum_{j=1}^{N} w_{ij} x_{k,j} + \varepsilon_i. \tag{7.7}$$

We assume SDM to be a comprehensive model for global spatial regression because it nests many of the models (LeSage and Pace, 2009). Apart from these global regression models, geographically weighted regression (GWR) and multi-scale geographically weighted regression (MGWR) refer to a range of specifications that focus on the spatial non-stationarity of relationships, where localized parameters are used to capture the spatial heterogeneous effects (Fotheringham *et al.*, 2017).

Many specialized estimation methods have been developed for spatial regression models, because the autocorrelation and collinearity between variables make it problematic to apply ordinary least squares (OLS) methods to estimate the model parameters (Anselin, 1988). The representative methods include maximum likelihood estimation (MLE) (Ord, 1975), instrumental variables (IV) (Kelejian and Prucha, 1998), general method of moments (GMM) (Kelejian and Prucha, 1999), and Bayesian methods (LeSage, 1997). Back-fitting algorithms, which calibrate the optimal parameters through an iteration manner, are also found to be computational effective when the model incorporates complex spatial weights and a large number of parameters (Fotheringham *et al.*, 2017).

7.4.1 MODELING SPATIAL ASSOCIATION RELATIONSHIPS

Spatial regression models, however, are somehow constrained by the paradigm of spatial econometrics and statistics. Applications are limited by prior assumptions, such as the linearity of regression model and the normality of data distributions (Anselin, 2010; Kelejian and Prucha, 2007). Traditional spatial regression models deal with the formal mathematical expression of spatial effects through predetermined functional forms, linear combinations of variables often the case, while the non-linear nature of spatial relationships has been overlooked (Fischer, 1998). The potential of modeling nonlinear spatial relationships in neural networks, is discussed in some recent works by combining the coefficients of ordinary least squares (OLS) with artificial neural networks parameters (Dai *et al.*, 2022; Du *et al.*, 2020; Wu *et al.*, 2021a). Another drawback is that the ad-hoc spatial dependence structure incorporated in the spatial weights matrix can only be defined among the locations where both independent and dependent variables are observed, without considering the other locations where the dependent variable is unobserved. This prerequisite enforces the specification of spatial relationships to be within the observed location set, which is not an ideal solution for transductive spatial prediction under missing data (Mennis and Guo, 2009).

Notably, encouraging pieces of evidence have been found in GeoAI research that graph convolutional neural networks (GCNNs) is a promising paradigm for spatial regression, thanks to its propagation mechanisms, spatial locality learning nature, and the semi-supervised training strategy (Defferrard *et al.*, 2016; Kipf and Welling, 2017). As shown in Figure 7.7, GCNNs as a variant of the CNN framework, is naturally suitable to build a conceptual mapping between the graph structure and the spatial weights matrix for irregularly distributed spatial units and is capable of capturing complex spatial lagged effects via the multi-layer graph convolutional filters across feature channels. This new spatial regression logic has been adopted in several

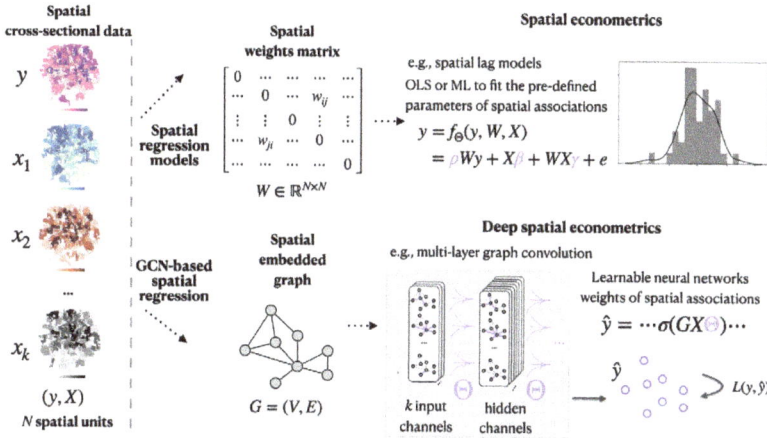

Figure 7.7 Traditional spatial lag models vs. graph convolutional neural networks in spatial regression. (*Source*: Adapted from Zhu *et al.* (2022).)

latest urban studies to understand complex association relationships based on irregular geographic units. Yan *et al.* (2019) proposed a two-layer GCNNs architecture to perform a binary classification on vectorized building patterns. Liu and De Sabbata (2021) introduced a semi-supervised approach to classify geo-located social media posts into multiple categories. Latest works started to use GCNNs to predict the continuous value of urban characteristics (Hu *et al.*, 2021; Xiao *et al.*, 2021; Zhu *et al.*, 2021, 2020b) as an interpolation task, the experiments were designed similar to the typical spatial regression workflow and the emphases were on the different spatial weighting measures for more accurate regression modeling. GCNN modules can also be integrated into temporal deep learning frameworks such as recurrent neural network (RNN) and long-short term memory (LSTM) models, in order to achieve better forecasting accuracies in areas such as disaster warning (Chen *et al.*, 2019b) and traffic prediction (Bai *et al.*, 2021; Bui *et al.*, 2021; Zhang *et al.*, 2020; Zhao *et al.*, 2019).

7.4.2 SPATIAL REGRESSION GRAPH CONVOLUTIONAL NEURAL NETWORKS

Given these knowledge on how GCNNs could imitate a typical spatial regression analysis (Section 7.2.1), Zhu *et al.* (2021) proposed spatial regression convolutional neural networks (SRGCNNs) as a deep learning paradigm to conduct spatial regression and prediction on spatial multivariate distributions. The major contribution of SRGCNNs is to formalize GCNNs in the context of spatial regression, which enables geographers to intuitively understand the linkages between graph convolution mechanisms and the key concepts in spatial regression models. As illustrated in Figure 7.8, the framework of SRGCNNs is similar to the traditional spatial regression analysis in all four steps: SRGCNNs starts by collecting the location-referenced data on the

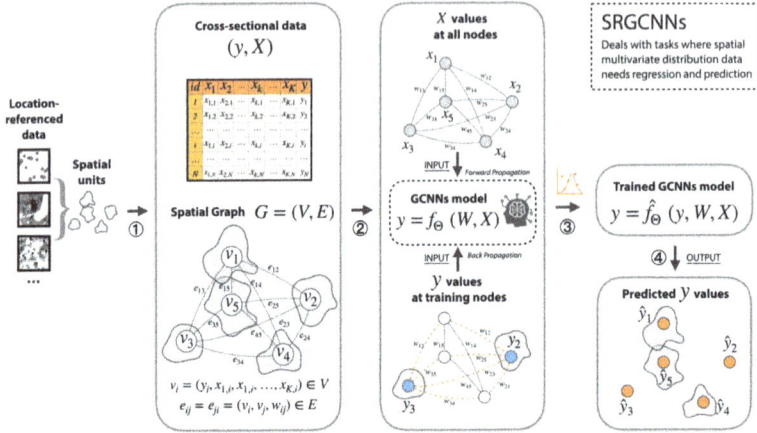

Figure 7.8 Spatial regression graph convolutional neural networks (SRGCNNs) as a deep learning paradigm for the regression analysis of spatial multivariate distributions. (*Source*: Adapted from Zhu *et al.* (2021).)

N spatial units and building a spatial graph that encodes the spatial weights matrix and cross-sectional data (step 1). Then, we initialize all nodes with the observed X values. These values are forward propagated through a specific GCNNs architecture, whilst the y values sampled at the training nodes are input to the last GCNNs layer to calculate the output errors and to enable the back propagation of the model (step 2). Next, SRGCNNs optimizes parameters in its GCNN model following a semi-supervised learning strategy (step 3), which explicitly considers the spatial weights among all the N spatial units, even though the y values are observed only at M ($\leqslant N$) training nodes. Profited by the propagation mechanism and spatial locality nature of GCNNs, SRGCNNs optimizes a GCNN model to achieve the most credible approximation of spatial relationships, and as the output, to predict the unobserved y values (step 4). Details including commonalities and differences between SRGCNNs and traditional spatial regression models can be found in Zhu *et al.* (2021).

In the following section, we will display the experiments based on two implementations of the SRGCNNs, i.e., the basic model and the geographically weighted model, on predicting social media check-in activities at the POIs level.

7.4.3 CASE STUDY: SOCIAL MEDIA CHECK-INS ON POIs

7.4.3.1 Data Description and Feature Selection

The urban region within the fifth ring road of Beijing, China, is selected as the study area. The region has an area of 668.72 km^2, a giant metropolis where the fast-developing economy has bred rich urban facilities and diverse human activities. This study area is considered the most complex and populous region in Beijing regarding socioeconomic vitality such as dining, residence, transportation and business. The dataset was collected from a social media platform named Sina Weibo that contains over 868 million annual check-in records in the year of 2014 on the POIs in Beijing.

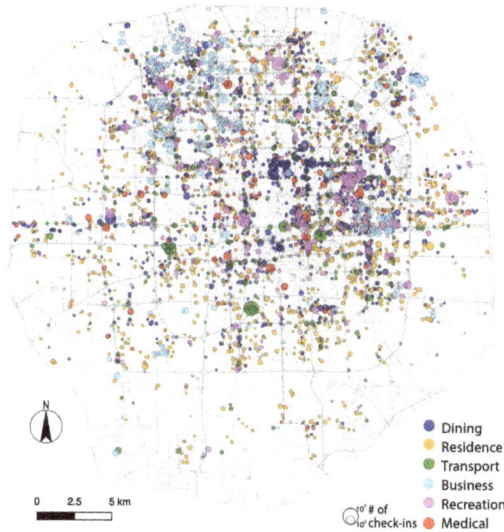

Figure 7.9   Study area and the POI check-in data. (*Source*: Adapted from Zhu *et al.* (2021).)

Within the study area, we selected the POIs with over 100 annual check-in records as the basic spatial analysis units (4,636 POIs in total) in the case study, as is shown in Figure 7.9. By selecting only the POIs with over 100 annual check-ins, inactive POIs with less check-ins are left out to balance the data distribution. All data are transformed into the logarithmic numbers of the check-in activities to the base 10 as the feature values, denoted as log_{10}(# of check-ins), in order to re-duce the skewness of check-in numbers. Raw sub-types of POIs are classified into six dominant categories (corresponding POI numbers are denoted by #) according to their functional types, i.e., dining (#=1,923), residence (#=1,117), transport (#=351), business (#=524), recreation (#=528) and medical (#=133). POIs are more clustered in the northern and eastern areas of the study area. Residential facilities (orange) are scattered around the ring roads, active business POIs (cyan) are more concentrated in CBD areas in the northwest and the east, while recreation (pink) and dining (blue) POIs are scattered in the whole area, with a diverse range of check-in numbers.

As for the experiments, we treated POIs' types as the independent categorical variables (X), the layout structure of POIs as the spatial support to build the spatial weights matrix and graph (W), while the logarithmic check-in numbers are the de-pendent variable values (y) to be predicted at the POI level. The loss L between the ground-truthing y and predictions $\hat{y}$ is computed only for the training POIs, and L is then used for the back propagation and gradient decent in the model. A final output $\hat{y}$ for all POIs can be obtained after sufficient training epochs.

More specifically, independent variable values are encoded as one-hot vectors $X \in \mathbb{R}^{N \times C}$, where $N = 4,636$ is the number of all POIs and $C = 6$ denotes the six type channels ordered by dining, residence, transport, business, recreation and medical. For example, if the POI with $id = 1$ is labeled as dining, then the corresponding one-hot vector passed to the graph convolution would be $x_1 = [1,0,0,0,0,0]$. The input features are output as $\hat{y} = [\hat{y_1}, \hat{y_2}, \ldots, \hat{y_N}] \in \mathbb{R}^N$ after graph convolutions and layer-wise transformations. We use ReLU as the activation function in the hidden layer. No activation function is used before the output layer, since we would like the regression model to approximate accurate numbers of y instead of generating classified labels. No drop-out is applied because we hope to maintain the spatial structure as well the attributes at every location to make comprehensive predictions.

7.4.3.2   Models and Settings

Spatial regression models can vary in terms of whether they allow the parameters to change spatially. Given that, two SRGCNN-based models are tested:

Basic SRGCNN model: The logic of basic SRGCNN model can be simplified as the following layer-wise graph convolution:

$$X^{(l+1)} = W \times X^{(l)} \times \Theta^{(l)}. \tag{7.8}$$

Assuming N spatial units, input features $X^{(l)}$ with C_{in} channels at layer l and output features $X^{(l+1)}$ with C_{out} channels and layer $l + 1$, we have $X^{(l+1)} \in \mathbb{R}^{N \times C_{out}}$, $X^{(l)} \in \mathbb{R}^{N \times C_{in}}$, spatial weights matrix $W \in \mathbb{R}^{N \times N}$ and neural network layer-wise parameters $\Theta^{(l)} \in \mathbb{R}^{C_{in} \times C_{out}}$. The parameters $\Theta^{(l)}$ are shared among all locations, leading to a global spatial regression within a local-connected graph structure.

SRGCNN-GW model: Similar to how GWR extends global spatial regression, we propose the geographically weighted SRGCNN model (SRGCNN-GW) with a geographically weighted layer-wise graph convolution:

$$X^{(l+1)} = W \times X^l \otimes \Theta^{(l)}_{local} \times \Theta^{(l)}, \tag{7.9}$$

where $\Theta^{(l)}_{local} \in \mathbb{R}^{N \times C_{in}}$ contains the geographically weighted parameters. SRGCNN-GW calculates the element-wise product between input spatial features and the geographically weighted parameters $X^l \otimes \Theta^{(l)}_{local}$. Thus, the features are parameterized, with each spatial unit having an independent set of trainable parameters.

Both models are designed as two-layer neural networks and the spatial weights matrix and graph are initialized based on a categorical k nearest neighbor (k−NN) searching, that is, considering up to the kth nearest neighbors regarding each POI types to be the adjacent locations. We conducted the experiments based on multiple training ratios, i.e., 10%, 20%, 40%, 60%, 80% to evaluate the model performances. For each of the training ratio, fifty parallel simulations with randomly selected training POIs and test POIs were carried out to achieve stable results. For each experiment, we evaluated the model fitting of training data, and the accuracy of prediction on the test POIs. The predicted check-in values on the test POIs are visualized onto

0.5 *km* × 0.5 *km* geographical grid cells, a reasonable scale of the regular spatial grid for social media data analysis (Chen *et al.*, 2019a). The saturation of color in each cell denotes the corresponding summed check-in numbers of all test POIs within the cell's extent. By transforming points into raster cells, we hope to provide relatively clear spatial patterns for the visual comparison between real numbers y and predicted $\hat{y}$ on the test POIs.

Two common quantitative evaluation metrics were used to help measure the model performances regarding the discrepancy between y and $\hat{y}$: mean absolute percentage error (MAPE) on the test locations and mean square error (MSE) on the training locations. MAPE ranges from 0 to 1, with a higher value indicating a larger dissimilarity between y and $\hat{y}$. MSE measures the squared error of estimation, which is widely used for comparing the stability of prediction models. Assuming that there are M training POIs and $(N-M)$ test POIs, the calculation of MAPE on the test POIs is:

$$MAPE = \frac{100\%}{N-M} \sum_{i=1}^{N-M} \frac{|\hat{y}_i - y_i|}{y_i}, \tag{7.10}$$

and the MSE on training POIs is:

$$MSE = \frac{1}{M} \sum_{i=1}^{M} (\hat{y}_i - y_i)^2. \tag{7.11}$$

As for training the SRGCNN-based models, we set the loss function $L(\hat{y}, y)$ to be L2-loss, which has exactly the same mathematical form as MSE. We used Adam optimizer, where $\beta_1 = 0.5$ and $\beta_2 = 0.999$. The learning rate was $\eta = 10^{-3}$ for SRGCNN model and $\eta = 3 \times 10^{-4}$ for the SRGCNN-GW model. Training epochs were capped at 8,000, and we recorded the best results among all epochs. The experiments were implemented using PyTorch, a deep learning framework in python with GPU acceleration. All the benchmark linear spatial models were implemented using PySAL, an open-source python package designed to support spatial data science. The computational environment was a Linux server with one NVIDIA 1080TI GPU, a 2.40GHz Intel E5-2680 CPU, and 128 GB RAM.

7.4.3.3 Evaluation of the Results

Figure 7.10 illustrate the results of basic SRGCNN model on a sample experiment with only 10% sampling ratio. The randomly initialized training POIs (10%) and test POIs (90%) are visualized in Figure 7.10a. The real numbers of check-ins on the training POIs (black points) are used (y), while the numbers of check-ins on the test POIs (while points) are to be predicted ($\hat{y}$). The spatial distribution patterns of real and predicted check-ins on test POIs are visualized in Figures 7.10b and 7.10c, respectively. The colors from yellow to black denote the numbers from the minimum to the maximum in each 0.5 *km* × 0.5 *km* grid cell, the mapping between colors and logarithmic check-in numbers is based on the quantile method of six categories. As a result, the overall prediction accuracy obtained is MAPE≈11.75% in the sample

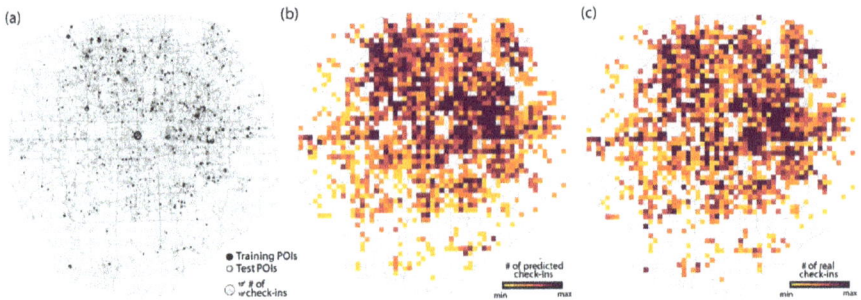

Figure 7.10   Spatial distribution patterns of check-ins on the test POIs in a sample experiment with 10% sampling ratio. (*Source*: Adapted from Zhu *et al.* (2021).) (a) Initialization of training (10%) and test (90%) POIs in the sample experiment. (b) Predicted spatial pattern of check-ins on the test POIs using SRGCNNs. (c) Real spatial pattern of check-ins on the test POIs.

experiment, based upon only the 10% sampled y and the comprehensive graph structure $W \in \mathbb{R}^{4636 \times 4636}$. The predicted spatial pattern of check-in numbers looks very similar to the real pattern. The rank-size relationships in most local regions are well reproduced. Local regions are distinguishable in the predicted pattern in terms of the spatial autocorrelation of check-in numbers. However, we can see that the clustered patches in Figure 7.10b are often slightly larger than those in Figure 7.10c. The reason for that is the weight-sharing graph convolution scheme introduces a smoothing effect, which tends to overestimate low-value locations and underestimate the high-value locations.

Furthermore, we conducted experiments based on all five training ratios: 10%, 20%, 40%, 60%, and 80%. For each ratio, we repeated fifty parallel simulations to investigate the accuracy and the stability of model performances. Figure 7.11 illustrates the statistical comparison between the basic SRGCNN model and SRGCNN-GW in box plots. Seen from Figure 7.11a, SRGCNN-GWs outperform SRGCNNs in terms of fitting the training data across all sampling ratios. Basic SRGCNN models have both larger median values and standard deviations of the training MSEs. However, Figure 7.11b proves that SRGCNN models actually perform better than SRGCNN-GW in terms of predicting the test data, with lower test MAPEs and smaller standard deviations. It is also interesting to see that the difference between basic SRGCNN models and SRGCNN-GW is more significant when the sampling ratio is low. The SRGCNN model, as a weight-sharing neural network, is more focused on prediction, while SRGCNN-GW, as a geographically weighted model, is more into data fitting. We may prefer to use the basic SRGCNN model when the sampling ratio is lower, while if the data is sufficiently sampled, SRGCNN-GW would be a better model. These findings are in accordance with spatial econometric models, where global spatial regression models are better at modeling and prediction, while GWR models are more powerful in fitting the sampled data for explanation.

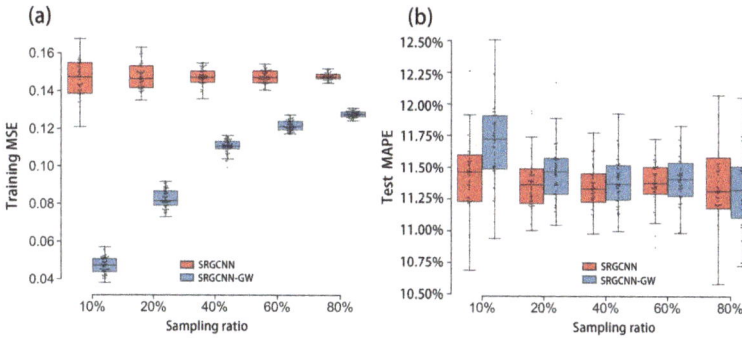

Figure 7.11 Statistical comparison between the basic SRGCNN model and SRGCNN-GW across different sampling ratios. (a) MSEs on the training POIs. (b) MAPEs on the test POIs.

To evaluate the prediction accuracy based on different sampling ratios, we selected two classical spatial regression models, i.e., SAR (Eq. 7.5) and SDM (Eq. 7.7), that consider spatial lagged effects, so as to be comparable with our proposed SRGCNN-based models. The model performances on predicting test locations are summarized in Table 7.2. We can see that compared with SAR and SDM, the SRGCNN-based models are significantly less sensitive to the sampling ratio (with lower MAPEs). As the sampling ratio goes down, traditional models exhibit obvious decreasing in accuracies. SDM achieves the best prediction accuracy when the sampling ratio $\geqslant 40\%$, but is less stable than the deep learning models (higher standard deviations). When the sampling ratio is lower than 40%, SRGCNNs and SRGCNN-GWs are much more accurate and stable.

Table 7.2

Model performances on spatial prediction.

	SAR		SDM		SRGCNN		SRGCNN-GW	
Ratio	MAPE	SD	MAPE	SD	MAPE	SD	MAPE	SD
10%	113.69%	57.47%	18.41%	10.79%	11.42%	0.30%	11.69%	0.32%
20%	97.94%	45.13%	12.49%	4.85%	11.38%	0.21%	11.46%	0.23%
40%	77.60%	26.08%	10.34%	1.69%	11.34%	0.18%	11.40%	0.20%
60%	50.19%	11.84%	9.94%	0.70%	11.38%	0.17%	11.40%	0.20%
80%	25.15%	2.67%	10.19%	0.38%	11.37%	0.33%	11.33%	0.32%

- MAPE: mean absolute percentage error; SD: standard deviation.
- Results are reported based on 50 simulations each sampling ratio.

7.5 DISCUSSIONS

7.5.1 GEOSPATIAL UNCERTAINTY

Geospatial uncertainty describes the differences between the geospatial data and the corresponding true phenomena or processes they represent. Since it is impossible to create a perfect representation of the infinitely complex real world, all geospatial data are subject to uncertainty (Goodchild, 2008). The core question is how to model the geospatial uncertainty and evaluate the impact of data uncertainty in practical applications and scientific modelings. It is a critical task shared by many domain fields where geospatial data are heavily involved and is closely related to the "replicability crisis" in scientific research (Kedron *et al.*, 2021).

Geospatial uncertainty inherits many of the characteristics of geospatial measurements themselves, including the complex spatial dependence structures and nonstationarity discussed above (Goodchild, 2020). Hence many of the previous discussions can be applied for geospatial uncertainty. In addition to the uncertainty of geospatial measurements (or *aleatoric uncertainty* in Bayesian modeling), the uncertainty of spatial prediction can also arise from the model itself (*model uncertainty* or *epistemic uncertainty*). Big models inevitably lead to big uncertainty. Bayesian models provide a powerful venue for uncertainty quantification. By assuming prior distributions on both the geospatial measurements and specification parameters. Bayesian spatial methods, such as Bayesian Kriging (Diggle *et al.*, 1998), can account for both the aleatoric and model uncertainty. However, as mentioned earlier, the performances of these traditional Bayesian methods are limited for complex heterogeneous cases due to strict assumptions.

The issue of model uncertainty becomes more prominent in deep learning-based methods because of the sheer complexity of the model and the amount of unknown parameters needed to be fitted. Despite the above mentioned progresses in GeoAI for spatial predictions, the discussions on uncertainty quantification remain scarce. This is partially due to the fact that most modern machine learning methods cannot adequately represent uncertainty. This has started to change following recent developments in Bayesian deep learning (BDL) (Gal, 2016; Gal and Ghahramani, 2016). BDL integrates the advances of Bayesian models and deep learning, and inherits the advantages of both sides: the former for uncertainty modeling and the latter for the representation of complex patterns. Many new methods and applications have been developed to take advantages of the integration; see Wang and Yeung (2020) for a recent review. Similarly, the integration of GeoAI methods and Bayesian framework might offer a new venue for geospatial uncertainty modeling.

7.5.2 TRANSFERABILITY AND GENERALIZATION

GeoAI model's transferability and generalization across space are weak due to the spatial heterogeneity Goodchild and Li (2021). To address this issue, geospatial knowledge-informed models have been developed and are generalizable for both natural and man-made features Li *et al.* (2021). Spatial-heterogeneity-aware deep learning architectures have shown promising results in spatial prediction tasks Xie *et al.*

(2021). GeoAI models should take the advantage of generalization capability across geographic scales from traditional spatial analysis and the automatic spatiotemporal feature extraction capability from deep neural networks. Further investigations are needed on the potential to perform cross-city, cross-modality, cross-event, and cross-scale spatial predictions with the help of transfer learning (Jean *et al.*, 2016) and physics-informed AI models (Raissi *et al.*, 2019).

7.5.3 INTERPRETABILITY AND EXPLAINABILITY

To open the AI model "black boxes", great efforts have been made in the AI community to increase the interpretability and explainability of deep learning models, such as the layer-wise relevance propagation (LRP) to assess the feature importance in classification tasks and the attention mechanisms to explain the relevant context in neural networks (Li *et al.*, 2021). Future development of GeoAI-based spatial prediction should keep interpretability and explainability in mind, especially how to incorporate spatial principles (e.g, spatial dependence) and geographic knowledge to advance explainable models (See also Chapter 9). One direction might be the incorporation of spatiotemporal-LRP and attention weights in assessing the relevance of geographic contexts in spatially explicit neural network models (Yan *et al.*, 2017). In addition, adding traditional diagnostic techniques from geostatistics such as distance-based and directional semivariograms into deep learning models, can help assess the impacts of spatiotemporal dependencies on interpolation tasks. Moreover, the inclusion of causal inference capabilities (Pearl, 2019), including association, intervention and counterfactuals from econometrics would further enhance the intelligence of spatial prediction and interpolation methods.

7.6 SUMMARY

Spatial prediction methods represent a set of tools for obtaining accurate data of geographic variables from limited observations. As an emerging subfield of GI-Science that uses artificial intelligence and machine learning techniques for geographic knowledge discovery, GeoAI offers a novel and bold perspective on revisiting and improving current spatial prediction and interpolation methods. In this chapter, the GeoAI motivations of spatial data representation, spatial structure measuring and the spatial relationship modeling throughout the workflow of spatial prediction are presented in the context of leveraging AI techniques. This chapter reviewed GeoAI for spatial prediction and interpolation methods, with a particular focus on two major fields: geostatistics and spatial regression. Challenges are discussed around uncertainty, transferability and interpretability. Readers are now directed to Cressie (2015) for geostatistics and Darmofal (2015) for spatial regression, which are regarded as more standard reference on the two subjects, respectively.

ACKNOWLEDGMENTS

Di Zhu would like to acknowledge the support from Faculty Interactive Research Program at the Center for Urban and Regional Affairs, University of Minnesota (1801-10964-21584-5672018). Guofeng Cao would like to acknowledge the support from National Science Foundation (BCS-2026331).

BIBLIOGRAPHY

Anselin, L., 1988. *Spatial Econometrics: Methods and Models*. vol. 4. Springer Science & Business Media.

Anselin, L., 1992. Space and applied econometrics: introduction. *Regional Science and Urban Economics*, 22 (3), 307–316.

Anselin, L., 2010. Thirty years of spatial econometrics. *Papers in Regional Science*, 89 (1), 3–25.

Anselin, L. and Bera, A.K., 1998. Spatial dependence in linear regression models with an introduction to spatial econometrics. *Statistics Textbooks and Monographs*, 155, 237–290.

Anselin, L. and Rey, S.J., 2014. *Modern Spatial Econometrics in Practice: A Guide to Geoda, Geodaspace and Pysal*. GeoDa Press LLC.

Arbia, G., 2014. *A primer for spatial econometrics with applications in r*. Springer.

Arjovsky, M., Chintala, S., and Bottou, L., 2017. Wasserstein Generative Adversarial Networks. *In*: *Proceedings of the 34th International Conference on Machine Learning*, July. PMLR, 214–223.

Bai, J., *et al.*, 2021. A3t-gcn: Attention temporal graph convolutional network for traffic forecasting. *ISPRS International Journal of Geo-Information*, 10 (7), 485.

Besag, J., 1974. Spatial interaction and the statistical analysis of lattice systems. *Journal of the Royal Statistical Society: Series B (Methodological)*, 36 (2), 192–225.

Bui, K.H.N., Cho, J., and Yi, H., 2021. Spatial-temporal graph neural network for traffic forecasting: An overview and open research issues. *Applied Intelligence*, 1–12.

Burrough, P.A., McDonnell, R.A., and Lloyd, C.D., 2015. *Principles of Geographical Information Systems*. Oxford University Press.

Caers, J., 2002. Multiple-point geostatistics : A quantitative vehicle for integrating geologic analogs into Stanford University, Stanford Center for Reservoir Forecasting. 1–24.

Cao, G., Kyriakidis, P.C., and Goodchild, M.F., 2011a. A multinomial logistic mixed model for the prediction of categorical spatial data. *International Journal of Geographical Information Science*.

Cao, G., Kyriakidis, P., and Goodchild, M., 2013. On spatial transition probabilities as continuity measures in categorical fields. *arXiv preprint arXiv:1312.5391*.

Cao, G., Kyriakidis, P.C., and Goodchild, M.F., 2011b. Combining spatial transition probabilities for stochastic simulation of categorical fields. *International Journal of Geographical Information Science*, 25 (11), 1773–1791.

Cao, G., Yoo, E.H., and Wang, S., 2014. A statistical framework of data fusion for spatial prediction of categorical variables. *Stochastic Environmental Research and Risk Assessment*, 28 (7), 1785–1799.

Carle, S.F. and Fogg, G.E., 1996. Transition probability-based indicator geostatistics. *Mathematical Geology*, 28 (4), 453–476.

Chen, L., *et al.*, 2019a. Quantifying the scale effect in geospatial big data using semi-variograms. *PloS One*, 14 (11), e0225139.

Chen, R., *et al.*, 2019b. A hybrid CNN-LSTM model for typhoon formation forecasting. *GeoInformatica*, 23 (3), 375–396.

Chiles, J.P. and Delfiner, P., 1999. *Geostatistics: Modeling Spatial Uncertainty*. vol. 136. Wiley-Interscience.

Chu, X., *et al.*, 2021. Twins: Revisiting the design of spatial attention in vision transformers. *Advances in Neural Information Processing Systems*, 34, 9355–9366.

Cressie, N., 2015. *Statistics for Spatial Data*. John Wiley & Sons.

Dai, Z., *et al.*, 2022. Geographically convolutional neural network weighted regression: a method for modeling spatially non-stationary relationships based on a global spatial proximity grid. *International Journal of Geographical Information Science*, 0 (0), 1–22.

Dale, M.R. and Fortin, M.J., 2014. *Spatial Analysis: A Guide for Ecologists*. Cambridge University Press.

Darmofal, D., 2015. *Spatial Analysis for the Social Sciences*. Cambridge University Press.

Defferrard, M., Bresson, X., and Vandergheynst, P., 2016. Convolutional neural networks on graphs with fast localized spectral filtering. *In*: *Advances in Neural Information Processing Systems*. 3844–3852.

Diggle, P.J., Tawn, J.A., and Moyeed, R.A., 1998. Model-based Geostatistics. *Applied Statistics*, 47 (3), 299–350.

Draxl, C., *et al.*, 2015a. *Overview and Meteorological Validation of the Wind Integration National Dataset Toolkit*. National Renewable Energy Lab.(NREL), Golden, CO (United States).

Draxl, C., *et al.*, 2015b. The wind integration national dataset (wind) toolkit. *Applied Energy*, 151, 355–366.

Du, Z., *et al.*, 2020. Geographically neural network weighted regression for the accurate estimation of spatial non-stationarity. *International Journal of Geographical Information Science*, 34 (7), 1353–1377.

Fischer, M.M., 1998. Computational neural networks: a new paradigm for spatial analysis. *Environment and Planning A*, 30 (10), 1873–1891.

Fischer, M.M. and Wang, J., 2011. *Spatial Data Analysis: Models, Methods and Techniques*. Springer Science & Business Media.

Fotheringham, A.S., Yang, W., and Kang, W., 2017. Multiscale geographically weighted regression (mgwr). *Annals of the American Association of Geographers*, 107 (6), 1247–1265.

Gal, Y., 2016. Uncertainty in Deep Learning. *PhD Thesis*.

Gal, Y. and Ghahramani, Z., 2016. Dropout as a Bayesian approximation: Representing model uncertainty in deep learning. *In*: *International Conference on Machine Learning*. 1050–1059.

Gao, S., 2020. A review of recent researches and reflections on geospatial artificial intelligence. *Geomatics and Information Science of Wuhan University*, 45 (12), 1865–1874.

Goodchild, M.F., 2004a. Giscience, geography, form, and process. *Annals of the Association of American Geographers*, 94 (4), 709–714.

Goodchild, M.F., 2004b. The validity and usefulness of laws in geographic information science and geography. *Annals of the Association of American Geographers*, 94 (2), 300–303.

Goodchild, M.F., 2008. Statistical perspectives on geographic information science. *Geographical Analysis*, 40 (3), 310–325.

Goodchild, M.F., 2020. How well do we really know the world? Uncertainty in GIScience. *Journal of Spatial Information Science*, (20), 97–102.

Goodchild, M.F. and Li, W., 2021. Replication across space and time must be weak in the social and environmental sciences. *Proceedings of the National Academy of Sciences*, 118 (35).

Goovaerts, P., 2005. Geostatistical analysis of disease data: estimation of cancer mortality risk from empirical frequencies using poisson kriging. *International Journal of Health Geographics*, 4 (1), 1–33.

Griffith, D.A. and Paelinck, J.H.P., 2011. *Non-standard Spatial Statistics and Spatial Econometrics*. Springer Science & Business Media.

Hu, S., *et al.*, 2021. Urban function classification at road segment level using taxi trajectory data: A graph convolutional neural network approach. *Computers, Environment and Urban Systems*, 87, 101619.

Janowicz, K., *et al.*, 2020. Geoai: spatially explicit artificial intelligence techniques for geographic knowledge discovery and beyond. *International Journal of Geographical Information Science*, 34 (4), 625–636.

Jean, N., *et al.*, 2016. Combining satellite imagery and machine learning to predict poverty. *Science*, 353 (6301), 790–794.

Karras, T., *et al.*, 2018. Progressive growing of gans for improved quality, stability, and variation. 26.

Kedron, P., *et al.*, 2021. Reproducibility and replicability: Opportunities and challenges for geospatial research. *International Journal of Geographical Information Science*, 35 (3), 427–445.

Kelejian, H.H. and Prucha, I.R., 1998. A generalized spatial two-stage least squares procedure for estimating a spatial autoregressive model with autoregressive disturbances. *The Journal of Real Estate Finance and Economics*, 17 (1), 99–121.

Kelejian, H.H. and Prucha, I.R., 1999. A generalized moments estimator for the autoregressive parameter in a spatial model. *International Economic Review*, 40 (2), 509–533.

Kelejian, H.H. and Prucha, I.R., 2007. The relative efficiencies of various predictors in spatial econometric models containing spatial lags. *Regional Science and Urban Economics*, 37 (3), 363–374.

King, J., Clifton, A., and Hodge, B.M., 2014. *Validation of Power Output for the Wind Toolkit*. National Renewable Energy Lab.(NREL), Golden, CO (United States).

Kipf, T.N. and Welling, M., 2017. Semi-supervised classification with graph convolutional networks. *In: International Conference on Learning Representations*.

Kitanidis, P.K. and Shen, K.F., 1996. Geostatistical interpolation of chemical concentration. *Advances in Water Resources*, 19 (6), 369–378.

Kleiber, W. and Nychka, D., 2012. Nonstationary modeling for multivariate spatial processes. *Journal of Multivariate Analysis*, 112, 76–91.

Klemmer, K. and Neill, D.B., 2021. Auxiliary-task learning for geographic data with autoregressive embeddings. *In*: *Proceedings of the 29th International Conference on Advances in Geographic Information Systems*. 141–144.

Kyriakidis, P.C., 2004. A geostatistical framework for area-to-point spatial interpolation. *Geographical Analysis*, 36, 259–289.

Lam, N.S.N., 1983. Spatial interpolation methods: a review. *The American Cartographer*, 10 (2), 129–150.

LeCun, Y., Bengio, Y., and Hinton, G., 2015. Deep learning. *Nature*, 521 (7553), 436–444.

Ledig, C., *et al.*, 2017. Photo-realistic single image super-resolution using a generative adversarial network. *In*: *Proceedings of the IEEE Conference on Computer Vision and Pattern Recognition*. 4681–4690.

Lehmann, A., Overton, J.M., and Leathwick, J.R., 2002. Grasp: generalized regression analysis and spatial prediction. *Ecological Modelling*, 157 (2-3), 189–207.

LeSage, J. and Pace, R.K., 2009. *Introduction to Spatial Econometrics*. Chapman and Hall/CRC.

LeSage, J.P., 1997. Regression analysis of spatial data. *Journal of Regional Analysis and Policy*, 27 (1100-2016-89650), 83–94.

LeSage, J.P. and Fischer, M.M., 2008. Spatial growth regressions: model specification, estimation and interpretation. *Spatial Economic Analysis*, 3 (3), 275–304.

Li, W. and Hsu, C.Y., 2020. Automated terrain feature identification from remote sensing imagery: a deep learning approach. *International Journal of Geographical Information Science*, 34 (4), 637–660.

Li, W., Hsu, C.Y., and Hu, M., 2021. Tobler's first law in geoai: A spatially explicit deep learning model for terrain feature detection under weak supervision. *Annals of the American Association of Geographers*, 111 (7), 1887–1905.

Lieberman-Cribbin, W., Draxl, C., and Clifton, A., 2014. *Guide to Using the Wind Toolkit Validation Code*. National Renewable Energy Lab.(NREL), Golden, CO (United States).

Lim, B., *et al.*, 2017. Enhanced deep residual networks for single image super-resolution. *In*: *Proceedings of the IEEE Conference on Computer Vision and Pattern Recognition Workshops*. 136–144.

Liu, P. and De Sabbata, S., 2021. A graph-based semi-supervised approach to classification learning in digital geographies. *Computers, Environment and Urban Systems*, 86, 101583.

Liu, Y., *et al.*, 2018. Improve ground-level pm2. 5 concentration mapping using a random forests-based geostatistical approach. *Environmental Pollution*, 235, 272–282.

Lu, B., *et al.*, 2014. Geographically weighted regression with a non-euclidean distance metric: a case study using hedonic house price data. *International Journal of Geographical Information Science*, 28 (4), 660–681.

Mennis, J. and Guo, D., 2009. Spatial data mining and geographic knowledge discovery—an introduction. *Computers, Environment and Urban Systems*, 33 (6), 403–408.

Miller, H.J., 2004. Tobler's first law and spatial analysis. *Annals of the association of American geographers*, 94 (2), 284–289.

Oliver, M.A. and Webster, R., 1990. Kriging: a method of interpolation for geographical information systems. *International Journal of Geographical Information System*, 4 (3), 313–332.

Ord, K., 1975. Estimation methods for models of spatial interaction. *Journal of the American Statistical Association*, 70 (349), 120–126.

Paciorek, C.J. and Schervish, M.J., 2006. Spatial Modelling Using a New Class of Nonstationary Covariance Functions. *Environmetrics (London, Ont.)*, 17 (5), 483–506.

Paelinck, J.H., *et al.*, 1979. *Spatial Econometrics*. vol. 1. Saxon House.

Parks, B.O., Steyaert, L.T., and Goodchild, M.F., 1993. *Environmental Modeling with gis*. Oxford University Press.

Pearl, J., 2019. The seven tools of causal inference, with reflections on machine learning. *Communications of the ACM*, 62 (3), 54–60.

Raissi, M., Perdikaris, P., and Karniadakis, G.E., 2019. Physics-informed neural networks: A deep learning framework for solving forward and inverse problems involving nonlinear partial differential equations. *Journal of Computational Physics*, 378, 686–707.

Reichstein, M., *et al.*, 2019. Deep learning and process understanding for data-driven earth system science. *Nature*, 566 (7743), 195–204.

Ripley, B., 1981. *Spatial Statistics*. New York: John Wiley & Sons.

Scheider, S. and Huisjes, M.D., 2019. Distinguishing extensive and intensive properties for meaningful geocomputation and mapping. *International Journal of Geographical Information Science*, 33 (1), 28–54.

Sengupta, M., *et al.*, 2018. The national solar radiation data base (nsrdb). *Renewable and Sustainable Energy Reviews*, 89, 51–60.

Shepard, D., 1968. A two-dimensional interpolation function for irregularly-spaced data. *In: Proceedings of the 1968 23rd ACM National Conference*. 517–524.

Solow, A.R., 1986. Mapping by simple indicator kriging. *Mathematical Geology*, 18 (3), 335–352.

Strebelle, S., 2002. Conditional simulation of complex geological structures using multiple-point statistics. *Mathematical Geology*, 34 (1), 1–21.

Tobler, W., 2004. On the first law of geography: A reply. *Annals of the Association of American Geographers*, 94 (2), 304–310.

Tobler, W.R., 1970. A computer movie simulating urban growth in the detroit region. *Economic Geography*, 46 (sup1), 234–240.

Ulyanov, D., Vedaldi, A., and Lempitsky, V., 2018. Deep Image Prior. *In: Proceedings of the IEEE Computer Society Conference on Computer Vision and Pattern Recognition*.

van den Boogaart, K.G. and Schaeben, H., 2002. Kriging of regionalized directions, axes, and orientations i. directions and axes. *Mathematical Geology*, 34 (5), 479–503.

Van Westen, C.J., Castellanos, E., and Kuriakose, S.L., 2008. Spatial data for landslide susceptibility, hazard, and vulnerability assessment: An overview. *Engineering Geology*, 102 (3-4), 112–131.

Wang, H. and Yeung, D.Y., 2020. A Survey on Bayesian Deep Learning. *ACM Computing Surveys*, 53 (5), 108:1–108:37.

Wang, Q., Shi, W., and Atkinson, P.M., 2016. Area-to-point regression kriging for pansharpening. *ISPRS Journal of Photogrammetry and Remote Sensing*, 114, 151–165.

Wang, X., *et al.*, 2018. Esrgan: Enhanced super-resolution generative adversarial networks. *In: Proceedings of the European Conference on Computer Vision (ECCV) Workshops*. 0–0.

Whittle, P., 1954. On stationary processes in the plane. *Biometrika*, 434–449.

Wu, S., *et al.*, 2021a. Geographically and temporally neural network weighted regression for modeling spatiotemporal non-stationary relationships. *International Journal of Geographical Information Science*, 35 (3), 582–608.

Wu, Y., *et al.*, 2021b. Inductive graph neural networks for spatiotemporal kriging. *In: Proceedings of the AAAI Conference on Artificial Intelligence*. vol. 35, 4478–4485.

Xiao, L., *et al.*, 2021. Predicting vibrancy of metro station areas considering spatial relationships through graph convolutional neural networks: The case of Shenzhen, China. *Environment and Planning B: Urban Analytics and City Science*, 48 (8), 2363–2384.

Xie, Y., *et al.*, 2021. A statistically-guided deep network transformation and moderation framework for data with spatial heterogeneity. *In: 2021 IEEE International Conference on Data Mining (ICDM)*. IEEE, 767–776.

Yan, B., *et al.*, 2017. From ITDL to Place2Vec: Reasoning about place type similarity and relatedness by learning embeddings from augmented spatial contexts. *In: Proceedings of the 25th ACM SIGSPATIAL International Conference on Advances in Geographic Information Systems*. 1–10.

Yan, X., *et al.*, 2019. A graph convolutional neural network for classification of building patterns using spatial vector data. *ISPRS Journal of Photogrammetry and Remote Sensing*, 150, 259–273.

Zhang, H., *et al.*, 2019. Self-attention generative adversarial networks. *In: International Conference on Machine Learning*. PMLR, 7354–7363.

Zhang, Y., *et al.*, 2020. A novel residual graph convolution deep learning model for short-term network-based traffic forecasting. *International Journal of Geographical Information Science*, 34 (5), 969–995.

Zhang, Y. and Yu, W., 2022. Comparison of dem super-resolution methods based on interpolation and neural networks. *Sensors*, 22 (3), 745.

Zhang, Y., Yu, W., and Zhu, D., 2022. Terrain feature-aware deep learning network for digital elevation model superresolution. *ISPRS Journal of Photogrammetry and Remote Sensing*, 189, 143–162.

Zhao, L., *et al.*, 2019. T-GCN: A temporal graph convolutional network for traffic prediction. *IEEE Transactions on Intelligent Transportation Systems*.

Zhu, A.X., *et al.*, 2018. Spatial prediction based on third law of geography. *Annals of GIS*, 24 (4), 225–240.

Zhu, D., *et al.*, 2020a. Spatial interpolation using conditional generative adversarial neural networks. *International Journal of Geographical Information Science*, 34 (4), 735–758.

Zhu, D., Gao, S., and Cao, G., 2022. Towards the intelligent era of spatial analysis and modeling. *In: Proceedings of the 5th ACM SIGSPATIAL International Workshop on AI for Geographic Knowledge Discovery.* 10–13.

Zhu, D., *et al.*, 2021. Spatial regression graph convolutional neural networks: A deep learning paradigm for spatial multivariate distributions. *GeoInformatica*, 1–32.

Zhu, D., *et al.*, 2020b. Understanding place characteristics in geographic contexts through graph convolutional neural networks. *Annals of the American Association of Geographers*, 110 (2), 408–420.

8 Heterogeneity-Aware Deep Learning in Space: Performance and Fairness

Yiqun Xie
University of Maryland, College Park

Xiaowei Jia
University of Pittsburgh

Weiye Chen
University of Maryland, College Park

Erhu He
University of Pittsburgh

CONTENTS

8.1 Introduction 152
8.2 Key Concepts 153
8.3 Problem Definitions 154
    8.3.1 Performance-Driven Problem Statement 154
    8.3.2 Fairness-Driven Problem Statement 155
8.4 Prior Works and Limitations on Heterogeneity-Aware Learning 156
8.5 Performance-Driven Design: Multi-Model Heterogeneity-Aware Learning. 157
    8.5.1 Summary of Key Ideas 157
    8.5.2 Representation: Hierarchical multi-task learning 157
    8.5.3 Statistically-Guided Transformation 159
    8.5.4 Spatial Moderation 161
    8.5.5 Experiments 161
8.6 Fairness-Driven Design: Single-Model Heterogeneity-Aware Learning 164
    8.6.1 The MAUP Dilemma in Spatial Fairness 164
    8.6.2 Summary of Key Ideas 165
    8.6.3 Space as a Distribution of Partitionings 165
    8.6.4 SPAD-based Stochastic Training 167
    8.6.5 Bi-level Fairness Enforcement 168
    8.6.6 Experiments 169
8.7 New Directions 172

DOI: 10.1201/9781003308423-8

8.8 Conclusions 173
     Bibliography 173

8.1 INTRODUCTION

Spatial data have been collected at massive scales with an ever-growing variety. Common examples of spatial data are GPS locations or trajectories, maps of crime incidents or disease cases (e.g., COVID-19), tweets with geo-tags, satellite imagery, drone imagery, geo-located videos form autonomous vehicles, LiDAR point clouds, and many more. These datasets have become foundations across major societal domains, including transportation, public health, public safety, climate, etc. The main distinction between spatial data and non-spatial data, as the name suggests, is the inclusion of the spatial context, which is often represented by geo-coordinates such as longitudes and latitudes. In general, spatial coordinates can be defined in any space (e.g., human bodies). Our discussion focuses on the geospatial coordinate, which is generalizable to any space where the spatial relationships between data points matter.

Challenges and research gaps introduced by spatial data are no stranger to the general computing fields. For example, last decades have witnessed disruptions and, more importantly, transformations created by spatial data to traditional database techniques, where new data types (e.g., geometry objects), indexing (e.g., the R-tree family) and query optimizations have been developed to bridge the gaps. These innovations later became building blocks in contemporary infrastructures (e.g., Oracle Spatial, PostGIS).

With the recent fast advances in machine learning and deep learning, two other intrinsic properties of spatial data have raised attention: spatial autocorrelation and heterogeneity. Similarly, these properties violate the common independent and identical distribution (i.i.d.) assumption (Atluri *et al.*, 2018; Goodchild and Li, 2021; Shekhar *et al.*, 2015) underlying most learning models. Spatial autocorrelation indicates that data samples are not independent (e.g., traffic, temperature, crop types), and the similarity between samples is often a function of their distance (e.g., intrinsic stationarity and variograms). Spatial heterogeneity undermines the identical distribution assumption, where data distribution $p(\mathbf{y}|\mathbf{X})$ or generation process $f : \mathbf{X} \rightarrow \mathbf{y}$ is non-stationary over space. In other words, samples with similar features $\mathbf{X}$ may correspond to different $\mathbf{y}$ depending on the locations. For example, our observations of real-world events or phenomena are often incomplete (e.g., many physical or social variables are unavailable or difficult to observe), but the unobservable variables are almost certainly not constants over space (Goodchild and Li, 2021). This makes spatial problems theoretically inapproximable by a single model with a single set of parameters. This also applies to deep neural networks, which although are universal approximators, cannot approximate different functions at the same time (e.g., two simple functions $y = x$ and $y = -x$, where the same x can lead to different results at different locations). Moreover, spatial coordinates should not be naively added as input features to machine/deep learning algorithms, which may otherwise lead to overfitting and limit model generalizability (Jiang *et al.*, 2017; Mai *et al.*, 2022; Shekhar *et al.*, 2015; Shekhar and Xiong, 2007).

Explicit consideration of these properties of spatial data has the revolutionizing potential, and the earlier rise of convolutional neural networks (CNNs) shows a motivating example. A key game-changer in CNNs is the use of convolutional layers, which – instead of connecting every pixel to every other pixel – only focus on nearby local pixels. This transition from full to local connections exactly mirrors the First Law of Geography: "Everything is connected to everything else (fully-connected layers), but nearby things are more relevant than distant things (convolutional layers)." However, spatial heterogeneity remains largely understudied, adding a layer of unpredictable and uncontrolled risks. If left unattended, resulting "hit or miss" solutions can be especially problematic for large-scale applications. This paper focuses on heterogeneity-aware deep learning for spatial data, and will cover recent advances in this domain with respect to both prediction performance (Xie *et al.*, 2021a) and fairness (Xie *et al.*, 2022b).

8.2   KEY CONCEPTS

As there are various definitions of heterogeneity in different contexts, we use the following concepts to formally define spatial heterogeneity in the machine learning context.

Definition 8.2.1 (Data generation process). Given input features $\mathbf{X}$ and labels $\mathbf{y}$ for supervised learning, a data generation process Φ can be considered as the true function $\Phi(\mathbf{X}) = \mathbf{y}$ or $\Phi : \mathbf{X} \to \mathbf{y}$.

Definition 8.2.2 (Heterogeneity). In this paper, a dataset $D = (\mathbf{X}, \mathbf{y})$ is considered heterogeneous if its data samples belong to multiple different data generation processes. This can be formally defined as:

$$\exists (\mathbf{X}', \mathbf{y}') \subset D \text{ and } (\mathbf{X}'', \mathbf{y}'') \subset D,$$
$$\text{such that } (\Phi' : \mathbf{X}' \to \mathbf{y}') \neq (\Phi'' : \mathbf{X}'' \to \mathbf{y}'')$$

where $(\mathbf{X}', \mathbf{y}') \cap (\mathbf{X}'', \mathbf{y}'') = \varnothing$. When heterogeneity is present in the data, we cannot use a single model or a single set of parameters to approximate the true functions.

Definition 8.2.3 (Spatial heterogeneity). Spatial heterogeneity is a fundamental property of spatial data, implying that the true functions $\Phi : \mathbf{X} \to \mathbf{y}$ are different for data in different locations or geographic regions. Denote S as the study area, and $S' \subset S$ and $S'' \subset S$ as its two non-overlapping sub-regions. Further denote Φ' and Φ'' as the true functions for data samples in S' and S'', respectively. Spatial heterogeneity can be then defined as a special case of heterogeneity:

$$\exists S' \text{ and } S'', \text{ such that } \Phi' \neq \Phi''$$

Example visualizations of spatially heterogeneous data are available in Sec. 8.3.1.

Definition 8.2.4. Partition p vs. Partitioning P. A partitioning P splits an input spatial domain S into m individual partitions p_i, i.e., $P = \{p_1, ..., p_i, ..., p_m\}$.

8.3 PROBLEM DEFINITIONS

In the following, we formally define the problems of performance-driven and fairness-driven learning under spatial heterogeneity (or distribution shift in space).

8.3.1 PERFORMANCE-DRIVEN PROBLEM STATEMENT

Performance-driven learning focuses on the overall prediction quality (e.g., measured by F1 scores for classification problems) of a deep learning model. Performance can also refer to computational performance (i.e., efficiency and scalability), which is out of scope for this paper. The following statement specifies the inputs, outputs and objectives:

Inputs:

- Features $\mathbf{X}$ and labels $\mathbf{y}$ in a spatial domain S;

- Spatial locations $\mathbf{L}$ of data samples;

- A user-selected deep learning model F (e.g., a 10-layer convolutional network);

- A significance level α (used to test if two data subsets are generated by the same process Φ);

Outputs:

- A space-partitioning $P = \{p_1, p_2, ...\}$ of S, where data in each spatial partition $p_i \in P$ belong to the same data generation process Φ_i;

- A spatialized deep network F_P of the input F based on P, which learns different parameters Θ_i for each partition $p_i \in P$.

Objectives:

- Maximizing overall/global prediction quality (e.g., F1-scores for classification, RMSE for regression, etc.).

Figure 8.1 shows an example of the inputs and outputs of the performance-driven version of the problem.

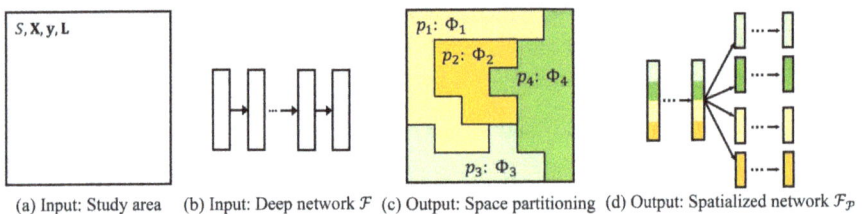

(a) Input: Study area (b) Input: Deep network $\mathcal{F}$ (c) Output: Space partitioning (d) Output: Spatialized network $\mathcal{F}_P$

Figure 8.1 Illustrations of inputs and outputs of performance-driven problems.

8.3.2 FAIRNESS-DRIVEN PROBLEM STATEMENT

Fairness-driven learning focuses on prediction quality parity over locations, where the goal is to reduce the variation of prediction quality (e.g., accuracy) in different sub-regions of the study area. Here we show a simplified version of the problem for illustration purposes with a single spatial partitioning P (Def. 8.2.4), and will introduce the complete version together with the methodology in Sec. 8.6.

Inputs:

- Features $\mathbf{X}$ and labels $\mathbf{y}$ in a spatial domain S;

- Spatial locations $\mathbf{L}$ of data samples;

- A user-selected deep learning model F (e.g., a 10-layer convolutional network);

- A prediction quality metric (e.g., F1-score);

- A space-partitioning P of S.

Outputs:

- Parameters Θ of the input deep network F (in this formulation Θ are the same for data samples at all locations in S).

Objectives:

- Fairness: Minimizing variation (e.g., variance) of prediction quality on data samples in different spatial partitions $p_i \in P$ (i.e., prediction quality parity).

Constraints:

- Prediction quality: Keeping the overall/global prediction quality (e.g., F1-scores for classification, RMSE for regression, etc.) similar to an F trained without any fairness objective.

There are two considerations to interpret this formulation:

(1) **The use of a single model for all locations:** In this formulation where location-based fairness is considered, we do not use different parameters for different locations because we do not want to discriminate data points by their locations. Analogically, in fairness-aware learning that concerns multiple genders or races, the group information (e.g., gender) is often considered as a sensitive attribute and a model is not allowed to bias the prediction based on that (Du *et al.*, 2021, 2020; Kilbertus *et al.*, 2018; Morales *et al.*, 2020). Thus, we operate under this constraint in this specific formulation.

(2) **Prediction quality vs. fairness:** Deep learning methods tend to have a huge number of parameters (e.g., millions, billions). This can lead to a large number of performance-wise symmetric solutions, i.e., many trained models with different parameters/weights Θ can yield very similar overall prediction quality. This leads to a large "degree of freedom" for the fairness of the model. This formulation aims to identify the fairest solution among these symmetric solutions. Figure 8.2 shows an example of spatial fairness.

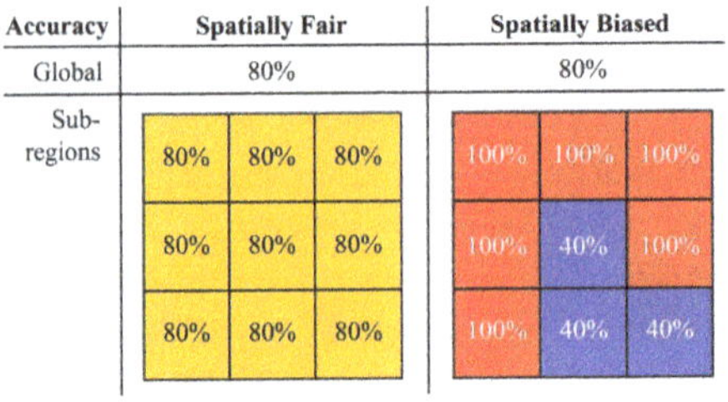

Accuracy	Spatially Fair			Spatially Biased		
Global	80%			80%		
Sub-regions	80%	80%	80%	100%	100%	100%
	80%	80%	80%	100%	40%	100%
	80%	80%	80%	100%	40%	40%

Figure 8.2   An example of spatial fairness.

8.4   PRIOR WORKS AND LIMITATIONS ON HETEROGENEITY-AWARE LEARNING

The wide adoption of convolutional kernels (He *et al.*, 2016; Krizhevsky *et al.*, 2012) in deep learning architectures have – to some degree – explicitly filled the missing representation to capture spatial autocorrelation (e.g., local connections and maintained spatial relationships between cells). However, the complex spatial heterogeneity challenge has not been sufficiently addressed. In a recent study, a spatial-variability aware neural network (SVANN) approach was developed (Gupta *et al.*, 2020, 2021), which requires the spatial footprints of heterogeneous processes to be known as an input. HeteroConvLSTM also uses a fixed space-partitioning (e.g., a grid) to learn local models and then combine results from nearby models during prediction (Yuan *et al.*, 2018). Similarly, a spatial incomplete multi-task deep learning (SIMDA) (Gao *et al.*, 2019) uses predefined partitionings to create local tasks, with additional autocorrelation losses to penalize nearby models with highly different parameters. These models require known spatial partitions as inputs where all samples in each partition follow the same process $\Phi : \mathbf{X} \rightarrow \mathbf{y}$, but such partitions are often unavailable in real applications. In addition, they require sufficient training samples for each unit partition. However, training data often have a skewed distribution over space and are limited in many sub-regions. Explicit spatial ensemble approaches aim to adaptively partition a dataset (Jiang *et al.*, 2019, 2017), but the algorithm and its variation are designed for two-class classification problems for two partitions; both training and prediction are performed separately for each partition. Outside recent literature on deep learning, a traditional approach for handling spatial heterogeneity is geographically-weighted regression (GWR) (Brunsdon *et al.*, 1999; Fotheringham *et al.*, 2017). However, GWR is mainly designed for inference and linear regression, and cannot handle complex prediction tasks commonly addressed by deep learning. Most existing methods also require dense training data across space to train models for individual partitions or locations. Finally, they cannot be applied to other regions outside the spatial extent of the training data.

More broadly in machine learning, transfer learning (Kaya *et al.*, 2019; Liu *et al.*, 2019; Ma *et al.*, 2019; Pan and Yang, 2009), domain adaptation (Ben-David *et al.*, 2007; Pan *et al.*, 2010; Wang and Deng, 2018), and meta learning (Finn *et al.*, 2017; Vilalta and Drissi, 2002) are also closely related topics to the heterogeneity problem.

They focus on generalizing models from one or multiple source domains to a new target domain with limited training labels. One strategy is to first learn a common feature representation, for example by auto-encoder (Finn *et al.*, 2016; Jean *et al.*, 2019) or contrastive learning (Chen *et al.*, 2020; Wei *et al.*, 2021), and then retrain a simple feature classifier for a new target domain. However, these methods assume the task domains are predefined and always available as inputs, which is not true for spatial data (e.g., Earth observation data) due to the unknown footprints of non-stationary processes. In addition, these methods' performance can degrade when data distributions in the spatial domain have large discrepancies.

8.5 PERFORMANCE-DRIVEN DESIGN: MULTI-MODEL HETEROGENEITY-AWARE LEARNING

This section discusses a model-agnostic Spatial Transformation And modeRation (STAR) framework for heterogeneity-aware learning in space (Xie *et al.*, 2021a, 2022a).

8.5.1 SUMMARY OF KEY IDEAS

The key ideas of the following contributions are listed in the following. We also provide a visual illustration of these ideas in Figure 8.3.

- We propose a spatial transformation approach to capture arbitrarily-shaped foot-prints of spatial heterogeneity at multiple scales during deep network training, and synchronously transform the network into a new "spatialized" architecture. The transformation is guided by a dynamic and learning-engaged generalization of multivariate scan statistic;

- We propose a spatial moderator to generalize the learned spatial patterns and transformed network architecture from the original region to new test regions;

- We implement the model-agnostic STAR framework using both snapshot and time-series based input network architectures (i.e., DNN, LSTM and LSTM-attention), and present the statistically guided transformation module for both classification and regression tasks.

8.5.2 REPRESENTATION: HIERARCHICAL MULTI-TASK LEARNING

As discussed in Sec. 8.5.1, heterogeneity in space is automatically captured using a hierarchical bi-partitioning process, which continues until no heterogeneity is detected in a child partition. To clarify notations, we formally define this hierarchy as follows:

Definition 8.5.1. Spatial hierarchy of processes H: A hierarchical representation of spatial heterogeneity (Xie *et al.*, 2021b). H uses a tree to represent the input spatial domain D; each node $H_j^i \in H$ is a partition of D, where i is the level in the hierarchy,

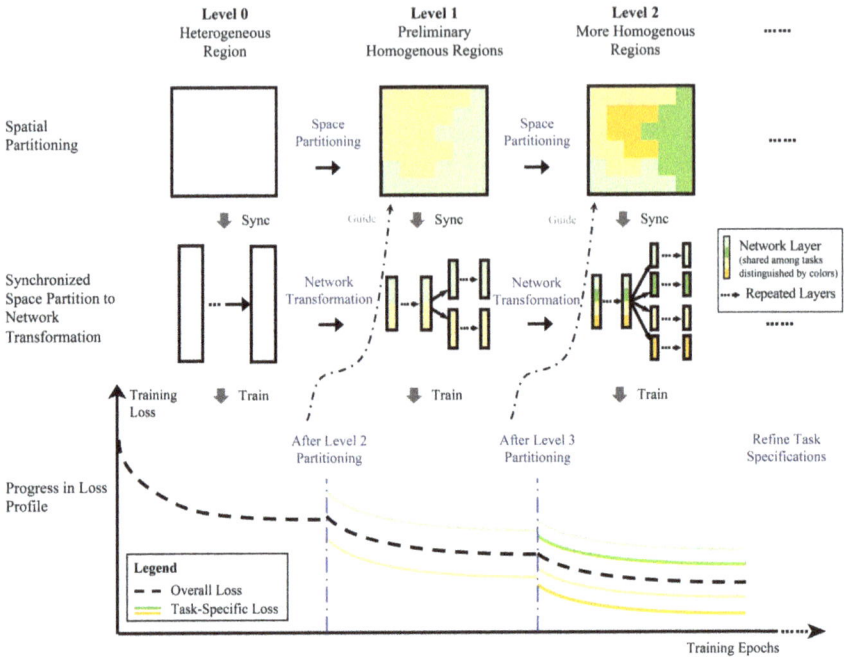

Figure 8.3 An illustration of key ideas in STAR. The three rows represent three aspects of the heterogeneity-aware learning process, where the first row finds the optimal space-partitioning to separate out heterogeneous processes, the second row synchronizes the partitioning into the deep network architecture by creating new branches, and finally the third row shows its impact on loss reduction.

and j is the unique ID for each partition at level-i. Here we use $H_j^i \in H$ instead of $p \in P$ (Def. 8.2.4) to more clearly denote a partition in the context of the spatial hierarchy H. Children of a partition H_j^i share the same lower-level processes (i.e., $\{\Phi\}$ at levels $i' < i$). Φ is homogeneous within a leaf-node and heterogeneous across leaf-nodes.

This definition of spatial hierarchy provides a natural way to represent both (1) the partitioning of data in the geographic space, and (2) the grouping of network layers in the deep learning model.

To illustrate this, Figure 8.4 (c) shows an example of spatial hierarchy H, where each node $H_j^i \in H$ can be considered as a spatial region with a spatial process Φ_j^i; here i is the level in the hierarchy and j is a unique ID of a node at this level. Based on this hierarchical representation of spatial partitions, Figure 8.4 (d) shows the deep network representation that synchronizes the structure of H, where each unique path from the input to output has the same architecture as the input deep network F. Using this representation, model parameters at each layer are shared by all leaf nodes branched out from the layer. This means nodes that share more common parent nodes

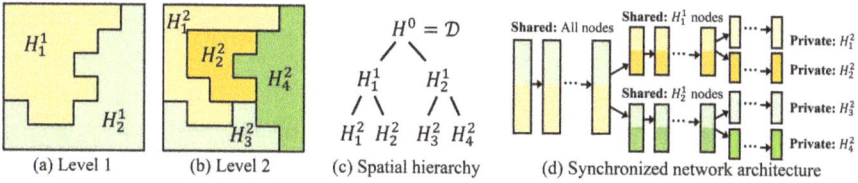

Figure 8.4 Spatial processes, hierarchy and network architecture with hierarchical weight-sharing.

in the spatial hierarchy H also share more common weights. Another intuitive interpretation is that spatial partitions that share the same parent H_j^i inherit the same higher level spatial process Φ_j^i. The learning at each leaf-node can be considered as a task in this multi-task learning context.

For the hierarchy-network synchronization (Figure 8.4), a final detail is the selection of the layer after which the following layers will be split into two parallel branches with private weights. We use an optional parameter β ($\beta \leqslant 1$; default to $1/2$) to control the proportion of the layers to keep as common layers among two child-branches (Xie *et al.*, 2021b).

8.5.3 STATISTICALLY-GUIDED TRANSFORMATION

Based on Def. 8.5.1, we separate data samples belonging to different spatial processes Φ using a hierarchical structure. Specifically, at each step (Figure 8.5), we identify an optimal space-bi-partitioning that maximizes the discrepancy of Φ between the partitions, and verify if the impact of separation is statistically significant for learning enhancement. The partitioning continues hierarchically until no significant heterogeneity can be recognized in new partitions. This transformation framework is a Dynamic and Learning-engaged generalization of the Multivariate Scan Statistic (DL-MSS) (Kulldorff *et al.*, 2007; Neill *et al.*, 2013; Xie *et al.*, 2022c). In the following we will discuss the three major components of DL-MSS.

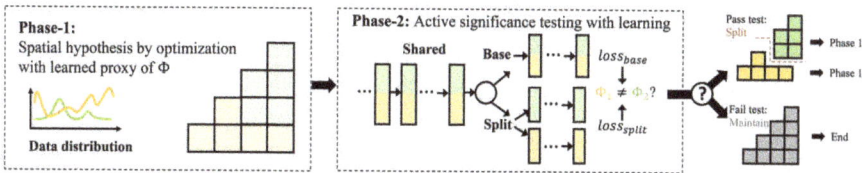

Figure 8.5 An illustrative example of the spatial transformation framework.

8.5.3.1 Space-Partitioning Optimization with Prediction Error Distribution

As spatial processes $\Phi : \mathbf{X} \to \mathbf{y}$ are not directly observable, we leverage the function approximation power of a deep network F and use its prediction error distribution for a partition $H_j^i \in H$ as a proxy of Φ_j^i. Intuitively, if all data belonging to a partition H_j^i are generated by a homogeneous Φ_j^i, we expect the prediction error made by a single model (in H_j^i) to follow a homogeneous distribution over space for each class; otherwise, the spatial process Φ_j^i is heterogeneous . To obtain statistics on error rates, locations are first aggregated into unit local groups using a grid, where all samples in a grid cell form a group; a user may also choose a different grouping strategy. Using classification as an example, denote $\hat{y}_{k,m}$ as the predicted labels for samples with class m (i.e., true labels are m) at a cell s_k. The number of misclassified samples of class m at s_k is then $e_{k,m} = |\hat{y}_{k,m} \neq m|$. Further, denote $n_{k,m}$ as the number of samples of class m at location s_k; and E_m and N_m as the number of misclassified and all samples of class m in the entire space. We identify an arbitrary set of cells $S = \{s_k\}$ that maximizes the error rate distribution discrepancy using Poisson-based likelihood ratio (Neill *et al.*, 2013):

$$
\begin{aligned}
S^* &= \arg\max_{S} \frac{\text{Likelihood}(H_1, S)}{\text{Likelihood}(H_0)} \\
&= \arg\max_{S} \prod_{s_k \in S} \prod_{m=1}^{m_{max}} \frac{\Pr(e_{k,m} \sim \text{Poisson}(q_m E(e_{k,m})))}{\Pr(e_{k,m} \sim \text{Poisson}(E(e_{k,m})))}
\end{aligned}
\tag{8.1}
$$

where the null hypothesis H_0 states that Φ_j^i is homogeneous, and H_1 states that there exists a set S where the rate of having an error is q_m times the expected rate under H_0; and $E(e_{k,m}) = E_m \cdot \frac{n_{k,m}}{N_m}$ is the expected number of misclassified samples at location s_k under H_0. The optimal set S^* can be efficiently solved using the Linear-Time Subset Scanning (LTSS) property (Neill, 2012; Xie *et al.*, 2022c) combined with coordinate ascent. More solution details and results on regression are available in (Xie *et al.*, 2021a).

8.5.3.2 Active Significance Testing with Learning

Once the optimal S^* is identified, the current node H_j^i will be temporarily split into two children H_{j1}^{i+1} and H_{j2}^{i+1}, where one child corresponds to S^* and the other for the rest of the space in H_j^i. To validate if the space-partitioning can lead to a statistically significant improvement on learning, we drop the traditional Monte-Carlo based descriptive test (Xie *et al.*, 2022c) and use a learning-engaged active test. Specifically, we carry out two training scenarios with and without the split (Figure 8.5) and perform an upper-tailed dependent T-test on the losses from the two scenarios (Xie *et al.*, 2021a). For the split scenario a new network branch will be created (e.g., adding a copy of the last L layers to the network) to allow private parameters (Figure 8.4(d)). If the impact of S^* is significant, we approve the new partitioning and network branch; otherwise, the transformation on H_j^i is terminated.

8.5.3.3   Spatial Transformation via a Dynamic and Learning-Engaged Hierarchy H

As both the space-partitioning and network parameters may be constantly updated during training, DL-MSS performs the first two key components as sub-routines for new partitions added to H to dynamically converge to the final H and heterogeneity-aware network F_H.

8.5.4   SPATIAL MODERATION

The spatial hierarchy H and "spatialized" deep network F_H learned and trained from the transformation step aim to capture heterogeneity in the spatial extent of the input $\mathbf{X}$ and $\mathbf{y}$. However, the partitions cannot be directly applied to a new spatial region. To bridge this gap, we propose a spatial moderator, which translates the learned network branches in F_H to prediction tasks in a

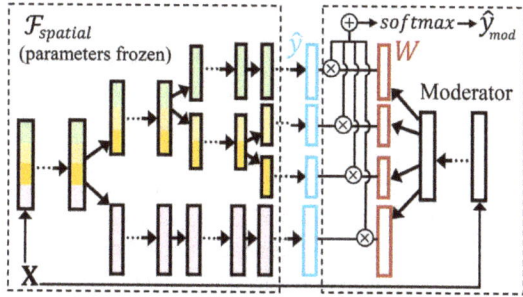

Figure 8.6   An example of the spatial moderator.

new region. The key idea of the spatial moderator is to learn and predict a weight matrix $\mathbf{W}$ for all branches in F_H (corresponding to all leaf-nodes in the spatial hierarchy H), and then use the weights to ensemble the branches' predictions to get the final result (Figure 8.6).

8.5.5   EXPERIMENTS

8.5.5.1   Datasets

California land-cover classification: We use multi-spectral data from Sentinel-2 satellites in two regions in Central Valley, California. Each region has a size of 4096×4096 (~ 6711 km^2 in 20m resolution). We first learn the spatial partitioning using the data from Region D_A and then use the moderator to transfer it to Region D_B. We use composite image series from May to October in 2018 (2 images/month) for time-series models, and one snapshot from August, 2018 for snapshot-based models. The labels are from the USDA Crop Data Layer (CDL). The training (and validation) set has 20% data at sampled locations in D_A, and 1% data in D_B is used for fine-tuning.

Boston COVID-19 human mobility prediction: Human mobility provides critical information to COVID-19 transmission dynamics models. We use the Boston COVID-19 mobility dataset from (Bao *et al.*, 2020), which includes data from US census, CDC COVID statistics, and SafeGraph mobility patterns. Here human mobility $\mathbf{y}$ is represented by the number of visits to points-of-interest (POIs; e.g., grocery stores, restaurants) and the counting is based on smartphone trajectories. The features include population, weekly COVID-19 cases and deaths, number of POIs, week and income. The dataset contains 12 weeks of data. we use the first 11 weeks for training/validation and the final week for testing.

8.5.5.2 Candidate Methods

For California land-cover classification, we have 12 candidate methods: (1) base F, spatially transformed F_H, and F_M (F_H with moderator), each for three base architectures: DNN (snapshot-based and fully-connected), LSTM and LSTM+Attention (Jia *et al.*, 2019); and (2) F integrated with the model-agnostic meta-learning (denoted as *meta*) (Finn *et al.*, 2017; Yao *et al.*, 2019) for fast adaptation in region D_B with the available 1% of data (D_A's data are clustered into tasks), for the three architectures.

For Boston COVID-19 human mobility regression, we use 3 examples of candidate methods for comparison, i.e., COVID-GAN (Bao *et al.*, 2020), DNN, and DNN$_H$ (spatially transformed version). As time-series is not used to construct additional features in (Bao *et al.*, 2020) and some features are aggregated to week-levels, we follow the same strategy and only use week IDs as features, which also allows a more direct comparison with COVID-GAN. In addition, as training and test samples are from the same set of spatial locations (timestamps are different), we directly use spatial transformation and skipped the moderator in this comparison.

8.5.5.3 Results

Land-cover classification: As shown in Figure 8.7, the "spatialized" network architectures overall achieved the highest F1-scores for different types of base models in both regions. For region D_A, the general trend is that the results of a base model F gradually improve with the addition of spatial transformation F_H, and the spatial moderator F_M. In addition, Figure 8.8(a) shows the hierarchical process of space-partitioning during spatial transformation for the first 3 levels. In the first level (largest scale), for example, H_1^1 is a mix of urban and suburban areas, whereas H_2^1 contains more rural and mountainous areas. Note that some partitions (e.g., H_3^2) are not further split, as determined by significance testing. We also visualize the loss reduction profile with the heterogeneity-awareness during the training in Figure 8.9. As we can see, the training achieved further progress with each optimized data partitioning by breaking the conflicts in $\Phi : \mathbf{X} \to \mathbf{y}$ between data samples, which otherwise causes the loss to stuck at a plateau. Finally, Figure 8.8(b) visualizes the weights

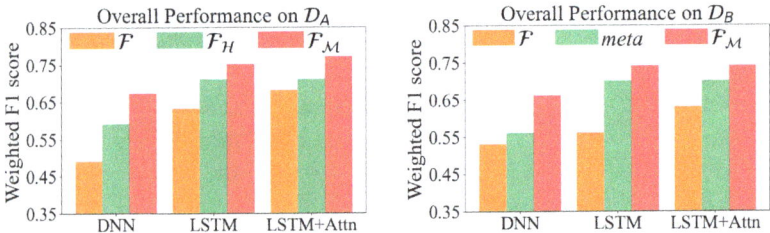

Figure 8.7 Weighted F1 scores for regions D_A and D_B.

predicted by the moderator for two example network branches in F_H for DNN for all locations in region D_B. For each branch, the weight is averaged over all classes in the predicted $\mathbf{W}$ at each location. As we can see, in the new region D_B, branch-2 is given higher weights for the left-side of the region, which is a mountainous area, whereas the weights for branch-7 shows the opposite spatial pattern.

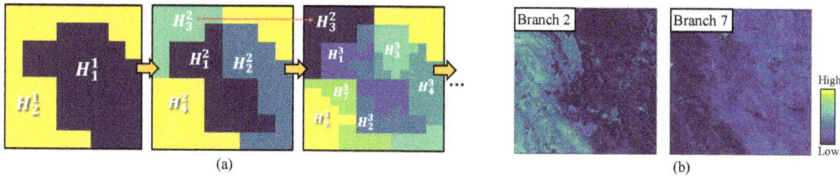

Figure 8.8 (a) Spatial hierarchy learned in region D_A (first 3 levels); (b) Learned branch weights across space (across samples). Each data point is represented by a pixel and the branch weights are estimated per pixel.

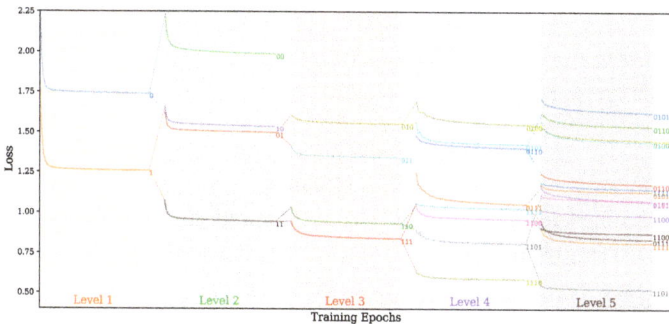

Figure 8.9 Partition-level loss reduction profile during training with heterogeneity-awareness. Dashed lines within breaks between adjacent levels indicate spatial process dependencies.

COVID-19 mobility regression: Figure 8.10 shows the results of the candidate methods. Several potential causes of the spatial heterogeneity here include different mobility patterns in the more populous downtown area versus the suburban regions, and several "hotspot" areas of POI visits that are a bit abnormal compared to the rest. As we can see, overall DNN_H's estimations align the best with the ground truth. DNN (the base model F used for DNN_H) substantially underestimates the total mobility, which could be a result of incorrect predictions on several mobility hotspots, whose patterns do not follow the global pattern.

In this example, DNN_H automatically identified three heterogeneous partitions (other splits are statistically insignificant) and branched out downtown, suburban and several mobility hotspots (our method allows a single partition to contain multiple disjoint large-footprints), greatly improving the prediction performance.

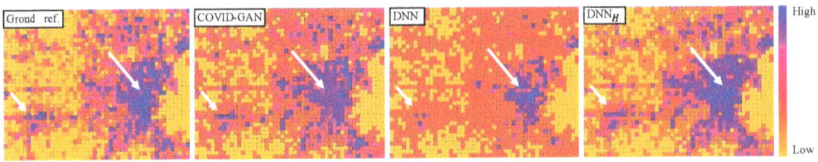

Figure 8.10   Visualization of human mobility maps (blue: high visits).

8.6 FAIRNESS-DRIVEN DESIGN: SINGLE-MODEL HETEROGENEITY-AWARE LEARNING

This section discusses a recent a fairness-aware learning framework for spatial data, which consists of a distributional representation of space and a bi-level learning framework for more robust fairness enhancements (Xie *et al.*, 2022b).

8.6.1 THE MAUP DILEMMA IN SPATIAL FAIRNESS

Groups are naturally needed to evaluate the fairness of a model (e.g., gender or racial groups in related AI-fairness problems). For spatial fairness, grouping of locations is also needed to calculate common metrics (e.g., accuracy) for fairness evaluation. Thus, we use space-partitioning P to generate location groups; in other words, each partition $p \in P$ is analogical to a group based on gender or race in related fairness studies. However, due to the MAUP dilemma in Def. 8.6.1, conclusions drawn from most – if not all – of common statistical measures are fragile to the variability in space-partitionings and scales. If this issue is ignored, then one may unintentionally or intentionally introduce additional bias, e.g., partisan gerrymandering (NPR, 2019).

Definition 8.6.1. Modifiable Areal Unit Problem (MAUP). MAUP states that statistical results and conclusions are sensitive to the choice of space partitioning P and scale. A change of scale (e.g., represented by the average area of $\{p_i \mid \forall p_i \in P\}$) always infers a change of P but not vice versa. MAUP is often considered as a dilemma as statistical results are expected to vary if different aggregations or groupings of locations are used.

In related work, while extensive fairness-aware learning methods have been developed, they largely focus on pre-defined categorical-attribute-based fairness (e.g., race and gender). Examples include regularization (Kamishima *et al.*, 2011; Serna *et al.*, 2020; Yan and Howe, 2019; Zafar *et al.*, 2017), sensitive category decorrelation (Alasadi *et al.*, 2019; Sweeney and Najafian, 2020; Zhang and Davidson, 2021), data collection/filtering strategies (Jo and Gebru, 2020; Steed and Caliskan, 2021; Yang *et al.*, 2020), and more (e.g., a recent survey (Mehrabi *et al.*, 2021)). In the context of spatial fairness, they can only handle a fixed space-partitioning and cannot address the MAUP challenge, where conclusions can be easily flipped due to the statistical sensitivity to the choice of partitionings.

Figure 8.11 shows an illustrative example of the effect of different partitionings on spatial fairness evaluation under MAUP. Without loss of generality, we consider partitionings P that follow a $s_1 \times s_2$ pattern (i.e., s_1 rows by s_2 columns) as examples. Figure 8.11 (a1) and (b1) show two example spatial distributions of prediction results (green: correct; red: wrong): (a1) has a large bias where the left side has 100% accuracy and the right side has 0%, and (b1) has a reasonably even distribution of each. However, as shown in Figure 8.11 (a2-3) and (b2-3), different partitionings or scales can lead to completely opposite conclusions, making fairness scores fragile in the spatial context.

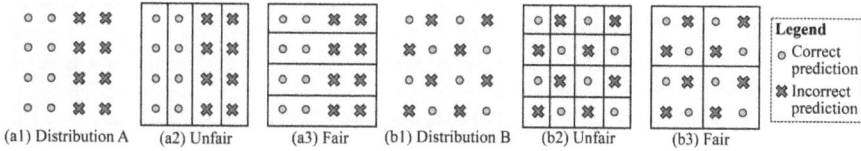

(a1) Distribution A   (a2) Unfair   (a3) Fair   (b1) Distribution B   (b2) Unfair   (b3) Fair

Figure 8.11   An examples showing the sensitivity of fairness evaluation to space-partitionings and scales.

8.6.2 SUMMARY OF KEY IDEAS

We explore new formulations and model-agnostic learning frameworks that are spatially-explicit and statistically-robust, with the following contributions:

- We propose a SPace-As-Distribution (SPAD) representation to formulate and evaluate the spatial fairness of learning models in continuous space, which mitigates the statistical sensitivity problems introduced by MAUP.

- We propose a SPAD-based stochastic strategy to efficiently optimize over an extensive distribution of candidate criteria for spatial fairness, which are needed to harness MAUP.

- We propose a bi-level player-referee training framework to enhance spatial fairness enforcement via adaptive adjustments of training priorities among locations.

8.6.3 SPACE AS A DISTRIBUTION OF PARTITIONINGS

Instead of relying on fragile scores calculated from a fixed partitioning or scale, we propose a SPace-As-Distribution (SPAD) representation to define spatial fairness. The idea is to go beyond a single partitioning or scale by treating space-partitionings at different scales $\{P\}$ as outcomes of a generative process governed by a statistical distribution. We use the $s_1 \times s_2$-type of partitioning as a concrete example in this work; our formulation and method do not depend on the type of partitioning. Following this, an example generative process may follow a joint two-dimensional distribution $Prob(s_1, s_2)$ where $s_1, s_2 \in \mathbb{Z}^+$, $s_1 \leqslant row_{max}$, $s_2 \leqslant col_{max}$. By default, one may assume a uniform distribution where $Prob(s_1, s_2) = (row_{max} \cdot col_{max})^{-1}$ (for

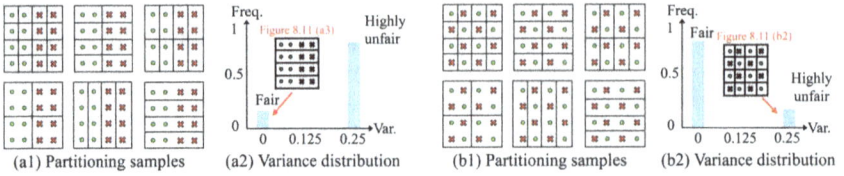

(a1) Partitioning samples (a2) Variance distribution (b1) Partitioning samples (b2) Variance distribution

Figure 8.12 Distributional representation by SPAD.

equal-size partitioning). This scheme also allows users to flexibly impose a different distribution or prior.

With the SPAD representation, spatial fairness becomes a distribution of scores, which can more holistically reflect fairness situations across a diverse set of partitions and scales. As an example, Figure 8.12 (a1) and (b1) show the same set of partitioning samples (different patterns and scales) overlaid on top of distributions A and B in Figure 8.11, respectively. The variance of accuracy across partitions for all 6 partitioning samples are aggregated in (a2) and (b2), where a lower variance means fairer results. As we can see, with the distributional extension, the majority of scores reflect our expected results on the fairness evaluation for distributions A and B, and the partitioning samples leading to unexpected results become outliers (highlighted by red arrows).

Once a distribution of scores is obtained from the SPAD representation, summary statistics can be conveniently used for fairness evaluation based on application preferences (e.g., mean). Finally, with SPAD, the formal formulation of spatial-fairness-aware learning is defined as follows:

$$\min_{\Theta} \int_{\Gamma} Prob(\Gamma) \cdot M_{fair}(F_{\Theta}, M_{perf}, P_{\Gamma}) d\Gamma \tag{8.2}$$

where F is an input deep network with parameters Θ; Γ parameterizes a space-partitioning P (e.g., number of rows and columns for $s_1 \times s_2$-partitionings); $Prob(\Gamma)$ is the probability of P_{Γ}; M_{perf} is a metric used to evaluate the performance of a model F (e.g., F1-score); and M_{fair} is a fairness measure (loss) that is defined as:

$$M_{fair}(F_{\Theta}, M_{perf}, P) = \sum_{p \in P} \frac{d(M_{perf}(F_{\Theta}, p), E_P)}{|P|} \tag{8.3}$$

where p is a partition in P (Def. 8.2.4), $d(\cdot, \cdot)$ is a distance measure (e.g., squared or absolute distance), $M_{perf}(F_{\Theta}, p)$ is the score (e.g., F1-score) of F_{Θ} on the training data of $p \in P$, $|P|$ is the number of partitions in P, and E_P is another key variable, which represents the mean performance of partitions $p \in P$ (weighted or unweighted by sample sizes). If $M_{perf}(F_{\Theta}, p)$ has a large deviation from the mean, the model F_{Θ} is potentially unfair across partitions. Finally, E_P here is calculated from a base model F_{Θ_0}, where parameters Θ_0 are trained without any consideration of spatial fairness:

$$E_P = \sum_{p \in P} \frac{M_{perf}(F_{\Theta_0}, p)}{|P|} \tag{8.4}$$

The benefit of using F_{Θ_0} to set the mean is that, ideally, we want to maintain the same level of overall model performance (e.g., F1-score without considering spatial fairness) while improving spatial fairness. Thus, this choice automatically takes the overall model performance into consideration as the objective function (Eq. (8.2)) will increase if F_Θ's overall performance diverges too far from it (e.g., a model that yields a F1-score of 0 on all partitions – which is fair but poor – will not be considered as a good candidate).

8.6.4   SPAD-BASED STOCHASTIC TRAINING

A direct way to incorporate the distributional SPAD representation into the training process – either through loss functions or the bi-level method to be discussed in the next section – is to aggregate results from all the partitionings $\{P\}$ for each iteration or epoch. However, this is computationally expensive and sometimes prohibitive. For example, the number of possible partitionings can be exponential to data size (e.g., the number of sample locations) when general partitioning schemes are considered (e.g., arbitrary, hierarchical, or $s_1 \times s_2$ partitionings with unequal-size cells). Even for equal-size $s_1 \times s_2$ partitionings, there can be easily over hundreds of candidates when large s_1 and s_2 values (e.g., 10, 40, or more) are used for large-scale applications.

Thus, we propose a stochastic training strategy for SPAD to mitigate the cumbersome aggregation. Considering SPAD as a statistical generative process G, in each iteration or epoch, we randomly sample a partitioning from G and use it to evaluate a fairness-related loss M_{fair} (e.g., Eq. (8.3)).

In particular, for equal-size $s_1 \times s_2$ partitionings, each time the generator may randomly sample (s_1, s_2) from a joint discrete distribution (Figure 8.13). In this way, the probability of each partitioning (Eq. (8.2)) is automatically taken into consideration during optimization over epochs. In addition, in scenarios where the difficulty of achieving fairness varies for different partitionings, the SPAD-based stochastic strategy may accelerate the overall convergence. It may first help a subset of partitionings reach good fairness scores faster without the averaging effect, which may in

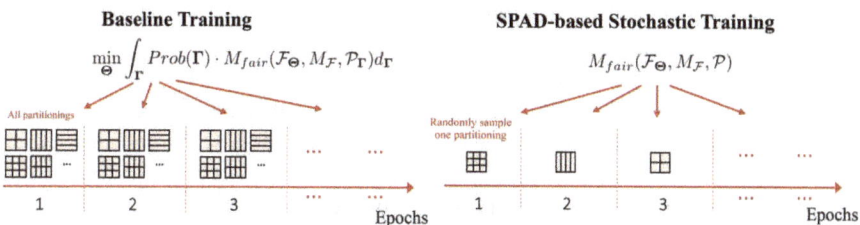

Figure 8.13   SPAD-based stochastic training strategy.

turn help related partitionings to move out local minima traps. In practice, we have three further recommendations for implementation:

- **Unconstrained initial training:** Ideally, we wish to maintain a high overall performance (e.g., F1-scores) while improving fairness across locations. However, it can be pre-mature to try to find a balance between the two objectives when the model still has a very poor overall performance (e.g., untrained). Hence, we keep fairness-related losses or constraints on-hold at the beginning, and optimize parameters by pure prediction errors till stable.

- **Epoch as a minimum unit:** Deep network training often involves mini-batches (i.e., a middle-ground between stochastic and batch gradient descent). As a result, the combined randomness of mini-batches and SPAD-based stochastic strategy may make the training unstable. Thus, using epoch as a minimum unit for changing partitioning samples can help reduce the superposed randomness.

- **Increasing frequency:** Extending the last point, denote k as the number of continuous epochs to train before a partitioning sample is changed. At the beginning of training, a biased model without any fairness consideration may need more epochs to make meaningful improvements, which means a larger k (e.g., 10) is preferred. In contrast, toward the end of the training, a large k can be undesirable as it may cause the model to overfit to a single partitioning at the finish. Thus, we recommend a decreasing k (finally $k = 1$) during training.

8.6.5   BI-LEVEL FAIRNESS ENFORCEMENT

A traditional way to incorporate fairness loss (e.g., Eq. (8.3)) is to add it as a term in the loss function, e.g., $L = L_{pred} + \lambda \cdot M_{fair}$, where L_{pred} is the prediction loss (e.g., cross-entropy or dice loss) and λ is a scaling factor or weight. This regularization-based formulation has three limitations when used for spatial-fairness enforcement: (1) Since deep learning training often uses mini-batches due to data size, it is difficult for each mini-batch to contain representative samples from all partitions $\{p_i \,|\, \forall p_i \in P\}$ when calculating M_{fair}. (2) To reflect true fairness over partitions, metrics M_{perf} used in M_{fair} in Eq. (8.3) are ideally exact functions such as precision, recall or F1-scores. However, since many of the functions are not differentiable as a loss function (e.g., with the use of arg max to extract predicted classes), approximations are often needed (e.g., threshold-based, soft-version), which introduce extra errors. Additionally, as such approximations are used to further derive fairness indicators (e.g., M_{perf}), the uncertainty created by the errors can be quickly accumulated and amplified; and (3) The regularization term M_{fair} requires another scaling factor λ, the choice of which directly impacts final output and varies from problem to problem.

   To mitigate these concerns, we propose a bi-level training strategy that disentangles the two types of losses with different purposes (i.e., L_{pred} and M_{fair}). Specifically, there are two levels of decision-making in-and-between epochs:

- **Partitioning-level** (P): Before each epoch, a referee evaluates the spatial fairness using Eq. (8.3) with exact metrics M_{perf} (e.g., F1-score); no approximation is needed as back-propagation is not part of the referee. The evaluation is performed on all partitions $p_i \in P$, guaranteeing the representativeness. Note that the model is evaluatable for the very first epoch because the fairness-driven training starts from a base model, as discussed in the previous section and explanations for Eq. (8.3). Based on an individual partition p_i's deviation $d(M_{perf}(F_\Theta, p_i), E_P)$ (a summand in M_{fair}'s numerator in Eq. (8.3)), we assign its learning rate η_i for this epoch as:

$$\eta_i = \frac{\eta_i' - \eta_{min}'}{\eta_{max}' - \eta_{min}'} \cdot \eta_{init} \qquad (8.5)$$

$$\eta_i' = \max(-(M_{perf}(F_\Theta, p_i) - E_P), 0) \qquad (8.6)$$

where η_{init} is the learning rate used to train the base model, $\eta_{min}' = \arg\min_{\eta_i'} \{\eta_i' \mid \eta_i' > 0, \forall i\}$, and $\eta_{max}' = \arg\max_{\eta_i'} \{\eta_i' \mid \forall i\}$. Different from Eq. (8.3), $d(\cdot, \cdot)$ is not used here as we need to know the sign of the difference (i.e., direction matters). The intuition is that, if a partition's performance measure is lower than the expectation E_p, its learning rate η_i will be increased (relatively to other partitions') so that its prediction loss will have a higher impact during parameter updates in this epoch. In contrast, if a partition's performance is the same or higher than the expectation, its η_i will be set to 0 to prioritize other lower-performing partitions. Positive learning rates after the update are normalized back to the range $[0, \eta_{init}]$ to keep the gradients more stable. This bi-level design also relieves the need for an extra scaling factor to combine the prediction and fairness losses.

- **Partition-level** (p): Using learning rates $\{\eta_i\}$ assigned by the referee, we perform regular training with the prediction loss L_{pred}, iterating over data in all individual partitions $p_i \in P$ in batches.

8.6.6 EXPERIMENTS

8.6.6.1 Candidate Methods

We evaluate our proposed method with the same dataset for California land-cover classification (the snapshot version) mentioned in Sec. 8.5.5.1. We consider the following candidate methods in our evaluation:

- **Base:** The base deep learning model (8-layer DNN) without consideration of spatial fairness.

- **Single:** Spatial fairness is evaluated and improved using a single space-partitioning P. Specifically, our experiment includes Single-(1,4) and Single-(4,1), which use 1×4 and 4×1 partitionings, respectively.

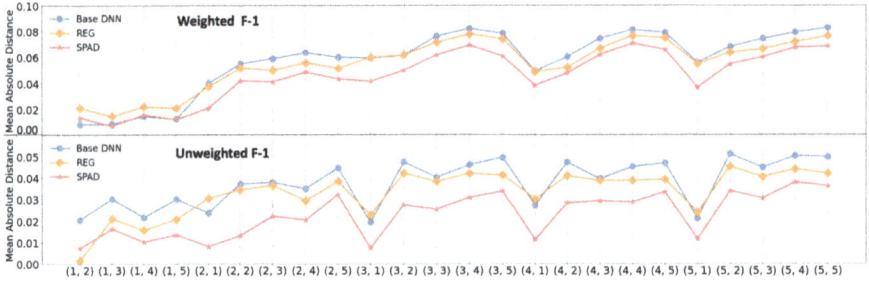

Figure 8.14 Fairness comparison amongst SPAD, REG and the base model over all the partitionings. The lines connecting the points are only used to make it visually easier to see the differences among the methods.

- **REG:** Spatial fairness is enforced using the SPAD representation by a regularization term; the inclusion of a regularizer is a common strategy in related work (Kamishima *et al.*, 2011; Yan and Howe, 2019). As F1-score is not differentiable, we use standard approximation via the threshold-based approach, which amplifies softmax predictions $\hat{y}$ over a threshold γ to 1 to suppresses others to 0 using $1 - \mathrm{ReLU}(1 - A \cdot \mathrm{ReLU}(\hat{y} - \gamma))$, where A is a sufficiently large number ($A = 10000$ in our tests). The scaling factor λ for the regularizer is set to 5.

- **SPAD:** The proposed approach using the SPAD representation with the stochastic and bi-level training strategies.

- **SPAD-GD:** SPAD without the stochastic strategy, which aggregates over gradients from all 24 partitionings before making parameter updates in each round.

8.6.6.2 Results

Comparison to the Regularization-based Method: We compare the fairness achieved by SPAD, the base DNN model (without considering fairness) and the REG method in Figure 8.14. For each partitioning P (x-axis), we report the mean of the absolute distances between F1-scores achieved on each partition p and the average performance over all partitions $\{p \in P\}$; both weighted and unweighted F-1 scores are considered. In Table 8.1, we summarize the overall performance (global F1-scores), the sum of mean absolute distance $S(d)_{\mathrm{mean}}$ and the sum of maximum absolute distance $S(d)_{\mathrm{max}}$ across all partitionings using weighted and unweighted F-1, respectively. Specifically, we have: $S(d)_{\mathrm{mean}} = \sum_{P_i \in \{P_1, P_2, \dots\}} \left(\sum_{p \in P_i} \frac{d(M_{perf}(F_\Theta, p), E_{P_i})}{|P_i|} \right)$ and $S(d)_{\mathrm{max}} = \sum_{P_i \in \{P_1, P_2, \dots\}} \left(\max_{p \in P_i} d(M_{perf}(F_\Theta, p), E_{P_i}) \right)$.

Figure 8.14 shows that both SPAD and REG achieve lower mean absolute distances over all space partitionings compared to the base model, confirming the effectiveness of the SPAD representation in improving the fairness. Comparing SPAD and REG, we can see that SPAD consistently outperforms REG in the experiments,

Table 8.1

Classification and fairness results by weighted (W.) F1 scores and unweighted (UW.) F1 scores.

Method	W. F1	$S(d)_{mean}$	$S(d)_{max}$	UW. F1	$S(d)_{mean}$	$S(d)_{max}$
Base DNN	0.572	1.379	3.799	0.377	0.906	1.808
REG	0.566	1.319	3.821	0.381	0.799	1.808
Single-(1,4)	0.576	1.356	3.666	0.362	0.627	1.392
Single-(4,1)	0.542	1.355	3.712	0.368	0.685	1.517
SPAD-GD	0.573	1.275	3.571	0.372	0.602	1.384
SPAD	0.573	**1.094**	**3.185**	0.374	**0.549**	**1.337**

which shows that the bi-level design is more effective in enforcing spatial fairness than regularization terms by improving sample representativeness, allowing the use of exact metrics (i.e., no need to use approximations of F1-scores for differentiability purposes), and eliminating the need for an extra scaling factor for the regularizer which may add extra sensitivity.

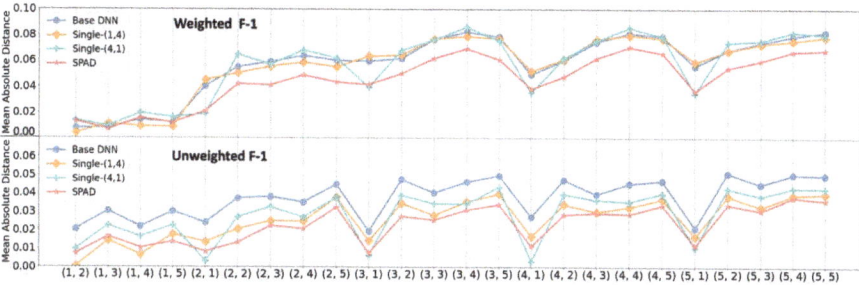

Figure 8.15 Fairness comparison amongst SPAD, Single-(1,4), Single-(4,1), and the base model over all the partitionings. The lines connecting the points are only used to make it visually easier to see the differences among the methods.

From the first column of Table 8.1, we can see that SPAD is able to maintain a similar overall/global classification performance compared to the base DNN, which does not have any fairness consideration. Meanwhile, the second and third columns in the table show that our method can significantly reduce the sums of mean and max absolute distance over all partitionings. This confirms that SPAD can effectively promote the fairness without compromising the classification performance.

Comparison to Partitioning-Specific Fairness-Aware Methods: Next, we compare SPAD with non-SPAD-based variants that only focus on a single partitioning to verify SPAD's robustness across a diverse set of partitionings. Figure 8.15 shows the fairness performance of partition-specific methods Single-(1,4) and Single-(4,1). The overall trend is that SPAD achieves better spatial fairness in most partitionings by modeling space-partitionings as a distribution. In addition, we can also observe that Single-(4,1) obtains a better fairness result for the given partitioning (4,1), and similarly, Single-(1,4) performs better for (1,4). However, their fairness improvements are limited for other partitionings. This conforms to the expectation that partitioning-specific methods are able to reach further improvements on a given P, but cannot generalize well to the others. Table 8.1 (rows 3-4) shows the weighted and unweighted F1-scores achieved by Single-(1,4) and Single-(4,1). The numbers confirm that the methods also have similar global F1-scores since our design takes the overall performance into account (Eqs. (8.3) and (8.4)). However, they produce larger values of $S(d)_{mean}$ and $S(d)_{max}$, which again confirms the benefits of SPAD.

Validation of Stochastic Training Strategies: Finally, we validate the effectiveness of the SPAD-based stochastic training strategy. We first compare to the SPAD-GD approach, which aggregates gradients from all partitionings in each epoch. Compared to our SPAD-based stochastic approach, the aggregation in SPAD-GD leads to a heavier computational load and requires longer training time (i.e., 2.5 hours vs. 9.5 hours using NVIDIA Tesla K80 GPU over two runs). Here we maintain the same number of parameter updates for the two methods, and the only difference is that each SPAD update is made by gradients from a sampled partitioning whereas each SPAD-GD update uses average gradients from all partitionings.

Interestingly, SPAD outperforms SPAD-GD for both weighted and unweighted scenarios (Table 8.1). One potential reason is that the added randomness from the stochastic sampling in SPAD may allow a better chance for the training to move out of local minima traps without the averaging effects, especially when fairness is harder to achieve at the beginning for some partitionings.

8.7 NEW DIRECTIONS

There have been recent new developments in AI to further improve heterogeneity-aware learning and fairness. The training of the STAR framework has been extended with meta-learning (Xie *et al.*, 2023a), where meta-updates are used to learn better common weights shared by different branches. Li *et al.* (2023) proposed a point-to-region co-learning framework, which learns region-level (e.g., city-level) contexts to automatically adjust classification criteria in different cities. The advantage of this approach is that it can self-adapt to a new city (i.e., no need for labelled samples), where the classification criteria may not be the same as any of those learned from the training cities. Experimental results on high-resolution poverty mapping show improved generalizability of the method in situations where no sample is available from a new city. Unsupervised or "label-free" approaches have also been explored to address the spatial heterogeneity challenge at large scales (Xie *et al.*, 2023b), where physical characteristics (e.g., spatio-temporal dynamics) are used to guide the training process for each geographic region. In terms of fairness-driven methods, He *et al.* (2022) showed the importance of explicitly considering locational fairness in machine learning, which otherwise tends to create biased results in various forms and can be easily manipulated to create a false sense of fairness. He *et al.* (2023) also developed a new physics-guided machine learning formulation, which aims to maintain the locational fairness of a model when it is applied to a new time period for spatio-temporal prediction problems.

8.8 CONCLUSIONS

Heterogeneity or distribution shift in space is a fundamental challenge for deep learning in geospatial datasets and problems. It not only causes reduced prediction performance and generalizability of a model but also creates bias and unfairness in the prediction results. We formally define spatial heterogeneity in the machine learning context with two general formulations that focus on prediction performance and fairness, respectively. We further discuss two recent methods: (1) a spatial transformation and moderation framework, and (2) a SPAD-based bi-level learning framework, to address these challenges. Experiment results on land cover monitoring and human mobility projection confirm the effectiveness of the methods on improving prediction quality and locational fairness.

FUNDING

This material is based upon work supported by the National Science Foundation under Grant No. 2105133, 2126474 and 2147195; NASA under Grant No. 80NSSC22K1164 and 80NSSC21K0314; USGS grants G21AC10207, G21AC10564, and G22AC00266; Google's AI for Social Good Impact Scholars program; the DRI award at the University of Maryland; and Pitt Momentum Funds award and CRC at the University of Pittsburgh.

BIBLIOGRAPHY

Alasadi, J., Al Hilli, A., and Singh, V.K., 2019. Toward fairness in face matching algorithms. *In: Proceedings of the 1st International Workshop on Fairness, Accountability, and Transparency in MultiMedia.* 19–25.

Atluri, G., Karpatne, A., and Kumar, V., 2018. Spatio-temporal data mining: A survey of problems and methods. *ACM Computing Surveys (CSUR)*, 51 (4), 1–41.

Bao, H., *et al.*, 2020. Covid-gan: Estimating human mobility responses to covid-19 pandemic through spatio-temporal conditional generative adversarial networks. *In: Proceedings of the 28th International Conference on Advances in Geographic Information Systems.* 273–282.

Ben-David, S., *et al.*, 2007. Analysis of representations for domain adaptation. *Advances in Neural Information Processing Systems*, 19, 137.

Brunsdon, C., Fotheringham, A.S., and Charlton, M., 1999. Some notes on parametric significance tests for geographically weighted regression. *Journal of Regional Science*, 39 (3), 497–524.

Chen, T., *et al.*, 2020. A simple framework for contrastive learning of visual representations. *In: International Conference on Machine Learning.* PMLR, 1597–1607.

Du, M., *et al.*, 2021. Fairness via representation neutralization. *Advances in Neural Information Processing Systems*, 34, 12091–12103.

Du, M., *et al.*, 2020. Fairness in deep learning: A computational perspective. *IEEE Intelligent Systems*, 36 (4), 25–34.

Finn, C., Abbeel, P., and Levine, S., 2017. Model-agnostic meta-learning for fast adaptation of deep networks. *In*: *International Conference on Machine Learning*. PMLR, 1126–1135.

Finn, C., *et al.*, 2016. Deep spatial autoencoders for visuomotor learning. *In*: *2016 IEEE International Conference on Robotics and Automation (ICRA)*. IEEE, 512–519.

Fotheringham, A.S., Yang, W., and Kang, W., 2017. Multiscale geographically weighted regression (mgwr). *Annals of the American Association of Geographers*, 107 (6), 1247–1265.

Gao, Y., *et al.*, 2019. Incomplete label multi-task deep learning for spatio-temporal event subtype forecasting. *In*: *Proceedings of the AAAI Conference on Artificial Intelligence*. vol. 33, 3638–3646.

Goodchild, M.F. and Li, W., 2021. Replication across space and time must be weak in the social and environmental sciences. *Proceedings of the National Academy of Sciences*, 118 (35).

Gupta, J., *et al.*, 2020. Towards spatial variability aware deep neural networks (svann): A summary of results. *In*: *ACM SIGKDD Workshop on Deep Learning for Spatiotemporal Data, App. & Sys.*

Gupta, J., *et al.*, 2021. Spatial variability aware deep neural networks (svann): A general approach. *ACM Trans. Intell. Syst. Technol.*, 12 (6).

He, E., *et al.*, 2022. Sailing in the location-based fairness-bias sphere. *In*: *Proceedings of the 30th International Conference on Advances in Geographic Information Systems*. 1–10.

He, E., *et al.*, 2023. Physics guided neural networks for time-aware fairness: An application in crop yield prediction. *In*: *AAAI Conference on Artificial Intelligence*.

He, K., *et al.*, 2016. Deep residual learning for image recognition. *In*: *CVPR*. 770–778.

Jean, N., *et al.*, 2019. Tile2vec: Unsupervised representation learning for spatially distributed data. *In*: *Proceedings of the AAAI Conference on Artificial Intelligence*. vol. 33, 3967–3974.

Jia, X., *et al.*, 2019. Spatial context-aware networks for mining temporal discriminative period in land cover detection. *In*: *SDM*. SIAM, 513–521.

Jiang, Z., *et al.*, 2019. Spatial ensemble learning for heterogeneous geographic data with class ambiguity. *ACM Trans. on Intelligent Sys. and Tech. (TIST)*, 10 (4).

Jiang, Z., *et al.*, 2017. Spatial ensemble learning for heterogeneous geographic data with class ambiguity: A summary of results. *In*: *Proceedings of the 25th ACM SIGSPATIAL International Conference on Advances in Geographic Information Systems*. 1–10.

Jo, E.S. and Gebru, T., 2020. Lessons from archives: Strategies for collecting sociocultural data in machine learning. *In*: *Proceedings of the 2020 Conference on Fairness, Accountability, and Transparency*. 306–316.

Kamishima, T., Akaho, S., and Sakuma, J., 2011. Fairness-aware learning through regularization approach. *In*: *2011 IEEE 11th International Conference on Data Mining Workshops*. IEEE, 643–650.

Kaya, A., *et al.*, 2019. Analysis of transfer learning for deep neural network based plant classification models. *Computers and Electronics in Agriculture*, 158, 20–29.

Kilbertus, N., *et al.*, 2018. Blind justice: Fairness with encrypted sensitive attributes. *In: International Conference on Machine Learning*. PMLR, 2630–2639.

Krizhevsky, A., Sutskever, I., and Hinton, G.E., 2012. Imagenet classification with deep convolutional neural networks. *Advances in Neural Information Processing Systems*, 25, 1097–1105.

Kulldorff, M., *et al.*, 2007. Multivariate scan statistics for disease surveillance. *Statistics in Medicine*, 26 (8), 1824–1833.

Li, Z., *et al.*, 2023. Point-to-region co-learning for poverty mapping at high resolution using satellite imagery. *In: AAAI Conference on Artificial Intelligence*.

Liu, X-P., *et al.*, 2019. Risk assessment using transfer learning for grassland fires. *Agricultural and Forest Meteorology*, 269, 102–111.

Ma, J., *et al.*, 2019. Improving air quality prediction accuracy at larger temporal resolutions using deep learning and transfer learning techniques. *Atmospheric Environment*, 214, 116885.

Mai, G., *et al.*, 2022. A review of location encoding for geoai: methods and applications. *International Journal of Geographical Information Science*, 36 (4), 639–673.

Mehrabi, N., *et al.*, 2021. A survey on bias and fairness in machine learning. *ACM Computing Surveys (CSUR)*, 54 (6), 1–35.

Morales, A., *et al.*, 2020. Sensitivenets: Learning agnostic representations with application to face images. *IEEE Transactions on Pattern Analysis and Machine Intelligence*, 43 (6), 2158–2164.

Neill, D.B., 2012. Fast subset scan for spatial pattern detection. *Journal of the Royal Statistical Society*, 74 (2), 337–360.

Neill, D.B., McFowland III, E., and Zheng, H., 2013. Fast subset scan for multivariate event detection. *Statistics in Medicine*, 32 (13), 2185–2208.

NPR, 2019. Supreme court rules partisan gerrymandering is beyond the reach of federal courts.

Pan, S.J., *et al.*, 2010. Domain adaptation via transfer component analysis. *IEEE Transactions on Neural Networks*, 22 (2), 199–210.

Pan, S.J. and Yang, Q., 2009. A survey on transfer learning. *IEEE Transactions on Knowledge and Data Engineering*, 22 (10), 1345–1359.

Serna, I., *et al.*, 2020. Sensitiveloss: Improving accuracy and fairness of face representations with discrimination-aware deep learning. *arXiv preprint arXiv:2004.11246*.

Shekhar, S., Feiner, S.K., and Aref, W.G., 2015. Spatial computing. *Communications of the ACM*, 59 (1), 72–81.

Shekhar, S. and Xiong, H., 2007. *Encyclopedia of gis*. Springer Science & Business Media.

Steed, R. and Caliskan, A., 2021. Image representations learned with unsupervised pre-training contain human-like biases. *In: Proceedings of the 2021 ACM Conference on Fairness, Accountability, and Transparency*. 701–713.

Sweeney, C. and Najafian, M., 2020. Reducing sentiment polarity for demographic attributes in word embeddings using adversarial learning. *In: Proceedings of the 2020 Conference on Fairness, Accountability, and Transparency*. 359–368.

Vilalta, R. and Drissi, Y., 2002. A perspective view and survey of meta-learning. *Artificial Intelligence Review*, 18 (2), 77–95.

Wang, M. and Deng, W., 2018. Deep visual domain adaptation: A survey. *Neurocomputing*, 312, 135–153.

Wei, Z., *et al.*, 2021. Large-scale river mapping using contrastive learning and multi-source satellite imagery. *Remote Sensing*, 13 (15), 2893.

Xie, Y., *et al.*, 2023a. Harnessing heterogeneity in space with statistically-guided meta-learning. *Knowledge and Information Systems*, 65, 2699–2729.

Xie, Y., *et al.*, 2021a. A statistically-guided deep network transformation and moderation framework for data with spatial heterogeneity. *In*: *2021 IEEE International Conference on Data Mining (ICDM)*. IEEE, 767–776.

Xie, Y., *et al.*, 2022a. Statistically-guided deep network transformation to harness heterogeneity in space (extended abstract). *In*: L.D. Raedt, ed. *Proceedings of the Thirty-First International Joint Conference on Artificial Intelligence, IJCAI-22*, 7, 5364–5368.

Xie, Y., *et al.*, 2022b. Fairness by "where": A statistically-robust and model-agnostic bi-level learning framework. *Proceedings of the AAAI Conference on Artificial Intelligence*, 36 (11), 12208–12216.

Xie, Y., *et al.*, 2021b. Spatial-net: A self-adaptive and model-agnostic deep learning framework for spatially heterogeneous datasets. *In*: *Proceedings of the 29th International Conference on Advances in Geographic Information Systems*. 313–323.

Xie, Y., *et al.*, 2023b. Auto-cm: Unsupervised deep learning for satellite imagery composition and cloud masking using spatio-temporal dynamics. *In*: *AAAI Conference on Artificial Intelligence*.

Xie, Y., Shekhar, S., and Li, Y., 2022c. Statistically-robust clustering techniques for mapping spatial hotspots: A survey. *ACM Computing Surveys (CSUR)*, 55 (2), 1–38.

Yan, A. and Howe, B., 2019. Fairst: Equitable spatial and temporal demand prediction for new mobility systems. *In*: *Proceedings of the 27th ACM SIGSPATIAL International Conference on Advances in Geographic Information Systems*. 552–555.

Yang, K., *et al.*, 2020. Towards fairer datasets: Filtering and balancing the distribution of the people subtree in the imagenet hierarchy. *In*: *Proceedings of the 2020 Conference on Fairness, Accountability, and Transparency*. 547–558.

Yao, H., *et al.*, 2019. Learning from multiple cities: A meta-learning approach for spatial-temporal prediction. *In*: *The World Wide Web Conference*. 2181–2191.

Yuan, Z., Zhou, X., and Yang, T., 2018. Hetero-convlstm: A deep learning approach to traffic accident prediction on heterogeneous spatio-temporal data. *In*: *Proceedings of the 24th ACM SIGKDD International Conference on Knowledge Discovery & Data Mining*. 984–992.

Zafar, M.B., *et al.*, 2017. Fairness beyond disparate treatment & disparate impact: Learning classification without disparate mistreatment. *In*: *Proceedings of the 26th International Conference on World Wide Web*. 1171–1180.

Zhang, H. and Davidson, I., 2021. Towards fair deep anomaly detection. *In*: *Proceedings of the 2021 ACM Conference on Fairness, Accountability, and Transparency*. 138–148.

9 Explainability in GeoAI

Ximeng Cheng, Marc Vischer, Zachary Schellin
Department of Artificial Intelligence, Fraunhofer Heinrich Hertz
Institute, Germany

Leila Arras
Department of Artificial Intelligence, Fraunhofer Heinrich Hertz
Institute, Germany; Berlin Institute for the Foundations of
Learning and Data, Germany

Monique M. Kuglitsch
Department of Artificial Intelligence, Fraunhofer Heinrich Hertz
Institute, Germany

Wojciech Samek
Department of Artificial Intelligence, Fraunhofer Heinrich Hertz
Institute, Germany; Department of Electrical Engineering and
Computer Science, Technische Universität Berlin; Berlin Institute
for the Foundations of Learning and Data, Germany

Jackie Ma
Department of Artificial Intelligence, Fraunhofer Heinrich Hertz
Institute, Germany

CONTENTS

9.1 Introduction 178
    9.1.1 Explainable AI 178
    9.1.2 The benefits of applying XAI methods 179
    9.1.3 Applying XAI methods in GeoAI 180
9.2 XAI Methods 181
    9.2.1 Local explanation methods 182
    9.2.2 Global explanation methods 185
    9.2.3 Further XAI methods and challenges 185
9.3 XAI Applications in GeoAI 186
    9.3.1 Use Case 1: Explain model decisions for rainfall-runoff prediction 187
    9.3.2 Use Case 2: Knowledge discovery for traffic analysis 189
    9.3.3 Use Case 3: Model improvement for remote sensing classification. 191
9.4 Discussion and Conclusion 194
    Bibliography 195

DOI: 10.1201/9781003308423-9

9.1   INTRODUCTION

Rich data resources have prompted the development of data-driven artificial intelligence (AI) techniques such as deep learning (Hinton *et al.*, 2006; Hinton and Salakhutdinov, 2006; LeCun *et al.*, 2015). These techniques have shown great promise in solving various types of tasks, fostering their adoption for geospatial applications and the emergence of a new sub-discipline of AI and geoscience [i.e., geospatial artificial intelligence (GeoAI) (Janowicz *et al.*, 2020; Reichstein *et al.*, 2019)]. However, the massive amount of parameters and complexity of structures inherent to neural networks, make it difficult for researchers and users to understand the basis for nonlinear model decisions (Lapuschkin *et al.*, 2019). This, among others, has given deep learning models the reputation of being "black boxes" (Castelvecchi, 2016; Erhan *et al.*, 2009). Consequently, this has led to discussions within the GeoAI community about the suitability of deep learning methods and, occasionally, limited the application of these methods. One approach that researchers have used to overcome this issue is through combining a priori knowledge (e.g., on geoscience processes) and traditional methods with general AI models to design spatially-explicit GeoAI methods (Liu and Biljecki, 2022). Another approach is leveraging explainable artificial intelligence (XAI) methods (Mamalakis *et al.*, 2022b; McGovern *et al.*, 2019; Samek *et al.*, 2019). For instance, XAI has been applied in GeoAI studies based on in situ, remotely sensed, and socially sensed data (Cheng *et al.*, 2021a; Hilburn *et al.*, 2021; Mayer and Barnes, 2021; Xing *et al.*, 2020). In this chapter, we give a compact overview of basic concepts in the field of XAI, in particular, we focus on (1) a brief introduction to XAI, (2) highlight several popular XAI methods, and finally (3) show some applications of XAI methods in GeoAI.

9.1.1   EXPLAINABLE AI

The field of XAI is an active research area in which experts from different communities discover novel ways to make opaque AI models more transparent. There are various methodologies to increase the model transparency and ultimately the usefulness of an explanation in practice depends on the type-, purpose-, and user of the model.

One of the very prominent types of approaches is the use of feature importance methods that compute attribution maps intending to visualize important features in the input data that have contributed to the model's decision. These methods are popular among deep learning models and prominent examples are, for instance, gradient-based methods such as *Integrated Gradients* (Sundararajan *et al.*, 2017) and backward decomposition-based methods such as *Layer-wise Relevance Propagation (LRP)* (Bach *et al.*, 2015). As already mentioned, depending on the model to be explained, XAI approaches are based on different principles, and different taxonomies for categorizing XAI exist. A common taxonomy categorizes XAI and the corresponding explanation as *local* and *global*. The former provides the task-specific input feature importance on individual samples while the latter provides the impact of input features on the entire model (Lundberg *et al.*, 2020). Based on this taxonomy, Section 9.2 provides an introduction to diverse popular XAI methods.

Providing an attribution map to interpret the decisions made by the AI model is often not sufficient. *Explanation assessment* determines the quality of the explanations generated by XAI approaches. According to Miller (2019), a good explanation should be human-friendly and meet quality criteria from psychology and cognitive science. Other researchers advocate for advanced XAI methods, such as *Concept Relevance Propagation (CRP)* (Achtibat *et al.*, 2022) and *Concept Recursive Activation FacTorization (CRAFT)* (Fel *et al.*, 2022b), which aim to provide human-understandable concept-based explanations. In general, no ground truth explanations are available for a given task and model. For simple and intuitive tasks, such as recognizing objects in images, humans might be able to judge whether an explanation reflects a correct understanding of the task. However, it is nearly impossible to judge the explanations for more complex tasks or tasks for which the researchers lack expertise. In such cases, multiple XAI methods can be used to cross-validate the explanations (Cheng *et al.*, 2021a). An alternative approach is to use diverse metrics to quantitatively evaluate explanations for faithfulness, robustness, localization, complexity, and randomization (Hedström *et al.*, 2023).

9.1.2   THE BENEFITS OF APPLYING XAI METHODS

XAI methods can yield valuable insights into the decisions made by AI models. Furthermore, if the explanations generated by XAI methods are consistent with human expert knowledge, then it suggests that the model has learned the expected knowledge which could increase the user's confidence in applying the AI model. Alternatively, the explanations could demonstrate that the model learned an unexpected strategy to make decisions (e.g., Clever Hans, a horse named Hans was once thought to be able to perform basic math operations. But in the end, it was found that the answers from Hans were given by observing human reactions `https://en.wikipedia.org/wiki/Clever_Hans`). In this scenario, the insights could hint toward novel knowledge or can be used by developers to find and solve problems in the model training. Figure 9.1 shows an example of a comparison of expected and unexpected strategies for a classification task. In GeoAI, XAI methods have been applied to explain the decisions made by trained models in the studies of remote sensing images and human activity data (Xing *et al.*, 2020; Zhang *et al.*, 2019).

XAI methods can be used as a magnification lens to help gain an understanding of the domain itself, which has tremendous value if the domain is largely unexplored. Deep learning, for instance, can mine features from huge training datasets and outperform humans in various tasks such as image classification (Russakovsky *et al.*, 2015), speech recognition (Assael *et al.*, 2016), and playing computer games (Berner *et al.*, 2019). The training process of these AI models requires little human intervention, which means that the model has the potential to develop novel data-based strategies which are not intuitive to humans but are critical to the specific task. In this case, XAI methods can transform the learned strategies into a human-understandable form and help humans better perform such tasks. This has been seen in the case of Go. AlphaGo beats the master of Go, with players now trying to learn new strategies from AI (Silver *et al.*, 2016). AI has also been demonstrated useful

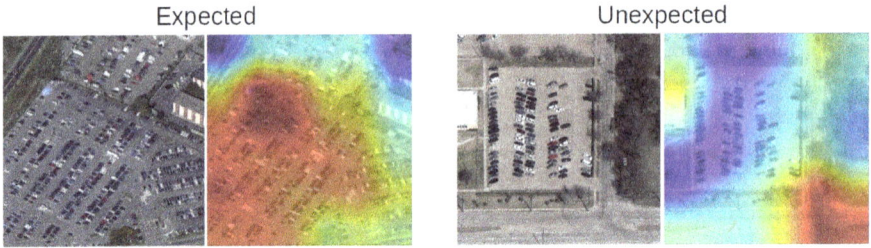

Figure 9.1   An example of XAI revealing strategies to make decisions. A model was trained to classify scenes based on remote sensing images. For each pair, the left one is a raw image and the right one is the corresponding explanation by XAI. A warm color denotes an important pixel that contributes to the task. Here, people may expect that many vehicles are the key feature to be classified as a parking lot (left). The use of other features such as the way to the parking lot is unexpected (right).

for discovering physical concepts (Iten *et al.*, 2020). In GeoAI, Cheng *et al.* (2021a) captured the task-specific importance of spatio-temporal units by XAI methods and successfully applied the new knowledge to perform efficient information extraction, cf. also Section 9.3.2. Li (2022) extracted spatial effects by using the *SHapley Additive exPlanations (SHAP)* approach (Lundberg and Lee, 2017) and compared them with the results of spatial statistical models. In Başağaoğlu *et al.* (2022), interpretable and explainable AI approaches were used to understand the rationale behind hydro-climatic predictions and to derive new insights into the involved processes. Section 9.3 delves further into this topic.

In addition to gaining insights into the model's decisions, XAI methods can also help improve AI models by alerting researchers of problems in model training. Based on the local explanation of samples, researchers can identify outliers and artifacts in the training data through manual inspection or clustering methods (Lapuschkin *et al.*, 2019), and then reorganize the training data, e.g., purposely augment the training data to remove the reliance on the model on such wrong features (Anders *et al.*, 2022). Furthermore, when domain knowledge is available, one can [e.g., by using the *Right for the Right Reasons (RRR)* approach from Ross *et al.* (2017)] introduce expert knowledge into the model training by changing the loss function and guide the AI model to leverage this expert domain knowledge. Cheng *et al.* (2022) demonstrated that the latter approach can improve deep learning models based on leaf image data. For more information about XAI-based model improvement, we refer to Anders *et al.* (2022) and Weber *et al.* (2023). Combining XAI with human expert knowledge to improve AI models (Fel *et al.*, 2022a) constitutes a promising direction for GeoAI research. The latter differs from the spatially-explicit GeoAI modeling approach that integrates geographical knowledge into model design or data organization before training (Liu and Biljecki, 2022).

9.1.3   APPLYING XAI METHODS IN GEOAI

Although the application of XAI methods in GeoAI research is still relatively new (McGovern *et al.*, 2019), there is great potential in leveraging established AI techniques from other domains with similar tasks, such as image classification or image segmentation. It is however worth noting that GeoAI has some characteristics that distinguish it from other domains. These characteristics are related to the data, models, and explanation requirements.

A lot of studies using XAI involve image data (e.g., explaining medical diagnoses). Indeed explanations based on image data are more intuitive and easier to interpret for humans compared to non-visual or unstructured data. In GeoAI however, data can range from remote sensing or street view images to traffic trajectories and time series, which requires the XAI methods to be applicable to multiple types of data.

XAI methods shall consider additional characteristics (e.g., spatial autocorrelation and scale effect) inherent to geographical research. As dependencies between individual sample points exist in spatio-temporal data, XAI approaches assuming the independence of input features might not be adequate. Further, there is great potential for extending and applying existing XAI methods to incorporate expert domain knowledge from GeoAI.

When it comes to explanations for GeoAI models, additional domain expertise is also often required to understand the raw outcomes of XAI methods. The pixel-level attribution maps, e.g., would need extensions to object- or concept-level explanations to deliver meaningful interpretations in GeoAI (Huang *et al.*, 2022a). Besides, the spatial heterogeneity of geographical studies can lead to explanations of the same model that differ across regions (e.g., big cities vs. small towns). This creates an additional layer of complexity when assessing local explanations and attempting to scale from local to global explanations.

9.2 XAI METHODS

As we have mentioned in Section 9.1 there exist various types of XAI methods (e.g., gradient-based and decomposition-based) and even different classes of XAI methods (e.g., local and global explanations). In Figure 9.2 we give an overview of some prominent methods and their categorization that we will also briefly present in the following. For a more detailed review, we refer to the works of Samek *et al.* (2021) and Holzinger *et al.* (2022) and the references therein.

In the following we denote by the function:

$$f : \mathbb{R}^d \longrightarrow \mathbb{R}$$
$$x \mapsto f(x) = f_L \circ \ldots \circ f_1(x)$$

an already trained machine learning model implemented as a Deep Neural Network with L layers each of the form:

$$f_l : \mathbb{R}^{n_{l-1}} \to \mathbb{R}^{n_l}$$
$$a^{l-1} \mapsto a^l = f_l(a^{l-1})$$

constituting chainable mappings between the real-valued input features x of dimension d, learned hidden representations a^l at each layer l of dimension n_l (with $a^0 = x$), and ultimately the scalar real-valued model output (e.g., the prediction for some target class in case of a classifier).

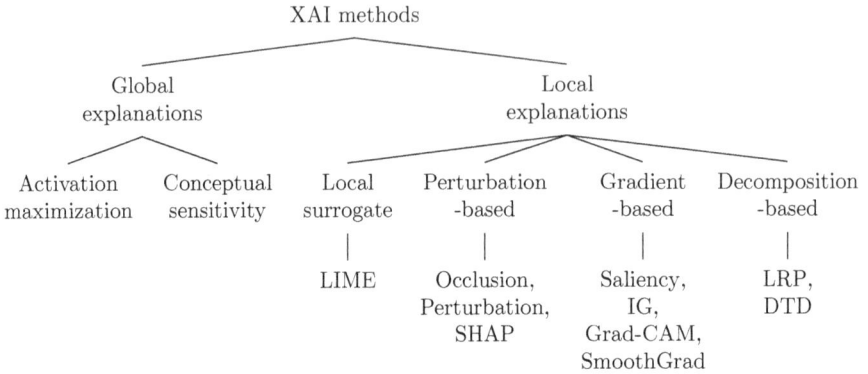

XAI methods

Global explanations   Local explanations

Activation maximization	Conceptual sensitivity	Local surrogate	Perturbation -based	Gradient -based	Decomposition -based
		LIME	Occlusion, Perturbation, SHAP	Saliency, IG, Grad-CAM, SmoothGrad	LRP, DTD

Figure 9.2   Structural overview and the XAI methods.

9.2.1 LOCAL EXPLANATION METHODS

Local explanation methods are XAI methods that are designed to make individual predictions interpretable to the user. This is most often done via attribution maps, where for a given prediction each input feature gets assigned a *relevance score* or a *feature importance* value. In the event of an image classifier, the output of an XAI method could, for instance, visualize the relevant input pixels inside the image that explain the decision for a specific class object. Since deep learning based models, in particular, deep neural networks, have shown incredible performance in image classification there has been a lot of research efforts in explaining such models, which has resulted in various types of local explanation methods based on different principles:

Local surrogate: The idea behind local surrogate based XAI methods is to replace the complex model f with a simpler function locally, that can be explained more easily. One prominent method of this type is *Local Interpretable Model Agnostic Explanations (LIME)* (Ribeiro *et al.*, 2016), where the surrogate model is a sparse linear model represented by its weights v, and which is found by minimizing a locally weighted squared loss of the form:

$$\min_{v} \int [f(\xi) - v^T \xi]^2 \cdot \pi_x(\xi) \cdot dp_x(\xi). \tag{9.1}$$

In practice, the optimization problem is solved by sampling data points ξ in the neighborhood of the point to explain x and weighting them by a proximity function $\pi_x(\xi)$. The LIME approach is model-agnostic and has been applied in many different applications, however, it does not directly solve the explanation problem in the original model. Instead, it requires fitting an ad-hoc model for each prediction, which is computationally expensive. Besides due to the sampling involved in the explanation process, some randomness is introduced, in other words, the explanation is not deterministic.

Perturbation-based: One natural way to find relevant features is to repeatedly delete features in the input signal x (e.g., in the case of an image classifier, replace portions of the input image by a constant-valued square around the considered pixel) and measure the effect on the output $f(x)$, typically by measuring the prediction difference, which is known as the occlusion sensitivity (Zeiler and Fergus, 2014). The higher the change in prediction, the more relevant the feature. Instead of deleting features, one could also replace them with a value sampled from the neighborhood of that feature (or e.g., in images blur pixels), and optimize the perturbation such that it affects the prediction the most, which is known as perturbation analysis (Fong and Vedaldi, 2017; Zintgraf *et al.*, 2017). Another XAI method that could be viewed as perturbation-based is the SHAP method (Lundberg and Lee, 2017). It is inspired by game theory and assigns importance to input features by estimating the change in the expected model prediction when conditioning the prediction on subsets of input features. In Matin and Pradhan (2021) the authors have used SHAP to study the feature importance of a multi-layer perceptron for building-damage classification using remote sensing images. Based on the feature analysis, irrelevant and redundant features were removed to increase the model performance. As the LIME method, SHAP is typically not deterministic and expensive to compute as it is also based on sampling, although some more efficient approximations for deep neural networks exist such as Deep SHAP.

In summary, for the perturbation-based explanation methods, one could represent the perturbation for a given input feature x_i roughly as a mask matrix M_i multiplied with the input x, resulting in a feature importance for feature i of the form:

$$f(x) - f(M_i x). \tag{9.2}$$

While such methods are model-agnostic, they require multiple perturbation steps or involve sampling-based approximations to explain a single prediction, thus they are computationally costly. Additionally, perturbing the inputs might introduce some artifacts and unreliable model predictions, if the perturbed inputs are too far away for the training data distribution.

Gradient-based: Gradient-based XAI methods involve partial derivatives of the model's prediction function w.r.t. the input x. A simple variant is to use directly the gradient of the prediction function $\nabla f(x)$ as the explanation, which is denoted as input saliency (Simonyan *et al.*, 2014), or in the case of a convolutional neural network, use a linear combination of the feature map activations in the last convolutional layer weighted by the average feature maps' gradients for that layer, which is known as the *Gradient-weighted Class Activation Mapping (Grad-CAM)* (Selvaraju *et al.*, 2017). Grad-CAM was also used, for instance, in McGovern *et al.* (2019) for tornado prediction, where the influence of grid cells for tornado prediction based on radar images was analyzed. While such gradient-based explanations are simple and cheap to compute (they require only one standard gradient backward pass through the neural network), they do not quantify the contributions of input features to the *actual* prediction, but instead indicate how sensitive the prediction function is to a small *change* in the input feature's value. Another popular gradient-based method is *Integrated Gradients (IG)* (Sundararajan *et al.*, 2017). IG integrates the gradient $\nabla f(x)$ along the segment $[\tilde{x}, x]$, where $\tilde{x}$ represents a baseline input that shall represent a neutral state without information. More precisely, using IG the importance of a feature x_i is computed by numerically approximating the following integral:

$$(x - \tilde{x})_i \int_0^1 [\nabla f(\tilde{x} + t(x - \tilde{x}))]_i \, dt. \tag{9.3}$$

The resulting feature importance fulfills several interesting properties of an explanation, in particular, the sum of all feature importances is equal to the prediction function difference $f(x) - f(\tilde{x})$ between the original input x and the baseline input $\tilde{x}$, such that feature importances can be readily interpreted as contributions to the model's output. However, in practice, gradient-based explanations are known to be noisy and subject to high variance, which is also why some smoothing techniques have been introduced such as SmoothGrad (Arras *et al.*, 2022; Smilkov *et al.*, 2017). Though, the latter introduces additional hyperparameters in the explanation problem that require careful fine-tuning, which is not always feasible when no a priori quality criterion or ground truth explanation is available for a given task.

Decomposition-based: The last type of local explanation method is based on a backward decomposition of the prediction function $f(x)$. Such a technique was initially introduced as LRP (Bach *et al.*, 2015) and originally designed as a message-passing propagation algorithm based on a layer-wise conservation principle. Subsequently, the LRP technique was theoretically justified and extended through the mathematical framework of *Deep Taylor Decomposition (DTD)* (Montavon *et al.*, 2017). This approach exploits the multi-layer structure of the model:

$$f(x) = f_L \circ \ldots \circ f_1(x), \tag{9.4}$$

to derive hidden representation relevances at each layer (one relevance value per neuron in each hidden representation a^l). To obtain the relevances in the input domain one redistributes the model output $f(x)$ layer by layer until the input layer is reached.

More precisely, given the relevances R_j^l of the neurons indexed by j at layer l (i.e., the relevances of the neurons a_j^l), one can obtain the relevances R_i^{l-1} of the neurons indexed by i at the layer below $l - 1$ (i.e., the relevance of the neurons a_i^{l-1}) by summing up shares of relevances representing the contributions of the hidden representations in the forward pass, through the generic propagation rule:

$$R_i^{l-1} = \sum_j \frac{a_i^{l-1} \cdot \rho(w_{ij})}{\sum_{i'} a_{i'}^{l-1} \cdot \rho(w_{i'j})} R_j^l, \tag{9.5}$$

where w_{ij} are the neural network connection weights between neurons i in layer $l - 1$ and neurons j in layer l, and the function $\rho(\cdot)$ is a mapping that depends on the type of layer and input domain. Several instances of the above rule have been proposed, for an overview and practical recommendations on their applications we refer to Montavon *et al.* (2019, 2018).

The explanations delivered by LRP and DTD verify a layer-wise, as well as an overall, conservation principle, i.e., they sum up to the model's output $f(x)$ for the considered data point x, and thus can be interpreted as feature contributions to the output. Such approaches have been successfully applied to computer vision neural networks (Arras *et al.*, 2022; Lapuschkin *et al.*, 2019; Yeom *et al.*, 2021), as well as to time series models (Arras *et al.*, 2019a,b; Poerner *et al.*, 2018; Yang *et al.*, 2018). In GeoAI, LRP was empirically validated through a synthetic task (Mamalakis *et al.*, 2022a), as well as applied to geospatial data, e.g., on geospatial fields of sea surface temperature where it was able to identify physically meaningful patterns of climate variability (Toms *et al.*, 2020). One limitation of LRP is that one has to carefully apply the relevance propagation rules depending on the given model's structure, layers and input domain, and for new types of layers new rules need to be designed, hence the approach is not model-agnostic. The advantages are that the explanation is deterministic and cheap to compute: it requires only a single backward pass through the model.

9.2.2 GLOBAL EXPLANATION METHODS

In contrast to local explanation methods that try to explain AI models based on local individual decisions, global explanation methods try to find concepts that are globally important for the model to make a successful prediction.

Activation maximization: In its simplest form *Activation Maximization* (Erhan *et al.*, 2009; Simonyan *et al.*, 2014) tries to find input data that maximizes a particular neuron activity a_i^l at layer l. This can be written as a simple optimization problem:

$$x^* = \text{argmax}_x a_i^l(x). \tag{9.6}$$

The resulting input gives an idea of what this neuron is particularly sensitive to.

Conceptual sensitivity: The idea behind conceptual sensitivity is to measure how much a particular concept affects the model output. One popular method in this category is *Testing with Concept Activation Vectors (TCAV)* (Kim *et al.*, 2018). First a concept activation vector v_C^l is computed by training a linear classifier that separates the activations a^l of data points containing a known concept C from data points not containing this concept. Then the concept activation vector v_C^l for the concept C at layer l is used to test the gradient-based sensitivity of the model's output toward this concept via a directional derivative:

$$\lim_{\varepsilon \to 0} \frac{f(a^l(x) + \varepsilon v_C^l) - f(a^l(x))}{\varepsilon}. \tag{9.7}$$

9.2.3 FURTHER XAI METHODS AND CHALLENGES

In Sections 9.2.1 and 9.2.2, we have presented prominent local and global explanation methods, in particular in conjunction with convolutional neural networks. However, a lot more XAI methods have been proposed recently that differ from the previously discussed methods. We continue with some of these methods and refer the interested reader to works such as Samek *et al.* (2021) and Holzinger *et al.* (2022).

- *TreeExplainer:* In Lundberg *et al.* (2020) the authors have introduced *TreeExplainer* which is an XAI method that can be used to explain (ensembles of) decision trees by measuring the input contribution to single tree decisions as well as interaction effects among features.

- *Explaining Graphs:* Another type of model that is gaining more attention in recent times is the graph neural network. It is therefore natural to extend the known concepts to these types of models. One example is *GraphLIME* (Huang *et al.*, 2022b) which follows the idea of LIME to use an interpretable model, in particular, it uses supervised nonlinear feature selection to explain node classification. Other extensions are for instance *GraphLRP* (Schnake *et al.*, 2020).

- *Glocal XAI:* The expression "glocal" is a mixture of the two words "global" and "local" and goes beyond the analysis of which features were learned by the model and where these are to be found. One such method is CRP which has been recently proposed in Achtibat *et al.* (2022), it combines both approaches to find global concepts for individual samples.

As already mentioned, the field of XAI is very active and there are many different methods and directions to foster transparency [including causality (Rieckmann *et al.*, 2022; Schölkopf, 2022) and other concepts]. However, a more elaborated and complete overview is beyond the scope of this chapter. We refer the interested reader to, for instance, Samek *et al.* (2021) and Holzinger *et al.* (2022).

Finally, we comment on one of the challenging problems for XAI methods: the evaluation and comparison of methods. There is a subjective human-based approach toward the validation, e.g., through visual inspection and comparisons with expert domain knowledge. This is the standard approach and it can also lead to further explanation improvements, e.g., by including domain knowledge in the XAI method or even in particular aspects of the modeling. Other quantitative criteria could be the computational complexity, e.g., measured by run time or arithmetic operations. Within the XAI community, further objective quantitative measures have also been proposed such as *pixel flipping* and *relevance mass accuracy* (Arras *et al.*, 2022; Bach *et al.*, 2015; Samek *et al.*, 2021). A comprehensive toolbox for quantitative evaluation of explanations has been recently published (Hedström *et al.*, 2023).

9.3   XAI APPLICATIONS IN GEOAI

In Section 9.2, we have given an overview of popular XAI methods, in particular, many of them are also already used in geo-applications. For example, Grad-CAM was used in tornado prediction (McGovern *et al.*, 2019), SHAP in building-damage classification (Matin and Pradhan, 2021), and LRP in climate variability (Toms *et al.*, 2020). In this section, we want to present further use cases and discuss the usage of XAI in GeoAI in more detail.

The primary goal in many GeoAI studies is to investigate and improve the model's prediction performance on specific tasks (e.g., how a model can increase the accuracy of traffic volume forecasting or land use classification). However, applying

XAI methods to understand *why* a model performs as it does bears further development potential in GeoAI (Cheng *et al.*, 2021a; Li, 2022; Liu and Biljecki, 2022; Xing and Sieber, 2021). As mentioned in Section 9.1, XAI methods can help explain model decisions, discover task-related knowledge, and contribute to model improvement. Here we provide three examples of how XAI has provided these services in GeoAI studies. Section 9.3.1 aims to predict rainfall-runoff in Germany based on physical environmental data such as digital elevation maps, soil features, and hydrogeological features, as well as precipitation data. In this use case an XAI method was used to explain the decision of a trained model, and we illustrate how the resulting feature importance can be interpreted in a physically meaningful way. Section 9.3.2 uses an XAI method to discover new knowledge related to human activities through exploiting taxi trajectories data collected in Beijing, China. Section 9.3.3 deals with a model that classifies scenes based on remote sensing images, which is a common task in GeoAI. An XAI method was used to help improve the model by identifying problems in the training process.

9.3.1   USE CASE 1: EXPLAIN MODEL DECISIONS FOR RAINFALL-RUNOFF PREDICTION

In this use case, we will illustrate the usefulness of the LRP approach (Arras *et al.*, 2017; Bach *et al.*, 2015) for understanding a neural network model of physical systems for rainfall-runoff prediction in Germany.

A key step for the prediction of flood events consists in modeling the effect of precipitation and water level in rivers and lakes [see primer in Beven (2011), overview in Sitterson *et al.* (2018), and history in Peel and McMahon (2020)]. The weather forecast might predict heavy rainfall, but will this cause the water levels of nearby water bodies to rise, or will the water trickle into the soil and evaporate into the air? Apart from meteorological forcing parameters, such as precipitation, temperature, and potential evapotranspiration, a variety of geographical factors (in this context referred to as ancillary features) contribute to this system: anthropogenic factors, such as soil sealing or agricultural formation of the landscape, affect the interaction between precipitation and the land surface. Soil and rock types determine the speed of trickling and subsurface transport of water. The recent history of precipitation determines how much water can be retained in soil and canopy. Orography has an obvious effect on trickling and surface runoff, and lastly, the routing of water along river networks can compound several non-critical events into a disaster. Luckily, information on all these factors is available in the form of weather forecasts, maps, and river gauging data.

When using models based on differential equations to express these relationships, it is up to the human modeler to decide on factors to consider, choose a suitable format or representation, and somehow include them in the equations. Over the last century, much effort has gone into creating sophisticated, detailed systems of equations, ultimately yielding performant operational models (Van Der Knijff *et al.*, 2010). In recent years, neural network models have been employed to tackle the rainfall-runoff modeling task (Kratzert *et al.*, 2018). In principle, these models are capable of discovering arbitrarily complex dynamics and processing vast amounts of data (e.g., remote sensing images) that could not be handled by traditional approaches. On the downside, neural networks are considered black-box models, which means that their increased capabilities come at the expense of not being readily interpretable. Below we demonstrate that LRP (Arras *et al.*, 2017; Bach *et al.*, 2015) allows overcoming this limitation and yields explanations of the model's decision-making process that are human-interpretable and in line with the physical understanding of the systems involved.

We would like to discuss the practical advantages first that LRP offers to the neural network modeler: runoff dynamics vary greatly over regions, so models are usually calibrated on relatively homogeneous and contingent areas, such as a medium-sized river basin. The model parameters are thus fitted to a relatively small amount of observations and need to be re-calibrated when extending the model to a new study area. This makes transferring models to remote or underdeveloped areas that lack costly modern measurement infrastructures (river gauges) difficult, as no data is available to re-calibrate the model for this region (Blöschl, 2006). In contrast, neural network models don't require to be "tailored" to a specific hydro-meteorological situation because they are expressive enough to capture several dynamics within a single model. To leverage this advantage, we can train them on all the data at our disposal. Providing additional information about the land surface, such as the aforementioned orographical, soil, hydrological, and land cover maps, allows the neural network model to differentiate between these different dynamics in a purely data-driven way (Kratzert *et al.*, 2019). As LRP explains which features are actually being used, this can help us to get an idea of how well the model will generalize to a given unfamiliar area.

For this use case, we employ data from Zink *et al.* (2017), collected daily between the years 1951 and 2010 at a spatial resolution of 4km, which covers the area of Germany. We trained a long short-term memory (LSTM) based model with five dynamic features: precipitation, potential evapotranspiration, minimum, maximum, and average temperature, as well as 102 static ancillary features taken from a hydro-geological map (HÜK) (Dorhöfer *et al.*, 2001), a soil map (BÜK) (Krug *et al.*, 2015), a land cover map (CLC) (GeoBasis-DE / BKG, 2012), and a digital elevation map (DEM) (GeoBasis-DE / BKG, 2021). For our analysis of feature importance, we will focus on the ancillary features. The task is to predict the daily generated runoff on a 4km × 4km grid. Our trained model reaches a test prediction performance in the mean spatial Nash–Sutcliffe Efficiency (NSE) of 0.6307.

Figure 9.3 shows which of the land surface features are used by the model to differentiate between different local hydro-meteorological dynamics. Following hydrological intuition, total elevation and elevation gradients are assigned high importance. Somewhat surprisingly, soil type (BÜK) has generally low importance, whereas rock features (HÜK) are more relevant. Land use features, especially for urban areas that imply a high level of soil sealing, are also detected as very relevant. Overall, the relative importance within and across different types of map information aligns with physical intuition. This underlines the credibility of LRP and demonstrates that a "black box" neural network together with LRP can create physically plausible and human-interpretable insights, that might prove further valuable when modeling a complex or poorly understood system for the first time.

Figure 9.3   LRP provides a task-specific relevance score for each ancillary input feature contained in the digital elevation map (DEM), soil map (BÜK), hydrogeological map (HÜK), and land cover map (CLC). Different scores are attributed within and across the ancillary input maps. Note that while relevance scores are calculated locally for each location, we show the mean over all locations to provide a first oversight. In terms of actionable insights, relevance scores suggest omitting soil type information or reducing the dimension of the representation.

It is generally held that neural networks extrapolate poorly to out-of-distribution samples. LRP can serve the modeler to gain a coarse estimate of how similar a new region is concerning the modeling task: as different regions are processed by the neural network, we can use their "signature" of LRP relevance scores to cluster the regions. Figure 9.4 shows that simple K-means clustering of the relevance scores yields spatially continuous regions that are based on a compound of elevation, soil and rock type, and urbanization. Performing the same clustering on raw input values yields uninformative clusters dominated by elevation. Our data-driven findings again qualitatively agree with hydro-meteorologically homogeneous regions. Relevance scores are geometrically interpretable (they can be treated as vectors and compared to one another), so comparing the relevance "signature" of a new sample to the previously built clusters provides insight into how novel the sample actually is and how much we can trust the network's prediction on the new sample.

9.3.2   USE CASE 2: KNOWLEDGE DISCOVERY FOR TRAFFIC ANALYSIS

As highlighted by Mamalakis *et al.* (2022b) and Roscher *et al.* (2020), a benefit of XAI methods is to reveal connections between input variables and model outputs (i.e., "explain to discover"). Such connections can provide insights into the task at hand and inspire additional research. For instance, Toms *et al.* (2021) applied the LRP method to understand neural network predictions of the climate system and were able to identify climate variability links between non-adjacent regions. In a study by Cheng *et al.* (2021a), XAI methods were applied to discover knowledge related to taxi trips. Below, we highlight how this study was conducted.

Cheng *et al.* (2021a) used Beijing, China, as the research area and the research data consists of the origin and destination (OD) points of taxi trajectories. This type of data has been widely used to investigate human activities and urban environments (Cheng *et al.*, 2021b; Liu *et al.*, 2012). Urban human activities and environmental features (e.g., commuting behaviors and land use) will affect taxi trips. This results

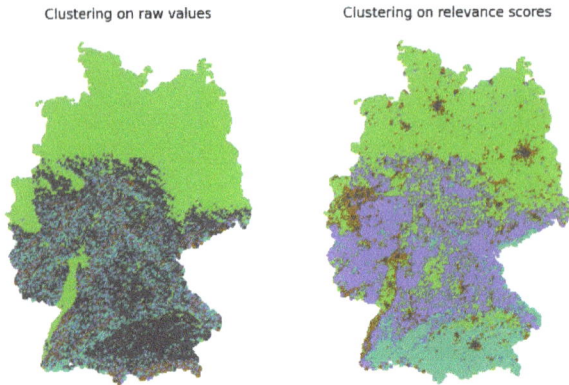

Figure 9.4 Standard K-means clustering on the raw input values (left) and the relevance scores (right). Clusters are denoted by different colors.

in patterns that differ in both temporal and spatial dimensions (e.g., on weekdays and weekends/holidays). Cheng *et al.* (2021a) explored knowledge related to taxi travels, in particular for which time slots and spatial units the characteristics of taxi travels are significant for distinguishing between the two periods of weekdays and weekends/holidays. Based on the raw taxi trajectory data, origin point volumes were extracted according to a 1km² grid square and half hour as the spatial resolution and temporal resolution. The daily spatio-temporal distributions of origin point volumes with a size of $30 \times 30 \times 48$ were labeled as *weekday* and *weekend/holiday* and served as input data for a 5-layer convolutional neural network. The classification accuracy of the trained model on the test data was 98%, indicating that the model learned the knowledge of taxi travel for the two periods.

To reveal the learned reasoning strategies, Cheng *et al.* (2021a) applied the LRP approach (Bach *et al.*, 2015) to obtain local explanations. According to the fixed location of spatio-temporal units in each sample, local explanations for this task can be transformed into a global explanation (i.e., contribution value of each spatio-temporal unit). Figure 9.5 displays the normalized unit contribution in temporal and spatial dimensions. For the XAI analysis the spatio-temporal units were first sorted according to their normalized contribution value from large to small, and then the cumulative contribution proportions in spatial and temporal dimensions were calculated. The units with a cumulative proportion of more than 60% were considered as *important* units. In the temporal dimension, *important* time slots (8:00 AM to 5:30 PM) align with office hours. In the spatial dimension, northern Beijing has higher contribution values than southern Beijing. These high-contribution areas are concentrated at major roads (e.g., the 3rd Ring Road) and junctions [e.g., labeled as specific zones in Figure 9.5(b)].

Then the newly acquired knowledge was applied to conduct a time series clustering task on two types of data: a time series of average daily origin point volumes of each spatial unit at a half-hour resolution with 48 elements, and a time series of average daily origin point volumes with 20 elements for the *important* time slots only according to the obtained temporal contribution. The clustering results of the two experiments only differed for 4.98% of the spatial units. This suggests that, for this time series clustering task, the XAI-derived knowledge could be used for more efficient information extraction.

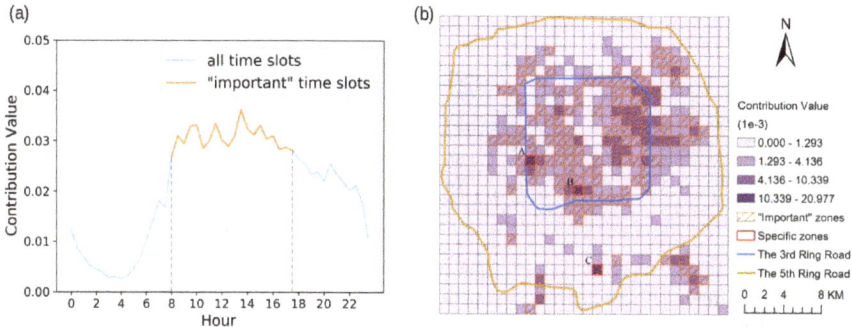

Figure 9.5   Normalized contribution value in (a) temporal and (b) spatial dimensions. The specific zones displayed in (b) are three transportation junctions in Beijing (A: Beijing West Railway Station; B: Beijing South Railway Station; and C: Beijing Nanyuan Airport). (*Source:* Image adopted from Cheng *et al.* (2021a).)

For the above case, a common way in geographical studies to learn the contribution of spatio-temporal units is through analyzing data directly, e.g., results could also be achieved by comparing the average activity distributions between weekdays and weekends/holidays. However, characteristics such as spatio-temporal dependence between units make the task more complex. Figure 9.6 shows that the contribution values obtained by XAI are different from the results of direct data analysis. Though there are geographical methods to deal with this dependence between units, it is often necessary to manually set parameters (e.g., the radius of neighbor definition). Fortunately, AI techniques can assist with automatically mining data even for complex geographical tasks. Through XAI, it is even possible to understand yet undiscovered new relationships learned by the AI model.

9.3.3   USE CASE 3: MODEL IMPROVEMENT FOR REMOTE SENSING CLASSIFICATION

Insights generated through XAI can help researchers identify problems (e.g., poor-quality data) in model training so that the AI model can be improved (Mamalakis *et al.*, 2022b). For instance, Cheng *et al.* (2022)'s study of leaf images integrated XAI feedback and used annotation matrices to introduce expertise into model training. This enabled them to improve the classification accuracy on three tasks (in particular, distinguishing between real and fake leaves, identifying disease leaves, and classifying plant species). Besides, Schramowski *et al.* (2020) demonstrated that

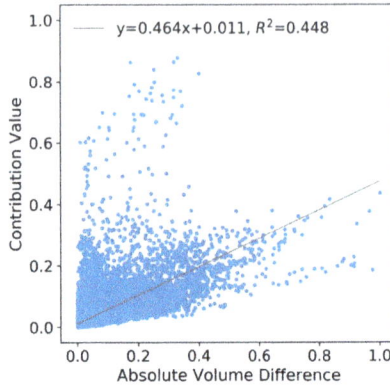

Figure 9.6 Linear regression of the spatio-temporal unit contribution value and the absolute volume difference between the two periods. It shows a non-linear relationship between XAI explanations and the results of direct data analysis, indicating that XAI can help discover complex knowledge related to a specific task. (*Source:* Image adopted from Cheng *et al.* (2021a).)

explanatory interactive learning which adds researchers to the training loop can avoid the "Clever Hans" situation. Inspired by these examples, we conducted an experiment based on remote sensing images. The results of this experiment suggest that XAI methods can also support model improvement in GeoAI.

The purpose of the experiment is to classify remote sensing scenes using a deep learning model. The images are derived from the open-source data set *RSSCN7* (Zou *et al.*, 2015), which are categorized based on their content into the labels: *grass*, *farm*, *industry* (including commercial regions), *lake* (including the greater catchment), *forest*, *resident*, and *parking areas*. *RSSCN7* consists of 2800 images (400 per category), which were collected from Google Earth and have the same size (400 × 400) with four spatial resolutions. The original data were randomly divided into three sets: train data (240 × 7 samples), validation data (80 × 7 samples), and test data (80 × 7 samples). The distribution of samples from each category is balanced in each set. The convolutional neural network *AlexNet* (Krizhevsky *et al.*, 2017) was trained and applied for the classification task. After 100 epochs of training, the classification accuracy on the test data was 81.43%. These results are consistent with those from a previous study that also used *RSSCN7* (Zou *et al.*, 2015).

For a seven-category classification task, the current accuracy (~80%) leaves space for improvement. Therefore, we used a confusion matrix (Figure 9.7) to better understand the classification results. However, there was a tendency for samples in some categories to be confused: *grass* and *farm*, *industry* and *resident*, and *industry* and *parking*. To determine the cause of this confusion, we applied the Grad-CAM method (Selvaraju *et al.*, 2017) to the test data. Figure 9.8 displays the misclassified input samples and provides explanations for each pair of confused categories. It is shown that these categories are sometimes difficult to distinguish by human visual inspection (e.g., *grass* and *farm*, and *industry* and *resident*). Furthermore, the samples within these categories can contain more than one type of object (e.g., an industrial building and a corresponding parking lot). Model explanations of the misclassified samples in Figure 9.8 are also reasonable. Therefore, the quality of the original data set seems to be responsible for the model performance on image classification.

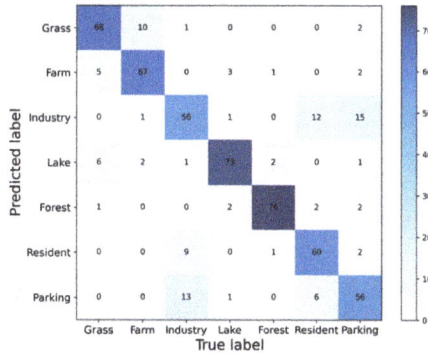

Figure 9.7 Confusion matrix of the classification results on the test data. *Grass*, *farm*, *industry*, *lake*, *forest*, *resident*, and *parking* are the seven categories. The value is the number of corresponding samples.

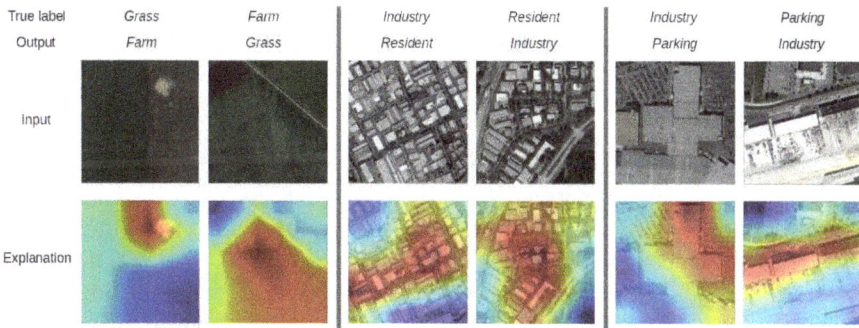

Figure 9.8 Examples of misclassified samples and corresponding explanations for three pairs of confused categories. A warmer color in the explanations indicates a more important pixel for the classification task.

Based on the samples and corresponding model explanations, the model accuracy can be ascribed to sample confusion. In such a situation, it is possible to relabel the samples with multiple labels and retrain the model for a multi-label classification task. Alternatively, it is possible to evaluate the uncertainty of the model classification (e.g., a sample is 60% likely to be *industry*). However, as relabeling samples and retraining a model takes time and can introduce artificial errors, we resorted to observing the full model output for the seven classes, not only considering the class with the maximum value. The larger the value associated with the class, the more likely the sample was to be classified into the corresponding category. The top-1, top-2, and top-3 classification accuracies for the testing data are 81.43%, 95.89%, and 98.57%, respectively. Compared with the top-1 accuracy, the top-2 accuracy increased by 14.46% and surpassed 95%. This suggests that the misclassification is mainly caused by the two-category confusion, which is consistent with Figures 9.7 and 9.8. To highlight the difference between confused samples and not-confused samples, we selected the most distinguishable (i.e., those for which the output value difference percentage of the top two categories is the largest) and the least distinguishable correctly classified samples for each class in the test data (displayed in Figure 9.9). Most of the samples in the top row have a higher spatial resolution and contain a single object, whereas the samples in the bottom row are difficult to interpret with the naked eye.

Figure 9.9 Examples of correctly classified samples of seven categories on the test data. The most distinguishable samples are in the top row. The least distinguishable samples are in the bottom row.

For this experiment, the insights derived by the XAI method suggest that the imperfect model accuracy is not caused by an inappropriate classification strategy from the model. Rather, to increase the model accuracy, we need to clean or relabel the data rather than debug the model structure and/or parameters. This example shows that XAI methods can help improve GeoAI models based on remote sensing images. However, the XAI methods could benefit from further development by integrating domain expertise into training. Furthermore, it would be useful to have a standardized workflow—including XAI methods—for improving GeoAI models.

9.4 DISCUSSION AND CONCLUSION

GeoAI refers to research that develops AI methods for geospatial tasks. Meanwhile, XAI methods aim to make AI models more transparent, e.g., by explaining model decisions. This book chapter presented popular XAI methods and several use cases to highlight how XAI methods can be used for different geo-applications. In particular, these use cases demonstrate that XAI methods can provide insights into GeoAI model decisions, discover task-related knowledge, or lead to model improvement. To exploit XAI in the best possible way, it is important to mention that the development of such methods has to be adapted to the given application. For instance, in the context of GeoAI, there is a variety of different data (e.g., sensor data and demographic data) and corresponding data structures (e.g., dynamic time series and static descriptions). This data can often be embedded into a spatio-temporal space to describe a particular event (e.g., traffic accident or flood). This underlying structure can be exploited to develop XAI methods that agree with existing knowledge or might even be used to validate data-driven explanations. On the other hand, XAI methods bear the potential to discover new task-specific dependencies and structures that were not visible by a mere statistical data analysis approach. Further advanced XAI methods such as CRP (Achtibat et al., 2022) may be particularly interesting in this context, as such methods bridge local and global explanations, and may find contextualized explanations for individual data points that are linked to concepts and structures that would not have been found if the data points were analyzed separately.

Another way to enhance explainability in GeoAI is through spatially-explicit GeoAI models. It consists in extending general AI models with geographical knowledge through changing network structures and using specific ways of training. Compared with XAI methods, spatially-explicit GeoAI models are the most direct way to introduce expertise and control model training. However, also these models suffer from the black box predicament: the rationale for spatially-explicit GeoAI model decisions may remain unknown under the guidance of geographical knowledge. In summary, a combination of spatially-explicit GeoAI models and XAI methods may constitute an optimal solution to increase the explainability of GeoAI. That is, researchers shall design specific network structures and control the training based on geographical knowledge, and then are advised to use XAI methods to explain and improve the trained models.

FUNDING

This chapter is supported by the Data- and AI- supported early warning system (DAKI-FWS) project (No. 01MK21009A), Federal Ministry for Economic Affairs and Climate Action (BMWK), Germany, and the German Research Foundation (No. DFG KI-FOR 5363).

BIBLIOGRAPHY

Achtibat, R., et al., 2022. From "Where" to "What": Towards human-understandable explanations through concept relevance propagation. arXiv preprint arXiv:2206.03208.

Anders, C.J., et al., 2022. Finding and removing Clever Hans: Using explanation methods to debug and improve deep models. Information Fusion, 77, 261–295.

Arras, L., et al., 2019a. Explaining and Interpreting LSTMs. In: Explainable AI: Interpreting, Explaining and Visualizing Deep Learning. Lecture Notes in Computer Science, vol. 11700. 211–238.

Arras, L., *et al.*, 2017. Explaining recurrent neural network predictions in sentiment analysis. *In*: *Proceedings of the EMNLP'17 Workshop on Computational Approaches to Subjectivity, Sentiment & Social Media Analysis*. 159–168.

Arras, L., *et al.*, 2019b. Evaluating recurrent neural network explanations. *In*: *Proceedings of the 2019 ACL Workshop BlackboxNLP: Analyzing and Interpreting Neural Networks for NLP*. 113–126.

Arras, L., Osman, A., and Samek, W., 2022. CLEVR-XAI: A benchmark dataset for the ground truth evaluation of neural network explanations. *Information Fusion*, 81, 14–40.

Assael, Y.M., *et al.*, 2016. LipNet: End-to-end sentence-level lipreading. *arXiv preprint arXiv:1611.01599*.

Bach, S., *et al.*, 2015. On pixel-wise explanations for non-linear classifier decisions by layer-wise relevance propagation. *PloS ONE*, 10 (7), e0130140.

Başağaoğlu, H., *et al.*, 2022. A review on interpretable and explainable artificial intelligence in hydroclimatic applications. *Water*, 14 (8), 1230.

Berner, C., *et al.*, 2019. Dota 2 with large scale deep reinforcement learning. *arXiv preprint arXiv:1912.06680*.

Beven, K.J., 2011. *Rainfall-Runoff Modelling: The primer*. John Wiley & Sons.

Blöschl, G., 2006. Rainfall-runoff modeling of ungauged catchments. *In*: *Encyclopedia of Hydrological Sciences*. John Wiley & Sons, Ltd, Ch. 133.

Castelvecchi, D., 2016. Can we open the black box of AI? *Nature News*, 538 (7623), 20.

Cheng, X., Doosthosseini, A., and Kunkel, J., 2022. Improve the deep learning models in forestry based on explanations and expertise. *Frontiers in Plant Science*, 1531.

Cheng, X., *et al.*, 2021a. A method to evaluate task-specific importance of spatio-temporal units based on explainable artificial intelligence. *International Journal of Geographical Information Science*, 35 (10), 2002–2025.

Cheng, X., *et al.*, 2021b. Multi-scale detection and interpretation of spatio-temporal anomalies of human activities represented by time-series. *Computers, Environment and Urban Systems*, 88, 101627.

Dorhöfer, G., Hannappel, S., and Voigt, H.J., 2001. Die hydrogeologische Übersichskarte von Deutschland (HÜK 200). *Die hydrogeologische Übersichskarte von Deutschland (HÜK 200)*, 47 (3 4), 153–159.

Erhan, D., *et al.*, 2009. Visualizing higher-layer features of a deep network. *University of Montreal*, 1341 (3), 1.

Fel, T., *et al.*, 2022a. Harmonizing the object recognition strategies of deep neural networks with humans. *Advances in Neural Information Processing Systems (NeurIPS)*.

Fel, T., *et al.*, 2022b. CRAFT: Concept recursive activation factorization for explainability. *arXiv preprint arXiv:2211.10154*.

Fong, R.C. and Vedaldi, A., 2017. Interpretable explanations of black boxes by meaningful perturbation. *ICCV 2017*, 3449–3457.

GeoBasis-DE / BKG, 2012. CORINE Land Cover 5 ha, Stand 2012 (CLC5-2012). Accessed: (2023/02/28), Available from: `https://gdz.bkg.bund.de/index.php/default/corine-land-cover-5-ha-stand-2012-clc5-2012.html`.

GeoBasis-DE / BKG, 2021. Digitales Geländemodell Gitterweite 200 m. Accessed: (2023/02/28), Available from: `https://mis.bkg.bund.de/trefferanzeige?` `docuuid=eaaa67a1-5ecb-4e57-af38-b5f1d6d57e2a#metadata_info`.

Hedström, A., *et al.*, 2023. Quantus: An explainable AI toolkit for responsible evaluation of neural network explanations and beyond. *Journal of Machine Learning Research*, 24 (34), 1–11.

Hilburn, K.A., Ebert-Uphoff, I., and Miller, S.D., 2021. Development and interpretation of a neural-network-based synthetic radar reflectivity estimator using GOES-R satellite observations. *Journal of Applied Meteorology and Climatology*, 60 (1), 3–21.

Hinton, G.E., Osindero, S., and Teh, Y.W., 2006. A fast learning algorithm for deep belief nets. *Neural Computation*, 18 (7), 1527–1554.

Hinton, G.E. and Salakhutdinov, R.R., 2006. Reducing the dimensionality of data with neural networks. *Science*, 313 (5786), 504–507.

Holzinger, A., *et al.*, 2022. Explainable ai methods - a brief overview. *In: xxAI - Beyond Explainable AI*. Lecture Notes in Artificial Intelligence, vol. 13200. Springer, 13–38.

Huang, J., *et al.*, 2022a. ConceptExplainer: Interactive explanation for deep neural networks from a concept perspective. *IEEE Transactions on Visualization and Computer Graphics*, 29 (1), 831–841.

Huang, Q., *et al.*, 2022b. GraphLIME: Local interpretable model explanations for graph neural networks. *IEEE Transactions on Knowledge and Data Engineering*.

Iten, R., *et al.*, 2020. Discovering physical concepts with neural networks. *Physical Review Letters*, 124 (1), 010508.

Janowicz, K., *et al.*, 2020. GeoAI: Spatially explicit artificial intelligence techniques for geographic knowledge discovery and beyond. *International Journal of Geographical Information Science*, 34 (4), 625–636.

Kim, B., *et al.*, 2018. Interpretability beyond feature attribution: Quantitative testing with concept activation vectors (TCAV). *In: ICML*. vol. 80, 2673–2682.

Kratzert, F., *et al.*, 2018. Rainfall–runoff modelling using long short-term memory (LSTM) networks. *Hydrology and Earth System Sciences*, 22 (11), 6005–6022.

Kratzert, F., *et al.*, 2019. Towards learning universal, regional, and local hydrological behaviors via machine learning applied to large-sample datasets. *Hydrology and Earth System Sciences*, 23 (12), 5089–5110.

Krizhevsky, A., Sutskever, I., and Hinton, G.E., 2017. ImageNet classification with deep convolutional neural networks. *Communications of the ACM*, 60 (6), 84–90.

Krug, D., Stegger, U., and Eberhardt, E., 2015. Bodenübersichtskarte 1:200.000 (BÜK200) – Status 2015. *In: Jahrestagung der DBG 2015: Unsere Böden - unser Leben*, München.

Lapuschkin, S., *et al.*, 2019. Unmasking Clever Hans predictors and assessing what machines really learn. *Nature Communications*, 10 (1), 1–8.

LeCun, Y., Bengio, Y., and Hinton, G., 2015. Deep learning. *Nature*, 521 (7553), 436–444.

Li, Z., 2022. Extracting spatial effects from machine learning model using local interpretation method: An example of SHAP and XGBoost. *Computers, Environment and Urban Systems*, 96, 101845.

Liu, P. and Biljecki, F., 2022. A review of spatially-explicit GeoAI applications in Urban Geography. *International Journal of Applied Earth Observation and Geoinformation*, 112, 102936.

Liu, Y., *et al.*, 2012. Urban land uses and traffic 'source-sink areas': Evidence from GPS-enabled taxi data in Shanghai. *Landscape and Urban Planning*, 106 (1), 73–87.

Lundberg, S.M., *et al.*, 2020. From local explanations to global understanding with explainable AI for trees. *Nature Machine Intelligence*, 2 (1), 56–67.

Lundberg, S.M. and Lee, S.I., 2017. A unified approach to interpreting model predictions. *Advances in Neural Information Processing Systems*, 30.

Mamalakis, A., Barnes, E.A., and Ebert-Uphoff, I., 2022a. Investigating the fidelity of explainable artificial intelligence methods for applications of convolutional neural networks in geoscience. *Artificial Intelligence for the Earth Systems*, 1 (4), e220012.

Mamalakis, A., Ebert-Uphoff, I., and Barnes, E.A., 2022b. Explainable artificial intelligence in meteorology and climate science: Model fine-tuning, calibrating trust and learning new science. *In: International Workshop on Extending Explainable AI Beyond Deep Models and Classifiers*. Springer, 315–339.

Matin, S.S. and Pradhan, B., 2021. Earthquake-induced building-damage mapping using Explainable AI (XAI). *Sensors*, 21 (13), 4489.

Mayer, K.J. and Barnes, E.A., 2021. Subseasonal forecasts of opportunity identified by an explainable neural network. *Geophysical Research Letters*, 48 (10), e2020GL092092.

McGovern, A., *et al.*, 2019. Making the black box more transparent: Understanding the physical implications of machine learning. *Bulletin of the American Meteorological Society*, 100 (11), 2175–2199.

Miller, T., 2019. Explanation in artificial intelligence: Insights from the social sciences. *Artificial Intelligence*, 267, 1–38.

Montavon, G., *et al.*, 2019. Layer-wise relevance propagation: An overview. *In: Explainable AI: Interpreting, Explaining and Visualizing Deep Learning*. Lecture Notes in Computer Science, vol. 11700. Springer, 193–209.

Montavon, G., *et al.*, 2017. Explaining nonlinear classification decisions with deep Taylor decomposition. *Pattern Recognition*, 65, 211–222.

Montavon, G., Samek, W., and Müller, K.R., 2018. Methods for interpreting and understanding deep neural networks. *Digital Signal Processing*, 73, 1–15.

Peel, M.C. and McMahon, T.A., 2020. Historical development of rainfall-runoff modeling. *Wiley Interdisciplinary Reviews: Water*, 7 (5).

Poerner, N., Roth, B., and Schütze, H., 2018. Evaluating neural network explanation methods using hybrid documents and morphosyntactic agreement. *In: Proceedings of the 56th Annual Meeting of the Association for Computational Linguistics*. 340–350.

Reichstein, M., *et al.*, 2019. Deep learning and process understanding for data-driven Earth system science. *Nature*, 566 (7743), 195–204.

Ribeiro, M.T., Singh, S., and Guestrin, C., 2016. "Why should I trust you?": Explaining the predictions of any classifier. *In: Proceedings of the 22nd ACM SIGKDD International Conference on Knowledge Discovery and Data Mining*, KDD '16. 1135–1144.

Rieckmann, A., *et al.*, 2022. Causes of Outcome Learning: a causal inference-inspired machine learning approach to disentangling common combinations of potential causes of a health outcome. *International Journal of Epidemiology*, 51 (5), 1622–1636.

Roscher, R., *et al.*, 2020. Explainable machine learning for scientific insights and discoveries. *IEEE Access*, 8, 42200–42216.

Ross, A.S., Hughes, M.C., and Doshi-Velez, F., 2017. Right for the right reasons: Training differentiable models by constraining their explanations. *In*: *Proceedings of the Twenty-Sixth International Joint Conference on Artificial Intelligence*. 2662–2670.

Russakovsky, O., *et al.*, 2015. ImageNet large scale visual recognition challenge. *International Journal of Computer Vision*, 115 (3), 211–252.

Samek, W., *et al.*, 2019. *Explainable AI: Interpreting, Explaining and Visualizing Deep Learning*. Lecture Notes in Computer Science, vol. 11700. Springer Nature.

Samek, W., *et al.*, 2021. Explaining deep neural networks and beyond: A review of methods and applications. *Proceedings of the IEEE*, 109 (3), 247–278.

Schnake, T., *et al.*, 2020. Higher-order explanations of graph neural networks via relevant walks. *IEEE Transactions on Pattern Analysis and Machine Intelligence*, 44, 7581–7596.

Schölkopf, B., 2022. *Causality for Machine Learning*, 1st ed. New York, NY, USA: Association for Computing Machinery, 765–804.

Schramowski, P., *et al.*, 2020. Making deep neural networks right for the right scientific reasons by interacting with their explanations. *Nature Machine Intelligence*, 2 (8), 476–486.

Selvaraju, R.R., *et al.*, 2017. Grad-CAM: Visual explanations from deep networks via gradient-based localization. *In*: *Proceedings of the IEEE International Conference on Computer Vision*. 618–626.

Silver, D., *et al.*, 2016. Mastering the game of Go with deep neural networks and tree search. *Nature*, 529 (7587), 484–489.

Simonyan, K., Vedaldi, A., and Zisserman, A., 2014. Deep inside convolutional networks: Visualising image classification models and saliency maps. *In*: *Proceedings of the International Conference on Learning Representations (ICLR)*. ICLR.

Sitterson, J., *et al.*, 2018. An overview of rainfall-runoff model types. *International Congress on Environmental Modelling and Software*.

Smilkov, D., *et al.*, 2017. SmoothGrad: removing noise by adding noise. *In*: *ICML*.

Sundararajan, M., Taly, A., and Yan, Q., 2017. Axiomatic attribution for deep networks. *In*: *Proceedings of the 34th International Conference on Machine Learning - Volume 70*, ICML'17. 3319–3328.

Toms, B.A., Barnes, E.A., and Ebert-Uphoff, I., 2020. Physically interpretable neural networks for the Geosciences: Applications to Earth system variability. *Journal of Advances in Modeling Earth Systems*, 12 (9).

Toms, B.A., Barnes, E.A., and Hurrell, J.W., 2021. Assessing decadal predictability in an Earth-system model using explainable neural networks. *Geophysical Research Letters*, 48 (12), e2021GL093842.

Van Der Knijff, J., Younis, J., and De Roo, A., 2010. LISFLOOD: a GIS-based distributed model for river basin scale water balance and flood simulation. *International Journal of Geographical Information Science*, 24 (2), 189–212.

Weber, L., *et al.*, 2023. Beyond explaining: Opportunities and challenges of XAI-based model improvement. *Information Fusion*, 92, 154–176.

Xing, J. and Sieber, R., 2021. Integrating XAI and GeoAI. *In: GIScience 2021 Short Paper Proceedings*. UC Santa Barbara: Center for Spatial Studies.

Xing, X., *et al.*, 2020. Mapping human activity volumes through remote sensing imagery. *IEEE Journal of Selected Topics in Applied Earth Observations and Remote Sensing*, 13, 5652–5668.

Yang, Y., *et al.*, 2018. Explaining therapy predictions with layer-wise relevance propagation in neural networks. *In: 2018 IEEE International Conference on Healthcare Informatics (ICHI)*. IEEE, 152–162.

Yeom, S.K., *et al.*, 2021. Pruning by explaining: A novel criterion for deep neural network pruning. *Pattern Recognition*, 115, 107899.

Zeiler, M.D. and Fergus, R., 2014. Visualizing and understanding convolutional networks. *In: Computer Vision–ECCV 2014: 13th European Conference*. 818–833.

Zhang, F., *et al.*, 2019. Social sensing from street-level imagery: A case study in learning spatio-temporal urban mobility patterns. *ISPRS Journal of Photogrammetry and Remote Sensing*, 153, 48–58.

Zink, M., *et al.*, 2017. A high-resolution dataset of water fluxes and states for Germany accounting for parametric uncertainty. *Hydrology and Earth System Sciences*, 21 (3), 1769–1790.

Zintgraf, L.M., *et al.*, 2017. Visualizing deep neural network decisions: Prediction difference analysis. *In: ICLR 2017*.

Zou, Q., *et al.*, 2015. Deep learning based feature selection for remote sensing scene classification. *IEEE Geoscience and Remote Sensing Letters*, 12 (11), 2321–2325.

10 Spatial Cross-Validation for GeoAI

Kai Sun
GeoAI Lab, Department of Geography, University at Buffalo

Yingjie Hu
GeoAI Lab, Department of Geography, University at Buffalo

Gaurish Lakhanpal
GeoAI Lab, Department of Geography, University at Buffalo;
Adlai E. Stevenson High School

Ryan Zhenqi Zhou
GeoAI Lab, Department of Geography, University at Buffalo

CONTENTS

10.1 Introduction 201
10.2 Four Main Spatial CV Methods 203
    10.2.1 Clustering-based spatial CV 204
    10.2.2 Grid-based spatial CV..205
    10.2.3 Geo-attribute-based spatial CV...205
    10.2.4 Spatial leave-one-out CV 206
    10.2.5 Summary...207
10.3 Examples Based on Real-World Data 207
    10.3.1 Example 1: Predicting domestic violence using random forest........208
    10.3.2 Example 2: Predicting obesity prevalence using deep neural network 210
10.4 Conclusions 212
    Bibliography..213

10.1 INTRODUCTION

Cross-validation (CV) has been widely used in GeoAI research to evaluate the performance of machine learning models. Often, a labeled data set is split into a number of subsets, with one subset held out for validation and the remaining subsets used for model training; the training-and-validation process is then iterated multiple times until all subsets of data are used for training and validation (Browne 2000). The

DOI: 10.1201/9781003308423-10

commonly used k-fold CV splits data into k subsets (or k folds), and uses one subset for validation and k-1 subsets for model training. The process is then repeated k times, and the obtained k results are averaged to compute a final score for the machine learning model (Burman 1989). When k equals to the total number of data instances n in the data set, k-fold CV becomes leave-one-out CV, in which only one data instance is held out for validation each time and the process is repeated n times. As an evaluation technique, CV often provides more robust evaluation scores for machine learning models since the models are trained and evaluated on different subsets of data (Géron 2019). Accordingly, CV avoids the potential bias when only one particular subset of the data is used for validation.

While increasing evaluation robustness, CV could lead to another type of bias when applied to geographic data. CV typically randomly splits data into training and validation subsets, and such a random CV approach can generate many training and validation data instances that are spatially close. As widely discussed in the spatial study literature, spatially close data instances are likely to share similar attribute values due to the existence of spatial dependency (Tobler 1970; Diniz-Filho, Bini, and Hawkins 2003; Fotheringham 2009; Hijmans 2012). A model trained on such training data, therefore, can be considered as having already "peeked" into the nearby validation data, given their spatial closeness and likely attribute similarity. Consequently, we may obtain overly optimistic evaluation results that do not represent the true performance of the model (Roberts et al. 2017; Pohjankukka et al. 2017; Karasiak et al. 2019; Meyer and Pebesma 2022; Karasiak et al. 2022).

Spatial cross-validation (spatial CV) is a spatially explicit CV method that can help address this evaluation bias by splitting data spatially rather than randomly. Spatial CV uses strategies, such as spatial buffering and grid-based splitting (Valavi et al. 2019), to create subsets of data that are more independent from each other, thereby decreasing the spatial autocorrelation between training and validation data. As machine learning models have been increasingly applied to geographic data, researchers have developed a number of methods and software packages for spatial CV. However, these methods and software packages are scattered in the literature across multiple disciplines, including ecology (Roberts et al. 2017; Meyer et al. 2018; Valavi et al. 2019), remote sensing (Brenning 2012; Karasiak et al. 2019; Karasiak 2020), GIScience (Pohjankukka et al. 2017; Crosby, Damoulas, and Jarvis 2020; Brenning 2022), and computer science (Airola et al. 2019; Da Silva, Parmezan, and Batista 2021). Consequently, it is difficult for researchers new to spatial CV to quickly grasp the main methods and existing software packages from this multidisciplinary literature. This book chapter fills this gap by presenting four main spatial CV methods identified from the literature, and uses two examples based on real-world data to demonstrate these spatial CV methods by comparing them with random CV.

Before we move on to discussing spatial CV methods, it is worth noting that spatial CV is not always better than random CV, when machine learning models are applied to geographic data. This can be understood based on two common situations in GeoAI research: within-area prediction and between-area prediction (Roberts et al. 2017; Goodchild and Li 2021). For within-area prediction or interpolation (in Figure 10.1(a)), a model is trained on the known values in the study area and is then used to predict unknown values in the same area. In such a case, random CV is in fact preferred since the real test data can be spatially close to known training data too, and we want to leverage the existence of spatial autocorrelation to generate more accurate predictions. For between-area prediction or extrapolation (in Figure 10.1(b)), a model is trained on known values in one geographic area and is then used to predict unknown values in another area (which could be adjacent to the training data area as shown in Figure 10.1(b) or be further away). In such a case, spatial CV is likely to provide a more realistic evaluation of model performance since the spatially-split validation data can better mimic the real test scenario that the trained model will be used for (Crosby, Damoulas, and Jarvis 2020; Meyer and Pebesma 2021). In short, whether we should choose spatial CV or random CV depends on which strategy better mimics the real application scenario of the trained machine learning model.

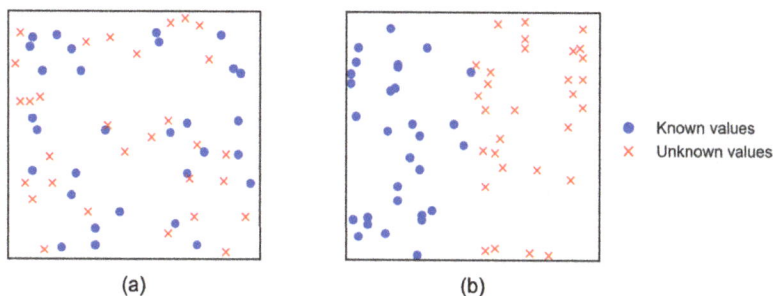

(a)   (b)

● Known values
✕ Unknown values

Figure 10.1 An illustration of two common situations under which machine learning models are applied to geographic data: (a) within-area prediction or *interpolation*; (b) between-area prediction or *extrapolation*.

The remainder of this chapter is organized as follows. Section 10.2 presents four main spatial CV methods and software packages that implement these methods. In Section 10.3, we use two examples based on real-world data to demonstrate the use of these four spatial CV methods in comparison with random CV. Finally, Section 10.4 concludes this chapter.

10.2 FOUR MAIN SPATIAL CV METHODS

While various spatial CV methods have been developed, they can be organized into four main types: *clustering-based spatial CV, grid-based spatial CV, geo-attribute-based spatial CV, and spatial leave-one-out CV*. These four methods share the common idea of spatially splitting training and validation data while differing in their specific ways of splitting the data. A number of software packages have been developed to implement these spatial CV methods. Given the popularity of the R programming language in the ecology research community where spatial CV has received much attention, a lot of these packages are developed in R, which include:

- *sperrorest*: https://cran.r-project.org/web/packages/sperrorest (Brenning 2012)

- *spatialsample*: https://cran.r-project.org/web/packages/spatialsample

- *blockCV*: https://cran.r-project.org/web/packages/blockCV (Valavi et al. 2019)

- *ENMeval*: https://cran.r-project.org/web/packages/ENMeval (Muscarella et al. 2014)

- *Mlr3spatiotempcv*: https://cran.r-project.org/web/packages/mlr3spatiotempcv (Schratz et al. 2021)

- *CAST*: https://cran.r-project.org/web/packages/CAST (Meyer et al. 2018)

There also exist two Python-based spatial CV packages, which are:

- *spacv*: https://github.com/SamComber/spacv

- *MuseoToolbox*: https://museotoolbox.readthedocs.io (Karasiak 2020)

Each package implements its own set of spatial CV methods, which may overlap with some other packages. In the following, we present each of the four main spatial CV methods and the software packages that have implemented the method. Interested readers can choose the software package based on the spatial CV methods needed and their preferred programming language.

10.2.1 CLUSTERING-BASED SPATIAL CV

One method to split data spatially is to perform spatial clustering. Based on the geospatial coordinates of the data and leveraging a clustering algorithm (e.g., K-means clustering), we can split data into several subsets with data instances within the same subset being spatially continuous and those in different subsets being relatively separated from each other. Figure 10.2 illustrates this clustering-based spatial CV using K-means clustering (k = 3). With the split data in Figure 10.2(b), one can then use, e.g., the data in Cluster 1 for validation and the data in Cluster 2 and Cluster

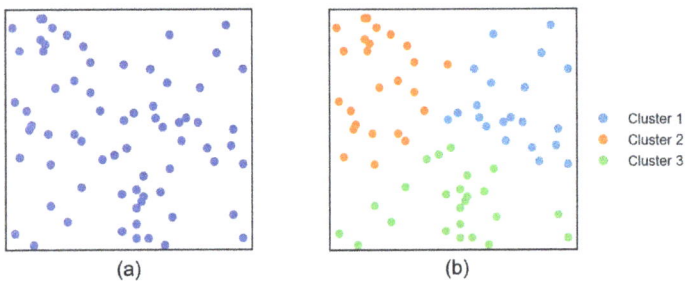

Figure 10.2 An illustration of clustering-based spatial CV: (a) original training data; (b) training data split into three folds using K-means clustering.

3 for model training, and then iterate this process. In addition, we are not limited to using only one cluster for validation but can use multiple clusters too. For example, we could split the data into ten clusters using a clustering algorithm, and then use three clusters for validation and seven clusters for model training.

Clustering-based spatial CV has been implemented in the following packages: *spatialsample* (R), *sperrorest* (R), *Mlr3spatiotempcv* (R), *blockCV* (R), and *spacv* (Python). K-means is the most commonly used clustering algorithm in these packages. The package *spatialsample* also implements another clustering algorithm, hierarchical clustering, for clustering-based spatial CV.

10.2.2   GRID-BASED SPATIAL CV

Another method to spatially split data is to use spatial grids. By dividing the entire geographic area of the data into n rows and m columns, we can split the data into $n \times m$ grid cells. Figure 10.3 illustrates grid-based spatial CV using nine grid cells. We can then use the data in one grid cell for validation and those in the remaining grid cells for model training, and then iterate this process. Similar to clustering-based spatial CV, data from multiple grid cells can be used for validation too. In addition to squares or rectangles, grid cells can be in other shapes too, such as hexagons which have often been used in GIScience research (Montello, Friedman, and Phillips 2014; Gao et al. 2017).

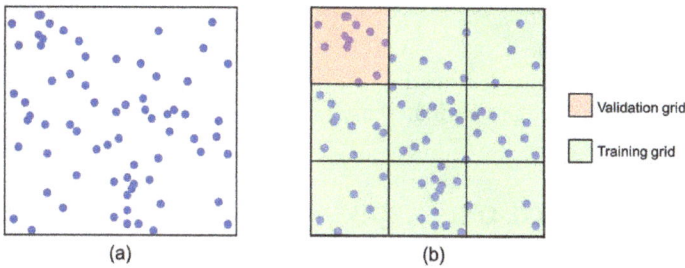

Figure 10.3   An illustration of grid-based spatial CV: (a) original training data; (b) training data split into nine grid cells.

Grid-based spatial CV has been implemented in the following packages: *ENMeval* (R), *spatialsample* (R), *sperrorest* (R), *Mlr3spatiotempcv* (R), *blockCV* (R), and *spacv* (Python). These packages implement rectangle-based grids, and two of them, *spatialsample* and *spacv*, also implement hexagonal grids. In addition, *sperrorest* provides an extra feature that allows users to specify a minimum number of data points inside a grid cell, and grid cells with smaller numbers of data instances will be merged into their neighboring cells.

10.2.3   GEO-ATTRIBUTE-BASED SPATIAL CV

When a geo-attribute is available, such as an attribute containing county names or city district names, the data can be split based on the spatial regions defined by the geo-attribute. Figure 10.4 illustrates this spatial CV using four geographic regions. Since we are using pre-defined geographic regions to split the data, the evaluation results can be communicated relatively easily using the names of the geographic regions. For example, we can clearly describe the geographic regions whose data are used for model training and the region(s) whose data are used for validation in each iteration of the CV.

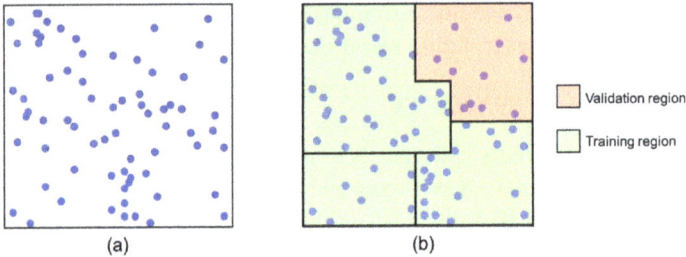

(a)   (b)

Validation region

Training region

Figure 10.4 An illustration of geo-attribute-based spatial CV: (a) original training data; (b) training data split based on four geographic regions defined by the geo-attribute.

Geo-attribute-based spatial CV has been implemented in *Mlr3spatiotempcv* (R) and *spacv* (Python). *Mlr3spatiotempcv* requires the user to specify a target attribute, while *spacv* expects the user to provide polygons related to the geo-attribute. In addition, we can also use typical spatial data processing packages, such as *GeoPandas* (Python), to implement geo-attribute-based spatial CV by splitting data based on the geo-attribute.

10.2.4 SPATIAL LEAVE-ONE-OUT CV

The fourth spatial CV method is an extension of the classic leave-one-out CV. Instead of simply using one data instance for validation and the remaining data for model training, spatial leave-one-out CV creates a buffer zone surrounding the validation data. Data falling into the buffer zone will not be used for model training. Figure 10.5 illustrates spatial leave-one-out CV. Note that this method is not limited to using one validation data instance only in each iteration, but can include more data within the same local region. This extension is called spatial leave-one-disc-out (Schratz et al. 2021), where a disc is a circular local region for validation, a buffer zone is created around the disc, and the remaining data are used for model training.

Spatial leave-one-out CV has been implemented in the following packages: *CAST* (R), *Mlr3spatiotempcv* (R), *blockCV* (R), *sperrorest* (R), *spatialsample* (R), *spacv* (Python), and *museotoolbox* (Python). The extended spatial leave-one-disc-out CV has been implemented in *Mlr3spatiotempcv* and *sperrorest*. In addition, the technique of adding a buffer zone to further increase the independence between training and validation data is not limited to spatial leave-one-out CV only, but has been implemented in the three other spatial CV methods as well.

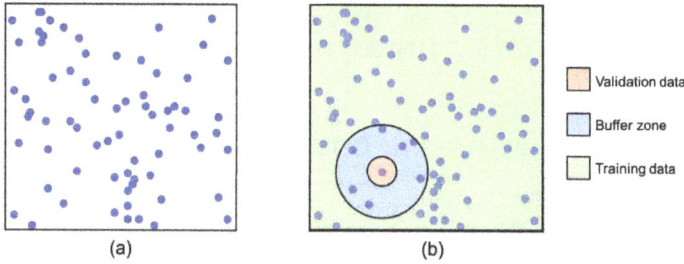

Figure 10.5   An illustration of spatial leave-one-out CV: (a) original training data; (b) training data split into validation data, a buffer zone (data in which are not used for model training), and training data.

10.2.5   SUMMARY

We have presented four main spatial CV methods and their corresponding software package implementations. In practice, the decision of which spatial CV method to use depends on which method best mimics the real application scenario of the trained model. For example, if the trained model will be used for making predictions in pre-defined administrative regions, then a geo-attribute-based spatial CV may be more suitable. By contrast, if the trained model will be used to make predictions in new regions defined by grid cells (or tiles), then a grid-based spatial CV may be preferred. The principle of mimicking the real application scenario also guides our choice between spatial CV and random CV (e.g., interpolation). Each spatial CV method involves certain parameters, such as the number of clusters, the size of the grids, and the distance of the buffer zone. There is no single best way for specifying the parameter values, and existing research generally recommends considering the effect of spatial autocorrelation and domain knowledge related to the machine learning task (Karasiak et al. 2019; Valavi et al. 2019; Brenning 2022).

10.3   EXAMPLES BASED ON REAL-WORLD DATA

In this section, we use two examples based on real-world data to demonstrate the use of the four spatial CV methods and how they may affect the model evaluation results compared with using random CV. To quantify the performance of the trained models, we use two metrics, R^2 and Root Mean Square Error (RMSE) calculated using the following two equations:

$$R^2 = 1 - \frac{\sum_{i=1}^{N} (y_i - \hat{y}_i)^2}{\sum_{i=1}^{N} (y_i - \bar{y})^2} \tag{10.1}$$

$$RMSE = \sqrt{\frac{1}{N} \sum_{i=1}^{N} (\hat{y}_i - y_i)^2} \tag{10.2}$$

where y_i is the observed value of the ith data instance; $\hat{y}_i$ is the predicted value made by the model; $\bar{y}$ is the average of the dependent variable; and N is the total number of data instances.

10.3.1 EXAMPLE 1: PREDICTING DOMESTIC VIOLENCE USING RANDOM FOREST

In the first example, we leverage the domestic violence data in the city of Chicago from our previous study (Chang et al. 2022), and use a random forest model to predict domestic violence based on socioeconomic and demographic variables. The geographic unit of analysis for this example is census block groups (CBGs), and the time period of the data is from December 31, 2018 to January 6, 2020.

The dependent variable is the domestic violence rate at the CBGs level. The global Moran's I of the domestic violence data is 0.648 ($p < 0.001$). The independent variables are 19 socioeconomic and demographic variables in five categories: (1) race and ethnicity, (2) age, (3) social disadvantage level, (4) residential instability, and (5) urbanicity. More details about the dependent and independent variables are available in the original article (Chang et al. 2022).

We use a random forest (RF) model to predict domestic violence rate based on the 19 independent variables. The RF model has two major hyperparameters to be set: `n_estimators` controlling the number of trees in the RF model, and `max_features` controlling the number of features to be considered when a node of a tree is split. Here, we use the hyperparameter values identified from our previous study by setting `n_estimators` to 80 and `max_features` to $\sqrt{n_v}$ where n_v is the number of independent variables.

10.3.1.1 Data Partition Using Spatial and Random CV

We use the four spatial CV methods and a random CV method to split the data into training and validation sets. The parameters of these five CV methods are set as follows. The data splitting result of each method is visualized in Figure 10.6.

- *Random CV*: We employ the traditional 10-fold CV by randomly splitting all CBGs into 10 subsets.

- *Clustering-based spatial CV*: We use K-means clustering to split the whole data set also into 10 clusters (i.e., k = 10).

- *Grid-based spatial CV*: We use a 3×3 grid to split the data into 9 grid cells. Among these 9 grid cells, 7 of them contain data and therefore a 7-fold cross-validation is used.

- *Geo-attribute-based spatial CV*: Using a geo-attribute of community names in Chicago, we split the data (2,146 CBGs) to 96 communities.

- *Spatial leave-one-out CV*: We set the buffer distance to 0.05 quantile of all distances among CBGs. One CBG is used for each iteration of the validation process.

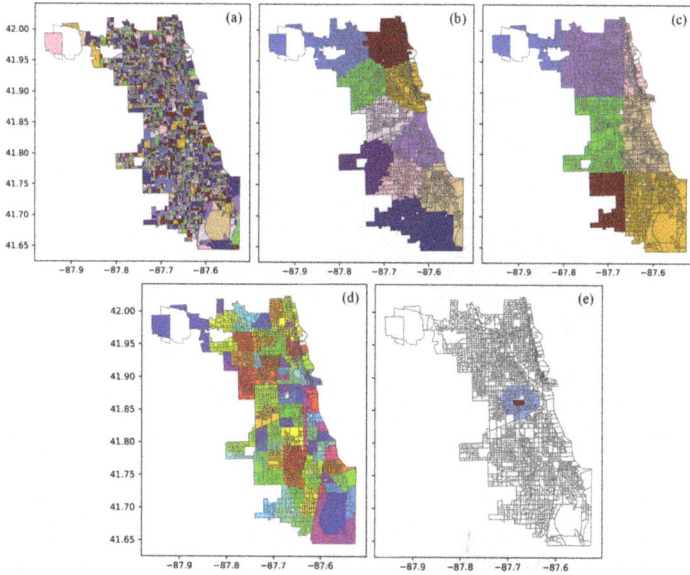

Figure 10.6 Data splitting results for CBGs in Chicago: (a) random CV; (b) clustering-based spatial CV; (c) grid-based spatial CV; (d) geo-attribute-based spatial CV; (e) spatial leave-one-out CV.

In Figure 10.6, each color in subfigures (a)-(d) indicates one subset of data, and subfigure (e) shows one CBG used for validation (in red color) and its surrounding buffer zone (in blue). As can be seen in subfigure (a), the training and validation data generated by random CV are scattered across the entire study area and can be spatially adjacent to each other. By contrast, the training and validation data generated by clustering-based, grid-based, and geo-attribute-based spatial CV (in subfigures (b), (c), and (d)) are more spatially separated from each other. In subfigure (c), i.e., grid-based spatial CV, we can also clearly see the horizontal and vertical lines formed by the 3×3 grid.

10.3.1.2 Results and Discussion

Table 10.1 presents the evaluation results using different CV methods. The hyperparameters of the RF model are fixed during the entire evaluation process to ensure that the performance difference only comes from the way that the data is split.

As can be seen, different performance scores are obtained when different CV methods are used for splitting the data. The performance scores by themselves do not suggest which CV method should be ideally used; instead we should choose the CV method that best mimics the real application scenario of the trained model. Overall, the results in Table 10.1 show that the highest performance scores are obtained using random CV, while lower performance scores are obtained using the four spatial CV methods. The highest performance scores from random CV are likely due to the many spatially adjacent training and validation data instances generated by random

Table 10.1

Performance of the RF model using random CV and four spatial CV methods.

CV method	R2	RMSE
Random CV	0.5952	8.9398
Clustering-based spatial CV	0.5443	9.4853
Grid-based spatial CV	0.5643	9.2752
Geo-attribute-based spatial CV	0.5667	9.2501
Spatial leave-one-out CV	0.5470	9.4575

CV, which allows the RF model to "peek" into nearby validation data during the training process. The performance scores obtained based on random CV are about 5% - 8% higher than those obtained via the four spatial CV methods, but are not radically different in this example. This suggests that the scores of random CV in this example are not entirely invalid even when spatial CV should ideally be used, but will likely present an overestimate of the model performance. If the real application scenario of the trained RF model is to predict domestic violence rate in CBGs of Chicago (i.e., the same study area), the evaluation scores obtained from random CV can still reflect the true performance of the model; if the application scenario is to predict domestic violence rate in nearby cities or regions, using random CV could lead to a 5% - 8% overestimate of the model performance.

10.3.2 EXAMPLE 2: PREDICTING OBESITY PREVALENCE USING DEEP NEURAL NETWORK

In this second example, we use the obesity prevalence data in New York City (NYC) from another recent study (Zhou et al. 2022), and use a deep neural network (DNN) model to predict obesity prevalence. The geographic unit of analysis is census tract and the time period of the data is the entire year of 2018.

The dependent variable is the obesity prevalence at the census tract level, and its global Moran's I is 0.740 ($p < 0.001$). The original data were obtained from the PLACES project of the Centers for Disease Control and Prevention (CDC). The independent variables are 21 socioeconomic and demographic factors related to race and ethnicity, gender, age, education, income, and housing conditions. More details about the dependent and independent variables are available in the original article (Zhou et al. 2022).

We use DNN in this example for predicting obesity prevalence. The hyperparameters of the model are set based on our previous study, and the model has 9 hidden layers with 160, 208, 160, 160, 256, 32, 240, 96, and 208 neurons for them respectively.

10.3.2.1 Data Partition Using Spatial and Random CV

Same as the last example, we use the four spatial CV methods and a random CV method to split the data. The parameters of these five CV methods are set as follows.

The splitting results are visualized in Figure 10.7.

- *Random CV*. We use the same classic 10-fold random CV as used in the previous example.

- *Clustering-based spatial CV*. We use the same K-means clustering method (k=10) to split the data.

- *Grid-based spatial CV*. We split the NYC area using a 3 × 3 grid, and 8 of the grid cells contain census tracts.

- *Geo-attribute-based spatial CV*. NYC consists of five boroughs. We split data into these five boroughs.

- *Spatial leave-one-out CV*. Similar to the previous example, the 0.05 quantile of distances among all census tracts is used as the buffer distance.

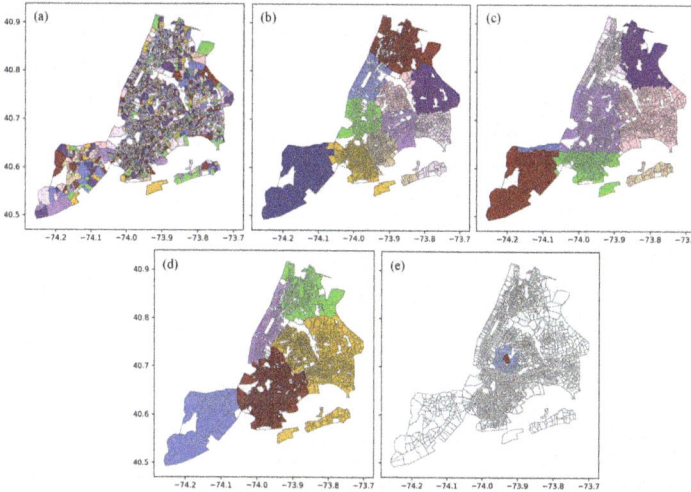

Figure 10.7 Data splitting results for census tracts in NYC: (a) random CV; (b) clustering-based spatial CV; (c) grid-based spatial CV; (d) geo-attribute-based spatial CV; (e) spatial leave-one-out CV.

10.3.2.2   Results and Discussion

Table 10.2 presents the evaluation results using different CV methods. Similar to the previous example, the hyperparameters of the DNN model are fixed during the entire evaluation process to ensure that the performance difference only comes from the way that the data is split.

As can be seen from Table 10.2, the overall pattern of the obtained performance scores holds, i.e., the scores from random CV are the highest compared with the

Table 10.2

Performance of the DNN model using random CV and four spatial CV methods.

CV method	R2	RMSE
Random CV	0.8692	2.1287
Clustering-based spatial CV	0.7244	3.0899
Grid-based spatial CV	0.7466	2.9624
Geo-attribute-based spatial CV	0.6613	3.4250
Spatial leave-one-out CV	0.8083	2.5766

scores obtained via the four spatial CV methods. As shown in Figure 10.7(a), random CV again generates many training and validation census tracts that are spatially adjacent, allowing the DNN model to "peek" into nearby validation census tracts during the training process. The performance scores from random CV are 8% - 31% higher. Note that the spatial autocorrelation in the obesity prevalance data is higher than that in the domestic violence data used in the previous example. This higher spatial autocorrelation likely contributes to the higher overestimate of performance scores from random CV. If the real application scenario of the trained model is to predict into other nearby geographic regions not in the training data, using random CV in this case could present a relatively large overestimate of the model performance.

10.4 CONCLUSIONS

Cross-validation is an evaluation approach that has been widely used in GeoAI research. In this chapter, we have discussed the concept of spatial CV and why we may need it over random CV under certain situations (e.g., extrapolation). We then present four main spatial CV methods identified and synthesized from the multidisciplinary literature, which are: *clustering-based spatial CV, grid-based spatial CV, geo-attribute-based spatial CV, and spatial leave-one-out CV*. We also list the software packages that have implemented these methods and their corresponding programming languages to inform the choice of interested readers. Two examples based on real-world data are used to demonstrate the different performance scores obtained via spatial CV and random CV. While the two examples involve different topics, study areas, and machine learning models, the obtained results share a similar pattern, i.e., the performance scores obtained via random CV tend to be higher than performance scores obtained via spatial CV methods. While random CV can still provide some extent of evaluation when spatial CV should ideally be used, we need to be cautious about these scores and know that those evaluation scores from random CV are likely to be overestimates of the true model performance. The ultimate goal of model evaluation is to provide realistic estimates that can reflect the true performance of a model when it is put into use. We hope that these spatial CV methods and their software packages can help expand the model evaluation toolbox of researchers in GeoAI and spatial data science.

BIBLIOGRAPHY

Airola, A., *et al.*, 2019. The spatial leave-pair-out cross-validation method for reliable auc estimation of spatial classifiers. *Data Mining and Knowledge Discovery*, 33, 730–747.

Brenning, A., 2012. Spatial cross-validation and bootstrap for the assessment of prediction rules in remote sensing: The r package sperrorest. *In: 2012 IEEE International Geoscience and Remote Sensing Symposium*. IEEE, 5372–5375.

Brenning, A., 2023. Spatial machine-learning model diagnostics: a model-agnostic distance-based approach. *International Journal of Geographical Information Science*, 37 (3), 584–606.

Browne, M.W., 2000. Cross-validation methods. *Journal of Mathematical Psychology*, 44 (1), 108–132.

Burman, P., 1989. A comparative study of ordinary cross-validation, v-fold cross-validation and the repeated learning-testing methods. *Biometrika*, 76 (3), 503–514.

Chang, T., *et al.*, 2022. The role of alcohol outlet visits derived from mobile phone location data in enhancing domestic violence prediction at the neighborhood level. *Health & Place*, 73, 102736.

Crosby, H., Damoulas, T., and Jarvis, S.A., 2020. Road and travel time cross-validation for urban modelling. *International Journal of Geographical Information Science*, 34 (1), 98–118.

Da Silva, T.P., Parmezan, A.R., and Batista, G.E., 2021. A graph-based spatial cross-validation approach for assessing models learned with selected features to understand election results. *In: 2021 20th IEEE International Conference on Machine Learning and Applications (ICMLA)*. IEEE, 909–915.

Diniz-Filho, J.A.F., Bini, L.M., and Hawkins, B.A., 2003. Spatial autocorrelation and red herrings in geographical ecology. *Global ecology and Biogeography*, 12 (1), 53–64.

Fotheringham, A.S., 2009. "the problem of spatial autocorrelation" and local spatial statistics. *Geographical Analysis*, 41 (4), 398–403.

Gao, S., *et al.*, 2017. A data-synthesis-driven method for detecting and extracting vague cognitive regions. *International Journal of Geographical Information Science*, 31 (6), 1245–1271.

Géron, A., 2022. *Hands-on Machine Learning with Scikit-Learn, Keras, and Tensorflow*. "O'Reilly Media, Inc.".

Goodchild, M.F. and Li, W., 2021. Replication across space and time must be weak in the social and environmental sciences. *Proceedings of the National Academy of Sciences*, 118 (35), e2015759118.

Hijmans, R.J., 2012. Cross-validation of species distribution models: removing spatial sorting bias and calibration with a null model. *Ecology*, 93 (3), 679–688.

Karasiak, N., 2020. Museo toolbox: A python library for remote sensing including a new way to handle rasters. *Journal of Open Source Software*, 5 (48), 1978.

Karasiak, N., *et al.*, 2022. Spatial dependence between training and test sets: another pitfall of classification accuracy assessment in remote sensing. *Machine Learning*, 111 (7), 2715–2740.

Karasiak, N., *et al.*, 2019. Statistical stability and spatial instability in mapping forest tree species by comparing 9 years of satellite image time series. *Remote Sensing*, 11 (21), 2512.

Meyer, H. and Pebesma, E., 2021. Predicting into unknown space? estimating the area of applicability of spatial prediction models. *Methods in Ecology and Evolution*, 12 (9), 1620–1633.

Meyer, H. and Pebesma, E., 2022. Machine learning-based global maps of ecological variables and the challenge of assessing them. *Nature Communications*, 13 (1), 2208.

Meyer, H., *et al.*, 2018. Improving performance of spatio-temporal machine learning models using forward feature selection and target-oriented validation. *Environmental Modelling & Software*, 101, 1–9.

Montello, D.R., Friedman, A., and Phillips, D.W., 2014. Vague cognitive regions in geography and geographic information science. *International Journal of Geographical Information Science*, 28 (9), 1802–1820.

Muscarella, R., *et al.*, 2014. Enm eval: An r package for conducting spatially independent evaluations and estimating optimal model complexity for maxent ecological niche models. *Methods in Ecology and Evolution*, 5 (11), 1198–1205.

Pohjankukka, J., *et al.*, 2017. Estimating the prediction performance of spatial models via spatial k-fold cross validation. *International Journal of Geographical Information Science*, 31 (10), 2001–2019.

Roberts, D.R., *et al.*, 2017. Cross-validation strategies for data with temporal, spatial, hierarchical, or phylogenetic structure. *Ecography*, 40 (8), 913–929.

Schratz, P., *et al.*, 2021. mlr3spatiotempcv: Spatiotemporal resampling methods for machine learning in r. *arXiv preprint arXiv:2110.12674*.

Tobler, W.R., 1970. A computer movie simulating urban growth in the detroit region. *Economic Geography*, 46 (sup1), 234–240.

Valavi, R., *et al.*, 2018. blockcv: An r package for generating spatially or environmentally separated folds for k-fold cross-validation of species distribution models. *Biorxiv*, 357798.

Zhou, R.Z., *et al.*, 2022. Deriving neighborhood-level diet and physical activity measurements from anonymized mobile phone location data for enhancing obesity estimation. *International Journal of Health Geographics*, 21 (1), 1–18.

Section III

GeoAI Applications

11 GeoAI for the Digitization of Historical Maps

Yao-Yi Chiang
Computer Science and Engineering Department, University of
Minnesota

Muhao Chen
Information Sciences Institute & Computer Science Department,
University of Southern California

Weiwei Duan
Computer Science Department, University of Southern California

Jina Kim
Computer Science and Engineering Department, University of
Minnesota

Craig A. Knoblock
Information Sciences Institute, University of Southern California

Stefan Leyk
Department of Geography, University of Colorado, Boulder

Zekun Li, Yijun Lin, Min Namgung
Computer Science and Engineering Department, University
of Minnesota

Basel Shbita
Information Sciences Institute & Computer Science Department,
University of Southern California

Johannes H. Uhl
Institute of Behavioral Science & Cooperative Institute for Research in
Environmental Sciences (CIRES), University of Colorado, Boulder

CONTENTS

11.1 Introduction 218
11.2 Extracting and Linking Geographic Features from Scanned Historical
    Maps 219

DOI: 10.1201/9781003308423-11

11.2.1 Exploiting Prior Information for Automated Geographic Feature
        Extraction from Scanned Historical Maps 219
11.2.2 Turning the Extracted Features into Linked Spatiotemporal Data ... 223
11.3 Recognizing and Linking Text in Scanned Historical Maps 228
11.3.1 Creating Synthetic Maps for Training Text Spotters 228
11.3.2 PostOCR for Map Text 231
11.3.3 Generating Linked Metadata from Map Text 232
11.4 Generating Large, Long-Term Historical Data from Historical Map Series .233
11.4.1 Historical Map Data Acquisition 234
11.4.2 Examples of Large-Scale Geographic Feature Extraction...............234
11.5 Ready-to-Use Map Processing System and Pre-Trained Model 239
11.5.1 The mapKurator System 239
11.5.2 SpaBERT 240
11.6 Conclusion 243
Bibliography 244

11.1   INTRODUCTION

Historical maps are a unique source of geohistorical information, documenting the natural and anthropogenic characteristics of past landscapes. Such geohistorical information for the periods before the 1970s (before the launch of the Landsat program) only exists, with few exceptions, on printed map sheets. For instance, the U.S. Geological Survey (USGS) produced over 178,000 topographic map sheets between 1884 and 2006. These maps have exceptional scientific, cultural, and societal value because they commonly hold high-resolution data collected and mapped at high (and well-documented) accuracy standards. Also, the complete USGS historical topographic map series provides the most continuous, pre-satellite imagery for different key natural and human landscape features from their inception through the early 2000s for most of the US. In addition, many other diverse historical maps and map series exist, such as the Sanborn maps (in the US) or the Ordnance Survey maps (in the UK).

Many of these maps have been scanned in the last decade and are now publicly available. These scanned maps portray 100+ years of detailed evolution of continental-scale landscapes, including landscape change, urban growth and sprawl, persistent underdevelopment of other urbanized areas, coastline erosion and modification, draining of swamps, growth and shrinkage of forest cover, and changes in river courses. This kind of geographic information depicted in historical maps can enable fundamentally new scientific research in the social and natural sciences covering large, national-scale areas and long time periods that would not be possible otherwise. However, the geographic information embedded in the map scans is not easily discoverable, accessible, or interpretable. To date, researchers review and analyze these maps, oftentimes in a manual, error-prone, and labor-intensive process (Chiang *et al.*, 2020, 2014), limiting these efforts in scope and geographic areas.

Our long-term research goal is to develop new automated technologies for over-coming challenges in extracting and linking geographic information from historical maps and map series, aiming to make these maps searchable and their content read-ily usable (Chiang, 2015). Specifically, we build machine learning models that can generalize well to handle a wide variety of maps and also make the trained models, training data, analyses, and findings easily discoverable, shareable, and reproducible.

This chapter presents an overview of AI technologies in methods and systems for processing historical maps to generate valuable data, information, and knowledge. Individual sections summarize our recently published research results in various domains, including the semantic web, big data, data mining, machine learning, doc-ument understanding, natural language processing, remote sensing, and geographic information systems.

The remainder of the chapter is structured as follows. Section 11.2 describes our recent work for training machine learning technologies to extract and link ge-ographic features from historical maps when limited data annotations are available. Section 11.3 presents methods for spotting text content in historical maps and link-ing map text to external knowledge bases to create linked metadata. Section 16.3.2 shows approaches for analyzing historical map series and systematically generating extensive, long-term geographic data (e.g., road networks and urban extent) from the map series. Section 11.5 presents a ready-to-use map processing system for gener-ating metadata from text content in large varieties of scanned historical maps and a pre-trained language model for linking and typing map text. Section 16.4 concludes the chapter.

11.2 EXTRACTING AND LINKING GEOGRAPHIC FEATURES FROM SCANNED HISTORICAL MAPS

The focus of this section is on training machine learning models to extract geographic features from scanned historical maps using limited human-annotated training data, structuring the extracted data into spatiotemporal linked data, and finally linking those data to existing knowledge bases. To achieve this goal, we present techniques in Section 11.2.1 that utilize prior information to reduce the number of manual an-notations needed for geographic feature extraction. In Section 11.2.2, we outline a method for transforming the extracted geographic features into linked spatiotemporal data and connecting them with existing knowledge bases on the web.

11.2.1 EXPLOITING PRIOR INFORMATION FOR AUTOMATED GEO-GRAPHIC FEATURE EXTRACTION FROM SCANNED HISTORICAL MAPS

Cartographic depictions of geographic features in scanned historical topographic maps offer valuable insights into the evolution of natural features and human activities. Efficient extraction of this information requires training a semantic segmentation system, which needs a large amount of annotated data. Using exter-nal vector data covering the same areas as the topographic maps can help to generate annotated training data automatically, but the annotations can be subject to misalignment and other types of errors (Uhl and Duan, 2020). For example, the green lines in Figure 11.1 are external vector data directly overlaid on a map. The desired annotation locations are where the rasterized vector lines (green) and features in the map image coincide. Misalignments between topographic maps and external vector data can be caused by different publication years, generalization, and spatial reference systems. Misaligned and false annotations can confuse a semantic segmentation model, leading to inaccurate extraction of feature locations in the map.

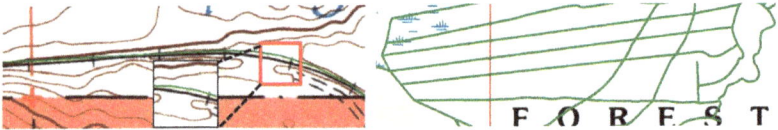

Figure 11.1   The green lines are the external vector data overlaid with the target map. The left figure shows the misaligned annotations for railroads. We magnify the misalignment in the red box. The right figure shows the incorrect annotations (false positives) for waterlines.

There are two types of existing approaches to handling misaligned and false annotations in image understanding models. The first approach aims to correct the annotations before model training, but these methods can only address significant annotation errors. For instance, our earlier vector-to-raster algorithm (Duan *et al.*, 2017) utilizes the color information of map symbols to align the annotations with target geographic features. However, maps are complex documents and pixels near the desired features can also belong to the same color range, leading to false annotations. The second type of approach assumes that the annotations are noisy and employs novel loss functions (Bhushan Damodaran *et al.*, 2018; Kang *et al.*, 2021; Wu *et al.*, 2019) to reduce segmentation errors by utilizing additional information from the input image. For instance, the roads detection model (Wu *et al.*, 2019) utilizing OpenStreetMap (OSM) annotations includes a normalized cut loss (Ncut) that calculates the similarities between annotated pixels based on both color (RGB) and spatial (XY) information. The idea is that true road pixels should have similar colors and be close to each other. This assumption allows Ncut to classify falsely annotated pixels as non-road to maximize the similarity between road pixels in color and space. However, this method is not reliable when false annotations are not minor nor similar to each other.

To address these challenges, we have developed the Label Correction Algorithm (LCA) (Duan *et al.*, 2021b). LCA effectively leverages the geometry of the external vector data and shapes and colors of features in the rasterized input maps to minimize false annotations. LCA is a variant of the Chan-Vese (CV) algorithm (Chan and Zhu, 2005), which aims to iteratively segment pixels in an image into foreground and background areas according to an objective function that calculates the color homogeneity within each area separately. (Song and Chan (2002) provide detailed proof about the correctness of fast optimization of the CV.) In addition to color homogeneity, LCA aims to find connected groups of map pixels similar to the target object's shape. These pixel groups are the annotations (examples of the desired objects) for training a semantic segmentation model for geographic feature extraction. Figure 11.2 shows the integration of LCA into a geographic feature extraction system.

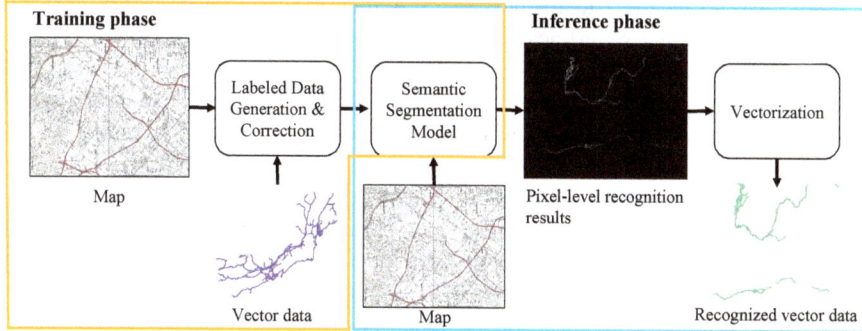

Figure 11.2 An example workflow of LCA incorporated in a geographic feature extraction system.

Specifically, LCA first derives the pixels-of-interest (PxOI) representing potential annotations by buffering the external vector data and overlaying them on the map image. The idea is that although directly overlaying two data sources could result in feature misalignment, the alignment errors are usually no more than a few pixels and can fall into the buffered area. Then LCA iteratively identifies correct annotations within the PxOI using two optimization goals: First, the colors of the identified annotations should be homogeneous. This goal is similar to the one in the CV algorithm. Second, LCA compares the shapes of the identified annotations in the map to their closest geographic features in the external vector data and uses the shape similarity to refine the annotation pixels. A simple way to achieve this goal is to maximize the intersecting area between the identified annotations in the map and the desired geographic features in the external vector data. However, since the map and the external vector data come from different sources, their shapes are not always the same. Hence, inspired by feature alignment approaches (e.g., (Duan *et al.*, 2017)), LCA includes a novel method to allow using a local affine transformation on the external vector data to approximate the shape similarity between external vector features and shapes of map pixel groups instead of directly overlaying the two datasets and calculating the intersecting area. LCA's optimization process minimizes an objective function toward these two goals by updating pixels into annotated and non-annotated areas iteratively. Also, in each iteration, LCA updates the PxOI and the local affine transformation to refine the search space. (See Duan *et al.* (2021b) for details of the objective function and the updating process.)

Once LCA generates the annotations for an input map using the external vector data, we can train a semantic segmentation model using the annotations (the yellow box in Figure 11.2). During the inference phase, the semantic segmentation model classifies map pixels into the target category and non-target category to detect pixels of the geographic features in the map based on the external vector data (the blue box in Figure 11.2). Then the system converts the pixel-level recognition results into vector data using the vectorization method proposed in the SpaceNet Road Detection and Routing Challenge.[1]

We have tested the effectiveness of LCA for extracting two linear geographic objects: railroads and waterlines, in the scanned USGS topographic maps covering Bray, California, published in 2001, and Louisville, Colorado, published in 1965. The maps contain diverse geospatial objects, including railroads, waterlines, roads, lakes, mountains, and wetlands. The external vector data for annotation is the vector-formatted map data from the USGS published in 2018, and the semantic segmentation model is DeepLabV3+ (Chen *et al.*, 2018). Our experimental results, verified with manual annotations, show that LCA significantly improves the annotation quality (1% to 7% improvement in F1 scores) compared with the annotations from simply overlaying the external vector data with the input map or using our previous alignment algorithm called vector-to-raster (Duan *et al.*, 2017). Also, the results show that the geographic feature extraction results based on LCA's annotations to train DeepLabV3+ significantly outperform the DeepLabV3+ models trained with the annotations from the two baseline approaches, especially measured by the Average Path Length Similarity (ALPS) scores (19% to 30% improvement). The APLS score is a commonly used metric for comparing linear geographic features. Low APLS scores indicate that the detected geographic features of continuous lines may have many false gaps. (See Duan *et al.* (2021b) for details of the experimental results.)

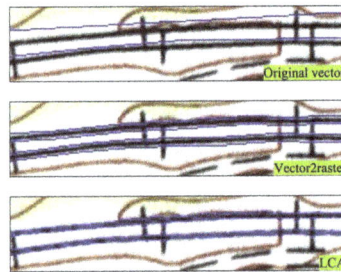

Figure 11.3   Annotations visualization. The areas within the blue outlines are the annotated railroads (target) objects in the Bray map.

Figure 11.3 shows the annotations in the Bray map from the three tested methods. The pixels along the blue lines are the annotated railroad pixels. As expected, the annotations using the vector-to-raster algorithm are more accurate than directly overlaying the original vector data with the map image. However, these annotated

---

[1]https://github.com/SpaceNetChallenge/RoadDetector/blob/master/albu-solution/src/skeleton.py

locations are still a few pixels off from the railroad pixels on the map, possibly because the pixel colors near the railroad edges are inconsistent. In contrast, the annotations from LCA are precisely at the railroad edges, showing the effectiveness of combining shape and color information for accurate annotation.

In summary, LCA is an automated technique that accurately annotates geographic features in historical map images. LCA leverages spatial information, such as feature locations and shapes, from external vector data to minimize annotation errors that often exist in traditional methods. This results in a significant reduction of manual annotating efforts and enables more accurate extraction of geographic features from historical maps.

11.2.2 TURNING THE EXTRACTED FEATURES INTO LINKED SPATIOTEMPORAL DATA

Linked geospatial data has received increased attention in recent years as researchers and practitioners have begun to explore the wealth of geospatial information on the Web (Athanasiou *et al.*, 2014; Bone *et al.*, 2016; Li *et al.*, 2020). We need more than just the extracted vector data from individual maps to better support analytical tasks and understand how particular map features change over time (e.g., long-term changes in railroad networks). We need to tackle several challenges, including (i) the generation and interlinking of geospatial entities ("building block" geometries originating from the vector data) that constitute the desired features (e.g., railroad lines) and can exist across different map editions, (ii) contextualizing and linking the generated entities to external resources to enable data enrichment and allow users to uncover additional information that does not exist in the original map sheets, and (iii) representing the resulting data in a structured and semantic output that can be easily interpreted by humans and machines, and adheres to the principles of the Semantic Web and Linked Data while enabling an incremental addition of map sheets over time with minimal cost and backward compatibility.

Overall, given geographic vector data extracted from multiple map editions of the same region (e.g., by using the method described in Section 11.2.1), we want to automatically construct a knowledge graph depicting all the geographic features that represent the original data, their relations (interlinks), and their semantics across different points in time.

11.2.2.1 Generating Geospatial Entities from Multi-temporal Vector Data

A geographic feature is represented using a collection of "building block" geometries. Given a set of geographic features about the same region from different points in time, each generated "building block" is essentially the largest geographic feature part that is either shared across different points in time or unique for a specific point in time. This task can be classified as a common entity matching/linking and entity "partitioning" task. These blocks can be either lines or areas in the case of linear geographic features or polygon geographic features, respectively. Figure 11.4 shows an example in which extracted features from two map editions (A in red, B outlined in blue) result in three building blocks (A' in red, B' in blue, and AB in green). Each building block geometry is encoded as well-known text (WKT) representation (MULTILINE or MULTIPOLYGON format) and corresponds to a geospatial entity in the resulting KG.

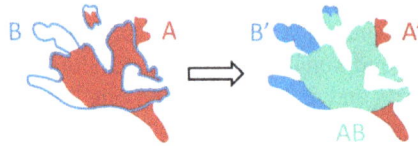

Figure 11.4   Illustration of geometry partitioning for a polygon geometry input of two map sheets: each color/outline represents a different building block. A single building block may contain disconnected areas.

We generate building block geometries of the geographic features (e.g., railroad networks) across time (originating in different map editions) in a granular and efficient fashion by utilizing a spatially-enabled database service to simplify handling data manipulations of geospatial objects. PostGIS is a powerful PostgreSQL extension for spatial data storage and query. It offers various functions to manipulate and transform geographic objects in databases.

The partitioning task is addressed by employing several PostGIS Application Programming Interface (API) calls over the geometries of the lines or polygons in the database. In the case of line geometries, we buffer each line segment to create two-dimensional areas before applying any geospatial operation. Specifically, for each extracted feature in a given map edition, we create the initial "building block" geometries to create the (initial) "leaf" nodes in the semantic graph. These nodes represent the most fine-grained building blocks computed so far. With each insertion of an additional map sheet, we refine the set of "leaf" nodes to include the most updated and fine-grained blocks. In each step, we compute the common and distinct parts of the leaves with the newly added feature by computing their intersection and difference. Finally, we subtract the union of the resulting geometries of the leaf node intersections from the newly added feature to find the latest added unique part. In (Shbita *et al.*, 2022), we describe the linking-partitioning algorithm in detail.

11.2.2.2   Contextualizing the Geo-data via Geo-entity Linking

Volunteered geographic information platforms (Goodchild, 2007) are used for collaborative mapping activities with users contributing geographic data. OpenStreetMap (OSM) is one of the most pervasive and representative examples of such a platform and broadly operates with a certain, albeit somewhat implicit, ontological perspective of place and geography. OSM suggests a hierarchical set of tags that users can attach to its basic data structures to organize their map data. These tags correspond to geographic feature types that can later be efficiently queried (e.g., wetland, railroad) and attributed to any geospatial entity, which can have more granular building blocks with similar attributes. Additionally, a growing number of OSM entities are linked to corresponding Wikipedia articles, Wikidata (Vrandečić and Krötzsch, 2014) entities, and feature identifiers in the USGS Geographic Names Information Service (GNIS) database.

To enrich the generated geospatial entities (i.e., building block geometries) with an external resource and link them to a corresponding entity, we employ a simple entity linking mechanism with OSM. This is again a task of entity matching; this time, it is with an entity in an external knowledge base. The method is based on reverse geocoding, which is the process of mapping the latitude and longitude measures of a point or a bounding box to an address or a geospatial entity.

The geo-entity linking process is illustrated in Figure 11.5 and described in detail in (Shbita *et al.*, 2022). The method requires pre-defining a given and known feature type (e.g., wetland) as an input. The procedure generates a global bounding box for each building block by finding its northmost, southmost, eastmost, and westmost coordinates. The OSM service takes the resulting box as input to execute a reverse-geocoding API call that locates instances of the given type on the external knowledge base. As a heuristic, we randomly sample a small number of coordinate pairs (Points), corresponding to the number of entities composing the block; thus, we gain more confidence in the detected instances with respect to the number of times they matched. Finally, we select the candidate with the highest number of matches.

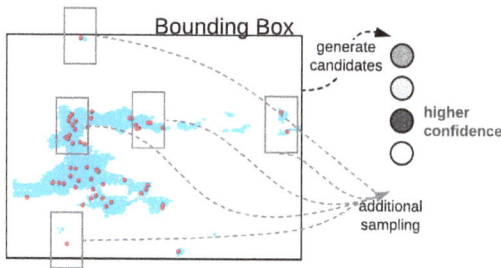

Figure 11.5 The method for acquiring external knowledge base instances. An example building block is seen in blue. Sample points of detected instances are seen in red.

Each resulting instance is used in later stages to enrich the blocks with additional semantics and metadata from the external knowledge base. Figure 11.6 shows an example of a scanned topographic map (seen in the background), which we used to extract its corresponding wetland vector data using the technologies in Duan *et al.* (2021a) alongside two OSM instances detected using the geo-entity linking method. The labels in Figure 11.6 (i.e., Four Mile Cove Ecological Preserve and Six Mile Cypress Slough Preserve) correspond to the name attribute of each entity. These labels are part of a set of attributes we use to augment the resulting data with information that did not exist in the original dataset.

Figure 11.6 An example of two OSM instances (enclosed in purple) and their `name` labels detected using the geo-entity linking method over a scanned topographic map.

11.2.2.3 Mapping the Data and Building a Spatiotemporal Knowledge Graph

As a generic model or framework, the Resource Description Framework (RDF) can be used to publish geographic information. One of the strengths of RDF is its structural flexibility, making it particularly well-suited for rich and varied forms of metadata. However, it has no specific features for encoding geometry, which is central to geographic information. The OGC GeoSPARQL (Battle and Kolas, 2011) standard defines a vocabulary for representing geospatial data on the Web and is designed to accommodate systems based on qualitative spatial reasoning and systems based on quantitative spatial computations. To provide a representation with valuable semantic meaning and universal conventions for the resulting data, we define a semantic model that builds on GeoSPARQL. Figure 11.7 shows the semantic model used to map the extracted and retrieved data. Each instance has at least one property of type `dcterms:date` to denote the different points in time in which the building block exists. `dcterms` stands for the Dublin Core Metadata Initiative metadata model, as recommended by the World Wide Web Consortium (W3C).[2] In order to describe the geometries in a compact and human-readable way, we use the WKT format. The `geo:asWKT` property is defined to link geometry with its WKT serialization. The resulting RDF data can be easily utilized by downstream applications via simple SPARQL queries.

The spatiotemporal data, resulting from the previous steps, is mapped into the semantic model to generate the final RDF graph and produce a structured standard ontologized output in the form of a knowledge graph that humans and machines can easily interpret as linked data. In order to encourage the reuse and application of the generated data in a manageable manner, one needs to make sure that the linked data publication process is robust and maintainable.

This hierarchical structure of the data (vertically built on top of the "building blocks") and its metadata management allows us to avoid an update across all the existing published geographic vector data (in linked data). Instead, we handle the

---

[2]https://www.w3.org/

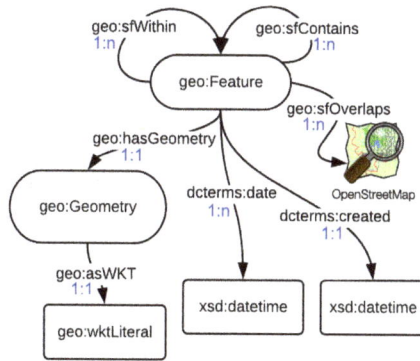

Figure 11.7　Semantic model for the linked data. Each node's cardinality is shown in blue next to each edge. In GeoSPARQL, the class type `geo:Feature` represents the top-level feature type. It is a superclass of all feature types. In our model, each instance of this class represents a single building block extracted from the original vector data. By aggregating a collection of instances of the class `geo:Feature` with a property of type `geo:sfWithin` we can construct a full representation for the geometry of a specific geographic feature in a given point in time.

computations incrementally once a new representation of the feature from a subsequent map edition is introduced by performing computations only with "leaf" nodes.

Figure 11.8 shows a visualized result of a SPARQL query that identifies parts of a wetland that were present in 1961 but are not present anymore in 2018 (in dark blue) following the processing of two map sheets from a swamp area in California. The results match the contemporary ground truth data (the light blue "stains" in the map's background depict the current swamp) and demonstrate the usefulness of our approach.

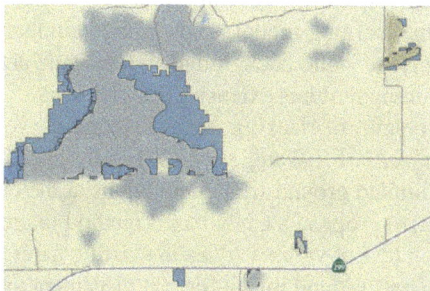

Figure 11.8　Example of wetland changes over time visualized via a SPARQL query: the parts of the Big Swamp in Bieber (CA) that are present in 1961 but are not present in 2018 are marked in dark blue, emphasizing its decline throughout time.

11.3   RECOGNIZING AND LINKING TEXT IN SCANNED HISTORICAL MAPS

Map text include names of places, labels for physical features, descriptions of areas or landmarks, and other important details. This information can play a crucial role in helping users to better find relevant maps and interpret the information presented on the map using valuable additional context and detail. For example, the names of places and other text on a historical map can provide insights into the historical, political, and cultural context in which the map was created.

This section describes the approach that automatically creates unlimited map images for training text spotting models (Section 11.3.1), exploits existing data for correcting the recognized text (Section 11.3.2), and links the recognized map text to external knowledge bases (Section 11.3.3).

11.3.1   CREATING SYNTHETIC MAPS FOR TRAINING TEXT SPOTTERS

Automatically localizing and recognizing text (i.e., text spotting) on historical maps facilitates generating useful map metadata and analyzing map content at scale. An end-to-end text spotting pipeline usually consists of a text detector and a recognizer. In general, the detector localizes the text instances within an image using a box or polygon representation, and the recognizer decodes the characters of the detected text instances. Recent end-to-end text spotting approaches (Huang *et al.*, 2022; Liu *et al.*, 2020; Zhang *et al.*, 2022) have achieved impressive performance on detecting and recognizing text with arbitrary shapes by taking advantage of computer vision techniques, e.g., convolutional neural networks (CNN) (Liu *et al.*, 2018) and vision transformers (Carion *et al.*, 2020; Xia *et al.*, 2022; Zhu *et al.*, 2020).

However, text spotting on scanned historical maps is challenging. Directly adopting existing text spotting models for historical maps typically results in poor performance. First, most text spotting methods deal with out-of-domain datasets, e.g., scene images (Long *et al.*, 2021) and documents (Chen *et al.*, 2021), where text are horizontal and compact. Nevertheless, text labels on historical maps have various rotation angles and letter spacings, causing failures in these spotting models. Second, training a machine learning model needs a large number of samples (i.e., text annotations) with diverse text and background styles, while annotating text labels in historical maps for training requires extensive manual work.

To solve the problem of lacking text annotations for training, synthetic datasets (Jaderberg *et al.*, 2014; Wang *et al.*, 2012) are cheap and scalable alternatives that generate unlimited ground-truth annotations at the character level or word level. Gupta *et al.* (2016) propose a CNN-based method to create synthetic images, SynthText, by blending text of various styles in existing natural scenes. The method identifies regions on a background image for text placement using segmentation and aligns text to the estimated region according to its boundary. SynthText provides annotations in minimum rectangle bounding boxes for characters and words. Liu *et al.* (2020) further create word-level polygon annotations based on SynthText for arbitrary shape text spotting. However, these synthetic images are significantly different from scanned historical maps regarding text and background styles. Li *et al.* (2021) propose to create synthetic maps for text detection on historical maps, which exploits a style transfer model (i.e., CycleGAN) to convert contemporary map images (e.g., OSM) into a historical style (e.g., Ordnance Survey) and place text labels upon them. However, the synthetic map images contain solely one historical background style. Also, the dataset provides only text bounding polygons (i.e., no transcription), which is applicable for end-to-end text spotting on diverse historical maps.

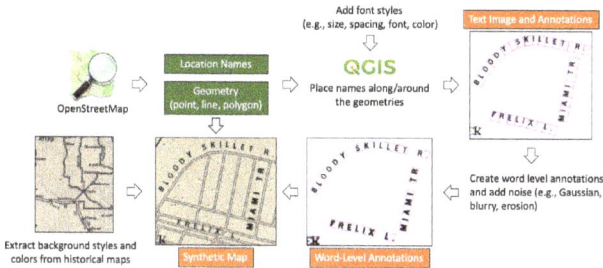

Figure 11.9 The workflow of generating synthetic map images with word-level annotations.

Inspired by (Li *et al.*, 2021), we have developed an approach that automatically generates unlimited historical-styled map images with text annotations, named SynMap, for training text spotters for historical maps. Figure 11.9 shows the workflow of generating SynMap images and annotations. First, we extract location names and geometries (e.g., points, lines, and polygons) from existing geographic data sources, such as OSM. The location names can be in any language (e.g., Chinese, Japanese). Second, we generate text images and word-level annotations by (1) using QGIS API[3] to place location names (or text labels) with various styles (e.g., font size, color, letter spacing) on a blank canvas. For point features, the text labels are placed around the point. For line features, the text labels follow the curvature of the line (e.g., road). For polygon features, the text labels are curved along the polygon boundary (e.g., lake); (2) obtaining text images and character bounding boxes using QGIS API. For example, the text label "BLOODY SKILLET R." in Figure 11.9 has 17 bounding boxes (including spaces); (3) automatically creating word-level bounding polygons for text labels as annotations. For example, we remove the bounding boxes of spaces and group the remaining characters into three words: "BLOODY", "SKILLET" and "R.". The control points of bounding polygons are sampled evenly from a polynomial curve fitting on the original bounding box coordinates. Then, we generate realistic scanned map background by extracting background profile around map text from the target historical maps in the David Rumsey Historical Map Collection[4] (Figure 11.10). Specifically, we first run text detection on these maps and draw rectangle bounding boxes with some buffer size around the detection results to

---

[3] https://api.qgis.org/api/
[4] https://www.davidrumsey.com/

retrieve pixels around texts (Figure 11.10 (a)). We construct grid cells of 8×8 pixels to preserve texture and color and adopt K-means to cluster these cells into background and foreground (Figure 11.10 (b) shows background cells). We use K-means again to cluster background cells into several groups as the map style profile. Figure 11.10 (c) shows two style groups in the given map. Finally, we use QGIS API to paint geographic features of various types using pre-defined cartographic rules, and merge them with the text and background images to generate synthetic map images.

The spotting results on the David Rumsey Map Collection show that the state-of-the-art text spotting models (e.g., TESTR (Zhang *et al.*, 2022)) trained with SynMap significantly improve its text detection and recognition performance on historical maps. For example, Figure 11.11 shows two examples of text spotting results on historical map images. We observe that the model can correctly detect and recognize words with (1) wide letter spacing (e.g., "COUNTY", "GREEN" in the left image) and (2) large rotation angles (e.g., "GOSHUTE", "TOANO" in the right image), demonstrating the effectiveness of training with SynMap for text spotting on historical maps.

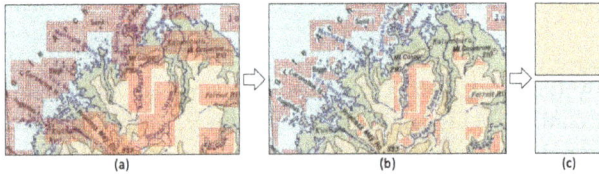

Figure 11.10   The process of creating background profile for producing synthetic map background by extracting the background color and texture from real historical maps.

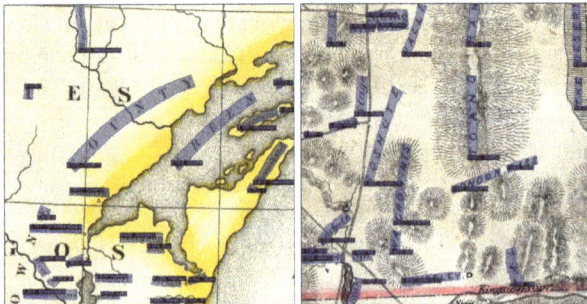

Figure 11.11   Two examples of text detection and recognition results on historical map images. The image size is 1000×1000 pixels.

11.3.2 POSTOCR FOR MAP TEXT

Existing optical character recognition (OCR) engines can help convert text content in historical maps to searchable metadata (e.g., Sections 11.3.1, 11.3.3, and 11.5.1). However, the retrieved text information often shows partially or incorrectly recognized words due to complicated backgrounds, such as text labels appearing on the dark background, lines across words, or large spacing between characters. Therefore, it is still challenging for modern OCR technologies to exploit the retrieved text from historical maps to generate meaningful metadata.

To conduct post-OCR processing, previous work utilizes lexical-based approaches with the linguistic context or applying language models to correct document OCR results. The lexical-based methods are based on the Levenshtein edit distance (LED) (Levenshtein, 1966), which can automatically generate replacement words for OCRed text using statistical methods (e.g., the n-gram approach (Chiron *et al.*, 2017)). Another method is to utilize neural networks, which apply pretrained BERT (Nguyen *et al.*, 2020) for error token detection only. In addition, fine-tuned Bidirectional and Auto-regressive Transformers (BART) (Soper *et al.*, 2021), pretrained on multiple denoising tasks, can run error detection and correction together. Although these language models show promising results in post-OCR for documents (i.e., 1-Dimensional (1D) word sequences), they cannot directly work on OCRed map text distributed in a 2-Dimensional (2D) space.

To preserve the original 2D space in post-OCR processing, we present a novel approach (Namgung and Chiang, 2022) that exploits the spatial arrangement of map text using a contextual language model, BART (Lewis *et al.*, 2020). Our post-OCR processing, which incorporates spatial arrangement in map text with BART, can directly consider spatial relations of map text for correction. For example, "Mississippi" and "River" constitute the place phrase "Mississippi River" (linguistic relation), and near "highway", there are likely to exist intersected "road" to enter the "highway" (spatial relation). We first obtain OCRed map text from the synthetic map images (see Section 11.3.1). Then, we convert the OCRed map text into 1D pseudo sentences and pair them with the ground truth text to fine-tune BART. To formulate these pseudo sentences, we group the map text into several spatial clusters (e.g., using K-means, DBScan). This is similar to the typical document structure analysis that divides a document (e.g., a scanned newspaper or research paper) into multiple meaningful paragraphs. We then sort the text labels in each cluster using their 2D coordinate, mimicking the natural reading order to constitute 1D pseudo sentences. The pseudo sentences help BART capture common map OCR errors by leveraging linguistic information between words in a place name and spatial information between place names (e.g., information from nearby words on a map).

For model training, we first fine-tune the BART-base model using the ICDAR dataset (Chiron *et al.*, 2017) inspired by (Soper *et al.*, 2021) to learn common English OCR error patterns. The dataset contains 5M English characters with 27,414 sentences in monographs and 11,561 sentences in periodicals. Next, we fine-tune the model using pseudo sentences (1D) constructed from the synthetic map (2D) to enable BART to capture the map OCR error patterns. We perform our experiment on both synthetic and historical map images in the state of Minnesota (MN) and Illinois (IL). The comparison includes a total of 9 models: the proposed method (i.e., cluster and fine-tuned with ICDAR data + Synthetic map) and its variations in the sentencing (i.e., Order by centering and by Euclidean) and fine-tuning (i.e., ICDAR only and Synthetic map only) processes. Our proposed sentencing method (cluster) fine-tuned with ICDAR data + Synthetic map improves recall significantly for IL and MN synthetic maps, 26% and 32.1%, respectively, while the best lexical method has 4% and 7% improvement. This shows the effectiveness of the proposed spatial technologies. (See Namgung and Chiang (2022) for details about the post-OCR approach and experiment.)

11.3.3 GENERATING LINKED METADATA FROM MAP TEXT

Historical maps typically exist as scanned images without searchable metadata. Existing approaches to making historical maps searchable rely on tedious manual work (including crowd-sourcing) to generate the metadata (e.g., geolocations and keywords). Optical character recognition (OCR) software could alleviate the required manual work, but the recognition results are individual words instead of location phrases (e.g., "Black" and "Mountain" vs. "Black Mountain").

This section presents an end-to-end pipeline, mapKurator, to address the real-world problem of finding and indexing historical map images. mapKurator automatically processes historical map images to extract their text content and generates metadata linked to large external geospatial knowledge bases. The linked metadata in the RDF (Resource Description Framework) format support complex queries for finding and indexing historical maps, such as retrieving all historical maps covering mountain peaks higher than 1,000 meters in California. The approach and experimental results are detailed in (Li *et al.*, 2020).

mapKurator includes three major modules for generating map metadata. In the first module, mapKurator produces complete location phrases from the input map. In the second module, mapKurator geolocates the place phrases and determines the approximate geolocation of the map. In the last module, mapKurator matches the georeferenced place phrases to entities on the LinkedGeoData to generate linked metadata for the input map. The location phrase generation module takes full map images as input and runs text detection and recognition as a preprocessing step which yields the recognized text content and their bounding boxes. Both textual predictor and visual predictor work toward pairwise linkage prediction, whereas textual predictor considers text-related information, and visual predictor considers image-related information. Text-related information can be derived from the text region, such as location, orientation angle, word embedding, etc. At the same time, the visual information comes from the neighborhood of a specific text region on the image. After pairwise linkage prediction, we construct a graph, $G = \{N, E\}$, whose nodes, N, are separate text regions and edges, E, are predicted linkages. We construct the location phrases by computing the connected components and sorting their elements.

For the geolocalization module, mapKurator employs the Google Geocoding API to convert the extracted phrases from the previous module into a set of candidate coordinates. Since the extracted text can sometimes be erroneous, the mapped candidate coordinates are noisy. To address this problem, mapKurator applies the DBSCAN algorithm to cluster the coordinates and choose the cluster's center with the most significant number of members as the geolocalization output. This is based on the intuition that extracted location phrases from maps should have a high spatial affinity (i.e., located near each other). DBSCAN is capable of removing noises by treating them as outliers. Consequently, with the members from the most significant cluster and the geo-coordinate of the cluster center, we can tell "true-mappings" apart from noises and associate the map location phrases with external databases.

The two previous modules can produce different types of metadata of the map image, and those metadata can help facilitate geospatial queries after being organized into a linked database. We use common vocabularies (see Section 11.2.2 for the semantic model) and the RDF format to represent the automatically generated metadata from historical maps and link the metadata to the existing LinkedGeoData database using the inferred geographical coordinates obtained from the previous modules. This schema provides information about the map image type, coordinates of the map center, location names in the map image, and their associated information from other sources (e.g., GeoLinkedData).

11.4 GENERATING LARGE, LONG-TERM HISTORICAL DATA FROM HISTORICAL MAP SERIES

The methods presented herein support large-scale applications using georeferenced historical maps that cover large spatial extents and are available at one or more points in time. While there are increasing numbers of whole historical map collections systematically scanned, georeferenced, and archived, the programmatic and efficient access to these map datasets and their contents often constitutes a critical bottleneck in map processing systems and may impede their application at scale. One example of an accessible historical map collection is the USGS Historical Topographic Map Collection (HTMC), which encompasses over 178,000 map sheets from 1884 to 2006 at various map scales (Allord et al., 2014). The maps in the HTMC are available as GeoTIFF format through AWS S3 buckets[5] and are accompanied by detailed metadata on each map sheet available online in CSV format.[6] Other examples of accessible georeferenced historical map collections include the British Ordnance Survey maps,[7] as well as historical maps available as Web Map Service (WMS) for several European countries (e.g., Spain[8] or Belgium[9]). As the large-scale, programmatic acquisition of historical map data is still a major issue in many regions of the world, only a few studies have focused on content extraction at the country level, e.g., (Hosseini et al., 2022).

---

[5]https://prd-tnm.s3.amazonaws.com/StagedProducts/Maps/HistoricalTopo/GeoTIFF/
[6]https://thor-f5.er.usgs.gov/ngtoc/metadata/misc/
[7]https://maps.nls.uk/os/
[8]http://www.ign.es/wms/minutas-cartograficas?request=GetCapabilities&service=WMS
[9]https://www.geopunt.be/catalogus/datasetfolder/2d7382ea-d25c-4fe5-9196-b7ebf2dbe352

11.4.1 HISTORICAL MAP DATA ACQUISITION

Herein, we focus on the USGS HTMC, for which several tools exist for bulk access to historical map sheets. For example, we developed and published the MapProcessor, a Python script that enables the user to select historical map sheets by region, time period, and map scale and returns a mosaic raster file in VRT format of the map sheets of interest.[10] Similarly, the USGS-topo-tiler allows for the selection of multiple map sheets, which will be returned in MosaicJSON format.[11]

Based on the metadata available for the USGS HTMC, it is possible to analyze different aspects of the historical map collection (Uhl *et al.*, 2018), visualize and track topographic mapping efforts over time (Uhl, 2021), and analyze the interactions between land development and topographic mapping activities over more than 100 years (Uhl *et al.*, 2022b). Moreover, these metadata enable the analysis of map coverage over time for each map scale (Figure 11.12a). For the two most commonly used map scales (i.e., 1:24,000 and 1:62,500), we used the MapProcessor tool to create historical map composites that consist of over 50,000 individual historical map sheets (covering most of the conterminous US) each of which represents the earliest map at its respective location that contains green color (i.e., woodland tint, to depict forest areas) (Figure 11.12b). It is worth noting that the temporal reference (i.e., edition or acquisition year) of these early map sheets varies between 1940 and 1980, approximately. These temporal variations are driven by the dynamics of the historical mapping and map-making process and represent a significant issue for large-scale information extraction from historical maps and applications using the extracted spatial data. Even in smaller subregions (Figure 11.12 c,d), the temporal reference of individual map sheets may vary by up to 20 years or more.

11.4.2 EXAMPLES OF LARGE-SCALE GEOGRAPHIC FEATURE EXTRACTION

This section describes two exemplary large-scale geographic feature extraction methods, focusing on historical road networks (Section 4.2.1) and historical urban areas (Section 4.2.2). Unlike the supervised geographic feature extraction methods discussed in Section 11.2.1, these methods are fully unsupervised and use modern geospatial ancillary data to support training and sampling.

11.4.2.1 Extraction of Historical Road Networks

Historical road network data are of great interest for urban planning and transportation research; however, they are rarely available (Burghardt *et al.*, 2022). To address this shortcoming, we have developed the Historical Road Network Extractor (Hironex) (Uhl *et al.*, 2022a), publicly available as a Python tool.[12] Hironex requires vector data of the modern road network (e.g., from OpenStreetMap available via OSMNx (Boeing, 2017)), as well as a georeferenced historical map from the year

---

[10]https://github.com/johannesuhl/mapprocessor
[11]https://github.com/kylebarron/usgs-topo-tiler
[12]https://github.com/johannesuhl/hironex

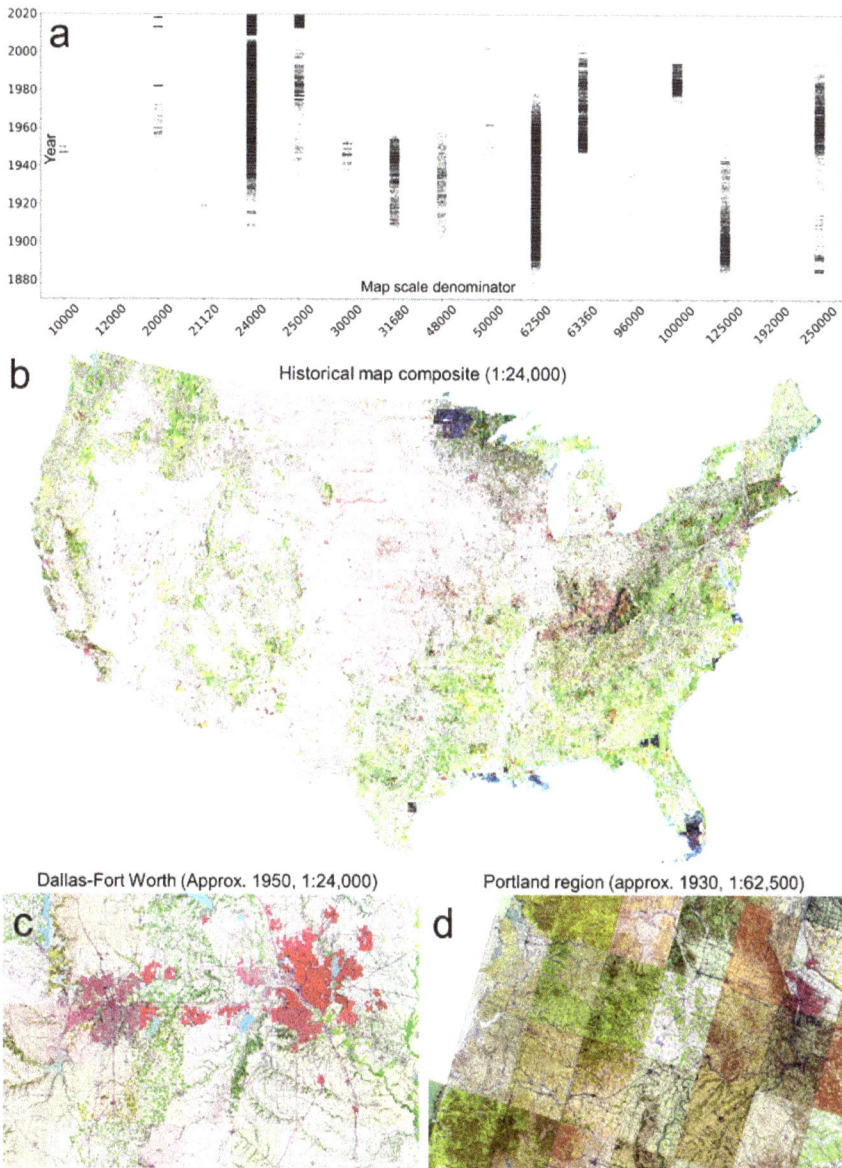

Figure 11.12 Large-scale acquisition of historical maps: (a) metadata analysis, showing the commonly used map scales over time (*Source:* (Uhl *et al.*, 2022b)), (b) historical map composite for the conterminous United States, consisting of over 50,000 historical maps of scale 1:24,000, each representing the earliest map containing woodland tint within the respective quadrangle; (c) Map composite for the cities of Dallas / Fort Worth around 1950 (at scale 1:24,000), and (d) Map composite for the region West of Portland, around 1930 (at scale 1:62,500).

T. Under the assumption that road symbols in the historical map approximately spatially coincide with the modern road (if it existed in *T*), the tool samples the color values of the scanned map image near the modern road vector data and calculates the directional grey value variance perpendicular to the modern road axis. A linear object in the map, parallel to the modern road axis, will result in a high directional variance when such a perpendicular sample is collected. Thus, high variance, found at various sampling locations along the modern road axis, can be interpreted as a proxy for existing road symbols in the underlying historical map (or as a higher likelihood that the road existed in *T*). The identified instances of high variance are registered for each modern road vector in the study area (Figure 11.13a,b). This method is fast and can be applied to large regions, as shown for the Bay area, which encompasses over 200,000 road segments (Figure 11.13c,d). Finally, a 1D-clustering method (Wang and Song, 2011) is applied to this continuous metric to identify the clusters of coinciding "historical" and "more recent" roads, which can be done for each point in time at which historical maps are available (Figure 11.13e).

11.4.2.2 Extraction of Historical Urban Areas

The quantitative analysis of the long-term evolution of cities and urban areas has long been impeded by the lack of historical spatial data on urban extents and built-up areas (Leyk *et al.*, 2020). We have developed a method to address this shortcoming by integrating color information from historical maps and remote-sensing-derived, modern data on built-up areas (Uhl *et al.*, 2021). Using our recently developed MapProcessor tool, we obtained composites of the earliest available maps per quadrangle for several metropolitan areas in the US (Figure 11.14a,d). We also used built-up areas from the Global Human Settlement Layer (GHSL) (Florczyk *et al.*, 2019), available for the time period 1975 to 2014. Our pipeline first downsamples the historical map composite (with a native resolution of 1m-5m) to the 30m grid of the GHSL, by averaging the color values, for each color channel. Then, a color space segmentation is performed, using a simple k-means clustering in combination with the Elbow method (Figure 11.14b) to identify existing groups.

We assume urban areas to grow over time rather than shrink. Hence, for each historical cluster, we calculate the area proportion within the modern urban areas as reported in the GHSL. We then use a decision function that assigns the historical cluster (or clusters, depending on the configuration) that has the largest share with the modern urban areas as the "urban" clusters. The areas of these clusters are then corrected by removing false positives, i.e., pixels of the urban clusters that are located outside of the modern urban areas (Figure 11.14c). This method is extremely fast and effective and can easily be applied to large spatial extents, such as the Boston metropolitan area (Figure 11.14d,e,f). The resulting geographical extents represent the urban areas in 1890 within the Boston area. These analysis-ready raster data enable long-term urban growth analyses and many other assessments involving changes in urban extents. For comparison, we show the 2014 built-up areas from the GHSL, which are used as ancillary data but also illustrate more recent urban extents and the urban growth that the Boston area has undergone from 1890 to 2014 (Figure 11.14g).

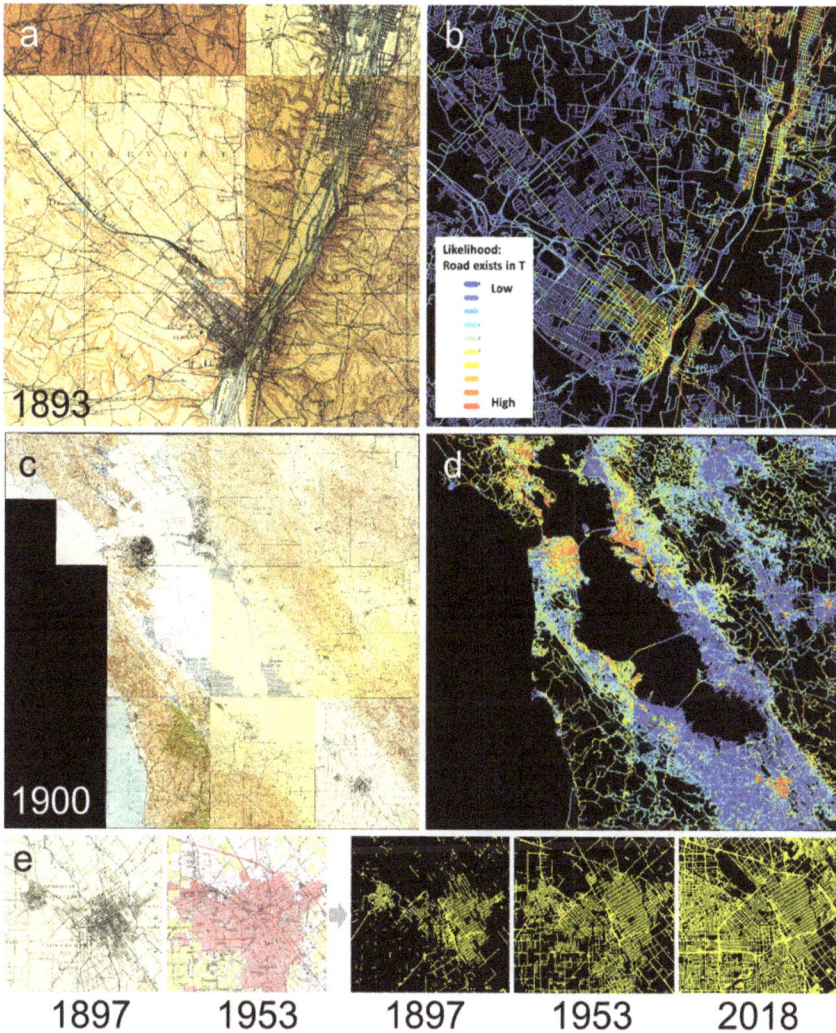

Figure 11.13 Illustrating the unsupervised extraction of the historical road networks: (a) map composite from 1893 covering the Albany (NY) region, (b) modern road network data for the same region attributed with a high likelihood of existence in 1893, (c) and (d) show a historical 10-map composite and the modern road network with likelihood attributes, respectively, for the Bay area (CA), (e) illustrating the multi-temporal application of the method, obtaining approximations of the historical road network for 1897 and 1953 for comparison to the 2018 road network data, shown for Santa Clara and San Jose (CA). *Source* (a)-(d): (Uhl *et al.*, 2022a).

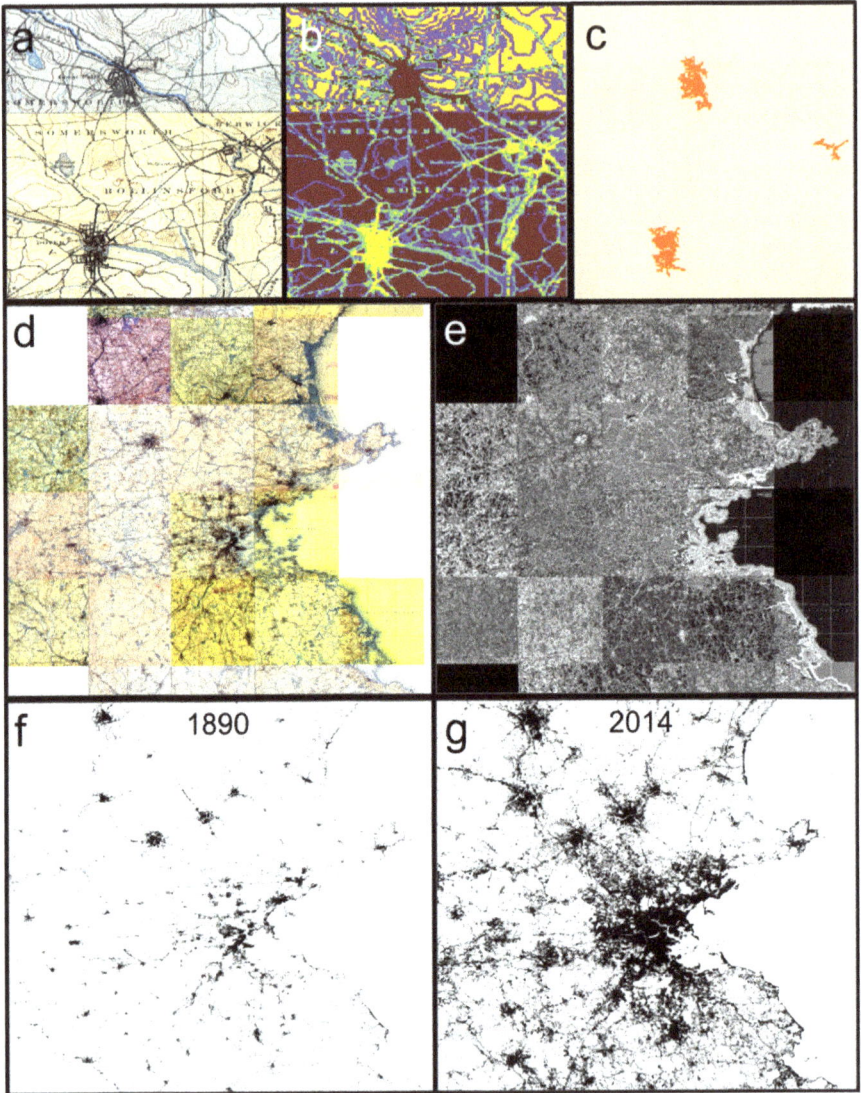

Figure 11.14 Illustrating the unsupervised extraction of urban areas from historical maps. (a) Historical map composite from around 1890, (b) Color clustering results after downsampling to the target resolution, (c) cluster identified as urban area based on decision rules that involve contemporary ancillary data. Panels (d)-(f) Results after the method was scaled up to over 20 map sheets for the Boston area: (d) historical map composite, (e) Cluster analysis results, (f) extracted urban areas, and (g) built-up areas from the GHSL in 2014 for comparison. *Source* (a)-(c): (Uhl *et al.*, 2021).

11.5 READY-TO-USE MAP PROCESSING SYSTEM AND PRE-TRAINED MODEL

This section showcases a map processing system for text recognition from historical maps and a pre-trained language model for linking and typing map text. The map processing system and pre-trained models are open-source and readily usable.

11.5.1 THE MAPKURATOR SYSTEM

To enable advanced search queries allowing users to retrieve relevant historical maps, data curators of various backgrounds, including geographers, historians, and librarians, have put substantial efforts into generating comprehensive metadata for individual map scans using map content and additional data sources. However, creating and maintaining metadata of scanned maps requires expert knowledge and extensive manual work (Rumsey and Williams, 2002).

There have been many attempts to convert text labels on maps to analytic-ready data, including crowdsourced projects (e.g., GB1900 [13]) and machine learning model-based approaches (Li *et al.*, 2018, 2020; Weinman *et al.*, 2019). Existing automatic methods could facilitate finding relevant maps and studying their content, but they often require ad-hoc post-processing to improve and integrate the results with other data sources. Thus, we present a set of ready-to-use intelligent technologies called the mapKurator system, which enables accessing automated approaches for extracting and linking text labels from historical map scans and generates a standard output file. The mapKurator system is a readily usable system derived from the mapKurator pipeline described in Section 11.3.3 and incorporated with the SynthMap and post-OCR technologies described in Sections 11.3.1 and 11.3.2, respectively.

The overall mapKurator system's capabilities include automatic processes of (1) detecting and recognizing text from maps and (2) linking map text to their corresponding entities in external knowledge bases historical gazetteers. The automatically generated results are in GeoJSON format, which can be utilized across various geographic information systems (e.g., QGIS). The GeoJSON result has features that refer to each map text. Each feature contains the geocoordinates of the bounding polygon (the "coordinates" field) and several properties, including the recognized text label (the "text" field), and the unique identifiers of matched entities in the external knowledge bases (e.g., "osm˙id" field). Example GeoJSON results are on the mapKurator website.[14]

The mapKurator system processes a map image in several steps (Figure 11.15). First, *mapKurator* slices a map image into smaller patches and processes each patch separately (M1). As deep learning models for image recognition typically rescale an input image into a small, fixed-size image, slicing the map image into small patches prevents over-downscaling and preserves the scan resolution. The system employs a text spotter, a one-stage, end-to-end trainable model to predict a bounding polygon and transcribe the text for each map text instance (e.g., TESTR (Zhang *et al.*, 2022)

---

[13] https://geo.nls.uk/maps/gb1900/
[14] https://knowledge-computing.github.io/mapkurator-doc/

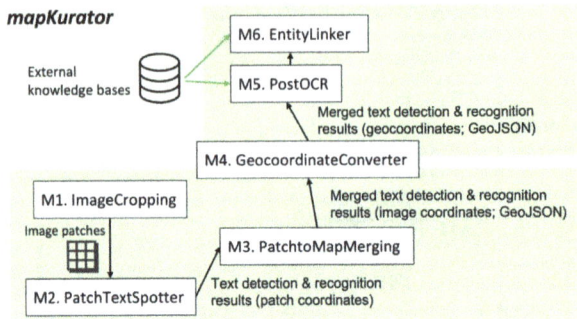

Figure 11.15   The architecture of the mapKurator system consisting of five modules (abbreviated as M#).

trained with SynthMap, Section 11.3.1) (M2). Then the system merges the results from small patches and generates a GeoJSON file in image coordinates (M3). Moreover, if the ground control points are available for georeferencing the input map, the system converts the image coordinates of bounding polygons to geocoordinates (M4). By leveraging external datasets, the system post-processes the recognized text results of the text spotter (M5) (i.e., post-OCR in Section 11.3.2). Lastly, the system takes the extracted map text and its location (i.e., geocoordinates of bounding polygon) to find and match the relevant entities in the external datasets (M6).

In summary, the mapKurator system is a complete map processing system that consists of ready-to-use intelligent technologies. As there are lots of open datasets that contain rich geospatial information (e.g., WikiData, OpenStreetMap, LinkedGeoData, and GeoNames), map text linked to external entities can support comprehensive queries and broaden access to individual maps. The proposed system allows users to quickly generate useful data from large numbers of historical maps for in-depth analysis of the map content.

11.5.2   SPABERT

Characterizing geo-entities is vital to various applications, such as geo-entity typing and geo-entity linking, while a key challenge is to capture the spatially varying context of an entity. Specifically, we humans contextualize a geo-entity by a reasonable surrounding neighborhood learned from experience and, from the neighborhood, relate other relevant geo-entities based on their name and spatial relations (e.g., distance) to the geo-entity.

Recently, the research community has seen a rapid advancement in pre-trained language models (PLMs) (Devlin *et al.*, 2019; Lewis *et al.*, 2020; Liu *et al.*, 2019; Sanh *et al.*, 2019), which supports strong contextualized language representation abilities (Lan *et al.*, 2020) and serves as the backbones of various natural language processing (NLP) systems (Rothe *et al.*, 2020; Yang *et al.*, 2019). However, it is challenging to adopt existing PLMs or their extensions to capture geo-entities' spatially varying semantics. First, geo-entities exist in the physical world. Their spatial relations (i.e., distance and orientation) do not have a fixed structure that can help contextualization. Second, existing language models (LMs) pre-trained on general domain corpora (Devlin *et al.*, 2019; Liu *et al.*, 2019) require fine-tuning for domain adaptation to handle names of geo-entities.

To tackle these challenges, we present SPABERT, an LM that captures the spatially varying semantics of geo-entity names for entity representation (Li *et al.*, 2022). SPABERT is built upon a pretrained BERT and further trained to produce contextualized geo-entity representations given large geographic datasets. In the following, we will introduce how SPABERT encodes a spatial neighborhood to obtain a contextualized representation of a geo-entity and how SPABERT is pre-trained.

Contextualizing Geo-entities Let g_i be a geo-entity within a geographic dataset. We assume that the name and location of the geo-entity are known. SPABERT aims to generate a contextualized representation for each geo-entity g_i given its spatial context. We call the "geo-entity of interest" the pivot entity, and the spatial context is represented by a set of neighbor entities whose distance to the pivot entity is within some certain threshold. For a pivot, p, SPABERT first linearizes its neighboring geo-entities' names to form a BERT-compatible input sequence, called a *pseudo sentence*. The pseudo sentence starts with the pivot name followed by the names of the pivot's neighboring geo-entities, ordered by their distance to the pivot in ascending order (as seen in Figure 11.16).

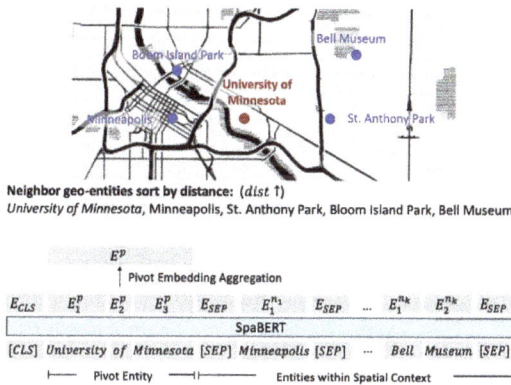

Figure 11.16 Overview for generating the pivot entity representation, E^p (red circle: pivot entity; blue circle: neighboring geo-entities). SPABERT sorts and concatenate neighboring geo-entities by their distance to pivot in ascending order to form a pseudo-sentence. [CLS] is prepended at the beginning. [SEP] separates entities. SPABERT generates token representations and aggregates representations of pivot tokens to produce E^p.

SPABERT sums two types of position embeddings in addition to the token embeddings in the pseudo sentences. The sequence position embedding represents the token order, the same as the original position embedding in BERT and other Transformer LMs. Also, SPABERT incorporates a *spatial coordinate embedding* mechanism, which seeks to represent the spatial relations between the pivot and its neighboring geo-entities. Specifically, for the 2D space, SPABERT generates normalized distance $dist_x^{n_k}$ and $dist_y^{n_k}$ between each neighboring geo-entity n_k along and the pivot along the horizontal and vertical direction. SPABERT encodes $dist_x^{n_k}$ and $dist_y^{n_k}$ using a continuous spatial coordinate embedding layer with real-valued distances as input to preserve the continuity of the output embeddings. Let $\mathscr{S}_{n_k}$ be the spatial coordinate embedding of the neighbor entity n_k, M be the embedding's dimension, and $\mathscr{S}_{n_k} \in \mathbf{R}^M$. We define $\mathscr{S}_{n_k}$ as:

$$\mathscr{S}_{n_k}^{(m)} = \begin{cases} sin(dist^{n_k}/10000^{2j/M}), m = 2j \\ cos(dist^{n_k}/10000^{2j/M}), m = 2j+1 \end{cases}$$

where $\mathscr{S}_{n_k}^{(m)}$ is the m-th component of $\mathscr{S}_{n_k}$. $dist^{n_k}$ represents $dist_x^{n_k}$ and $dist_y^{n_k}$ for the spatial coordinate embeddings along the horizontal and vertical directions. Also, SPABERT uses DSEP, a constant numerical value larger than $max(dist_x^{n_k}, dist_y^{n_k})$ for all neighboring entities to differentiate special tokens from entity name tokens.

Pre-training We train SPABERT with two tasks to adapt the pretrained BERT backbone to geo-entity pseudo sentences. One task is the masked language modeling (MLM) (Devlin *et al.*, 2019), for which SPABERT needs to learn how to complete the full names of geo-entities from pseudo sentences with randomly masked subtokens using the remaining subtokens and their spatial coordinates (i.e., partial names and spatial relations between subtokens). In addition, we hypothesize that given common spatial co-occurrence patterns in the real world, one can use neighboring geo-entities to predict the name of a geo-entity. Therefore, we propose and incorporate a masked entity prediction (MEP) task, which randomly masks all subtokens of an entity name in a pseudo sentence. For MEP, SPABERT relies on the masked entity's spatial relation to neighboring entities to recover the masked name.

Both MLM and MEP have a masking rate of 15% (without masking [CLS] and [SEP]). The sequence position and spatial coordinates are not masked. In our previous paper (Li *et al.*, 2022), we provide additional details about these techniques.

Geo-entity Typing In this task, we aim to predict the geo-entity's semantic type (e.g., transportation and healthcare). We stack a softmax prediction head on top of the final-layer hidden states (i.e., geo-entity embeddings) to perform the classification. The dataset is collected from Open Street Map (OSM)[15] in London, UK and California, US with 9 class types. (Li *et al.*, 2022) contains the details about dataset statistics. Table 11.1 summarizes the geo-entity typing results on the OSM test set with fine-tuned baselines and SPABERT. The last column shows the micro-F1 scores (per-instance average). The remaining columns are the F1 scores for individual classes. Among baseline models, BERT has the highest F1, and SPABERT outperforms all the context-aware LM baselines.

---

[15]https://www.openstreetmap.org/

Geo-entity Linking The goal of this task is to link the same geo-entity in different geographical databases. Specifically, we link geo-entities from the United States Geological Survey (USGS) Historical Topographic Maps Collection to the Wikidata knowledge base. (Li *et al.*, 2022) contains the details about dataset statistics. We directly use the contextualized geo-entity embeddings from SPABERT and calculate the embedding similarity based on cosine distance to match the corresponding geo-entities from separate data sources. Table 11.2 shows the geo-entity linking results. In particular, among the baseline models, SimCSE$_{BERT}$ has the highest score for MRR and R@1. SPABERT has significantly higher R@5 and R@10 than SimCSE. Also, SPABERT outperforms all baselines on MRR, R@5, and R@10. Additionally, SPABERT outperforms its backbone model BERT.

Table 11.1

Geo-entity typing with the state-of-the-art LMs and SpaBERT. Column names are the OSM classes. Bold: the highest scores in each column. Underlined: the highest scores among baselines.

Classes →	Edu.	Ent.	Fac.	Fin.	Hea.	Pub.	Sus.	Tra.	Was.	Micro Avg
BERT	.674	.634	.763	.929	.856	.872	.856	.862	.678	.835
RoBERTa	.626	.627	.605	.951	**.869**	.818	.838	.850	.475	.820
SpanBERT	.633	.589	.608	.916	.859	.882	.824	.867	.735	.819
LUKE	.648	.608	.598	.945	.857	.867	.854	.851	.517	.825
SimCSE$_{BERT}$	.623	.590	.504	.925	.867	.852	.857	.810	.470	.810
SimCSE$_{RoBERTa}$	.621	.629	.499	.951	.841	.853	.828	.856	.500	.814
SPABERT	**.674**	**.653**	.680	**.959**	.865	**.900**	**.883**	**.888**	.703	**.852**

Table 11.2

Geo-entity linking result. Bold and underlined numbers are the highest scores in each column and the highest in the baselines, respectively.

Model	MRR	R@1	R@5	R@10
BERT	.400	.289	.559	.635
RoBERTa	.326	.232	.446	.540
SpanBERT	.164	.138	.201	.213
LUKE	.306	.188	.440	.547
SimCSE$_{BERT}$	.453	**.371**	.547	.628
SimCSE$_{RoBERTa}$	.227	.188	.264	.301
SPABERT	**.515**	.338	**.744**	**.850**

11.6   CONCLUSION

This chapter presented a broad overview of the recent advances in methods, systems, and data related to processing historical maps to generate valuable data, information, and knowledge. We hope the chapter will shed light on the future development of

AI technologies for processing historical maps and benefit the broad community of historical map users and researchers from various disciplines.

ACKNOWLEDGMENTS

This material is based upon work supported in part by NVIDIA Corporation, the National Endowment for the Humanities under Award No. HC-278125-21 and Council Reference AH/V009400/1, the National Science Foundation of United States Grant IIS 2105329, a Cisco Faculty Research Award (72953213), the University of Minnesota, Computer Science & Engineering Faculty startup funds, and the generous gift from David and Abby Rumsey to the Knowledge Computing Lab at the University of Minnesota. We thank all our collaborators who contribute datasets and valuable knowledge in library science and digital humanities.

BIBLIOGRAPHY

Allord, G.J., Fishburn, K.A., and Walter, J.L., 2014. Standard for the us geological survey historical topographic map collection. *US Geological Survey: Reston, VA, USA*, 3.

Athanasiou, S., *et al.*, 2014. Geoknow: making the web an exploratory place for geospatial knowledge. *ERCIM News*, 96 (12-13), 119–120.

Battle, R. and Kolas, D., 2011. Geosparql: enabling a geospatial semantic web. *Semantic Web Journal*, 3 (4), 355–370.

Bhushan Damodaran, B., *et al.*, 2018. An entropic optimal transport loss for learning deep neural networks under label noise in remote sensing images. *arXiv e-prints*, arXiv–1810.

Boeing, G., 2017. Osmnx: New methods for acquiring, constructing, analyzing, and visualizing complex street networks. *Computers, Environment and Urban Systems*, 65, 126–139.

Bone, C., *et al.*, 2016. A geospatial search engine for discovering multi-format geospatial data across the web. *International Journal of Digital Earth*, 9 (1), 47–62.

Burghardt, K., *et al.*, 2022. Road network evolution in the urban and rural united states since 1900. *Computers, Environment and Urban Systems*, 95, 101803.

Carion, N., *et al.*, 2020. End-to-end object detection with transformers. *In: European Conference on Computer Vision*. Springer, 213–229.

Chan, T. and Zhu, W., 2005. Level set based shape prior segmentation. *In: 2005 IEEE Computer Society Conference on Computer Vision and Pattern Recognition (CVPR'05)*. IEEE, vol. 2, 1164–1170.

Chen, L.C., *et al.*, 2018. Encoder-decoder with atrous separable convolution for semantic image segmentation. *In: Proceedings of the European Conference on Computer Vision (ECCV)*. 801–818.

Chen, X., *et al.*, 2021. Text recognition in the wild: A survey. *ACM Computing Surveys (CSUR)*, 54 (2), 1–35.

Chiang, Y.Y., 2015. Querying historical maps as a unified, structured, and linked spatiotemporal source: Vision paper. SIGSPATIAL '15, New York, NY, USA. Association for Computing Machinery. Available from: `https://doi.org/10.1145/2820783.2820887`.

Chiang, Y.Y., *et al.*, 2020. *Using Historical Maps in Scientific Studies: Applications, Challenges, and Best Practices.* Springer.

Chiang, Y.Y., Leyk, S., and Knoblock, C.A., 2014. A survey of digital map processing techniques. *ACM Computing Surveys (CSUR)*, 47 (1), 1–44.

Chiron, G., *et al.*, 2017. Icdar2017 competition on post-ocr text correction. *In: 2017 ICDAR.* vol. 01, 1423–1428.

Devlin, J., *et al.*, 2019. BERT: Pre-training of deep bidirectional transformers for language understanding. *In: Proceedings of the 2019 Conference of the North American Chapter of the Association for Computational Linguistics: Human Language Technologies, Volume 1 (Long and Short Papers)*, June, Minneapolis, Minnesota. ACL, 4171–4186.

Duan, W., *et al.*, 2017. Automatic alignment of geographic features in contemporary vector data and historical maps. *In: Proceedings of the 1st Workshop on Artificial Intelligence and Deep Learning for Geographic Knowledge Discovery.* 45–54.

Duan, W., *et al.*, 2021a. Guided generative models using weak supervision for detecting object spatial arrangement in overhead images. *In: 2021 IEEE International Conference on Big Data (Big Data).* IEEE, 725–734.

Duan, W., *et al.*, 2021b. A label correction algorithm using prior information for automatic and accurate geospatial object recognition. *In: 2021 IEEE International Conference on Big Data (Big Data).* IEEE, 1604–1610.

Florczyk, A.J., *et al.*, 2019. Ghsl data package 2019. *Luxembourg, EUR*, 29788 (10.2760), 290498.

Goodchild, M.F., 2007. Citizens as sensors: the world of volunteered geography. *GeoJournal*, 69 (4), 211–221.

Gupta, A., Vedaldi, A., and Zisserman, A., 2016. Synthetic data for text localisation in natural images. *In: Proceedings of the IEEE Conference on Computer Vision and Pattern Recognition.* 2315–2324.

Hosseini, K., *et al.*, 2022. Mapreader: a computer vision pipeline for the semantic exploration of maps at scale. *In: Proceedings of the 6th ACM SIGSPATIAL International Workshop on Geospatial Humanities.* 8–19.

Huang, M., *et al.*, 2022. Swintextspotter: Scene text spotting via better synergy between text detection and text recognition. *In: Proceedings of the IEEE/CVF Conference on Computer Vision and Pattern Recognition.* 4593–4603.

Jaderberg, M., *et al.*, 2014. Synthetic data and artificial neural networks for natural scene text recognition. *arXiv preprint arXiv:1406.2227.*

Kang, J., *et al.*, 2021. Noise-tolerant deep neighborhood embedding for remotely sensed images with label noise. *IEEE Journal of Selected Topics in Applied Earth Observations and Remote Sensing*, 14, 2551–2562.

Lan, Z., *et al.*, 2020. ALBERT: A lite BERT for self-supervised learning of language representations. *In: 8th International Conference on Learning Representations, ICLR 2020, Addis Ababa, Ethiopia, April 26-30, 2020.* OpenReview.net. Available from: `https://openreview.net/forum?id=H1eA7AEtvS`.

Levenshtein, V., 1966. Binary codes capable of correcting deletions, insertions, and reversals. *In: Soviet Physics Doklady.* Soviet Union, vol. 10, 707–710.

Lewis, M., *et al.*, 2020. BART: Denoising sequence-to-sequence pre-training for natural language generation, translation, and comprehension. *In*: *Proceedings of the 58th Annual Meeting of the ACL*, July. ACL, 7871–7880.

Leyk, S., *et al.*, 2020. Two centuries of settlement and urban development in the united states. *Science Advances*, 6 (23), eaba2937.

Li, H., Liu, J., and Zhou, X., 2018. Intelligent map reader: A framework for topographic map understanding with deep learning and gazetteer. *IEEE Access*, 6, 25363–25376.

Li, Z., *et al.*, 2020. *An Automatic Approach for Generating Rich, Linked Geo-Metadata from Historical Map Images*. New York, NY, USA: Association for Computing Machinery, 3290–3298.

Li, Z., *et al.*, 2021. Synthetic map generation to provide unlimited training data for historical map text detection. *In*: *Proceedings of the 4th ACM SIGSPATIAL International Workshop on AI for Geographic Knowledge Discovery*. 17–26.

Li, Z., *et al.*, 2022. Spabert: A pretrained language model from geographic data for geo-entity representation. *In*: *EMNLP-findings*.

Liu, X., *et al.*, 2018. Fots: Fast oriented text spotting with a unified network. *In*: *Proceedings of the IEEE Conference on Computer Vision and Pattern Recognition*. 5676–5685.

Liu, Y., *et al.*, 2019. Roberta: A robustly optimized bert pretraining approach. *arXiv preprint arXiv:1907.11692*.

Liu, Y., *et al.*, 2020. Abcnet: Real-time scene text spotting with adaptive bezier-curve network. *In*: *Proceedings of the IEEE/CVF Conference on Computer Vision and Pattern Recognition*. 9809–9818.

Long, S., He, X., and Yao, C., 2021. Scene text detection and recognition: The deep learning era. *International Journal of Computer Vision*, 129 (1), 161–184.

Namgung, M. and Chiang, Y.Y., 2022. Incorporating spatial context for post-ocr in map images. *In*: *Proceedings of the 5th ACM SIGSPATIAL International Workshop on AI for Geographic Knowledge Discovery*. 14–17.

Nguyen, T., *et al.*, 2020. Neural machine translation with bert for post-ocr error detection and correction. *In*: *Proceedings of the ACM/IEEE JCDL*, New York, NY, USA. ACM, 333–336.

Rothe, S., Narayan, S., and Severyn, A., 2020. Leveraging pre-trained checkpoints for sequence generation tasks. *Transactions of the Association for Computational Linguistics*, 8, 264–280.

Rumsey, D. and Williams, M., 2002. *Historical Maps in GIS*. Cartography Associates.

Sanh, V., *et al.*, 2019. Distilbert, a distilled version of bert: smaller, faster, cheaper and lighter. *arXiv preprint arXiv:1910.01108*.

Shbita, B., *et al.*, 2022. Building spatio-temporal knowledge graphs from vectorized topographic historical maps. *Semantic Web*, (Preprint), 1–23.

Song, B. and Chan, T., 2002. A fast algorithm for level set based optimization. *UCLA Cam Report*, 2 (68).

Soper, E., Fujimoto, S., and Yu, Y., 2021. BART for post-correction of OCR newspaper text. *In*: *Proceedings of W-NUT 2021*, November, Online. ACL, 284–290.

Uhl, J.H., 2021. How the U.S. was mapped - visualizing 130 years of topographic mapping in the conterminous U.S. Available from: `https://doi.org/10.6084/m9.figshare.17209433.v3`.

Uhl, J.H. and Duan, W., 2020. Automating information extraction from large historical topographic map archives: New opportunities and challenges. *Handbook of Big Geospatial Data*, 509–522.

Uhl, J.H., *et al.*, 2018. Map archive mining: visual-analytical approaches to explore large historical map collections. *ISPRS International Journal of Geo-Information*, 7 (4), 148.

Uhl, J.H., *et al.*, 2022a. Towards the automated large-scale reconstruction of past road networks from historical maps. *Computers, Environment and Urban Systems*, 94, 101794.

Uhl, J.H., *et al.*, 2022b. Unmapped terrain and invisible communities: Analyzing topographic mapping disparities across settlements in the united states from 1885 to 2015. July. Zenodo. Available from: `https://doi.org/10.5281/zenodo.6789259`.

Uhl, J.H., *et al.*, 2021. Combining remote-sensing-derived data and historical maps for long-term back-casting of urban extents. *Remote Sensing*, 13 (18), 3672.

Vrandečić, D. and Krötzsch, M., 2014. Wikidata: a free collaborative knowledgebase. *Communications of the ACM*, 57 (10), 78–85.

Wang, H. and Song, M., 2011. Ckmeans. 1d. dp: optimal k-means clustering in one dimension by dynamic programming. *The R Journal*, 3 (2), 29.

Wang, T., *et al.*, 2012. End-to-end text recognition with convolutional neural networks. *In: Proceedings of the 21st International Conference on Pattern Recognition (ICPR2012)*. IEEE, 3304–3308.

Weinman, J., *et al.*, 2019. Deep neural networks for text detection and recognition in historical maps. *In: Proceedings of IEEE ICDAR '19*. IEEE, 902–909.

Wu, S., *et al.*, 2019. Road extraction from very high resolution images using weakly labeled openstreetmap centerline. *ISPRS International Journal of Geo-Information*, 8 (11), 478.

Xia, Z., *et al.*, 2022. Vision transformer with deformable attention. *In: Proceedings of the IEEE/CVF Conference on Computer Vision and Pattern Recognition*. 4794–4803.

Yang, S., *et al.*, 2019. Exploring pre-trained language models for event extraction and generation. *In: Proceedings of the 57th Annual Meeting of the Association for Computational Linguistics*, July, Florence, Italy. ACL, 5284–5294.

Zhang, X., *et al.*, 2022. Text spotting transformers. *In: Proceedings of the IEEE/CVF Conference on Computer Vision and Pattern Recognition*. 9519–9528.

Zhu, X., *et al.*, 2020. Deformable detr: Deformable transformers for end-to-end object detection. *arXiv preprint arXiv:2010.04159*.

12 Spatiotemporal AI for Transportation

Tao Cheng
SpaceTimeLab, Department of Civil, Environmental and
Geomatic Engineering, University College London

James Haworth
SpaceTimeLab, Department of Civil, Environmental and
Geomatic Engineering, University College London

Mustafa Can Ozkan
SpaceTimeLab, Department of Civil, Environmental and
Geomatic Engineering, University College London

CONTENTS

12.1 Introduction: Background on Spatiotemporal AI and Transportation 248
12.2 Data-driven Prediction of Traffic Variables 249
12.3 Optimization of Traffic Networks Using Reinforcement Learning 251
12.4 Computer Vision for Sensing Complex Urban Environments 253
12.5 Spatiotemporal AI in Transportation: Challenges and Opportunities 255
     Bibliography 255

12.1 INTRODUCTION: BACKGROUND ON SPATIOTEMPORAL AI AND TRANSPORTATION

The transportation sector has long had an interest in the application of computational techniques, centered on the field of intelligent transportation systems (ITS). Transportation systems, particularly vehicular traffic and transit systems, have been monitored using a range of technologies since the latter part of the 20th century, generating vast quantities of data. These data include: flows and volumes collected by static loop detectors; travel times collected by automatic number plate recognition (ANPR); trajectories from smart phones; and public transit trips from smart cards (Welch and Widita 2019). This data rich environment has led to sustained research in academia and industry into how they can be best used to improve the efficiency, safety and customer experience of transport systems. From the 1980s, researchers started applying statistical methods such as autoregressive integrated moving average (ARIMA) and related models to traffic forecasting. However, due to the nonlinearity, heterogeneity

DOI: 10.1201/9781003308423-12

and nonstationarity present in time series of transport data, research shifted toward machine learning approaches such as neural networks, kernel methods and random forests in the 1990s to 2000s. These methods were shown to provide improved performance across a range of tasks in ITS. Most recently, deep learning approaches have been applied to various tasks, in particularly Spatiotemporal AI methods that leverage the spatial and spatiotemporal correlation in transport data (Cheng, Zhang, and Haworth 2022). These methods make use of neural architectures that have been developed in the fields of computer science and AI, including recurrent neural networks (RNNs), convolutional neural networks (CNNs), graph convolutional neural networks (GCNNs), autoencoders and attention modules (Veres and Moussa 2020). Alongside this, AI methods are beginning to find application in the range of simulation and optimization problems in transportation, such as scheduling, routing and planning. In this chapter, we focus on the latest advances in Spatiotemporal AI for transportation in three areas: (1) data driven prediction of traffic variables; (2) optimization of traffic networks using reinforcement learning; and (3) computer vision for sensing complex urban environments.

12.2 DATA-DRIVEN PREDICTION OF TRAFFIC VARIABLES

Prediction of traffic variables such as flows, densities and travel times is an extremely important in topic in ITS. If future traffic conditions can be accurately predicted, then transport planners and individuals can take actions to smooth traffic and minimize congestion. For example, at the network level, signal timings can be adjusted based on short-term forecasts of vehicles arriving at intersections, which allows queuing traffic to be cleared more effectively. At the individual level, road users can be delivered information on predicted travel times to their destination on different routes and tailor their choices accordingly. For example, if a crash causes congestion on their usual route, they may take an alternative route or travel at a different time.

Although many different algorithms have been used for short-term traffic forecasting, the basic approach is typically the same. Historical values of traffic data are used to build a model to predict future values. Research began in the second half of the 20th century with statistical time series models that are applied to a single location, such as ARIMA models (Smith, Williams, and Oswald 2002; Williams and Hoel 2003). However, these fail to take into account the spatio-temporal propagation of traffic conditions. Recognizing this, research in the early 2000s shifted toward spatio-temporal models such as space-time (ST) ARIMA and related models (Cheng et al. 2014; Kamarianakis and Prastacos 2003), as well as various machine learning (ML) models (Karlaftis and Vlahogianni 2011); Vlahogianni et al. (2014) provided reviews of the progress during this period. Among the ML methods, ANNs have been applied in the traffic forecasting literature since the 1990s, starting with simple time series methods (Smith and Demetsky 1997) and progressing to spatio-temporal approaches. For example, Van Lint *et al.* (2005) used ANNs for freeway traffic prediction, with the structure of the hidden layer mimicking the spatio-temporal arrangement of data measurement locations. Recurrent neural networks (RNNs) such as Jordan Networks, Elman Networks and Long-short Term Memory Networks (LSTMs) have often been used due to their ability to process temporal information and overcome the vanishing gradient problem. For example, Zhao et al. (2017) used LSTMs for traffic forecasting, incorporating spatial and temporal information. While LSTMs are typically used for single time series, Zhao's approach uses a 2D-LSTM to incorporate spatial neighborhood. In a similar vein, researchers have combined LSTMs with CNNs to forecast traffic flow aggregated to regular grids (Ren et al. 2020; Wu and Tan 2016).

While grid-based CNNs may yield promising results, they break the network structure of road traffic, which results in a loss of information about traffic propagation. The latest work in Spatiotemporal AI has adopted deep learning architectures for network-based forecasting of spatio-temporal data, which can be referred to as Network SpaceTimeAI (Cheng, Zhang, and Haworth 2022). Ren *et al.* (2019) used deep spatio-temporal residual networks that construct locally connected neural network layers to model road network structure, demonstrating improved performance over the grid based methods developed in (Ren et al. 2020) in forecasting taxi flows in Chengdu, China. While Ren et al.'s (2019) approach does take into account spatio-temporal correlation, graph convolutional neural networks (GCNNs) have been the most popular approach due to the inherent suitability of their structure for modeling network data. A GCNN takes the principles of CNNs and applies them to graph data structures. This approach presents challenges: while images have a regular structure in which the number of neighbors of a pixel is fixed and the scanning direction can be easily defined (e.g., top left to bottom right), the same is not true for graphs, which can be of arbitrary arrangement and differing numbers of neighbors (S. Zhang et al. 2019). Therefore, one of the challenges of GCNNs is to define orderings and neighborhood structures that are suitable for CNN models.

There are two main approaches for GCNNs: spectral-based methods and spatial-based methods (S. Zhang et al. 2019). The former approach uses spectral convolution for filtering in the frequency domain and has been found to be more suitable to the large networks encountered in transportation. Yu *et al.* (2018) developed spectral-based spatio-temporal (ST) GCNNs for multi-step traffic forecasting using an undirected graph structure. The approach combines graph convolutions with gated temporal convolutions, which achieve efficiency gains over traditional recurrent architectures. The method is tested on traffic flow data from highways in Beijing, China and California, USA, outperforming a range of comparison models including LSTMs. Recognizing that Yu *et al.* (2018) neglects the direction of traffic flow in their undirected network, Zhang, Cheng, and Ren (2019) used a spectral-based GCNN to develop a graph deep learning model for short-term travel time forecasting named the spatial–temporal graph inception residual network (STGI-ResNet). The approach uses the Laplacian matrix of the graph to capture the graph's topology and pattern of edge weights, allowing it to model local spatial dependency. Using spectral convolution reduces the complexity to O(N), where N is the number of nodes in the graph, making the approach suitable for large networks. Short, medium and long-term temporal features are extracted for each location and combined with the spatial features using spatio-temporal graph convolution. The STGI-ResNet is tested against a range of benchmark algorithms including ARIMA, LSTM and Graph Convolutional-LSTM, demonstrating improved performance in forecasting urban travel times based on a DiDi taxi trajectory data in Chengdu, China.

In summary, transportation researchers have always been forward thinking in adopting the latest developments from ML and computer science to data driven forecasting of traffic variables. State-of-the-art results are obtained when those methods are designed with the properties of spatio-temporal network data in mind. In the context of Spatiotemporal AI and SpaceTimeAI, this is illustrated in the improved performance of methods that take into account network structure and traffic flow dynamics over grid-based approaches.

12.3 OPTIMIZATION OF TRAFFIC NETWORKS USING REINFORCEMENT LEARNING

Optimization is an important topic in transportation due to the inherent need to manage traffic and resources efficiently. Optimization problems can range from controlling signal timings at a single intersection to optimizing the timetables of urban transit systems. An optimization problem involves minimizing or maximizing a cost function according to a set of constraints. For example, in traffic signal optimization at an intersection, the objective is to minimize the delay for vehicles passing through the intersection subject to constraints such as the minimum and maximum green time for each signal (Liu and Xu 2012). These problems become more complex when multiple intersections are involved, for example across a city. If each intersection is treated in isolation then optimization of one location may lead to delays elsewhere. It is because of this inherent complexity that scholars have turned to Spatiotemporal AI based methods such as reinforcement learning (RL) for transport optimization.

Reinforcement learning is a ML technique inspired by the way that animals learn behaviors through positive and negative reinforcement of their actions. For example, if a dog owner gives their pet a treat when it performs an action, such as sitting, they are positively reinforcing that behavior, which is learned by repetition. In RL, a learner or agent takes actions within an environment to seek rewards, with the aim of finding sequences of actions that maximize reward (Sutton and Barto 2018). RL problems are usually formulated as a Markov Decision Process and solved by an algorithm such as Q-Learning. The most famous example of RL in computer science is Deep Mind's AlphaGo, which has been shown to beat top human players at Go, a notoriously complex board game (Silver et al. 2016). This success has led to its popularity in other application domains.

In transportation research, RL found early application in traffic signal optimization, which is an established field of ITS where RL has natural application (Miletic et al. 2022; Wei et al. 2021). In the past 5 years, the application of RL has widened to include decision-making systems, connected and autonomous vehicles, and vehicle routing (planning, routing, navigation).

Transport decision-making problems such as logistics, supply chain management and scheduling all involve optimization. For example, in parcel delivery the goal is to find the optimal allocation of finite resources (e.g., delivery vehicles) to deliver parcels across an environment (transport network) in the minimum time. This problem may be subject to constraints such as the maximum number of parcels per vehicle, the shift length of the drivers and the working zones of the vehicles. Because of its suitability to this type of problem, RL has gained popularity in this domain. For example, Kang, Lee, and Chung (2019) used RL for parcel delivery management, while Irannezhad *et al.* (2020) and Hassan *et al.* (2020) proposed RL based intelligent decision systems for optimizing freight supply and demand chains. In a similar vein, Vikharev and Liapustin (2019) and Yang et al. (2020) have applied RL to identify passengers' decisions in public transit systems, while Khadilkar (2019), Matos et al. (2021), and Wang and Sun (2020) applied RL to transport scheduling problems.

Within the broader goal of resource allocation on networks, vehicle routing has gained particular interest. Work on RL for routing began as early as 2010, when (Mostafa and Talaat 2010) and (Tamagawa, Taniguchi, and Yamada 2010) used Q-Learning to minimize travel times for freight. While the early studies focused on static problems, more recent work has shifted to optimization of shared systems such as taxi fleets and connected autonomous vehicles (CAVs). RL is a particularly attractive approach in these multi-agent systems (many vehicles) with dynamic conditions because it allows the properties of the agents, their level of knowledge, and their environment to be encoded. For example, CAVs can have knowledge of each other's position and communicate with one another, with the goal of minimizing total travel time across the network subject to constraints on the individual vehicle travel times. RL has been used in taxi fleet management to redistribute idle vehicles according to future demand, vehicle dispatch (Shi et al. 2019), and taxi carpooling (Jindal et al. 2018).

In the CAV setting, Grunitzki, Ramos, and Bazzan (2014) proposed a system in which vehicles can communicate and send information on network conditions. This mimics the real situation where users of navigation systems transmit their position and speed to a centralized system like Google live traffic maps. It was found that cooperation between agents to maximize system utility yields improved performance. Similar approaches have been taken for dynamic route selection in uncertain networks (Edrisi, Bagherzadeh, and Nadi 2022) and electric vehicle routing to minimize energy consumption (Basso et al. 2022). It is also worth noting that RL plays an important role in onboard technology in CAV because it allows algorithms to be trained through simulations in the absence of real-world data, which is challenging to acquire while CAVs are not widespread on public roads. In this context, RL has been used to train CAVs to carry out a range of tasks, including lane-changing (Wang et al. 2019), car-following (Zhou, Fu, and Wang 2020) and path planning (Chen et al. 2020).

While RL has found applications in transportation research, it has not yet replaced traditional methods for traffic simulation due to the inherent complexity in modeling large scale multi-agent systems like urban transport networks. Studies to date have often focused on small case studies (Walraven, Spaan, and Bakker 2016). Where city-wide applications exist, they are based on microscopic simulation models in which the signaling is optimized rather than the behavior of individual agents (Chu et al. 2020; H. Zhang et al. 2019). Therefore, they optimize the network's

response to traffic, rather than optimizing individual choices to minimize congestion. It is expected that multi-agent RL (MARL) will become more widespread in
transportation research in the coming years as algorithms improve and computational
power increases.

12.4 COMPUTER VISION FOR SENSING COMPLEX URBAN ENVIRONMENTS

Computer vision (CV) is a field of research "concerned with extracting information
about a scene by analyzing images of that scene" (Rosenfeld, 1988, p. 863). CV
uses computational methods to extract features from images and videos that could
be easily perceived by the human eye. The key tasks of computer vision include
object detection and tracking, semantic segmentation, image classification and action
detection. These tasks are summarized in Table 12.1, including an overview of their
use in transportation research.

The tasks listed in Table 12.1 have benefited from improvements in deep learning
algorithms in the past two decades. Most of the core tasks are accomplished using
CNNs of different types. A CNN is a deep neural network that uses convolutional op-
erators to extract features from images. The basic architectures are sometimes combined with other algorithms, e.g., for tracking objects between subsequent images.
Notable algorithms in each task are shown in Table 12.1. The list is non-exhaustive
and readers are directed to the cited review articles for detailed overviews of each
task.

Computer vision methods saw early adoption in transport with their use in automatic number plate recognition cameras (ANPR) for traffic monitoring. While most

Table 12.1

Main tasks of computer vision.

Task	Description	Algorithms	Transport applications
Image Classification	The classification of a whole image into a class	ResNet (He et al. 2016)	Weather and visual conditions classification
Object Detection	The localization of objects in an image	Region based CNN (RCNN); You Only Look Once (YOLO); Single Shot Detector (SSD) (Xiao et al. 2020)	Plate recognition; vehicle counting; infrastructure mapping (e.g., road signs and signals); fault detection (e.g., cracks, potholes)
Semantic Segmentation	Pixel level classification of an image into different classes	Fully Convolutional Network (FCN); U-Net; SegNet; Deeplab (Mo et al. 2022)	Environment segmentation, e.g., road, pavement, lane
Panoptic Segmentation	Pixel level classification of an image into semantic classes and objects	Mask R-CNN and JSIS-Net; Panoptic FPN (Li and Chen 2022)	Combines tasks of semantic segmentation and object detection
Object Tracking	Tracking of objects through sequences of frames	Deep SORT (Pal et al. 2021)	Tracking vehicles through intersections; CAV onboard systems
Action Detection	Detection of specific actions of objects or agents within a scene	Algorithm choice is dependent on action being detected (Yao, Lei, and Zhong 2019; H. Zhang et al. 2019)	Near-miss detection; driver impairment (e.g., tiredness)

ANPR systems were installed for enforcement purposes, such as detecting uninsured vehicles or enforcing congestion charging, many have been repurposed to record travel times, such as London's Congestion Analysis Project (LCAP) (Cheng, Haworth, and Wang 2011). This involves setting up networks of links formed of pairs of cameras. When a vehicle passes the first camera in a link the time is recorded, and if/when the vehicle is reidentified by a second camera the travel time is recorded as the difference between the two times. This type of network is constrained by the coverage of ANPR cameras so tends to be limited to major routes such as arterials and ring roads. While ANPR is able to differentiate between motorized modes by linking up to vehicle licensing databases, it is unable to detect road users that do not have plates such as bicycles, micromobility (e-bikes, e-scooters etc.) and pedestrians.

More recently, advances in CNNs have enabled vehicle detection and tracking from CCTV cameras. The use cases vary from simple counts passing through intersections broken down by mode, to extraction of risky interactions between road users. (Z. Yang and Pun-Cheng 2018) provides a review of the application areas.

One of the most promising application domains for computer vision in transportation is in safety analysis. Nowadays, many users of private transport modes use dash cams or action cameras to record their trips. This provides rich data on incidents and near misses, which are situations in which the driver or rider had to take action to avoid a crash, or felt destabilized or unsafe (Ibrahim et al. 2020). In the context of cycling, video data has been used in naturalistic studies to generate detailed risk factors associated with near misses, which required time consuming annotation. Spatiotemporal AI provides the opportunity for automated analysis of crashes and near-misses using CV. This involves deploying a pipeline of algorithms for detecting and tracking the road users involved (Ibrahim, Haworth, and Cheng 2021), detecting the weather and visual conditions (Ibrahim, Haworth, and Cheng 2019) and detecting the road surface conditions (Asad et al. 2022). Going beyond this, CV can also be used to detect risky incidents from video streams. For example, Ibrahim *et al.* (2021b) used a convolutional structure embedded with self-attention bidirectional LSTM blocks to detect cycling near-misses using optical flow of video streams from action cameras. At present, most of these algorithms are deployed retrospectively once the journey has been completed. However, great potential lies in deploying CV methods on edge devices for real-time detection of risk, including near-real time communication between road users (Ibrahim et al. 2020). This technology is referred to as vehicle-to-everything (V2X) communication and is an active area of research (Kiela et al. 2020; Wang et al. 2019).

Research in V2X is predicated on the assumption of future high penetration rates of CAVs, which will require a high degree of cooperation, as well as accurate and robust performance in real-time to avoid risk of collisions. CAVs use a range of sensors to detect and interpret their environment. Generally speaking, LiDAR is used to estimate the range of objects for collision avoidance and driver assistance, while video is used to detect relevant visual features. Such features could include lane markings (Tang, Li, and Liu 2021), vehicles (Fayyad et al. 2020), obstacles (X. Yu and Marinov 2020) and road signs (Wali et al. 2019). This is accomplished using computer vision algorithms such as object detection and tracking, which are based on CNN architectures.

12.5 SPATIOTEMPORAL AI IN TRANSPORTATION: CHALLENGES AND OPPORTUNITIES

This chapter has introduced how Spatiotemporal AI methods are being applied in transportation research in three broad application areas. The progress and opportunities for future research can be summarized as follows:

- **Data Driven Traffic Forecasting:** Progress has been made in adapting powerful Spatiotemporal AI models such as CNNs to transport network structures using GCNNs. The most promising models take-into-account the local spatial dependency between network nodes and the temporal correlation in traffic data to improve short-term traffic forecasting when compared with benchmark models. How to improve training efficiency and develop transferable generative spatiotemporal transport models is still challenging. Furthermore, algorithms that accommodate missing and multimodal data are still to be developed.

- **Optimization of traffic networks using reinforcement learning:** RL has been widely applied in traffic signal optimization and is being increasingly adopted for resource allocation problems such as logistics and timetabling. It also plays an important role in the development of onboard systems in CAVs. While recent work has applied RL to routing of vehicles in cooperative settings (e.g., taxi fleets and CAVs), this work remains in its infancy. In particular, while network-level RL of traffic signaling systems has been accomplished, MARL of individual driver behavior is lacking, especially when network structures change.

- **Computer vision for sensing complex urban environments:** Computer vision using CNNs and related architectures has been employed widely in transportation research, starting with ANPR, moving to traffic counting and segmentation of images based on visual features. The most recent work involves scene understanding, both of static images and videos, to quantify risk and ensure safe operation of CAVs. These systems will be tested and improved as the penetration rate of CAVs increases. Fruitful avenues of research lie in deploying lightweight Spatiotemporal AI algorithms on edge devices, particularly in the context of vulnerable road users.

BIBLIOGRAPHY

Asad, M.H., *et al.*, 2022. Pothole detection using deep learning: A real-time and ai-on-the-edge perspective. *Advances in Civil Engineering*, 2022, e9221211.

Basso, R., *et al.*, 2022. Dynamic stochastic electric vehicle routing with safe reinforcement learning. *Transportation Research Part E: Logistics and Transportation Review*, 157, 102496.

Chen, C., *et al.*, 2020. An intelligent path planning scheme of autonomous vehicles platoon using deep reinforcement learning on network edge. *IEEE Access*, 8, 99059–99069.

Cheng, T., Haworth, J., and Wang, J., 2012. Spatio-temporal autocorrelation of road network data. *Journal of Geographical Systems*, 14, 389–413.

Cheng, T., Zhang, Y., and Haworth, J., 2022. Network spacetime ai: Concepts, methods and applications. *Journal of Geodesy and Geoinformation Science*, 5 (3), 78.

Chu, T., *et al.*, 2019. Multi-agent deep reinforcement learning for large-scale traffic signal control. *IEEE Transactions on Intelligent Transportation Systems*, 21 (3), 1086–1095.

Edrisi, A., Bagherzadeh, K., and Nadi, A., 2022. Applying markov decision process to adaptive dynamic route selection model. *In*: *Proceedings of the Institution of Civil Engineers Transport*. vol. 175, 359–372.

Fayyad, J., *et al.*, 2020. Deep learning sensor fusion for autonomous vehicle perception and localization: A review. *Sensors*, 20 (15), 4220.

Grunitzki, R., de Oliveira Ramos, G., and Bazzan, A.L.C., 2014. Individual versus difference rewards on reinforcement learning for route choice. *In*: *2014 Brazilian Conference on Intelligent Systems*. IEEE, 253–258.

Hassan, L.A.H., Mahmassani, H.S., and Chen, Y., 2020. Reinforcement learning framework for freight demand forecasting to support operational planning decisions. *Transportation Research Part E: Logistics and Transportation Review*, 137, 101926.

Ibrahim, M.R., Haworth, J., and Cheng, T., 2019. Weathernet: Recognising weather and visual conditions from street-level images using deep residual learning. *ISPRS International Journal of Geo-Information*, 8 (12), 549.

Ibrahim, M.R., Haworth, J., and Cheng, T., 2021a. Urban-i: From urban scenes to mapping slums, transport modes, and pedestrians in cities using deep learning and computer vision. *Environment and Planning B: Urban Analytics and City Science*, 48 (1), 76–93.

Ibrahim, M.R., *et al.*, 2021b. Cyclingnet: Detecting cycling near misses from video streams in complex urban scenes with deep learning. *IET Intelligent Transport Systems*, 15 (10), 1331–1344.

Ibrahim, M.R., *et al.*, 2020. Cycling near misses: a review of the current methods, challenges and the potential of an ai-embedded system. *Transport Reviews*, 41 (3), 304–328.

Irannezhad, E., Prato, C.G., and Hickman, M., 2020. An intelligent decision support system prototype for hinterland port logistics. *Decision Support Systems*, 130, 113227.

Jindal, I., *et al.*, 2018. Optimizing taxi carpool policies via reinforcement learning and spatiotemporal mining. *In*: *2018 IEEE International Conference on Big Data*. 1417–1426.

Kamarianakis, Y. and Prastacos, P., 2003. Forecasting traffic flow conditions in an urban network: Comparison of multivariate and univariate approaches. *Transportation Research Record*, 1857 (1), 74–84.

Kang, Y., Lee, S., and Do Chung, B., 2019. Learning-based logistics planning and scheduling for crowdsourced parcel delivery. *Computers & Industrial Engineering*, 132, 271–279.

Karlaftis, M.G. and Vlahogianni, E.I., 2011. Statistical methods versus neural networks in transportation research: Differences, similarities and some insights. *Transportation Research Part C: Emerging Technologies*, 19 (3), 387–399.

Khadilkar, H., 2018. A scalable reinforcement learning algorithm for scheduling railway lines. *IEEE Transactions on Intelligent Transportation Systems*, 20 (2), 727–736.

Kiela, K., *et al.*, 2020. Review of v2x–iot standards and frameworks for its applications. *Applied Sciences*, 10 (12), 4314.

Li, X. and Chen, D., 2022. A survey on deep learning-based panoptic segmentation. *Digital Signal Processing*, 120, 103283.

Liu, Q. and Xu, J., 2012. Traffic signal timing optimization for isolated intersections based on differential evolution bacteria foraging algorithm. *Procedia-Social and Behavioral Sciences*, 43, 210–215.

Matos, G.P., *et al.*, 2021. Solving periodic timetabling problems with sat and machine learning. *Public Transport*, 13 (3), 625–648.

Miletic, M., *et al.*, 2022. A review of reinforcement learning applications in adaptive traffic signal control. *IET Intelligent Transport Systems*, 16 (10), 1269–1285.

Mo, Y., *et al.*, 2022. Review the state-of-the-art technologies of semantic segmentation based on deep learning. *Neurocomputing*, 493, 626–646.

Mostafa, T.S. and Talaat, H., 2010. Intelligent geographical information system for vehicle routing (igis-vr): A modeling framework. *In: 13th International IEEE Conference on Intelligent Transportation Systems*. IEEE, 801–805.

Pal, S.K., *et al.*, 2021. Deep learning in multi-object detection and tracking: state of the art. *Applied Intelligence*, 51, 6400–6429.

Ren, Y., *et al.*, 2020. A hybrid integrated deep learning model for the prediction of citywide spatio-temporal flow volumes. *International Journal of Geographical Information Science*, 34 (4), 802–823.

Ren, Y., Cheng, T., and Zhang, Y., 2019. Deep spatio-temporal residual neural networks for road-network-based data modeling. *International Journal of Geographical Information Science*, 33 (9), 1894–1912.

Rosenfeld, A., 1988. Computer vision: basic principles. *Proceedings of the IEEE*, 76 (8), 863–868.

Shi, D., *et al.*, 2019. Optimal transportation network company vehicle dispatching via deep deterministic policy gradient. *In: Wireless Algorithms, Systems, and Applications: 14th International Conference, WASA 2019, Honolulu, HI, USA, June 24–26, 2019*. Springer, 297–309.

Silver, D., *et al.*, 2016. Mastering the game of go with deep neural networks and tree search. *Nature*, 529 (7587), 484–489.

Smith, B.L. and Demetsky, M.J., 1997. Traffic flow forecasting: comparison of modeling approaches. *Journal of Transportation Engineering*, 123 (4), 261–266.

Smith, B.L., Williams, B.M., and Oswald, R.K., 2002. Comparison of parametric and nonparametric models for traffic flow forecasting. *Transportation Research Part C: Emerging Technologies*, 10 (4), 303–321.

Sutton, R.S. and Barto, A.G., 2018. *Reinforcement Learning: An Introduction*. MIT Press.

Tamagawa, D., Taniguchi, E., and Yamada, T., 2010. Evaluating city logistics measures using a multi-agent model. *Procedia-Social and Behavioral Sciences*, 2 (3), 6002–6012.

Tang, J., Li, S., and Liu, P., 2021. A review of lane detection methods based on deep learning. *Pattern Recognition*, 111, 107623.

Van Lint, J., Hoogendoorn, S., and van Zuylen, H.J., 2005. Accurate freeway travel time prediction with state-space neural networks under missing data. *Transportation Research Part C: Emerging Technologies*, 13 (5-6), 347–369.

Veres, M. and Moussa, M., 2019. Deep learning for intelligent transportation systems: A survey of emerging trends. *IEEE Transactions on Intelligent Transportation Systems*, 21 (8), 3152–3168.

Vikharev, S. and Liapustin, M., 2019. Reinforcement learning cases for passengers behavior modeling. *In*: *AIP Conference Proceedings*. AIP Publishing, vol. 2142, 170020.

Vlahogianni, E.I., Karlaftis, M.G., and Golias, J.C., 2014. Short-term traffic forecasting: Where we are and where we're going. *Transportation Research Part C: Emerging Technologies*, 43, 3–19.

Wali, S.B., *et al.*, 2019. Vision-based traffic sign detection and recognition systems: Current trends and challenges. *Sensors*, 19 (9), 2093.

Walraven, E., Spaan, M.T., and Bakker, B., 2016. Traffic flow optimization: A reinforcement learning approach. *Engineering Applications of Artificial Intelligence*, 52, 203–212.

Wang, J., *et al.*, 2019. A survey of vehicle to everything (v2x) testing. *Sensors*, 19 (2), 334.

Wang, J. and Sun, L., 2020. Dynamic holding control to avoid bus bunching: A multi-agent deep reinforcement learning framework. *Transportation Research Part C: Emerging Technologies*, 116, 102661.

Wei, H., *et al.*, 2021. Recent advances in reinforcement learning for traffic signal control: A survey of models and evaluation. *ACM SIGKDD Explorations Newsletter*, 22 (2), 12–18.

Welch, T.F. and Widita, A., 2019. Big data in public transportation: a review of sources and methods. *Transport Reviews*, 39 (6), 795–818.

Williams, B.M. and Hoel, L.A., 2003. Modeling and forecasting vehicular traffic flow as a seasonal arima process: Theoretical basis and empirical results. *Journal of Transportation Engineering*, 129 (6), 664–672.

Wu, Y. and Tan, H., 2016. Short-term traffic flow forecasting with spatial-temporal correlation in a hybrid deep learning framework. *arXiv preprint arXiv:1612.01022*.

Xiao, Y., *et al.*, 2020. A review of object detection based on deep learning. *Multimedia Tools and Applications*, 79, 23729–23791.

Yang, M., *et al.*, 2020. Inferring passengers' interactive choices on public transits via MA-AL: Multi-agent apprenticeship learning. *In*: *Proceedings of the Web Conference 2020*. 1637–1647.

Yang, Z. and Pun-Cheng, L.S., 2018. Vehicle detection in intelligent transportation systems and its applications under varying environments: A review. *Image and Vision Computing*, 69, 143–154.

Yao, G., Lei, T., and Zhong, J., 2019. A review of convolutional-neural-network-based action recognition. *Pattern Recognition Letters*, 118, 14–22.

Yu, B., Yin, H., and Zhu, Z., 2018. Spatio-temporal graph convolutional networks: a deep learning framework for traffic forecasting. *In*: *Proceedings of the 27th International Joint Conference on Artificial Intelligence*. 3634–3640.

Yu, X. and Marinov, M., 2020. A study on recent developments and issues with obstacle detection systems for automated vehicles. *Sustainability*, 12 (8), 3281.

Zhang, H., *et al.*, 2019a. A comprehensive survey of vision-based human action recognition methods. *Sensors*, 19 (5), 1005.

Zhang, S., *et al.*, 2019b. Graph convolutional networks: A comprehensive review. *Computational Social Networks*, 6 (1), 11.

Zhang, Y., Cheng, T., and Ren, Y., 2019c. A graph deep learning method for short-term traffic forecasting on large road networks. *Computer-Aided Civil and Infrastructure Engineering*, 34 (10), 877–896.

Zhao, Z., *et al.*, 2017. Lstm network: A deep learning approach for short-term traffic forecast. *IET Intelligent Transport Systems*, 11 (2), 68–75.

Zhou, Y., Fu, R., and Wang, C., 2020. Learning the car-following behavior of drivers using maximum entropy deep inverse reinforcement learning. *Journal of Advanced Transportation*, 2020 (4752651), 1–13.

13 GeoAI for Humanitarian Assistance

Philipe A. Dias
Geospatial Science and Human Security Division, Oak Ridge
National Laboratory

Thomaz Kobayashi-Carvalhaes and Sarah Walters
Geospatial Science and Human Security Division, Oak Ridge
National Laboratory

Tyler Frazier, Carson Woody, and Sreelekha Guggilam
Geospatial Science and Human Security Division, Oak Ridge
National Laboratory

Daniel Adams and Abhishek Potnis
Geospatial Science and Human Security Division, Oak Ridge
National Laboratory

Dalton Lunga
Geospatial Science and Human Security Division, Oak Ridge
National Laboratory

CONTENTS

13.1 Introduction 261
13.2 Agencies and Data Sources in Humanitarian Context.................................263
13.3 Population Mapping 265
    13.3.1 Gridded Population Modeling Datasets...266
13.4 Built Environment Characterization 269
13.5 Human-Centric Vulnerability Assessments and Risk Analysis....................270
    13.5.1 Identifying and characterizing drivers of international migration
    and internally displaced persons (IDP)...270
    13.5.2 Developing indices for human impacts and lifeline infrastructure
    provision 271
    13.5.3 Interdisciplinary Applications for Narrative and Sentiment Anal-
    ysis of IDPs and refugees..272
13.6 Agent-Based Models and Humanitarian Crises Response 273
    13.6.1 Synthetic Human Populations...274
    13.6.2 Modeled Built Environments..275

DOI: 10.1201/9781003308423-13

13.6.3 Simulating Human Movement...275
13.6.4 Current State of ABMs..276
13.7 Risks, Challenges and Opportunities for GeoAI in Humanitarian Assistance277
13.7.1 Issues on Using GeoAI in Humanitarian Assistance......................277
13.7.2 Resources & recommendations for rights-respecting GeoAI in Humanitarian Assistance 279
13.7.3 Open challenges and vision for the future 280
Bibliography...281

13.1 INTRODUCTION

The principal motivation of humanitarian assistance is to save lives and alleviate the suffering of a crisis-affected population. Such crises can range from short-term events to persistent issues, arising from causes such as conflict, violence, and natural or anthropogenic disasters. Thus, the scope of humanitarian action ranges from emergency relief to longer-term assistance and protection. Emergency relief corresponds to scenarios where rapid aid is needed to save lives in conflicts or disasters. It relates to the disaster response component of the disaster management cycle, which also includes mitigation, preparedness, and recovery. Meanwhile, indirect and longer-term humanitarian assistance includes, for example, informing decision-making through vulnerability assessments, population/environmental mapping, and simulations that support risk mitigation, reconstruction efforts, and planning toward sustainable development goals (Kuner *et al.*, 2020; Ortiz, 2020; Quinn *et al.*, 2018).

According to the Global Humanitarian Overview released by the United Nations (UN) in December 2022 (`https://www.unocha.org/2023gho`), 339 million individuals across the globe are currently in need for humanitarian assistance, with annual numbers increasing due to issues such as increased hunger levels and population displacements due to conflicts and epidemics. The climate crisis is already contributing to these numbers and pose prospects of further increasing them, given the increased likelihood of catastrophic events, water/food shortages, and the ensuing prospects of forced migration.

There is, thus, an ever-rising interest in leveraging modern technologies to assist decision-making surrounding humanitarian action at such a large scale. As a by-product of the digitization of our lives, we have witnessed astonishing growth in the amounts of data being collected worldwide. Combined with the capability of AI models of extracting meaningful information from such data in a timely and scalable manner, this context positions GeoAI tools with the potential for delivering significant breakthroughs to support humanitarian practices.

Building upon typologies and examples from multiple works on GIS and data-driven approaches for humanitarian applications (Ortiz, 2020; Quinn *et al.*, 2018; Verjee, 2005), we summarize below some main scenarios where GeoAI can benefit humanitarian assistance:

- *Cartography and humanitarian intelligence*: population mapping, land use characterization, infrastructure mapping, damage assessment, and mapping communication/transportation networks based on, for example, analysis of remote-sensing data;

- *Crisis simulation and impact models*: modeling of incidents (e.g., nuclear, biological, environmental), human dynamics, and simulating the impact of different types of responses;

- *Vulnerability and risk assessment*: integration of socioeconomic and environmental data to help identification of hot spots, highlighting populations at risk;

- *Surveying and monitoring*: detection of human rights violations, monitoring for protection of cultural heritage, detection/ early-warning of natural disasters (e.g., wildfires), crowd-sourcing efforts (e.g., humanitarian OSM), documentation of displacements and relevant events;

- *Sustainable development*: AI may act as an enabler for 79% of the targets agreed in the 2030 Agenda for Sustainable Development, with examples such as deep learning (DL) for assisting monitoring agricultural land (Zero Hunger goal), mapping of informal settlements and socioeconomic inequalities (Sustainable Cities goal), and mapping presence of renewable energy generation sources (Climate Action goal) (Persello *et al.*, 2022; Vinuesa *et al.*, 2020).

Humanitarian principles: Efforts toward humanitarian assistance shall take into consideration aspects summarized in the form of humanitarian principles. Derived from international humanitarian law, four core Humanitarian Principles guide humanitarian action: *humanity, neutrality, impartiality,* and *independence.* Described in Table 13.1 by United Nations High Commissioner for Refugees and Office for the Coordination of Humanitarian Affairs, these principles are highlighted in the Code of Conduct for the International Red Cross and Red Crescent Movement (ICRC) and have been taken up by the United Nations in General Assembly Resolutions 46/182 and 58/114. Adherence to the principles is central to the work of humanitarian organizations, enabling them to distinguish themselves from other actors whose activities are influenced by political, economic, military or other non-humanitarian objectives. In addition to the core principles, Table 13.1 highlights three other internationally recognized principles: i) *do no harm*, ii) *participation, empowerment & respect* of persons-of-concern, and iii) *sustainability & accountability* of humanitarian actions.

In the remainder of this chapter, we focus on six main components. First, Section 13.2 discusses the main agencies and data sources in the context of humanitarian assistance. Subsequent sections provide brief literature examples and case studies on four topics of cross-cutting importance for humanitarian assistance: Population Mapping (13.3), Built Environment Characterization (13.4), and Human-centric Vulnerability Assessments and Risk Analysis (13.5). While these sections address the questions of where individuals are located and how current environmental and socioeconomic factors make them vulnerable and at risk, Section 15.5.3 presents tools in the form of Agent-Based Models for simulation and forecasting of multiple scenarios. In light of the humanitarian principles described above, Section 13.7 discusses critical aspects for the development and usage of GeoAI tools that respect humanitarian

Table 13.1

Main humanitarian principles recommended to guide humanitarian action.

	Principle	Description
Core principles	Humanity	human suffering must be addressed wherever it is found. The purpose of humanitarian action is to protect life and health and ensure respect for human beings;
	Neutrality	humanitarian actors must not take sides in hostilities or engage in controversies of a political, racial, religious or ideological nature;
	Impartiality	humanitarian action must be carried out on the basis of need alone, giving priority to the most urgent cases of distress and making no adverse distinction on the basis of nationality, race, gender, religious belief, class or political opinion;
	Independence	humanitarian action must be autonomous from the political, economic, military or other objectives that any actor may hold with regard to areas where humanitarian action is being implemented;
Additional principles	Do no harm	prevent/mitigate any negative impact of humanitarian actions on the affected populations;
	Participation, empowerment & respect	engage and empower persons-of-concern in decisions that affect their lives, while representing them as dignified humans rather than hopeless objects;
	Sustainability & accountability	humanitarian actors shall use resources wisely and put mechanisms in place to assign responsibilities.

principles and human rights. Finally, we also discuss some of the most important prospects and needs for future research on GeoAI for humanitarian assistance.

13.2 AGENCIES AND DATA SOURCES IN HUMANITARIAN CONTEXT

During a humanitarian crisis, critical online data is one of the first resources to become inaccessible due to restricted and damaged servers. The opportune time to capture humanitarian data is before disaster strikes, and in cases where disaster can be forecasted, some of this data can be saved prior to its loss. Even so, there are still several types of data sources analysts can use in the event of a humanitarian disaster, each with their own potential advantages and possible disadvantages. These can be broadly categorized into *Authoritative* versus *"Crowd-Sourced"* sources, the first coming from international agencies such as the United Nations with their own dedicated team of researchers, and the second from volunteer-based open sources such as OpenStreetMap Points of Interest (POIs), social media data, individual journalists, and news publications. Table 13.2 provides a summary of some main data sources of relevance, with more detailed discussion in the paragraphs below.

Authoritative Datasets: Authoritative datasets can include State Statistical Bureaus, governmental agencies, and various United Nations organizations. Prior to a humanitarian crisis, collecting data from a country's own statistical bureau can provide baseline estimates of how the country operated before the crisis. State Statistical Agencies often provide yearly estimates on population and the general living conditions of the country. As this data comes from the country itself, it can be seen as an accurate and authoritative source, but it may contain certain biases due to different political environments. One major drawback is that the country profile pre-conflict

Table 13.2

List of main data sources of relevance for humanitarian assistance.

Data	Org	Use	URL
State Statistical Bureaus	State Governments	Country level statistics and datasets	`https://www.census.gov/programs-surveys/international-programs/about/related-sites.html`
UNITAR	United Nations	Training and learning services to enhance decision-making	`https://www.unitar.org`
UNOSAT	United Nations	Satellite imagery and emergency mapping services	`https://unosat.org`
The Humanitarian Exchange	United Nations	Open datasets across different organizations	`https://data.humdata.org`
The UN Refugee Agency	United Nations	Data on refugees and IDP camps	`https://www.unhcr.org/en-us/internally-displaced-people.html`
OpenStreetMap (OSM)	OSM	Volunteer POI Data	`https://www.openstreetmap.org`
Live Universal Awareness Map	Liveuamap	Live map using news sources/social media	`https://liveuamap.com`

rarely resembles the after-crisis version. Since having the baseline pre-conflict data is necessary to determine the post-crisis health of an area, although State Statistical Bureaus are one of the first places where data is either lost or restricted during a conflict, these state datasets are also some of the first sources saved before website servers become unavailable.

The United Nations also has many agencies and services that respond to humanitarian crises with data for analysts to use. Damage assessment maps created by UNITAR (United Nations Institute for Training and Research) using satellite imagery obtained from their UNOSAT (United Nations Satellite Center) division, are useful for tracking the damage in cities during conflicts, terrorist attacks, or environmental disasters. The United Nations also runs a data service called the Humanitarian Data Exchange through their Office for the Coordination of Humanitarian Affairs (OCHA). This is an open platform that shares data from hundreds of different crisis organizations in one place to make humanitarian data easier to find and access. It mainly uses openly sourced data such as baseline population estimations, damage assessments, data on impacted peoples, refugee information, response data from various organizations, and geospatial mapping services.

"Crowd-Sourced" Datasets: While authoritative datasets are often the first choice an analyst might use after a humanitarian crisis, they would be remiss in skipping over the "crowd-sourced" sources. While "crowd-sourced" datasets generally rely on volunteer-based information, they can provide an on the ground view of the humanitarian disaster, even if the data is not as easily verifiable. OpenStreetMap (OSM) and various POI services collect data on a volunteer basis but can give a local's understanding of the current conflict and the region impacted. While these points cannot always be accurately verified, they can provide useful reference points on which to base population assumptions and damage off.

News outlets also provide a wealth of information via "crowd-sourced" data. Along with useful pre- and post-population data, they allow for near-real time updates of the crisis as it unfolds. The Live Universal Awareness map harnesses these news updates to provide near instantaneous updates of different humanitarian crises around the world using local news sources and various social media platforms. For a broad use, it shows information on civilian protests, crime reporting and fires in parts of the world such as the United States. It also gives live updates on armed conflicts and terrorist attacks in several parts of the world. Using these independent mapping and data collection services to gather data, a better, more human-based understanding of humanitarian crises can be gained. This, along with datasets from authoritative sources, can provide a more robust view of a humanitarian crisis.

Case Study – Türkiye/Syria Earthquake: As of the writing of this chapter, a recent example of a humanitarian crisis and the scramble to collect humanitarian data were the recent earthquakes of Türkiye and Syria. Several of the resources listed above became useful tools to analyze the current crisis. Data from Türkiye's statistical bureaus and yearly population "yearbooks" helped provide baseline estimates as to what the country looked like before the earthquakes to compare to the situation after the subsequent devastation. The sources from the various United Nations agencies also provided damage assessment maps of regions of Türkiye and Syria hit by the earthquake. Imagery from UNOSAT was used to compare destruction before and after the earthquake.

Along with the use of the authoritative datasets, multiple POI datasets on building damage and disaster response were found using OSM and other volunteer-based data services. This provides an additional useful tool to estimate population assumptions and facility use in areas damaged by the earthquakes.

13.3   POPULATION MAPPING

High-resolution population data is a foundational component of humanitarian aid preparation and rapid response, also serving to inform policy development and implementation in the longer term. The ability to efficiently identify the number, location, distribution, and demographics of people at scale using current, high-resolution population data allows governments and aid-organizations to quickly identify areas of greatest impact to the most vulnerable populations. This data is necessary for determining the most optimal distribution of resources and personnel in the wake of unanticipated humanitarian disasters, and for providing assistance to refugees and internally displaced persons as a result of migration events.

The population modeling and mapping community provides several products that have been made openly available to users and provide gridded population distribution and counts at various scales. Advances in computer vision (CV) and machine learning (ML) have helped to drive further advancement of higher-resolution geospatial data products, many of which now serve as primary inputs for various applied GeoAI problems in humanitarian assistance contexts.

High-resolution population data products are typically modeled using either a top-down or a bottom-up modeling methodology, or some combination of the two, based on available input data. The standard high-resolution data product is represented in the form of a gridded raster, in which each cell is valued by a population count. Top-down modeling has been the leading methodology since the inception of global-scale human population modeling and, while other models are emerging in the field, at present it remains the primary methodology (Kuffer *et al.*, 2022). Top-down modeling involves the disaggregation and gridded placement of human population counts, as reported from country-level censuses, across a geographic landscape at a selected scale, most often at the nation-level. Population counts are assigned per grid based on a modeled weight that is representative of how suitable any given location is for human occupation. Under a top-down modeling approach, all population values are constrained by the reported census count, meaning when aggregated to any given administrative unit the resulting population counts will equate to the total census reported value.

In contrast, the bottom-up method was developed to support population modeling where the top-down approach is not necessarily feasible due to limited population information, especially in areas that have significant gaps or outdated data. This methodology utilizes micro-census data – i.e., census data that samples only a percentage of an area's population – that is then fused with other ancillary datasets, to find statistical relationships with relevant geographic and remotely sensed data, resulting in similar population distributions as the top-down approach (Wardrop *et al.*, 2018).

As further discussed in Section 13.4, many GeoAI-driven methods are key to both approaches, as the ability to automate the identification of a broad range of structures throughout the built environment, the mapping of building footprints, and distinguishing between various settlement patterns are essential components for efficiently producing high-quality products on a global scale (Wardrop *et al.*, 2018). For population mapping purposes, the segmented exterior boundaries of building footprints are used to produce ground floor area estimates that can be used to populate locations at finer resolution than, e.g., census data. In addition to footprints, quantification of building heights is also exploited to improve model performance, especially in densely developed urban areas where buildings can be considerably taller.

13.3.1 GRIDDED POPULATION MODELING DATASETS

Table 13.3 provides summary descriptions of many of the human population datasets currently available, with more details on several community state-of-the-art examples in the paragraphs below.

Global Human Settlement Layer – Population (GHS-POP) – Joint Research Center: The GHS-POP, developed by the Joint Research Center (JRC), is a spatial raster product that depicts the distribution of human population per cell across a selected geography. These modeled human population data are available globally, at a temporal extent from 1975 to 2030 in 5-year increments and have a resolution of 100*m*. The GHSL team produces GHS-POP by leveraging information derived from

Table 13.3

Gridded population modeling resources for Geo AI and humanitarian aid efforts.

Product	Organization	Scale	Resolution*	Dates	URL
Data for Good High Resolution Population Density Maps	Meta/CIESIN	More than 160 Countries	30 m	2018 to present	https://dataforgood.facebook.com
Demobase	U.S. Census Bureau and Partners	7 Countries	100 m	1998 to present	https://www.census.gov/geographies/mapping-files/time-series/demo/international-programs/demobase.html
Global Human Settlement Layer – Population (GHS-POP)	European Commission – Joint Research Center	Global	100 m – 1 km	1975-2030	https://ghsl.jrc.ec.europa.eu/
Gridded Population of the World (v4)	SEDAC/CIESIN	Global	1 km	2000, 2005, 2010, 2015, 2020	https://sedac.ciesin.columbia.edu/data/collection/gpw-v4
GRID3 Population Estimates	GRID3 and Partners	11 Countries	100 m	2018 to present	https://grid3.org
Global Rural Urban Mapping Project (GRUMPv1)	SEDAC/CIESIN	Global	1 km	1990, 1995, 2000	https://sedac.ciesin.columbia.edu/data/collection/grump-v1
High Resolution Settlement Layer	CIESIN, Meta, the World Bank	More than 140 Countries	30 m	2015	https://ciesin.columbia.edu/data/hrsl
History Database of the Global Environment of Population Grids (HYDEv3.2)	Copernicus Land Change Lab, Utrecht University	Global	85 km	10000 BCE to CE 2015	https://landuse.sites.uu.nl/hyde-project/
LandScan Global	ORNL	Global	1 km	1998 to present	https://LandScan.ornl.gov
LandScan HD	ORNL	25 Countries and Territories	90 m	2014 to present	https://LandScan.ornl.gov
LandScan USA	ORNL	United States	90 m	2014 to present	https://LandScan.ornl.gov
Living Atlas of the World	ESRI and WorldPop	Global	1 km	2000-2020	https://livingatlas.arcgis.com
WorldPop	WorldPop, University of Southampton	Global	1 km	2000 to present	https://www.worldpop.org

*Approximate resolution at the equator in metric

satellite imagery, census data, and voluntary-derived geographic data. GHS-POP is modeled using a top-down dasymetric mapping model to distribute census derived population counts across a geography using weights modeled as a part of the GHS-BUILT dataset (Schiavina *et al.*, 2022) production. GHS-POP conducts verification of population estimates by ensuring all disaggregated values equate to an authoritative aggregate count per administrative area via GEOSTAT by Eurostat. GHS-POP and the other GHSL data products are the core datasets used by the Atlas of the Human Planet, an annual scientific community report on human presence impact on Earth.

LandScan Global (LSG) and High Definition (LSHD) – ORNL: Oak Ridge National Laboratory established the LandScan Program in 1997 with the goal to address the need for improved estimates of population for consequence assessment. LandScan Global (LSG) is a top-down dasymetric mapping model that disaggregates census-derived population data across geographic landscapes of varying administrative levels, at approximately $1km$ (30-arc seconds) resolution. The LSG dataset provides coverage of all countries and is supplemented by specialty modeled outputs by LandScan USA (LSUSA) (Bhaduri *et al.*, 2007) and LandScan HD (LSHD) [1]. LSG receives an annual update in which a selection of countries receives subject domain expertise to revise population assignments.

In contrast, LandScan HD is a bottom-up human population model, available at approximately $90m$ (3-arc seconds) resolution, with coverage across 25 countries at the time of writing. Aspects of the methodology behind LSHD vary on a country per country basis, as the models are based on the adaptations of unique sociocultural activities and their representative data that serve as explanatory variables for the model. LSHD leverages the ongoing work of ORNL's Population Density Tables project (Stewart *et al.*, 2016) to derive micro census-like data and other ancillary information to serve as the response variable for the model, particularly in instances when micro census data is not available. For both LSG and LSHD, final population estimates are verified and aligned to those provided by the Central Intelligence Agency World Fact Book.

WorldPop - University of Southampton: The gridded population datasets from WorldPop – an interdisciplinary applied research group based at the University of Southampton, in the UK – are available at two scales: $100m$ for individual countries, or $1km$ for global and individual country products, both unconstrained and constrained. Their top-down modeled products have worldwide coverage, with partial coverage for their bottom-up modeled products developed with partners GRID3 and CIESIN for several nations, primarily focused on the African continent. The WorldPop group's global coverage products use a weighting layer for population disaggregation that is modeled using a Random Forest regressor that leverages various remotely sensed geographic data as covariate information. The bottom-up modeling has some variability in methods based on the adaptations of unique sociocultural activities that serve as covariates for the model and the various types of micro-census data available that serve as the dependent variable.

---

[1] https://LandScan.ornl.gov/

Considerations: Assessing modeling accuracy between all referenced data products can be challenging as each uses varying inputs and modeling methods. It is advised to review model inputs and select data products on a case-by-case basis, based on whichever authoritative population estimate source meets the corresponding needs. Additionally, many of the referenced organizations provide other geographic data resources that may provide helpful information beyond high-resolution population distribution/density. The POPGRID Data Collaborative (https://www.popgrid.org) is a noteworthy additional resource, as it brings together much of the population mapping community and associated products in a singular location.

13.4 BUILT ENVIRONMENT CHARACTERIZATION

Characterization of natural and built environment is a key source of information to support humanitarian actions. In addition to population mapping, it is an important component for vulnerability and risk assessment, as well as simulations. Thus, we briefly mention in the paragraphs below examples of research conducted at ORNL to illustrate how GeoAI techniques can be paired with high-quality remote sensing data for built environment characterization and land-use categorization.

The volunteered geographic information (VGI) from OpenStreetMap (OSM) is arguably the definitive source for describing transportation networks of roads at a global scale. OSM also provides building footprint data that can be particularly useful for densely populated areas, where valuable building attributions are often available. However, given the inconsistent quality of such VGI data across different scenarios, efforts combining AI and remote sensing data have been pursued.

In particular, the combination of deep CNNs with high-resolution optical imagery has been successfully demonstrated in several works for high-quality building footprint extraction and height estimation. The work being conducted by ORNL over the years reflects how advancements in the fields of CV/ML and Remote Sensing have been leveraged for such purposes. Yang *et al.* (2018) explored an encoder-decoder architecture with signed-distance labels to generate the first mapping of building footprints across the entire contiguous United States using a GPU cluster. ORNL's Human Geography group uses stereo-pair satellite images with spaceborne LiDAR to generate digital surface models and digital terrain models that are surface matched and registered to estimate building heights at a country scale.

As documented in Lunga *et al.* (2021, 2020), ORNL's GeoAI group has been further improving the quality and turnaround speed for generating high-quality building footprint maps at large-scale. Software resources such as Apache Spark, Horovod, and PyTorch Distributed Data Parallel have been crucial for scalability purposes, leveraging High Performance Computing (HPC) resources for model training and deployment with large data volumes. Moreover, the quality of the data products being generated has been ever increasing thanks to: i) the curation and annotation of large volumes of high-resolution optical imagery by in-house analysts with GIS expertise; ii) research and development on domain adaptation (Dias *et al.*, 2022b), model specialization (Lunga *et al.*, 2021), active learning and human-in-the-loop mechanisms

(Dias *et al.*, 2022a); iii) modern advancements in DL architectures such as attention mechanisms.

Several public organizations as well as private companies have made openly available built environment datasets generated with similar GeoAI-based approaches. Google's building footprint dataset (`https://sites.research.google/open-buildings/`) is particularly comprehensive in describing buildings throughout Africa, while Microsoft likewise shares building footprint data for many regions and countries throughout the world (`https://www.microsoft.com/en-us/maps/building-footprints`).

Similar strategies have been exploited for road extraction, urban and land-use/land-cover characterization. Within ORNL, building attribute data from OSM and geometric features from footprints polygons have been used as features by deep neural networks to categorize structures according to usage (e.g., residential, commercial, industrial) (Arndt and Lunga, 2020), with optical imagery allowing similar categorization of Urban Structural Units, i.e., partitioning urban areas into regions that exhibit homogeneous physical characteristics (Arndt and Lunga, 2021). Such categorizations allow improving occupancy estimates at global scale by ORNL's Human Geography group (Burris *et al.*, 2013; Stewart *et al.*, 2016).

13.5   HUMAN-CENTRIC VULNERABILITY ASSESSMENTS AND RISK ANALYSIS

While humanitarian assistance often requires tools for situational awareness and emergency response, spatial analytics in terms of local disaster risks and community resilience can enable preparedness to alleviate suffering and human loss more efficiently (Milán-García *et al.*, 2021). In fields such as industrial engineering and urban resilience, risk analysis and vulnerability assessments are used to preemptively allocate resources, harden infrastructure, and anticipate future conditions (Reddy, 2020). GeoAI approaches can adapt such frameworks to humanitarian contexts such as forced displacement, migration, food insecurity, insufficient sanitation, and more. As illustrated below, understanding the multi-dimensional drivers behind humanitarian crises is a highly relevant application space for big data and spatially explicit techniques.

13.5.1   IDENTIFYING AND CHARACTERIZING DRIVERS OF INTERNATIONAL MIGRATION AND INTERNALLY DISPLACED PERSONS (IDP)

Migration and forced displacement are both outcomes and causes of cultural conflict, violence, and stresses on critical infrastructure systems and institutions (Abel *et al.*, 2019; McAuliffe and Triandafyllidou, 2020). While sometimes a forced condition, migration is one potential adaptation strategy that individuals and households consider in response to loss of essential livelihoods, infrastructure services, and basic freedoms (Black *et al.*, 2011b). With climate change, migration is expected to increase as environmental anomalies impact agriculture, lifeline infrastructure (e.g., potable water), and natural resource-dependent economies (McLeman *et al.*, 2016).

However, climate migration is a multi-scale (from individual decisions to global sociological processes), and multi-dimensional (social, economic, political, environmental) complex process (Beine and Jeusette, 2021; Black *et al.*, 2011a). Such complexity has been at the center of recent theoretically informed methodological applications, yet novel adaptations and innovations are still needed (Beine and Jeusette, 2021).

GeoAI can be applied for humanitarian decision analysis and support for migration and IDPs, especially in terms of environmentally-driven contexts. For example, Carvalhaes *et al.* (2022a) used a Multi-scale Geographically Weighted Regression (MGWR) and Multi-Criteria Decision Analysis (MCDA) approach to identify drivers of international migration for 2019 and develop a geovisual index. Using annually published datasets, the study used total migrant stock by origin country as a response to calibrate "push factors" (i.e., stressors) along sociopolitical, environmental, and economic dimensions. The advantage of an MGWR approach is that variables can be identified as spatially non-stationary, meaning that statistical relationships are not constant through space and occur at varying spatial scales, from local (i.e., short bandwidth) to global (large bandwidth) for each covariate (Fotheringham *et al.*, 2004, 2017). Such geostatistical approaches can also be adapted by applying spatial weighting schemes (i.e., matrices) for multi-scale geographically weighted AI or ML frameworks.

13.5.2 DEVELOPING INDICES FOR HUMAN IMPACTS AND LIFELINE INFRASTRUCTURE PROVISION

Spatial assessments for social vulnerability typically capture unequally distributed impacts or provide a "yard stick" for how heterogeneous communities will adapt to or cope with natural hazards, but can also be extended to political violence and other hazards (Casali *et al.*, 2022). A common approach for vulnerability assessments and risk-based tools is to develop a geographic index that maps structural socioeconomic factors (e.g., educational opportunity, mobility, income) for disaster impacts or sustainable development goals (e.g., increases in mental health impacts, property damage). Casali *et al.* (2022) provide a useful review of related algorithms for machine learning applications.

Martínez-Rivera *et al.* (2022) introduce a geostatistical approach to develop a *Human Impacts Index* (HII) that measures the effect of Hurricane Maria in respect to human-centric disaster impacts in Puerto Rico (e.g., social, political, economic). The study applied a Treatment-Effect method using publicly available and spatially explicit panel data (i.e., structured and regularly published such as monthly or annual datasets) to quantify the marginal effect of a specific event (i.e., Maria) on human-centric measures such as changes in suicide rates, substance abuse, excess mortality, and unemployment. Methodologies like the HII are essentially modular in that the type of event and spatial unit of analysis are adaptable. For example, alternative events could include war, earthquakes, civil conflict, and others, while spatial units can be sub-national administrative boundaries, cities, and more.

A primary limitation of composite index approaches, however, is that they are generally static "snapshots" and highly reduced glimpses of complex human dynamics, which can be addressed with temporal frameworks or mixed methods that add analytical nuance or capture important contextual implications (Carvalhaes *et al.*, 2021; Spielman *et al.*, 2020). Panel data can help address these limitations, at least partially, by enabling temporal analysis that can be performed in a continuous or recursive manner, as well as regularly updated analytical models. In this way, GeoAI techniques can be used to both automate such indices (e.g., time series) or replicate model specifications with advanced algorithmic or big data techniques.

Recent applications are emerging that develop indices in a recurring fashion, especially those taking advantage of high-frequency data and common AI techniques. Knippenberg *et al.* (2019) leveraged monthly datasets for food security and welfare among populations living in shock-prone contexts to apply LASSO and Random Forest algorithms that reveal resilience-related factors such as household distance to drinking water in Malawi. de Blasio *et al.* (2022) show how algorithmic forecasts could support anti-corruption policy targeting with high accuracy data-driven predictions using municipality-level data on white-collar offences. Spatially explicit machine learning approaches can be applied to target factors like corruption as a driver of humanitarian crises such as civil unrest, war, and disasters. The studies above identify salient social, economic, and political factors to develop quantitative tools that help policy and decision-makers understand key dynamics, make informed adaptation plans, and better prepare for future humanitarian needs.

Socioeconomic assessments also extend into the global development space where the United Nations Sustainable Development Goals (SDG) are essential in guiding research and policy. Pandey *et al.* (2022) developed an *Infrastructure Provisioning Index* (IPI) using Nighttime Lights (VIIRS) satellite data to link unequal provision of infrastructure services to socioeconomic indicators in the Global South. Spatial variability in IPI across communities suggests disproportionate susceptibility to hardships (e.g., energy insecurity, frequent disruptions to potable water access) and resources to adapt amid sociopolitical violence and natural hazards (Cutter and Finch, 2008). Among other satellite products, nighttime lights imagery represents a recurring data source for automated analysis to map multi-scale infrastructural inequalities in space (e.g., international, national, rural-urban, intra-urban).

13.5.3   INTERDISCIPLINARY APPLICATIONS FOR NARRATIVE AND SENTIMENT ANALYSIS OF IDPs AND REFUGEES

Beyond data-driven applications, GeoAI can be applied to augment multi-method and interdisciplinary approaches that require processing and synthesizing qualitative data, where key areas include narrative and sentiment analysis, and transdisciplinary research frameworks. Carvalhaes *et al.* (2022b) demonstrate how quantitative and qualitative assessment approaches can be integrated given the complexity of socio-technical systems (i.e., the intertwined and dynamic nature of social processes and infrastructure resilience). Here, geographic perspectives provide powerful tools to anchor multiple frames of reference and data analysis approaches. Using concepts like *thick mapping*, which is set of concepts and methods that incorporate a multiplicity of place-based subjective records (Presner *et al.*, 2014), contemporary digital mapping capabilities enable web-based applications that combine index-based metrics for human impacts and vulnerability with personal stories and sentiments of specific events. For a practical example, see `https://varinaldi.shinyapps.io/triadGeo`.

Such approaches leverage text-based data, such as personal anecdotes describing disastrous events that can be collected via web-based tools (e.g., SenseMaker ^{TM}; Amazon MTurk) for hundreds to thousands of thick data points. As opposed to "big data", *thick data* have relatively few yet deeply descriptive records where topic modeling and natural language processing are applicable. In this way, GIScience is integrated with the digital humanities and participant involvement for more nuanced transdisciplinary assessments that account for contested perspectives, uncertainties, and weak signals. Additionally, incorporating narratives alongside metrics that otherwise reduce human impacts and factors to relatively simple quantitative metrics helps humanize research products with experiential data, participation, and empowerment. See Table 13.1 and Section 13.7 for further discussion.

GeoAI can also be leveraged as part of the research design process itself. For example, ML algorithms can be used to augment traditionally qualitative techniques like survey design to capture complex social concepts like social capital and agency. To assess women's agency in India, Jayachandran *et al.* (2023) leveraged ML to hone in on key close-ended survey questions that can be applied at large scales yet reveal rich insights. In this way, *integrated GeoAI* can fill gaps where interdisciplinary pursuits have met challenges and facilitate convergence in complex and multi-dimensional research domains like disaster risk, climate migration, and conflict assessment and forecasting (Li, 2020).

13.6 AGENT-BASED MODELS AND HUMANITARIAN CRISES RESPONSE

Simulating and predicting a humanitarian crisis at scale as a spatial process can enable stakeholder selection of sustainable future paths (Pagels, 1989). It is best achieved by focusing on the behavior of individuals, families and communities, as their behavior pertains to their networks of social relations, services and infrastructures as well as both their built and natural environments. A computational model designed to simulate agent and system behavior can be referred to as a multi-agent system, microsimulation or geosimulation, but we refer to this group simply as agent-based models (ABMs). As demonstrated by Batty (2007); Benenson and Torrens (2004); Epstein and Axtell (1996); Helbing and Molnar (1995); Nagel and Schreckenberg (1992); Waddell (2002), ABMs have become an increasingly popular approach to simulate complex, evolving and adapting economic, social and natural systems that involve emergent phenomenon and trend toward intractability (Mitchell, 1998; Newman, 2003). ABMs are well suited to simulate and plan in advance of a humanitarian crisis due to their capacity to model spatial phenomenon that emerge from the interaction between individuals and their response to an event or a series of events, such as one that threatens the health, safety or well-being of a community or large group of people.

ABMs generally result from case-by-case development where the final empirical implementation is intended to reflect a real-world scenario as closely as possible and have demonstrated effective simulation of many disaster related human activities including: the spread of infectious diseases; migration in response to crises, natural disasters, climate change, and persistent poverty. For example, the Netlogo-based Land Use Dynamic Simulator (LUDAS) was an ABM designed to simulate spatio-temporal dynamics of coupled human–landscape systems to better understand appropriate policies for sustainable management of land and water resources in the face of different climate change scenarios (Le *et al.*, 2010). Another example of an ABM developed for response to a natural disaster was the MATSim-based "Last Mile" project, which was developed to understand the potential impact of a tsunami on Padang, Indonesia and to create an early warning and evacuation information system (Taubenbock *et al.*, 2009). Whether characterized by static longer-term agent and system behavior or more dynamic, shorter-term intervals, ABMs have proven to be an effective tool for simulating a natural disaster and improve the ability of stakeholders to understand complex dynamics when planning for response.

13.6.1   SYNTHETIC HUMAN POPULATIONS

An important initial step when developing an ABM is the characterization of agents in a study area at the beginning timestamp of the simulation. For an ABM of a humanitarian disaster, this population dataset typically consists of synthetically generated persons and households that are described in terms of their demographic and economic attributes, but may also incorporate workplaces and businesses. Producing a synthetic population dataset that closely reflects the demographic, economic and social heterogeneity that characterizes the real population is significant for a probabilistic modeling system to accurately predict agent behavior. A current challenge in this context is capturing both the heterogeneity of households and persons as well as their actual locations. State-of-the-art methods generally fail to generate a synthetic population that is close-to-reality in descriptive heterogeneity at high spatial resolution.

Earliest efforts to generate synthetic population data were associated with transportation models and were likely driven by the format of available household surveys as well as the interest to preserve statistical disclosure limitations. Beckman *et al.* (1996) combined the US Census Bureau surveys with an iterative proportionate fitting algorithm to generate a synthetic population of persons and households from marginal totals at the block group level. In Alfons *et al.* (2011); Frazier and Alfons (2012); Munnich and Schurle (2003), complete observations provided in household surveys were exploited, such as those made available by the International Household Survey Network (IHSN) or the Demographic and Health Surveys Program (DHS). A multinomial logistic regression model was used to generate combinations which were not represented in the original survey but are likely to occur in the real population. Spatial accuracy was constrained by the uncertainty associated with each household survey being located within a $2km$ to $5km$ radius of a given cluster center-point. More recently, Alegana *et al.* (2015); Bosco *et al.* (2017); Gething *et al.* (2015); Utazi *et al.* (2018) used a hierarchical Bayesian model with DHS household survey observations to produce smoothed population surfaces combined with spatial measures of uncertainty. The resulting $1km^2$-large raster is effective at representing a population at a country level, but each dataset is generally limited to only a few variables from the DHS, due to the effect subsetting the survey observations has on validity.

Deep learning probabilistic estimators using multinomial logistic regressions, random forest or more recently gradient boosted trees have surpassed previous methods. Nonetheless, machine learning estimators cannot overcome limitations of model specification resulting from subsetting household surveys according to location and attributes. To assign synthetic populations as close as possible to their location, a spatial statistic, density ratio estimator is produced that takes advantage of both the high-resolution global population data that is available (such as LandScan Global at 100x100m resolution) as well as the survey design that is inherent to the DHS household sample. The resulting synthetic population is produced at approximately $1km^2$ with individual households matched to dwellings using a divisive hierarchical clustering algorithm allowing a dendrogram hierarchy to be read as a decision tree.

13.6.2   MODELED BUILT ENVIRONMENTS

In addition to the synthetic population, ABMs benefit from accurately modeled built environments that are used to learn how human behavior will interact during an event that threatens the health, safety or well-being of the community. A highly articulated built environment, in terms of buildings and their land uses as well as transportation networks and their functional classifications, is nearly always necessary for an ABM to productively simulate various humanitarian crises response scenarios. Openly available datasets of transportation networks and building footprints such as the ones reviewed in Section 13.4 are a few examples of resources that can be used in an ABM to simulate a study area's response to a humanitarian crisis.

It is worth noting that while humanitarian crises can range in scale of settlement size from villages to even the entire planet, it is highly unlikely that all of the needed built environment data will be available from open-source datasets. More likely is that some part of the study area will be populated with roads and streets from open sources, while the remainder will need to be generated in-house with techniques such as those reviewed in Section 13.4 to produce a modeled built environment that accurately reflects the existing built environment.

13.6.3   SIMULATING HUMAN MOVEMENT

Depending on the type of humanitarian crisis, ABMs that simulate human responses vary in spatio-temporal design from more static, longer-term, regional scale to dynamic, shorter-term, local scale. Many early ABMs for simulating crises and decisions associated with migration designed the agent decision-making process as a probability arising from survey or census data, often estimating the likelihood of deciding to move or stay, and then subsequently the location choice of the new destination. This probabilistic method was incorporated in (Tsegai and Le, 2010) for simulating response to long-term climate change impacts resulting from land use decisions, drought, and possible migration using variables from the Ghana Living Standard Survey 5 (GLSS5), to estimate district level spatial analysis of migration flows in their Netlogo based GH-LUDAS. Several early ABMs addressing longer-term humanitarian issues in both rural and urban contexts derive from the LUDAS ABM created by Le *et al.* (2010), including Feitosa *et al.* (2011); Frazier and Alfons (2012); Schindler (2013); Villamor *et al.* (2014). Later works from Wesolowski *et al.* (2013) and Garcia *et al.* (2014) used census data to model human movement or internal migration patterns, while Kraemer *et al.* (2019); Lai *et al.* (2019); Sallah *et al.* (2017) used CDR data from cellphones with gravity type spatial interaction models for modeling human behavior in the context of humanitarian crises in advance of GPS data.

Agent decision-making at higher spatio-temporal resolution has historically been associated with transportation activity models, and modified within the context of a particular humanitarian disaster. Traditionally, travel surveys have informed gravity type spatial interaction models to define the origin and destination of agents as the means of trip distribution, followed by routing across the network and optimization of learned routes. Several classical sources have informed ABM transport simulations including Wilson (1971) introduction of gravity models, Fotheringham (1983) formalization of basic implementation, Flowerdew and Aitkin (1982) method for fitting a Poisson gravity type spatial interaction model and Nagel and Schreckenberg (1992) use of an maximum entropy algorithm in a cellular automata model for trip routing. More recently, MATSim has been used in several humanitarian response contexts, including Burris *et al.* (2015); Durst *et al.* (2014); Lämmel *et al.* (2009); Wisetjindawat *et al.* (2013). In contrast to previous examples that primarily used location choice models, these MATSim simulations incorporated agent learning.

13.6.4 CURRENT STATE OF ABMs

Crooks *et al.* (2021) observe that big data and machine learning will dramatically impact geographically explicit agent-based model design, execution and evaluation. Authors note how reinforcement learning, genetic algorithms and random forests have been used for agents to learn from past experiences, and list several studies using neural or Bayesian networks to predict transportation activities, decision-making or agent choices about location and land use. In addition to these examples, there have been numerous examples where AI has been used to produce synthetic populations, modeling the built environment and simulating human movement.

In Miller (2021), Miller provides a thorough review of how real-time travel-related information and new mobility services are rapidly and radically transforming human mobility informatics. The current revolution in mobility informatics will disrupt both travel behavior and transportation system performance as well as the data available and the models using that data. The author surveys new mobility data sources including: cellphone trace data, transit transaction data, bluetooth sensor data, credit card transaction data and then also data fusion, statistical matching and imputation methods that advance a more comprehensive and complete representation of travel behavior. The survey by Miller (2021) is instructive as to the relationship between these newly available data sources and their use in deep learning ABMs for humanitarian crises response.

In Luca *et al.* (2020), the authors articulate a taxonomy of DL methods for human mobility tasks as related to the proliferation of digital mobility data – such as phone records, GPS traces and social media posts –, providing descriptions of the most relevant models within the context of disaster response. We refer the reader to Luca *et al.* (2020) for a thorough discussion on relevant metrics and DL techniques – ranging from RNNs, LSTMs, VAEs to modern Transformer-based architectures – that have been used for predicting next-location, crowd flow, trajectories, among other features relevant for ABMs.

Finally, it is worth that GPUs have been crucial to accelerate agent-based demand-responsive simulations. Saprykin *et al.* (2022) demonstrated that a MATSim-based human mobility model leveraging GPU acceleration was 9 times faster than existing state-of-the-art tools when running a 5.2 million agent simulation of human mobility throughout Switzerland.

13.7 RISKS, CHALLENGES AND OPPORTUNITIES FOR GEOAI IN HUMANITARIAN ASSISTANCE

The use of GeoAI tools to assist decision-making for humanitarian assistance presents risks and challenges that range from technical to ethical and legal issues. Beyond general ethical concerns related to AI, the risks regarding its application for humanitarian purposes can have particularly serious consequences considering the vulnerability of the populations involved (Dodgson *et al.*, 2020). In light of the humanitarian principles described in Section 13.1, we build upon Blumenstock (2018); Coppi *et al.* (2021); Dodgson *et al.* (2020); Pizzi *et al.* (2020) to highlight some of the main risks associated with GeoAI approaches for humanitarian action. We also describe some of the existing tools and strategies that can help mitigating such risks, concluding with a brief discussion on critical open challenges and promising directions for the near future of GeoAI for humanitarian assistance.

13.7.1 ISSUES ON USING GEOAI IN HUMANITARIAN ASSISTANCE

Lack of quality data: In humanitarian contexts, data might be often insufficient due to the need for fast response, the complexity of collection, political barriers for data access, as well as socioeconomic and other inequalities. For example, availability and quality of data from developing countries is often lower due to limited interest from major players, and low availability of local resources and expertise.

Biases: Biases risks of many types are present for data-driven humanitarian frameworks (Dodgson *et al.*, 2020), including: i) *historical biases*, e.g., due to gender/race inequalities; ii) *response & selection biases*, where the collected subset of data does not represent the full population; iii) *measurement biases*, where inappropriate proxy metrics are selected; and, iv) *aggregation bias*, where a "one-size-fits-all" approach is inaccurately applied to rather different groups. Since AI methods such as Deep Learning heavily rely on supervised learning from data provided by humans, they present a high risk of "reproducing, reinforcing, and amplifying patterns of marginalization, inequality and discrimination in society" (Leslie, 2019), menacing especially the humanitarian principle of impartiality (Pizzi *et al.*, 2020).

Privacy and data protection: The meaningful consent by affected individuals may be compromised by communication barriers (e.g., technical language), as well as time pressure and requirements imposed for receiving aid. Data-intensive models can augment such risks, as the incentives for fast and large data collection can lead to negligence in the process (Dodgson *et al.*, 2020). Concerns also include data leaks and risks of re-identification of subjects, since basic identifying information can be sufficient to expose vulnerable individuals, e.g., fleeing oppression (Pizzi *et al.*, 2020).

Generalization, opaqueness, and unpredictability: "Black-box" methods like Deep Neural Networks have the ability of extracting meaningful information from large datasets in ways that humans would often not be able to engineer. On the flip side, such model complexity and their common lack of uncertainty estimates and explainability poses difficulties for understanding when the models will fail and why. Moreover, GeoAI applications impose high generalization challenges, as models must often be deployed in "out-of-the-distribution" conditions since significant mismatches between training and deployment conditions can occur due to changes in location, weather, and sensor configuration, among many other variables.

Lack of contextual knowledge at the design phase: Several aspects act as barriers for domain experts to develop and understand AI systems: they often have limited AI knowledge, and the costs to obtain high-quality datasets and the computing hardware necessary for model development can be prohibitive and daunting (Dias and Lunga, 2022). Meanwhile, AI experts often lack the awareness of the importance of considering factors such as social and cultural aspects for domains like humanitarian assistance, which can lead to critically wrong outcomes (e.g., not understanding certain biases in the data).

Overuse of AI: While certain problems might be better suitable for simpler, more transparent techniques (e.g., binary trees), the success of AI in other domains and the influences from the private sector leading its development present the risk of AI overuse. Critically, overuse can be compelling to decision-makers as they might exploit AI as a "crutch" that adds a false impression of objectivity and neutrality to the decisions made, while compromising accountability processes (Pizzi *et al.*, 2020).

Accountability: From a social standpoint, model opaqueness combined with lack of technical literacy compromises the public's ability to know how their information is being used for decision-making and when their rights are potentially at risk. From a legal standpoint, opaqueness and the lack of traceable data and model pipelines compromises the ability of holding accountable those responsible for bad outcomes.

Improper deployment: Beyond model development and data acquisition, lack of expertise and other challenges such as inappropriate infrastructure also lead to risks at deployment level, which may include application of models in situations they are not intended for, or incorrect interpretation of results that can lead to failure to deliver aid, discrimination, and other human rights violations.

Decision-making concerns: All these factors combined lead to a range of ethical concerns for AI-informed decision-making that includes: *epistemic concerns*, as predictions might represent inconclusive, inscrutable, and/or misguided evidences for decision-making; *normative concerns*, as the actions based on such evidences might be unfair, and potentially introducing new ways of conceptualizing the world that could be problematic (e.g., population profiling) (Dodgson *et al.*, 2020).

13.7.2 RESOURCES & RECOMMENDATIONS FOR RIGHTS-RESPECTING GEOAI IN HUMANITARIAN ASSISTANCE

Human-rights and Ethics principles to benefit accountability: As discussed in Dodgson *et al.* (2020); Pizzi *et al.* (2020), outcomes from advancements in the field of AI ethics have been mostly in the form of "AI Code of Ethics", which are limited in several ways including: i) they are often developed by organizations with interests and priorities that do not necessarily reflect those of individuals affected by the decision-making; ii) they rely on self-regulation rather than offering universal accountability mechanisms. Based on consultations with experts in the humanitarian sector, Pizzi *et al.* (2020) advocate for leveraging international human rights laws in combination with ethics principles to build an operational baseline for rights-respecting AI for humanitarian action. Human rights are codified, more universal and well-defined than ethics principles, with laws and jurisprudence that offer a foundation for accountability (Dodgson *et al.*, 2020). Meanwhile, ethics principles can complement aspects not covered by law/jurisprudence yet. Authors highlight three main AI principles deriving from such combination of human rights and ethics: i) *non-discrimination*, ii) *transparency and explainability*, and iii) *accountability*. These are some of the key aspects discussed across the paragraphs below.

Oversight and incentives: Accountability requires *oversight mechanisms* to assess compliance with established principles/requirements. In addition to potential regulations grounded on human-rights laws, alternatives include certifications for recognizing organizations that abide to best practices, with transnational institutions (e.g., UN, IEEE) collaborating for their elaboration and implementation. Additional mechanisms include funding agencies, for example, imposing auditability as a prerequisite for grant eligibility. Recommendations also include: internal review boards combined with audits by external and independent reviewers, making reports public; AI human rights and ethics review boards that includes technical and non-technical staff, with powers that include halting projects as needed without fear of retaliation.

Capacity-building and knowledge sharing: These factors are crucial for inclusion of affected groups, public sector, developers and other involved organizations, which is key to ensure diversity and inclusion across all aspects of AI development and deployment. In turn, this benefits robustness to biases, non-discriminatory usage, as well as accountability since it is a challenge beyond the capabilities of any single organization when such complex technologies are involved.

Human-in-the-loop approaches: Highlighted as a promising avenue to benefit the understanding and verification of the outputs being provided by "opaque" AI-models, human-in-the-loop approaches also help ensuring a human decision-maker into every AI-supported decision, such that humans can be considered responsible for decisions made and their outcomes.

Data management protocols: Protocols on how to manage data are also important to ensure that outcomes and all steps that lead to decision-making for humanitarian action do respect humanitarian principles. Relevant resources highlighted in (Dodgson *et al.*, 2020) include: *the ICRC Handbook on Data Protection in Humanitarian Response*, which has a specific chapter on data protection challenges associated with AI and ML in the humanitarian sector; *the International Organization for Migration (IOM) Data Protection Manual*; and the *UN OCHA: Data Responsibility Guidelines* (`https://centre.humdata.org/the-ocha-data-responsibility-guidelines`). Risk Assessment Tools are additional relevant resources for risk management, such as the *UN Global Pulse's Risk, Harm and Benefits Assessment tool* (`https://www.unglobalpulse.org/policy/risk-assessment`).

13.7.3 OPEN CHALLENGES AND VISION FOR THE FUTURE

In light of recent developments in AI techniques, data sharing, and humanitarian challenges, the following paragraphs cast wider to highlight a few research priority directions that are promising but are likely to require multi-organizational partnerships (Lunga *et al.*, 2022).

Self-supervised learning from multimodal data: The ability to integrate the diverse and rich signals from different earth observation technologies, socioeconomic indicators, environmental sensors, social media and other sources could play a key role in knowledge discovery. Self-supervised learning (SSL) strategies are particularly well positioned to contribute to this direction, while also potentially drastically reducing the large need for labeled data that still hinders applications that require rapid response. Foundation Models (FMs) are a noteworthy example: trained on broad internet-scale data through SSL, such models present strong generalization capabilities that drastically reduce training requirements for adaptation to specific downstream tasks. While non-trivial, developing such capabilities for Earth Observation data could revolutionize cross domain applications including, e.g., humanitarian mapping tasks.

Scalability: The increased size of training data and model parameters directly inflate memory footprints. Humanitarian tasks often require rapid mapping of areas thousands of km^2 large, which is implied in extremely large data volumes. Already a significant challenge for modality- and task-specific models, this problem is further augmented for multimodal learning at the proportions of, e.g., foundation models. Scaling and optimization strategies, leveraging both data and model parallelism in tandem to reducing memory footprints during training as well as inferencing seem a prudent path forward to address this challenge.

Democratization of machine learning for humanitarian assistance: Developing open geospatial data infrastructures is a research direction particularly worthy investigating. GeoAI systems could benefit from platforms that are open, indexable, and standardized. The BigEarthNet platform (Sumbul *et al.*, 2019) and the Spatio-Temporal Asset Catalog (STAC) are good examples, but the field could further benefit from multi-organization collaboration to scale up such resources. The *development and sharing of generalizable pre-trained models* could also greatly lower the barrier for domain experts and organizations with less resources to explore AI models for, e.g., humanitarian applications. However, despite such reductions in needs for training data, large benchmark datasets will remain a crucial need to validate research results, particularly generalizability.

BIBLIOGRAPHY

Abel, G.J., *et al.*, 2019. Climate, conflict and forced migration. *Global Environmental Change*, 54, 239–249.

Alegana, V.A., *et al.*, 2015. Fine resolution mapping of population age-structures for health and development applications. *Journal of The Royal Society Interface*, 12 (105), 20150073.

Alfons, A., *et al.*, 2011. Simulation of close-to-reality population data for household surveys with application to EU-SILC. *Statistical Methods & Applications*, 20, 383–407.

Arndt, J. and Lunga, D., 2020. Sampling subjective polygons for patch-based deep learning land-use classification in satellite images. *In*: *IGARSS 2020 - 2020 IEEE International Geoscience and Remote Sensing Symposium*. 1953–1956.

Arndt, J. and Lunga, D., 2021. Large-scale classification of urban structural units from remote sensing imagery. *IEEE Journal of Selected Topics in Applied Earth Observations and Remote Sensing*, 14, 2634–2648.

Batty, M., 2007. *Cities and Complexity: Understanding Cities with Cellular Automata, Agent-Based Models, and Fractals*. Boston: The MIT Press.

Beckman, R.J., Baggerly, K.A., and McKay, M.D., 1996. Creating synthetic baseline populations. *Transportation Research Part A: Policy and Practice*, 30 (6), 415–429.

Beine, M. and Jeusette, L., 2021. A meta-analysis of the literature on climate change and migration. *Journal of Demographic Economics*, 87 (3), 293–344.

Benenson, I. and Torrens, P.M., 2004. *Geosimulation: Automata-Based Modeling of Urban Phenomena*. London: John Wiley & Sons, Ltd.

Bhaduri, B., *et al.*, 2007. Landscan usa: a high-resolution geospatial and temporal modeling approach for population distribution and dynamics. *GeoJournal*, 69 (1), 103–117.

Black, R., *et al.*, 2011a. The effect of environmental change on human migration. *Global Environmental Change*, 21, S3–S11.

Black, R., *et al.*, 2011b. Migration as adaptation. *Nature*, 478 (7370), 447–449.

Blumenstock, J., 2018. Don't forget people in the use of big data for development.

Bosco, C., *et al.*, 2017. Exploring the high-resolution mapping of gender-disaggregated development indicators. *Journal of The Royal Society Interface*, 14 (129), 20160825.

Burris, J.W., *et al.*, 2013. Uncertainty quantification techniques for population density estimates derived from sparse open source data. *In: Proc. SPIE 8747, Geospatial InfoFusion III*. vol. 8747, 254–258.

Burris, J.W., *et al.*, 2015. Machine learning for the activation of contraflows during hurricane evacuation. *In: 2015 IEEE Global Humanitarian Technology Conference (GHTC)*. vol. 1, 254–258.

Carvalhaes, T., *et al.*, 2022a. A spatially non-stationary approach to identify multi-dimensional push factors for international emigration. *Social Science Research Network*.

Carvalhaes, T., *et al.*, 2022b. Integrating spatial and ethnographic methods for resilience research: A thick mapping approach for hurricane maria in puerto rico. *Annals of the American Association of Geographers*.

Carvalhaes, T.M., *et al.*, 2021. An overview & synthesis of disaster resilience indices from a complexity perspective. *International Journal of Disaster Risk Reduction*, 57, 102165.

Casali, Y., Aydin, N.Y., and Comes, T., 2022. Machine learning for spatial analyses in urban areas: a scoping review. *Sustainable Cities and Society*, 85, 104050.

Coppi, G., Moreno Jimenez, R., and Kyriazi, S., 2021. Explicability of humanitarian ai: a matter of principles. *Journal of International Humanitarian Action*, 6 (1), 1–22.

Crooks, A., *et al.*, 2021. Agent-based modeling and the city: A gallery of applications. *In:* W. Shi, M.F. Goodchild, M. Batty, M.P. Kwan and A. Zhang, eds. *Urban Informatics*. Singapore: Springer, Ch. 46, 885–910.

Cutter, S.L. and Finch, C., 2008. Temporal and spatial changes in social vulnerability to natural hazards. *Proceedings of the National Academy of Sciences*, 105 (7), 2301–2306.

de Blasio, G., D'Ignazio, A., and Letta, M., 2022. Gotham city. Predicting 'corrupted' municipalities with machine learning. *Technological Forecasting and Social Change*, 184, 122016.

Dias, P., *et al.*, 2022a. Human-machine collaboration for reusable and scalable models for remote sensing imagery analysis. *Presented at the ICML 2022 Workshop on Human-Machine Collaboration and Teaming*.

Dias, P. and Lunga, D., 2022. Embedding ethics and trustworthiness for sustainable AI in earth sciences: Where do we begin? *In: IGARSS 2022 - 2022 IEEE International Geoscience and Remote Sensing Symposium*. 4639–4642.

Dias, P., *et al.*, 2022b. Model assumptions and data characteristics: Impacts on domain adaptation in building segmentation. *IEEE Transactions on Geoscience and Remote Sensing*, 60, 1–18.

Dodgson, K., *et al.*, 2020. A framework for the ethical use of advanced data science methods in the humanitarian sector. *Data Science and Ethics Group*.

Durst, D., Lämmel, G., and Klüpfel, H., 2014. Large-scale multi-modal evacuation analysis with an application to hamburg. *In: Pedestrian and Evacuation Dynamics 2012*. Springer, 361–369.

Epstein, J.M. and Axtell, R.L., 1996. *Growing Artificial Societies: Social Science from the Bottom Up*. Boston: The MIT Press.

Feitosa, F.F., Le, Q.B., and Vlek, P.L., 2011. Multi-agent simulator for urban segregation (masus): A tool to explore alternatives for promoting inclusive cities. *Computers, Environment and Urban Systems*, 35 (2), 104–115.

Flowerdew, R. and Aitkin, M., 1982. A method of fitting the gravity model based on the poisson distribution. *Journal of Regional Science*, 22 (2), 191–202.

Fotheringham, A.S., *et al.*, 2004. The Development of a Migration Model for England and Wales: Overview and Modelling Out-Migration. *Environment and Planning A: Economy and Space*, 36 (9), 1633–1672.

Fotheringham, A.S., Yang, W., and Kang, W., 2017. Multiscale Geographically Weighted Regression. *Annals of the American Association of Geographers*, 107 (6), 1247–1265.

Fotheringham, A., 1983. A new set of spatial-interaction models: The theory of competing destinations. *Environment and Planning A: Economy and Space*, 15 (1), 15–36.

Frazier, T. and Alfons, A., 2012. Generating a close-to-reality synthetic population of ghana. *Social Science Research Network*, 15.

Garcia, A.J., *et al.*, 2014. Modeling internal migration flows in sub-Saharan Africa using census microdata. *Migration Studies*, 3 (1), 89–110.

Gething, P., *et al.*, 2015. Creating spatial interpolation surfaces with DHS data. *DHS Spatial Analysis Reports*, No. 11, pp. 86.

Helbing, D. and Molnar, P., 1995. Social force model for pedestrian dynamics. *Physical Review E*, 51 (5), 4282–4286.

Jayachandran, S., Biradavolu, M., and Cooper, J., 2023. Using machine learning and qualitative interviews to design a five-question survey module for women's agency. *World Development*, 161, 106076.

Knippenberg, E., Jensen, N., and Constas, M., 2019. Quantifying household resilience with high frequency data: Temporal dynamics and methodological options. *World Development*, 121, 1–15.

Kraemer, M.U.G., *et al.*, 2019. Utilizing general human movement models to predict the spread of emerging infectious diseases in resource poor settings. *Scientific Reports*, 9 (5151).

Kuffer, M., *et al.*, 2022. The missing millions in maps: Exploring causes of uncertainties in global gridded population datasets. *ISPRS International Journal of Geo-Information*, 11 (7), 403.

Kuner, C., *et al.*, 2020. *Handbook on Data Protection in Humanitarian Action*. International Committee of the Red Cross.

Lai, S., *et al.*, 2019. Exploring the use of mobile phone data for national migration statistics. *Palgrave Communications*, 5 (34), 89–110.

Lämmel, G., Klüpfel, H., and Nagel, K., 2009. The matsim network flow model for traffic simulation adapted to large-scale emergency egress and an application to the evacuation of the indonesian city of padang in case of a tsunami warning. *In*: *Pedestrian Behavior*. Emerald Group Publishing Limited, 245–265.

Le, Q.B., Park, S.J., and Vlek, P.L., 2010. Land use dynamic simulator (ludas): A multi-agent system model for simulating spatio-temporal dynamics of coupled human–landscape system. *Ecological Informatics*, 5 (3), 203–221.

Leslie, D., 2019. Understanding artificial intelligence ethics and safety. *arXiv preprint arXiv:1906.05684*.

Li, W., 2020. Geoai: Where machine learning and big data converge in GIScience. *Journal of Spatial Information Science*, (20), 71–77.

Luca, M., *et al.*, 2020. Deep learning for human mobility: a survey on data and models. *CoRR*, abs/2012.02825.

Lunga, D., *et al.*, 2021. Resflow: A remote sensing imagery data-flow for improved model generalization. *IEEE Journal of Selected Topics in Applied Earth Observations and Remote Sensing*, 14, 10468–10483.

Lunga, D., *et al.*, 2020. Apache spark accelerated deep learning inference for large scale satellite image analytics. *IEEE Journal of Selected Topics in Applied Earth Observations and Remote Sensing*, 13, 271–283.

Lunga, D., *et al.*, 2022. GeoAI at ACM SIGSPATIAL: The New Frontier of Geospatial Artificial Intelligence Research.

Martínez-Rivera, W., *et al.*, 2022. A treatment-effect model to quantify human dimensions of disaster impacts: the case of hurricane maria in puerto rico. *Natural Hazards*, 1–36.

McAuliffe, M. and Triandafyllidou, A., 2020. *World Migration Report 2022*. Geneva: International Organization for Migration (IOM).

McLeman, R., *et al.*, 2016. Environmental migration and displacement: What we know and don't know. *In: Laurier Environmental Migration Workshop*.

Milán-García, J., *et al.*, 2021. Climate change-induced migration: a bibliometric review. *Globalization and Health*, 17 (1), 74.

Miller, E.J., 2021. Transportation modeling. *In*: W. Shi, M.F. Goodchild, M. Batty, M.P. Kwan and A. Zhang, eds. *Urban Informatics*. Singapore: Springer, Ch. 46, 911–931.

Mitchell, M., 1998. *An Introduction to Genetic Algorithms*. Boston: The MIT Press.

Munnich, R. and Schurle, J., 2003. On the simulation of complex universes in the case of applying the german microcensus. *DACSEIS Research Paper Series, No. 4, 2003*, No. 4, pp. 22.

Nagel, K. and Schreckenberg, M., 1992. A cellular automaton model for freeway traffic. *Journal de Physique*, 2 (12), 2221–2229.

Newman, M.E.J., 2003. The structure and function of complex networks. *SIAM Review*, 45 (2), 167–256.

Ortiz, D.A., 2020. Geographic information systems (gis) in humanitarian assistance: a meta-analysis. *Pathways: A Journal of Humanistic and Social Inquiry*, 1 (2), 4.

Pagels, H., 1989. *The Dreams of Reason*. USA: Simon & Schuster, Inc.

Pandey, B., Brelsford, C., and Seto, K.C., 2022. Infrastructure inequality is a characteristic of urbanization. *Proceedings of the National Academy of Sciences*, 119 (15), e2119890119.

Persello, C., *et al.*, 2022. Deep learning and earth observation to support the sustainable development goals: Current approaches, open challenges, and future opportunities. *IEEE Geoscience and Remote Sensing Magazine*, 10 (2), 172–200.

Pizzi, M., Romanoff, M., and Engelhardt, T., 2020. AI for humanitarian action: Human rights and ethics. *International Review of the Red Cross*, 102 (913), 145–180.

Presner, T., Shepard, D., and Kawano, Y., 2014. *Hypercities Thick Mapping in the Digital Humanities*. UCLA Previously Published Works. UCLA eScholarship.

Quinn, J.A., *et al.*, 2018. Humanitarian applications of machine learning with remote-sensing data: review and case study in refugee settlement mapping. *Philosophical Transactions of the Royal Society A: Mathematical, Physical and Engineering Sciences*, 376 (2128), 20170363.

Reddy, T.A., 2020. Resilience of complex adaptive systems: A pedagogical framework for engineering education and research. *ASME Journal of Engineering for Sustainable Buildings and Cities*, 1 (2), 021004.

Sallah, K., *et al.*, 2017. Mathematical models for predicting human mobility in the context of infectious disease spread: introducing the impedance model. *International Journal of Health Geographics*, 16 (42).

Saprykin, A., Chokani, N., and Abhari, R.S., 2022. Accelerating agent-based demand-responsive transport simulations with gpus. *Future Generation Computer Systems*, 131, 43–58.

Schiavina, M., *et al.*, 2022. GHSL Data Package 2022 GHS-BUILT-S R2022A - GHS built-up surface grid, derived from Sentinel-2 composite and Landsat, multitemporal (1975-2030).

Schindler, J., 2013. About the uncertainties in model design and their effects: An illustration with a land-use model. *Journal of Artificial Societies and Social Simulation*, 16 (4), 6.

Spielman, S.E., *et al.*, 2020. Evaluating social vulnerability indicators: criteria and their application to the social vulnerability index. *Natural Hazards*, 100 (1), 417–436.

Stewart, R., *et al.*, 2016. A bayesian machine learning model for estimating building occupancy from open source data. *Natural Hazards*, 81.

Sumbul, G., *et al.*, 2019. BigEarthNet: A Large-Scale Benchmark Archive For Remote Sensing Image Understanding. *CoRR*, arxiv.org/abs/1902.06148.

Taubenbock, H., *et al.*, 2009. ”last-mile” preparation for a potential disaster – interdisciplinary approach towards tsunami early warning and an evacuation information system for the coastal city of padang, indonesia. *Natural Hazards and Earth System Sciences*, 9 (4), 1509–1528.

Tsegai, D. and Le, Q.B., 2010. District-level spatial analysis of migration flows in ghana: Determinants and implications for policy. *ZEF-Discussion Papers on Development Policy*, No. 144, pp. 18.

Utazi, C., *et al.*, 2018. High resolution age-structured mapping of childhood vaccination coverage in low and middle income countries. *Vaccine*, 36 (12), 1583–1591.

Verjee, F., 2005. The application of geomatics in complex humanitarian emergencies. *Journal of Humanitarian Assistance*.

Villamor, G.B., *et al.*, 2014. Biodiversity in rubber agroforests, carbon emissions, and rural livelihoods: An agent-based model of land-use dynamics in lowland sumatra. *Environmental Modelling & Software*, 61, 151–165.

Vinuesa, R., *et al.*, 2020. The role of artificial intelligence in achieving the sustainable development goals. *Nature Communications*, 11 (1), 1–10.

Waddell, P., 2002. Urbansim: Modeling urban development for land use, transportation, and environmental planning. *Journal of the American Planning Association*, 68 (3), 297–314.

Wardrop, N., *et al.*, 2018. Spatially disaggregated population estimates in the absence of national population and housing census data. *Proceedings of the National Academy of Sciences*, 115 (14), 3529–3537.

Wesolowski, A., *et al.*, 2013. The use of census migration data to approximate human movement patterns across temporal scales. *PLOS ONE*, 8 (1), 1–8.

Wilson, A., 1971. A family of spatial interaction models, and associated developments. *Environment and Planning A: Economy and Space*, 3 (1), 1–32.

Wisetjindawat, W., *et al.*, 2013. Modeling disaster response operations including road network vulnerability. *Journal of the Eastern Asia Society for Transportation Studies*, 10, 196–214.

Yang, H.L., *et al.*, 2018. Building extraction at scale using convolutional neural network: Mapping of the united states. *IEEE Journal of Selected Topics in Applied Earth Observations and Remote Sensing*, 11 (8), 2600–2614.

14 GeoAI for Disaster Response

Lei Zou
Department of Geography, Texas A&M University

Ali Mostafavi
Department of Civil & Environmental Engineering, Texas A&M University

Bing Zhou, Binbin Lin, and Debayan Mandal
Department of Geography, Texas A&M University

Mingzheng Yang, Joynal Abedin, and Heng Cai
Department of Geography, Texas A&M University

CONTENTS

14.1 Introduction 288
14.2 The Paradigm of GeoAI for Disaster Response 289
    14.2.1 Remote Sensing Data...290
    14.2.2 Human Mobility Data 291
    14.2.3 Social Media Data..292
    14.2.4 Sensor Data..293
14.3 Case Study: Analyzing Twitter Data with NLP for Disaster Response 293
    14.3.1 Data Collection and Cleaning 293
    14.3.2 Intelligent Classification Model..294
    14.3.3 Intelligent Toponym Recognition Model...296
    14.3.4 Spatial Analysis 297
14.4 Challenges and Opportunities...299
    14.4.1 Data Quality and Availability 299
    14.4.2 Privacy Protection and Security..299
    14.4.3 Trustworthiness of GeoAI-based Solutions 300
    14.4.4 Practical Deployment and Maintenance 300
14.5 Conclusion..301
    Bibliography 302

DOI: 10.1201/9781003308423-14

14.1 INTRODUCTION

Over the past few years, the planet Earth has witnessed a series of catastrophic events, leaving indelible marks on human communities, ecosystems, and landscapes (Lam et al., 2018). Some recent major disasters include Hurricane Harvey in 2017, which tied with Katrina (2005) for the costliest ($125 billion) Atlantic hurricane in history (Mihunov et al., 2020), the COVID-19 pandemic from 2020 to 2023 that infected over 674 million people worldwide (Dong et al., 2020), the 2021 Winter Storm Uri, the most expensive natural disaster ($196.5 billion) in the United States (Lee et al., 2022), and the 2023 Turkey-Syria Earthquake, which claimed over 57,300 lives[1]. Under climate change and the growing population in hazard-prone areas, the frequency and intensity of disasters are projected to increase, posing significant challenges to disaster management efforts (Cai et al., 2018). Therefore, it is crucial to develop effective pathways for minimizing the impacts of disasters to build resilient communities that can withstand future challenges.

Disaster resilience can be strengthened through strategies in the four phases of disaster management: mitigation, preparedness, response, and recovery (Zou et al., 2018). Mitigation entails the implementation of measures aimed at minimizing the risks associated with disasters, including identifying hazard risks and enforcing building codes and zoning regulations. Preparedness focuses on planning, training, and conducting exercises to ensure that organizations and communities are equipped to respond to disasters. The response phase involves taking immediate actions to address disaster impacts, such as evacuation, search and rescue operations, and the identification of local needs and provision of assistance. Recovery aims to restore the community to its pre-disaster state or to a new state that is more resilient to future disasters. While disaster resilience can be improved through actions taken in any of the four phases, the unpredictable nature of disasters makes it difficult to foresee their occurrences and effects and prepare in advance, leaving disaster response as the last and most critical line of defense in saving lives, reducing infrastructural damages, lessening societal disturbances, and accelerating recoveries.

Efficient disaster response relies heavily on timely information describing local disaster impacts and survivors' needs to coordinate first responders and allocate resources to the most needed individuals and communities, but such information is difficult to obtain through traditional approaches. The burgeoning real-time or near-real-time geospatial big data collected from social media, web applications, cellphones, satellites, drones, cameras, and various sensors offer a novel channel to observe location-based, time-sensitive disaster information that can support effective disaster response (Zou et al., 2019). However, applying geospatial big data in disaster response is challenging due to technical difficulties in accurately extracting valuable, fine-grained disaster information from big, biased, and noisy data, and effectively incorporating them into disaster response practices.

The emergence of Geospatial Artificial Intelligence (GeoAI), a subfield of spatial data science that combines GIScience and AI for spatial data recognition, knowledge discovery, and spatial-temporal modeling, provides new opportunities

---

[1] https://reliefweb.int/report/syrian-arab-republic/syriaturkey-earthquakes-situation-report-7-march-8-2023

(Janowicz et al., 2020). Recent advances in GeoAI include the development of intelligent models to extract or encode location-based information from multimodal data, the creation of spatial-aware knowledge graphs for geographic question answering, and spatial-explicit machine learning and deep learning models to understand geographical phenomena. Combining GeoAI with geospatial big data is promising in generating valuable disaster-related information and new knowledge that can support disaster response missions more effectively.

This chapter endeavors to foster the convergence of GeoAI and disaster response research and practice by accomplishing three distinct objectives: (1) establishing a comprehensive paradigm that expounds upon the diverse applications of GeoAI toward enhancing disaster response efforts; (2) exhibiting the practical employment of GeoAI in disaster response through the analysis of social media data during Hurricane Harvey in 2017 with advanced Natural Language Processing (NLP) models; and (3) identifying the challenges associated with the complete realization of the potential of GeoAI in disaster response research and practice and propose possible solutions. The fulfillment of these objectives will serve to augment the extant corpus of knowledge pertaining to GeoAI and its critical role in disaster response, as well as underscore prospects for future research and practice in this domain.

14.2 THE PARADIGM OF GEOAI FOR DISASTER RESPONSE

Figure 14.1 presents a comprehensive paradigm that details the roles of GeoAI in supporting disaster response operations. The paradigm illustrates that various types of real-time or near real-time geospatial big data can be integrated with traditional GIS data (i.e., historical disaster records, built-environment, socioeconomic characteristics, and landscape features) to identify disaster impacts and needs, thereby facilitating disaster response missions. These impacts and needs include the evacuation

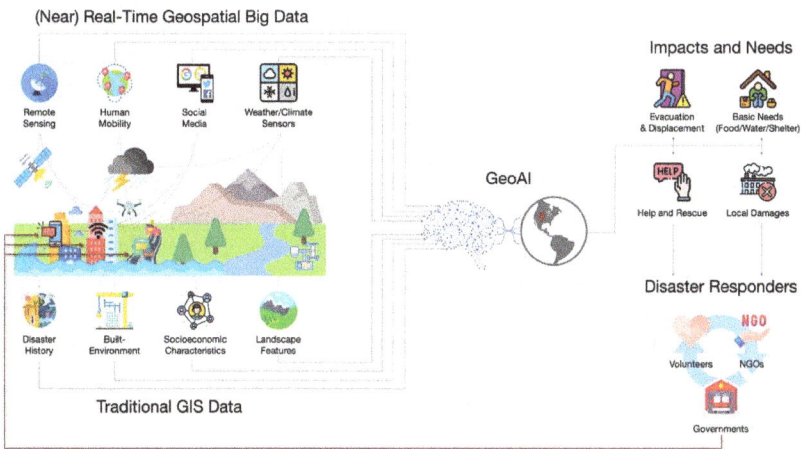

Figure 14.1   Conceptual framework of using GeoAI to support disaster response.

or displacement of residents in disaster-affected communities, survivors trapped in disaster-affected areas requiring rescue, basic necessities such as food, water, shelter, and medical assistance, and property and infrastructure damage. Armed with such information, first responders, including governmental organizations (such as police and fire departments, coast guards, and FEMA), non-governmental organizations (such as the Red Cross), and volunteers (such as the Cajun Navy and CrowdSource Rescue at `https://crowdsourcerescue.org`), can identify people and communities in need of evacuation or rescue, basic supplies, and infrastructure repairs and provide assistance. GeoAI serves as the engine that precisely translates geospatial data into near real-time disaster impacts and needs information in this paradigm. In the rest of this section, we exemplify the use of various GeoAI methods in disaster response through a review of previous investigations, categorized by four types of geospatial big data, including remote sensing, human mobility, social media, and sensor data.

14.2.1   REMOTE SENSING DATA

Remote sensing data, such as satellite images, drone footage, and street view photos, have been increasingly utilized to rapidly estimate disaster damage types, areas, and magnitudes and allocate rescue and repair forces in disaster response. In this context, two distinct approaches have been employed to estimate disaster damages using remote sensing images: multi-temporal techniques and mono-temporal techniques (Dong and Shan, 2013). The former involves comparing changes between pre- and post-event remote sensing data to identify damaged buildings or infrastructures, while the latter analyzes only post-event remote sensing data to identify disaster-induced damages. In both approaches, image processing techniques are applied to recognize objects such as buildings and roads and measure their damage sizes and severities. The use of GeoAI algorithms can significantly enhance the accuracy and efficiency of such image processing methods (Zhang et al., 2019).

Numerous studies have explored the potential of utilizing GeoAI methods, specifically deep learning-based image processing algorithms, for analyzing remote sensing data to estimate disaster damages. One such study by Zhao and Zhang (2020) utilized a Siamese neural network model based on the Mask Region-based Convolutional Neural Network (R-CNN) to evaluate disaster damages from pre- and post-disaster satellite imagery in two stages: building feature extraction and damage level classification. The results demonstrated that the proposed deep learning approach could accurately recognize collapsed buildings and classify them into four damage levels, with an accuracy 16 times higher than the baseline model. In another study, Nex et al. (2019) used an advanced CNN model to uniformly detect visible structural building damage caused by earthquakes from satellite, airborne, and unmanned aerial vehicle (UAV) images captured after disasters. Furthermore, Zhai and Peng (2020) collected post-disaster Google Street View photos from neighborhoods along coastal Florida affected by 2018 Hurricane Michael. They fine-tuned a deep convolutional network for visual recognition, the Visual Geometry Group-19 (VGG-19) model (Simonyan and Zisserman, 2015), to detect damaged building areas on street view photos and assess destruction degrees. The results indicate that processing post-disaster street view photos with GeoAI can estimate building-level disaster damage with an accuracy of 70%. Using street view photos can also reveal on-the-ground details that are not typically captured by satellite or aerial remote sensing images.

14.2.2   HUMAN MOBILITY DATA

Human mobility data refer to data that track the movements of individuals or groups of individuals. Such data can be collected from a variety of sources, including GPS-enabled mobile devices, transportation networks, social media platforms, and census surveys and have been used in diverse applications, e.g., tracking human behavioral changes during the COVID-19 pandemic (Gao et al., 2020; Hu et al., 2021). In the context of disaster response, human mobility data can be particularly useful for tracking the evacuation patterns of affected individuals. Monitoring evacuation compliance during disasters helps disaster responders locate communities that have failed to evacuate and provide necessary assistance. Additionally, decreased human mobility in disaster-affected regions, e.g., declined visitations to restaurants due to winter storms, can serve as an indicator of disaster damage. To analyze human mobility data for evacuation monitoring and damage estimation purposes, GeoAI methods can be employed to quantify human mobility changes, predict evacuation destinations, cluster similar evacuation patterns, and analyze the spatial relationships between different variables related to evacuation compliance or mobility decline, e.g., transportation infrastructure, land use, and socioeconomics.

Scholars have investigated the changes in individual/group-level mobility patterns before, during, and after disasters using different geospatial data and GeoAI approaches to infer evacuation behaviors and estimate disaster damage. One study analyzed GPS records of approximately 1.6 million anonymized users in Japan and discovered that human mobility following large-scale disasters correlate with their mobility patterns during normal times and are highly impacted by victims' social relationships, local disaster intensity, damage level, shelter locations, and news reporting (Song et al., 2014). They further modeled and predicted human evacuation behaviors under catastrophes in Japan using a deep learning model combining a deep belief network with a sigmoid regression layer (Song et al., 2017). Deng et al. (2021) obtained 30 million GPS records of roughly 150,000 unique, anonymous users from Cuebiq to quantitatively research Houston residents' evacuation behaviors during the 2017 Hurricane Harvey. By computing the evacuation rates and distances, they revealed that disadvantaged minority populations were less likely to evacuate than wealthier residents. They also discovered that solid social cohesion exists among evacuees from advantaged neighborhoods in their destination choices. Yabe and others quantified the economic impact of disasters on businesses using human mobility data from SafeGraph and a Bayesian causal inference approach (Yabe et al., 2020). Besides, social media activities of users sharing location information have also been utilized to track evacuation behaviors, as evidenced by Martin et al. (2017) and Jiang et al. (2019) who analyzed Twitter messages during the 2016 Hurricane Matthew in South Carolina and found that people in the same social networks tend to make similar evacuation decisions, and evacuated people tend to have larger long-term activity space and smaller long-term sentiment variances.

14.2.3 SOCIAL MEDIA DATA

Social media platforms such as Twitter, Facebook, and Weibo have emerged as popular mediums for users to disseminate information, seek assistance, and report on local conditions during times of disaster (Zou et al., 2023). The real-time nature of social media data, when analyzed in conjunction with location data, can provide valuable insights into the extent of damage and needs of the affected population. Since social media data typically comprise textual and visual content, incorporating GeoAI techniques, such as natural language processing (NLP) and image processing, can enable the accurate extraction of disaster-related information from social media platforms for use in disaster response efforts.

Apart from monitoring evacuations, social media has gained significant traction in the realm of emergency rescue, identifying local needs, and damage estimation, as evident from existing literature. Monitoring rescue requests, local needs, or damages during disasters via social media involves three steps: (a) classification - searching for messages requesting help, stating needs, or reporting damages, (b) location recognition - extracting their geographical information, and (c) location resolution – converting or standardizing the extracted geographical information into pairs of coordinates. Prior studies have integrated GeoAI, such as advanced natural language processing (NLP) models, into these steps to enhance the comprehension of social media message contents. For example, Zhou et al. (2022) employed the Bidirectional Encoder Representations from Transformers (BERT) model in the first step to identify rescue requests from Twitter during Hurricane Harvey and achieved an F1-score of 0.92. In the second step, Wang et al. (2020) developed a novel Twitter data toponym recognition model, NeuroTPR, based on pre-trained language models and the Long Short-Term Memory (LSTM) model, resulting in an F1-score of 0.73. Another study applied the Latent Dirichlet Allocation (LDA) method to identify topics discussed on Twitter during Hurricane Harvey and uncovered the spatial-temporal variations of topics relevant to local needs, e.g., gas and supplies, power, cellphone services, cable Internet, etc. (Mihunov et al., 2022).

Social media data can also be integrated with traditional GIS data to detect disaster-affected areas and estimate damages. For instance, Li et al. (2018) fused Twitter data with field water-level sensors and elevation data to visualize the extent of the 2015 South Carolina Floods in near-real-time. Huang and others improved upon this model by incorporating post-flood soil moisture information derived from satellite observations (Huang et al., 2018). Several studies have examined the value of incorporating social media data into rapid damage estimation using NLP and statistical and machine learning approaches (Kryvasheyeu et al., 2016; Mihunov et al., 2022; Wang et al., 2021; Yuan and Liu, 2020; Zou et al., 2018). Drawing from examples such as Hurricanes Sandy (2012), Isaac (2012), Matthew (2016), and Harvey (2017), these studies have found strong correlations between disaster-related

Twitter activities, i.e., disaster-related discussion intensity and sentiment, and economic losses across multiple spatial and temporal scales.

14.2.4 SENSOR DATA

Data obtained from weather, climate, geological, and human sensors can provide
critical information to assist in early warning in disaster response. These data can
aid disaster responders to identify or forecast affected individuals and alert them,
as well as formulate more effective response strategies. For instance, weather sensors can measure temperature, humidity, air pressure, wind speed, and precipitation,
which can be used to predict and track the progression of natural disasters such as
hurricanes, tornadoes, and floods. Similarly, seismic sensors can detect earthquakes
and aftershocks, enabling disaster responders to monitor the intensity and duration
of the event and determine where resources are most needed. Therefore, integrating
multiple sensor data sources with GeoAI-based analysis can provide more accurate
early warning and support disaster response.

The integration of sensor data and GeoAI, e.g., machine learning models, has
led to the implementation of several early-warning systems. One such system is provided by Grillo (https://grillo.io/), which offers a device that detects abnormal ground
vibrations and immediately verifies the occurrence of an earthquake to notify local
users (Ray et al., 2017). Another system, developed by the (British Geological Survey, 2023), uses an accelerometer-based sensor system to monitor slope instability
and landslide incidence. By measuring ground movement and density, an early warning message can be sent to the periphery through an application called Assessment
of Landslides using Acoustic Real-time Monitoring Systems (ALARMS).

14.3 CASE STUDY: ANALYZING TWITTER DATA WITH NLP FOR DISASTER RESPONSE

This section demonstrates the use of GeoAI in disaster response via a case study
of leveraging advanced NLP models to extract locations of help requests for human
beings and animals, infrastructure damages, and shelters and supplies from tweets
posted during the 2017 Hurricane Harvey. The extracted messages relevant to emergency rescue and infrastructure damages can be used for spatial-temporal analysis
and visualization to better understand the impact of the hurricane and coordinate
disaster response operations. The methodological framework and the tools used to
classify and geolocate those tweets can be inherited and applied in future hazardous
events, while the knowledge generated by analysis of the data can aid stakeholders
in making more informed decisions.

14.3.1 DATA COLLECTION AND CLEANING

Figure 14.2 presents the workflow of this case study with four primary steps: (1)
Twitter data collection and cleaning, (2) intelligent tweet classification, (3) intelligent geoparsing, and (4) spatial-temporal analysis. The first step is data collection

and cleaning. Harvey-related Twitter data from August 17, 2017, the day when Harvey was named, to September 7, 2017, two weeks after Harvey dissipated, were collected using Twitter Academic Research API v2, which offers certified researchers access to the full historical Twitter database with a monthly cap of 10 million tweets. We created a list of case-insensitive keywords about the disaster and known responding agencies and volunteer groups to identify an initial collection of tweets relevant to Hurricane Harvey response efforts: [harvey, hurricane, storm, flood, houston, txtf (Texas Task Force), coast guard, uscg (U.S. Coast Guard), houstonpolice, cajun navy, fema (Federal Emergency Management Agency), rescue] (Zou et al., 2023). Every tweet containing at least one of the keywords was retrieved. The initial collection resulted in 47.5 million tweets. The data were stored in a non-SQL database, MongoDB. The data collection process was supported by distributed systems so that several data-collecting streams can be initiated simultaneously with multiple research accounts to accelerate the data-collection process. The content of each tweet was cleaned by removing the contractions, URLs, punctuation, and emojis.

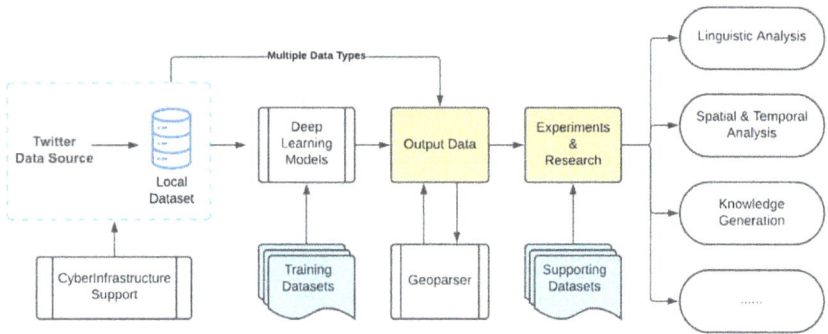

Figure 14.2   The workflow of Twitter data collection and processing with GeoAI.

14.3.2   INTELLIGENT CLASSIFICATION MODEL

The second step is to retrieve rescue and damage-related tweets from the raw data. This study developed an intelligent classification model to process the textual information from tweets and identify tweets falling into one of the four categories: (1) humanitarian help requests, (2) animal-related help requests, (3) reports of infrastructure malfunction, and (4) content that offers help or shelter information. This task can be accomplished by using deep learning models concatenating a pre-trained large language model (LLM) and a classifier, as delineated in Figure 14.3 (a). The pre-trained model learns the general language representations from existing corpus by converting texts into meaningful vectors or metrics, while the classifier targets at specific downstream tasks, e.g., text classification, based on the vectors or metrics. This approach allows the tweet classification model to benefit from the knowledge

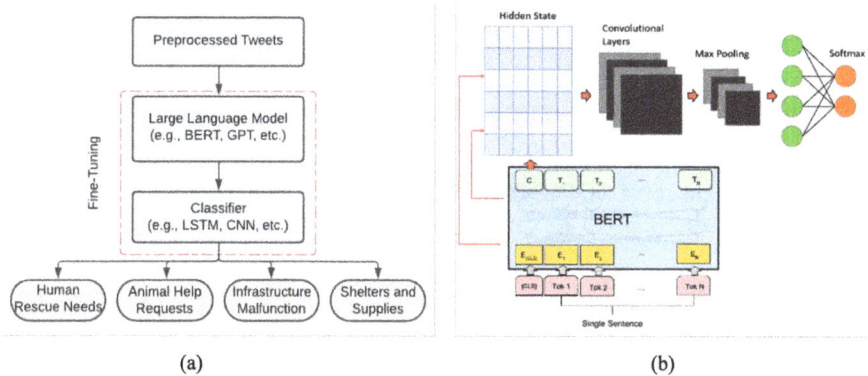

(a)   (b)

Figure 14.3   (a) The architecture of building an intelligent classification model; (b) the architecture of the proposed BERT-CNN model (*Source:* Zhou et al. 2022).

captured by large and complex pre-trained models and achieve the goal of detecting target categories from tweets.

This case study chose the Bidirectional Encoder Representations from Transformers (BERT) as the pre-trained LLM and the CNN as the classifier, referred to as the BERT-CNN model, suggested by Zhou et al. (2022). BERT is one of the latest pretrained language models for text understanding and significantly outperforms previous NLP models in numerous text analysis tasks (Devlin et al., 2018). A recent investigation has compared the performance of ten tweet classification models based on different pre-trained LLMs and classifiers (Zhou et al., 2022). The tested pre-trained language models include Global Vectors for Word Representation (GloVe), Embeddings from Language Models (ELMo), BERT and its modified versions (RoBERTa, DistilBERT, and ALBERT), and XLNet. Four classifier models were compared, including transformers, linear, non-linear, long short-term memory (LSTM) network, and CNN. The results demonstrated that the BERT-CNN model performed the best with its precision, recall, and F-1 score as 0.897, 0.933, and 0.919 (Zhou et al., 2022). The architecture of the BERT-CNN model adopted in this research is displayed in Figure 14.3b.

The training data were 8413 manually labeled tweets collected during Hurricanes Harvey and Irma in 2017. Hurricane Irma tweets were included in the training dataset to enhance the generalizability of the developed model. The dataset was comprised of 2461 tweets reporting humanitarian help needs, 193 help requests for animals, 687 messages reporting infrastructural damages, 229 posts sharing locations providing shelters or supplies, and 4843 other tweets. The dataset was randomly split into training, cross-validation, and testing data using a ratio of 7:1:2. The random state was stored to ensure the reproducibility of the model training process. A maximum of 20 training epochs were used with early stopping. The learning rate was initially set to 2e-5 for fine-tuning the model (Devlin et al., 2018). The cross-entropy loss function was applied, as it is commonly used for multiclass classification problems. The optimizer chosen to update the model weights was AdamW. To prevent the loss function from being trapped at a local optimum, a scheduler function was implemented to dynamically change the learning rate during training. Based on the testing dataset, the averaged precision, recall, and the F-1 score of the trained intelligent tweet classification model are 0.889, 0.886, and 0.887, respectively.

14.3.3 INTELLIGENT TOPONYM RECOGNITION MODEL

The third step geo-locates tweets, which can be achieved using three common attributes: geo-tags, user profile addresses, and text-mentioned locations. Despite geo-tags being the most reliable approach for determining user locations, only a small proportion (approximately 0.04%) of tweets contain geo-tags (Lin et al., 2022). Meanwhile, user profile addresses are usually at the city or coarser administrative levels and cannot predict user locations at hyperlocal scales. Instead, during hazard events, many users mentioned specific locations within their tweets when requesting rescue, reporting neighborhood needs or damages, or sharing shelter and supply information (Zou et al., 2023). Thus, it is imperative that these messages be included in spatial analysis to avoid information loss.

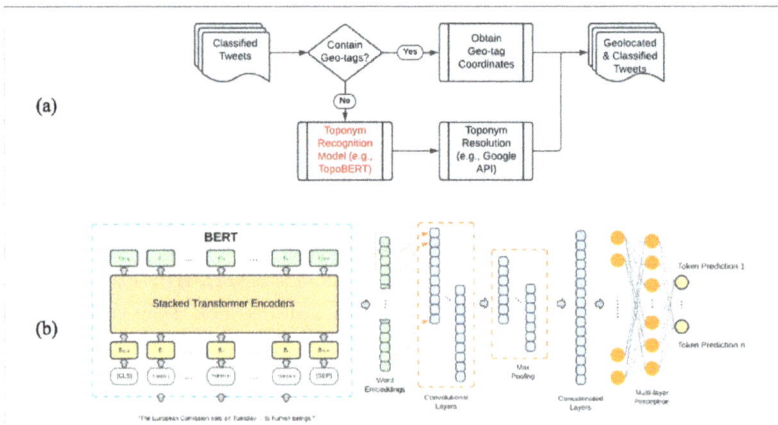

Figure 14.4   (a) The workflow of geolocating Twitter data; (b) the architecture of the TopoBERT model (*Source:* Zhou et al. 2023).

In this case study, both geo-tags and tweet-mentioned addresses were utilized to geolocate rescue and damage-related tweets, as depicted in Figure 14.4a. Geo-tagged Tweets were geolocated based on their geo-tag coordinates, while tweets without geo-tags were geolocated based on the tweet-mentioned addresses using toponym recognition and resolution tools. The process of recognizing place names from tweets (highlighted in red in Figure 14.4a) is an NLP task classifying words into place or non-place categories. This process can also benefit from the use of pre-trained LLMs. We employed a newly designed toponym recognition tool called TopoBERT, which is structured as a BERT model for language understanding with a one-dimensional

CNN model for word classification, as illustrated in Figure 14.4b (Zhou et al., 2023). The model was fine-tuned using four different open-source databases and achieved state-of-the-art performance with an F1-score of 0.865, outperforming the other five widely used baseline models by at least 18%, including Standford NER, SpaCy NER, BiLSTM-CRF, and NeuroTPR (Wang et al., 2020). More detailed information regarding the model training and validation can be found in Zhou et al. (2023). The toponym resolution was accomplished through the Google geocoding API.

14.3.4   SPATIAL ANALYSIS

The last step is to analyze the spatial patterns of geolocated tweets and interpret their values in disaster response and management practice. A total of 354,898 tweets were identified as one of the four categories, and 285,046 of them were successfully geolocated. Among the geolocated tweets, 106,104 (37.2%) are humanitarian rescue requests, 45,603 (16.0%) are posted asking for help for animals, 3576 (1.3%) are reports or complaints about infrastructural damages, and 129,763 (45.5%) are sharing information about shelters or supplies.

The geolocated tweets under the four categories that fall inside Texas were retained and visualized in Figure 14.5. Figure 14.5a shows the path of Hurricane Harvey and total tweet counts combining all four categories (referred to as total help-related tweets) at the county level. The counts were normalized by the population and the number indicates the tweet count per million people. Two clusters were identified. One centered in Aransas County (City of Rockport), where the hurricane made its first and second landfalls. The other one centered in Harris County (City of Houston), where the hurricane caused unprecedented precipitation and triggered severe flood.

To further reveal the geographical disparities of multi-dimensional impacts from Harvey on different communities and local needs, the tweeting frequencies per million people of the four categories were computed in the most affected areas at the county level, as displayed in Figure 14.2b, c, e, f, respectively. From Figure 14.5b, Harris County, Galveston County, and Jefferson County are the top 3 counties regarding normalized tweet counts which indicates that people in these areas might suffer from a lack of rescue forces and had to turn to social media to request help. Figure 14.5c displays slightly different patterns compared to Figure 14.5b and shows that Hardin County contains the greatest number of animal-related help requests, such as reports of lost animals and requesting rescue for abandoned animals. Figure 14.5e shows that Nueces County and Harris County are the two that stand out in terms of reports of infrastructure malfunction, while the rest of the counties exhibit few or no reports. This might be due to the fact that Hurricane Harvey triggered life-threatening damage that people tended to focus more on saving people's lives rather than reporting or complaining about infrastructure issues. Figure 14.2f shows that Harris County offers the largest amount of help or shelter information via Twitter.

The extraction of tweets not only allows for the identification of the impact of the disaster and the associated needs at the community level but also serves as a valuable guide for local disaster response efforts. Figure 14.5d illustrates the

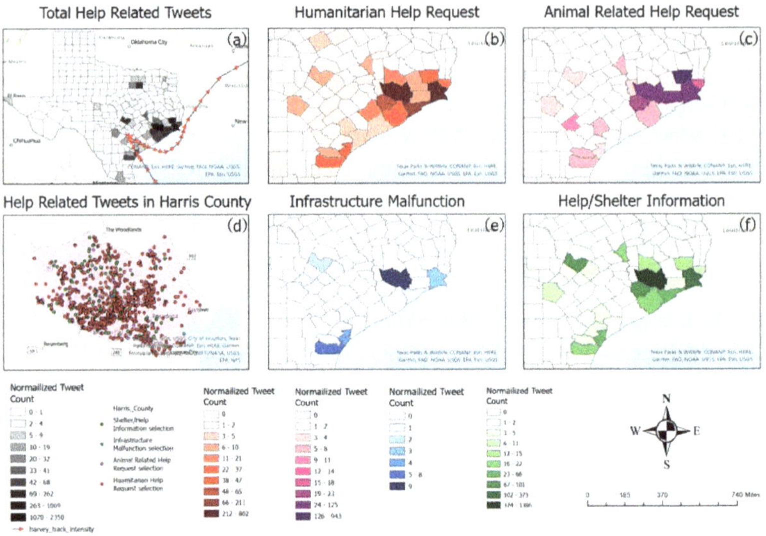

Figure 14.5   Spatial distributions of (a) all target tweets in Texas, (b) humanitarian rescue requests in affected areas, (c) animal-related help requests in affected areas, (d) all target tweet locations in Harris County, (e) reports of infrastructural malfunctions in affected areas, and (f) shelter and supplies tweets in affected areas.

geographic distribution of help-related tweets in all four categories within Harris County, depicted as points of varying colors. This real-time information can be utilized by first responders from governments and NGOs to promptly identify individuals and areas in need of assistance and to direct resources to those locations. Moreover, these organizations can also gather crucial data regarding local shelters and supplies that are not documented in official websites. Such timely information can be disseminated to the broader community residing in nearby regions that may be vulnerable to hazards but do not use social media during disasters.

Based on the data visualization and analysis of geolocated tweets, the most affected areas will be revealed so that policymakers can be more aware of the potential weaknesses of those areas and make future policies accordingly. In addition to visualizing the distribution of the data, several analyses can be conducted with extra supporting data. For example, a temporal analysis of tweet count will demonstrate the trend of social media usage and provide valuable insights into how and when society is impacted by hazardous events; correlation analysis with other data types such as social economic indices and damage reports can reveal their relationships and inform predictive models for rapid damage assessment; applying Natural Language Processing techniques like topic modeling in analyzing the content of the tweets can help researchers identify latent topics discussed or sentiment expressed in different regions so that community managers can prioritize most important issues in post-disaster recovery.

14.4 CHALLENGES AND OPPORTUNITIES

The paradigm proposed in this study, together with the presented case study, demonstrates the valuable potential of GeoAI and geospatial big data in improving disaster response efforts and contributing to the protection of communities and the preservation of human lives. Nevertheless, despite the numerous benefits that GeoAI provides, it is essential to consider several caveats when applying this technology.

14.4.1 DATA QUALITY AND AVAILABILITY

The foremost challenge facing the utilization of GeoAI in disaster response is the uncertainty surrounding data quality and availability, which hinders the creation of GeoAI models capable of addressing varying disaster scenarios and data inputs. GeoAI models require considerable amounts of high-quality geospatial training data to operate efficiently, which can be problematic in disaster response scenarios where data collection may be incomplete or difficult. This could be elucidated by the case study presented in Section 14.3. During the hurricane, not everyone in need of rescue sent a help request on Twitter, indicating that creating a GeoAI model to predict those in need of assistance using data harvested exclusively from Twitter may be biased and overlook digitally marginalized communities. Additionally, applying GeoAI models trained using data obtained from previous disaster events may overlook emerging disaster-related information that could be crucial for disaster response, but is absent in the training data.

One potential solution to this data challenge is to increase the quality, quantity, and diversity of training data used to create GeoAI models. Fortunately, the widespread use of open-source data and codes in the academic community has made this solution feasible. For example, the DesignSafe Data Depot (https://www.designsafe-ci.org/data/browser/) maintained by the Natural Hazards Engineering Research Infrastructure (NHERI) offers a shared online database of raw data and data products, including source codes and systems designed for disaster response and management. Through this portal, scholars can share or access valuable disaster-related data, source codes, and systems for developing and using GeoAI models for disaster research. Similarly, CrisisNLP (https://crisisnlp.qcri.org/) provides various manually labeled disaster-related text data voluntarily shared by researchers that can be used to create, compare, and improve NLP models for humanitarian aid-related text mining. These and numerous other open-source initiatives provide a valuable opportunity to reproduce, replicate, and refine GeoAI models to tackle the challenge of uncertain data quality and availability.

14.4.2 PRIVACY PROTECTION AND SECURITY

The second challenge is the concern of privacy protection and security when applying GeoAI in disaster response. The location-based nature of geospatial data presents a risk of exposing sensitive information about individuals, such as their home addresses or locations where they seek refuge during a disaster. The need of accelerating ethics in GIScience is also heighted in a recent publication (Nelson et al.,

2022). To address this issue, measures must be taken to protect personal identification information when training or applying GeoAI models. This could be solved by several methods, such as implementing encryption techniques and access controls, using anonymized data for model training, and adopting privacy-preserving techniques e.g., geomasking (Gao et al., 2019), when applying models for research purposes. For example, Zou et al. (2023) proposed a practical framework for using social media for emergency rescue, which highlights that only authorized users should access the information harvested from social media by GeoAI models. Another solution is to use federated learning techniques in training GeoAI models, where multiple parties can collaboratively train a model without sharing their raw data (Li et al., 2020). This is particularly useful when existing databases of rescue requests are held by different responding organizations that cannot directly share data due to privacy concerns.

14.4.3   TRUSTWORTHINESS OF GEOAI-BASED SOLUTIONS

The third challenge in using GeoAI for disaster response is ensuring the trustworthiness of GeoAI models and the results generated from them. The complexity and opacity of GeoAI models make it challenging to interpret the decision-making process, which can lead to potential biases or errors. Inaccurate or biased decisions can have severe consequences in disaster response scenarios. For instance, a misclassified rescue request may mean a life lost. To address this issue, several strategies can be implemented to promote the interpretability, transparency, and trustworthiness of GeoAI models. These strategies include incorporating explainable AI (XAI) techniques in practice to provide transparent and understandable explanations of modeled decisions, ensuring the transparency of data sources used for model training, adopting standards and guidelines when developing and deploying GeoAI models, and taking into account the concerns and perspectives of stakeholders in the building and using of GeoAI for disaster response. In addition to these strategies, leveraging advanced models such as GPT-based models for text mining can be an alternative approach to improve the accuracy and reliability of GeoAI solutions. By implementing these strategies, the trustworthiness of GeoAI models can be enhanced, enabling their effective use in disaster response scenarios.

14.4.4   PRACTICAL DEPLOYMENT AND MAINTENANCE

Finally, deploying and maintaining GeoAI-supported systems for disaster response is a significant practical challenge, mainly due to the previously mentioned obstacles. Successful implementation of GeoAI in disaster response requires collaboration among various stakeholders, including governmental organizations, NGOs, volunteers, data providers, researchers, and residents living in hazard-prone areas. Ideally, researchers should develop and continually upgrade GeoAI-based solutions for disaster response missions, incorporating feedback from disaster first responders to enhance the system's effectiveness. Disaster response organizations should integrate GeoAI-based solutions into their management system and maintain the system, which can be computationally intensive and require massive infrastructural support. During a hazard event, residents should have access to critical information on how to respond to disasters and seek assistance. At the same time, data providers should offer timely and accurate geospatial big data, while disaster practitioners analyze the data using GeoAI to identify critical information such as local needs and impacts. This information should then be shared securely with front-end first responders to ensure efficient allocation of personnel and resources. However, many disaster management practitioners lack the necessary infrastructural resources to implement and maintain GeoAI-based solutions in their disaster management systems. They also have limited technical expertise required to use GeoAI-based tools and outcomes for decision-making.

This challenge can be addressed through multiple pathways. Collaborative partnerships can help in sharing knowledge, expertise, and resources, resulting in improved decision-making in developing, maintaining, and applying GeoAI solutions, and, thus, effective response to disasters. The U.S. National Science Foundation (NSF) has launched several programs to support such collaborative partnerships, such as the Convergence Accelerator and the Civic Innovation Challenge. Meanwhile, training programs and capacity-building initiatives can be developed to build practitioners' skills and knowledge of applying GeoAI solutions in disaster response. Additionally, open-source, user-friendly GeoAI tools and platforms can be developed and shared to make it easier for non-experts to use GeoAI in disaster responses.

14.5 CONCLUSION

This chapter expounds on the roles of GeoAI in empowering smart disaster response through three objectives. Firstly, it proposes and explains a comprehensive paradigm of GeoAI for disaster response that specifies how GeoAI can extract disaster response-related information from multiple types of geospatial big data. Secondly, this chapter demonstrates the use of GeoAI in disaster response through a case study that analyzes social media data using advanced pre-trained language models. The case study employs two intelligent models, a classification model and a toponym recognition model, to classify social media data into four categories relevant to disaster response efforts and determine their locations. Both models leverage BERT, one of the most cutting-edge pre-trained language models, and achieved state-of-the-art performance with F1 scores of 0.919 and 0.865, respectively. The spatial-temporal patterns of the tweets of the four categories were analyzed to provide insights into the most damaged communities and the diverse impacts of the hurricane. Finally, this chapter discusses four challenges of practicing GeoAI in disaster response, including uncertain data quality and availability, privacy protection and security concerns, limited trustworthiness of GeoAI-based solutions, and difficulty in practical deployment and maintenance. Corresponding potential opportunities to resolve the four challenges are also proposed. Overall, the findings of this research will not only enhance the understanding of GeoAI and its critical role in disaster response but also highlight future research and practice opportunities in this field.

BIBLIOGRAPHY

British Geological Survey. (2023). ALARMS — *Assessment of Landslides using Acoustic Real-time Monitoring Systems*. British Geological Survey. Retrieved March 21, 2023, from https://www.bgs.ac.uk/geology-projects/geophysical-tomography/technologies/

Cai, H., N. S. Lam, Y. Qiang, L. Zou, R. M. Correll, and V. Mihunov (2018). A synthesis of disaster resilience measurement methods and indices. *International Journal of Disaster Risk Reduction 31*, 844–855.

Deng, H., D. P. Aldrich, M. M. Danziger, J. Gao, N. E. Phillips, S. P. Cornelius, and Q. R. Wang (2021). High-resolution human mobility data reveal race and wealth disparities in disaster evacuation patterns. *Humanities and Social Sciences Communications 8*(1), 1–8.

Devlin, J., M.-W. Chang, K. Lee, and K. Toutanova (2018). Bert: Pre-training of deep bidirectional transformers for language understanding. *arXiv preprint arXiv:1810.04805*.

Dong, E., H. Du, and L. Gardner (2020). An interactive web-based dashboard to track covid-19 in real time. *The Lancet Infectious Diseases 20*(5), 533–534.

Dong, L. and J. Shan (2013). A comprehensive review of earthquake-induced building damage detection with remote sensing techniques. *ISPRS Journal of Photogrammetry and Remote Sensing 84*, 85–99.

Gao, S., J. Rao, Y. Kang, Y. Liang, and J. Kruse (2020). Mapping county-level mobility pattern changes in the united states in response to covid-19. *SIGSpatial Special 12*(1), 16–26.

Gao, S., J. Rao, X. Liu, Y. Kang, Q. Huang, and J. App (2019). Exploring the effectiveness of geomasking techniques for protecting the geoprivacy of twitter users. *Journal of Spatial Information Science* (19), 105–129.

Hu, T., S. Wang, B. She, M. Zhang, X. Huang, J. Cui, J. Khuri, Y. Hu, X. Fu, X. Wang, et al. (2021). Human mobility data in the covid-19 pandemic: characteristics, applications, and challenges. *International Journal of Digital Earth 14*(9), 1126–1147.

Huang, X., C. Wang, and Z. Li (2018). A near real-time flood-mapping approach by integrating social media and post-event satellite imagery. *Annals of GIS 24*(2), 113–123.

Janowicz, K., S. Gao, G. McKenzie, Y. Hu, and B. Bhaduri (2020). Geoai: spatially explicit artificial intelligence techniques for geographic knowledge discovery and beyond. *International Journal of Geographical Information Science 34*(4), 625–636.

Jiang, Y., Z. Li, and S. L. Cutter (2019). Social network, activity space, sentiment, and evacuation: what can social media tell us? *Annals of the American Association of Geographers 109*(6), 1795–1810.

Kryvasheyeu, Y., H. Chen, N. Obradovich, E. Moro, P. Van Hentenryck, J. Fowler, and M. Cebrian (2016). Rapid assessment of disaster damage using social media activity. *Science Advances 2*(3), e1500779.

Lam, N. S.-N., Y. J. Xu, K.-B. Liu, D. E. Dismukes, M. Reams, R. K. Pace, Y. Qiang, S. Narra, K. Li, T. A. Bianchette, et al. (2018). Understanding the mississippi river delta as a coupled natural-human system: Research methods, challenges, and prospects. *Water 10*(8), 1054.

Lee, C.-C., M. Maron, and A. Mostafavi (2022). Community-scale big data reveals disparate impacts of the texas winter storm of 2021 and its managed power outage. *Humanities and Social Sciences Communications 9*(1), 1–12.

Li, L., Y. Fan, M. Tse, and K.-Y. Lin (2020). A review of applications in federated learning. *Computers & Industrial Engineering 149*, 106854.

Li, Z., C. Wang, C. T. Emrich, and D. Guo (2018). A novel approach to leveraging social media for rapid flood mapping: a case study of the 2015 south carolina floods. *Cartography and Geographic Information Science 45*(2), 97–110.

Lin, B., L. Zou, N. Duffield, A. Mostafavi, H. Cai, B. Zhou, J. Tao, M. Yang, D. Mandal, and J. Abedin (2022). Revealing the linguistic and geographical disparities of public awareness to covid-19 outbreak through social media. *International Journal of Digital Earth 15*(1), 868–889.

Martin, Y., Z. Li, and S. L. Cutter (2017). Leveraging twitter to gauge evacuation compliance: Spatiotemporal analysis of hurricane matthew. *PLoS One 12*(7), e0181701.

Mihunov, V. V., N. H. Jafari, K. Wang, N. S. Lam, and D. Govender (2022). Disaster impacts surveillance from social media with topic modeling and feature extraction: Case of hurricane harvey. *International Journal of Disaster Risk Science 13*(5), 729–742.

Mihunov, V. V., N. S. Lam, L. Zou, Z. Wang, and K. Wang (2020). Use of twitter in disaster rescue: lessons learned from hurricane harvey. *International Journal of Digital Earth 13*(12), 1454–1466.

Nelson, T., M. Goodchild, and D. Wright (2022). Accelerating ethics, empathy, and equity in geographic information science. *Proceedings of the National Academy of Sciences 119*(19), e2119967119.

Nex, F., D. Duarte, F. G. Tonolo, and N. Kerle (2019). Structural building damage detection with deep learning: Assessment of a state-of-the-art cnn in operational conditions. *Remote Sensing 11*(23), 2765.

Ray, P. P., M. Mukherjee, and L. Shu (2017). Internet of things for disaster management: State-of-the-art and prospects. *IEEE Access 5*, 18818–18835.

Simonyan, K. and A. Zisserman (2015). Very deep convolutional networks for large-scale image recognition. *arXiv preprint arXiv:1409.1556*.

Song, X., R. Shibasaki, N. J. Yuan, X. Xie, T. Li, and R. Adachi (2017). Deepmob: learning deep knowledge of human emergency behavior and mobility from big and heterogeneous data. *ACM Transactions on Information Systems (TOIS) 35*(4), 1–19.

Song, X., Q. Zhang, Y. Sekimoto, and R. Shibasaki (2014). Prediction of human emergency behavior and their mobility following large-scale disaster. In *Proceedings of the 20th ACM SIGKDD International Conference on Knowledge Discovery and Data Mining*, pp. 5–14.

Wang, J., Y. Hu, and K. Joseph (2020). Neurotpr: A neuro-net toponym recognition model for extracting locations from social media messages. *Transactions in GIS 24*(3), 719–735.

Wang, K., N. S. Lam, L. Zou, and V. Mihunov (2021). Twitter use in hurricane isaac and its implications for disaster resilience. *ISPRS International Journal of Geo-Information 10*(3), 116.

Yabe, T., Y. Zhang, and S. V. Ukkusuri (2020). Quantifying the economic impact of disasters on businesses using human mobility data: a bayesian causal inference approach. *EPJ Data Science 9*(1), 36.

Yuan, F. and R. Liu (2020). Mining social media data for rapid damage assessment during hurricane matthew: Feasibility study. *Journal of Computing in Civil Engineering 34*(3), 05020001.

Zhai, W. and Z.-R. Peng (2020). Damage assessment using google street view: Evidence from hurricane michael in mexico beach, florida. *Applied Geography 123*, 102252.

Zhang, B., Z. Chen, D. Peng, J. A. Benediktsson, B. Liu, L. Zou, J. Li, and A. Plaza (2019). Remotely sensed big data: Evolution in model development for information extraction [point of view]. *Proceedings of the IEEE 107*(12), 2294–2301.

Zhao, F. and C. Zhang (2020). Building damage evaluation from satellite imagery using deep learning. In *2020 IEEE 21st International Conference on Information Reuse and Integration for Data Science (IRI)*, pp. 82–89. IEEE.

Zhou, B., L. Zou, Q. Y. Hu, Yingjie, and D. Goldberg (2023). Topobert: Plug and play toponym recognition module harnessing fine-tuned bert. *arXiv preprint arXiv:2301.13631*.

Zhou, B., L. Zou, A. Mostafavi, B. Lin, M. Yang, N. Gharaibeh, H. Cai, J. Abedin, and D. Mandal (2022). Victimfinder: Harvesting rescue requests in disaster response from social media with bert. *Computers, Environment and Urban Systems 95*, 101824.

Zou, L., N. S. Lam, H. Cai, and Y. Qiang (2018). Mining twitter data for improved understanding of disaster resilience. *Annals of the American Association of Geographers 108*(5), 1422–1441.

Zou, L., N. S. Lam, S. Shams, H. Cai, M. A. Meyer, S. Yang, K. Lee, S.-J. Park, and M. A. Reams (2019). Social and geographical disparities in twitter use during hurricane harvey. *International Journal of Digital Earth 12*(11), 1300–1318.

Zou, L., D. Liao, N. S. Lam, M. A. Meyer, N. G. Gharaibeh, H. Cai, B. Zhou, and D. Li (2023). Social media for emergency rescue: An analysis of rescue requests on twitter during hurricane harvey. *International Journal of Disaster Risk Reduction 85*, 103513.

15 GeoAI for Public Health

Andreas Züfle
Emory University

Taylor Anderson, Hamdi Kavak, and Dieter Pfoser
George Mason University

Joon-Seok Kim
Oak Ridge National Laboratory

Amira Roess
George Mason University

CONTENTS

15.1 Introduction 305
15.2 Existing AI Models for Data-Driven Infectious Disease Spread Prediction . 306
15.3 Spatial and Temporal Heterogeneity of Health-Related Human Behavior ... 308
15.4 AI for Simulation of Health-Related Human Behavior in Agent-Based
     Models 311
15.5 A Use-Case of Using Agent-Based Modeling to Predict the Spread of an
     Infectious Disease in Fairfax, VA, USA 313
     15.5.1 Data Sets 313
     15.5.2 Population Initialization..314
     15.5.3 Representation of Disease Dynamics 314
     15.5.4 Foot Traffic Topic Extraction...316
     15.5.5 Dwell Time Distributions for POIs..317
     15.5.6 Simulation Results...318
15.6 Future Direction for GeoAI for Public Health: Prescriptive Analytics 319
     Bibliography..321

15.1 INTRODUCTION

Artificial intelligence (AI) models of infectious disease spread are critical decision-support systems to help explain the mechanisms of infectious disease spread, predict the number of cases and deaths, and prescribe effective policy guidelines. Ideally, by observing the *spatiotemporal behavior* of individuals, we can comprehend the uptake of preventative behaviors that help reduce transmission by wearing masks, reducing

DOI: 10.1201/9781003308423-15

mobility, limiting social interactions, or getting vaccinated. However, spatiotemporal human behavior is collectively shaped by the biological, environmental, cultural, sociological, and economic makeup of individuals and their spatial environments (Baldassare, 1978; Golledge, 1997). As a result, the degree to which individuals engage in preventative behaviors varies over space and time. Efforts to improve this understanding have been limited by the lack of reliable longitudinal spatiotemporal data containing observations that capture the change in human behavior in response to disease spread (Bharti, 2021).

This chapter summarizes some of the important findings of this community and identifies future research directions. We note that the term AI is used more generally in this chapter: Rather than narrowly focusing on neural networks we also include other models that extract patterns from large infectious disease spread related datasets, such as mechanistic models, agent-based models, regression models, and ensemble models. The remainder of this book chapter is organized as follows:

Section 15.2 provides a brief overview of data-driven approaches for infectious disease spread prediction before and after the COVID-19 pandemic.

Section 15.3 motivates the need for an AI solution in public health by exemplifying the heterogeneity of human behavior related to health in space and time. This heterogeneity means that health-related human behavior is difficult to capture by classic models that assume homogeneous human behavior.

Section 15.4 discusses research efforts toward leveraging AI to improve representations of human behavior in models of disease spread by leveraging AI for intelligent and realistic agent decision-making and using AI solutions for estimating simulation model parameters and patterns.

Section 15.5 describes a use case of using large-scale human mobility data and AI to realistically parameterize agent human mobility behaviors in an agent-based simulation of the 1.1 million population of a county in the United States.

Section 15.6 provides future research directions and ideas.

15.2 EXISTING AI MODELS FOR DATA-DRIVEN INFECTIOUS DISEASE SPREAD PREDICTION

Data-driven epidemic forecasting has been a very large research field in the last decade. A recent survey summarizes more than 300 publications in this field (Rodríguez et al., 2022). Even before the COVID-19 pandemic, this field was already a focus of the computing community (Marathe and Vullikanti, 2013) in the context of influenza-like illnesses (Adhikari et al., 2019). For example, ACM KDD has been organizing the International Workshop on Epidemiology meets Data Mining and Knowledge Discovery (EpiDaMiK) since 2018 (Adhikari et al., 2021, 2022). The COVID-19 pandemic has brought forth very large sets of human mobility data (Elarde et al., 2021; Gao et al., 2020; Qazi et al., 2020), which enabled new data-driven models. Existing data-driven models to predict the spread of infectious diseases include mechanistic models (Adiga et al., 2020), agent-based models (Pesavento et al., 2020; Venkatramanan et al., 2018), regression models (Ginsberg et al.,

2009), off-the-shelf sequential models (Volkova *et al.*, 2017), graph neural network models (Deng *et al.*, 2020; Wu *et al.*, 2018), density estimation models (Brooks *et al.*, 2018), ensemble models (Cramer *et al.*, 2022; Reich *et al.*, 2019), as well as many other types of models (Rodríguez *et al.*, 2022).

Propelled by the COVID-19 pandemic, researchers have sought interdisciplinary collaborations with researchers in epidemiology and the social sciences to fill this data availability gap and use data-driven approaches that seek to understand and model the relationship between spatiotemporal behavior and disease transmission. Here, we highlight three such collaborative efforts.

After the onset of the COVID-19 pandemic, the ACM SIGSPATIAL community focused its research efforts on understanding the spread of COVID-19 in two special issues on the topic (Züfle and Anderson, 2020; Züfle *et al.*, 2022b). In the SIGSPATIAL newsletter, Qazi et al. published a dataset of billions of tweets related to COVID-19 (Qazi *et al.*, 2020). This dataset was later updated in (Imran *et al.*, 2022). Gao *et al.* (2020) published a study on the change in human mobility during the COVID-19 pandemic with a dataset of human mobility during the pandemic based on SafeGraph foot-traffic data published in (Kang *et al.*, 2020). These two datasets were crucial to provide large human mobility datasets to understand the spread of COVID-19 and inform disease spread models. A mobility-informed epidemic simulation platform was published by Fan *et al.* (2020). Kim *et al.* (2020c) proposed a system to find a consensus among multiple simulations using an ensemble model. Bobashev *et al.* (2020) described an algorithm to rapidly detect COVID-19 clusters in space and time using a prospective space-time scan statistic and Bobashev et al. described a machine learning approach using mechanistic models for COVID-19 forecasting. Solutions for contact tracing were discussed in the context of privacy-preserving contact tracing (Xiong *et al.*, 2020) and an approach that moves the contact tracing functionality from individual users to facilities (Mokbel *et al.*, 2020).

The efforts of the newsletter led to dedicated workshops at the ACM SIGSPATIAL on Modeling and Understanding the Spread of COVID-19 (Anderson *et al.*, 2021b) and on Spatial Computing for Epidemiology (Anderson *et al.*, 2022, 2021a). Notable workshop publications related to AI and simulation include simulation-based infection risk estimation models (Agarwal and Banerjee, 2020; Kiamari *et al.*, 2020; Pechlivanoglou *et al.*, 2022), game-theory based COVID-19 simulation (Pejó and Biczók, 2020), a joint pandemic modeling and analysis platform (Thakur *et al.*, 2020), infectious disease case time series prediction (Aboubakr and Magdy, 2020; Susarla *et al.*, 2022), analysis of human behavior during the COVID-19 pandemic (Chen and McKenzie, 2021; Elsaka *et al.*, 2021; Ye and Gao, 2022), a tool for pandemic decision-making support (Lopes *et al.*, 2022), mapping and visualization of infectious disease data (Cabana *et al.*, 2022; Samet *et al.*, 2020), and real-time detection of COVID-19 clusters (Ajayakumar *et al.*, 2021).

The newsletter and workshops described above led to a special issue of ACM Transactions on Spatial Algorithms and Systems (TSAS) on Understanding the Spread of COVID-19 (Züfle *et al.*, 2022a). The main goal of this special issue was to facilitate interdisciplinary work that included social scientists and epidemiologists. Therefore, the call for papers read: *"This special issue intends to bring together transdisciplinary researchers and practitioners working in topics from multiple areas, including Spatial Data Scientists ... Mathematicians, Epidemiologists, Computational Social Scientists, Medical Practitioners, Psychologists, Emergency Response and Public safety, among others."* This special issue included extended versions of newsletter and workshop papers but also included many new research directions. Notable research directions toward understanding and simulating infectious diseases included infectious disease simulation (Azad *et al.*, 2022; Sydora *et al.*, 2022), infectious disease modeling (Burtner and Murray, 2022; Cardoso *et al.*, 2021; Coro and Bove, 2022; Lorch *et al.*, 2022), and infectious disease data analysis (Behera *et al.*, 2022; Fanticelli *et al.*, 2022; Mehrab *et al.*, 2021; Zakaria *et al.*, 2022).

15.3 SPATIAL AND TEMPORAL HETEROGENEITY OF HEALTH-RELATED HUMAN BEHAVIOR

Agent-based models (ABMs) are used to forecast disease spread trajectories and to support policymakers as they prepare for and respond to emerging and re-emerging infectious diseases (D'Orazio *et al.*, 2020; Hoertel *et al.*, 2020; Silva *et al.*, 2020). ABMs use a bottom-up approach to simulate disease dynamics among a population by representing the mobility, interactions, and subsequent transmission of infectious disease between individuals or "agents". ABMs expand upon traditional assumptions of the compartmental SIR model and its variations to include heterogeneity in the population, the spatial environment and the transmission likelihoods (Bian, 2004; Frias-Martinez *et al.*, 2011). Therefore, ABMs have been developed to simulate the spread of seasonal influenza (Anderson and Dragićević, 2020; Kim *et al.*, 2019, 2020b), H1N1 (Chao *et al.*, 2010; Halder *et al.*, 2010; Lee *et al.*, 2010), Ebola (Merler *et al.*, 2015; Siettos *et al.*, 2015), smallpox (Burke *et al.*, 2006), anthrax (Chen *et al.*, 2006), the pneumonic plague (Williams *et al.*, 2011), dengue (Karl *et al.*, 2014), and more recently COVID-19 (Hoertel *et al.*, 2020; Silva *et al.*, 2020).

However, despite significant advances in ABMs of disease spread, recent commentaries have pointed out that many still lack realistic representations of human behavior, a key driver of mobility, physical interaction, and subsequent disease transmission (Agusto *et al.*, 2020; Funk *et al.*, 2015; Holmdahl and Buckee, 2020; Manfredi and D'Onofrio, 2013; Ray *et al.*, 2020). Without this, ABMs are limited in making accurate forecasts, especially over longer prediction horizons. This limitation is best illustrated by examining some of the models included in the COVID-19 Forecast Hub (Ray *et al.*, 2020), a recent collaborative effort between the Center for Disease Control and Prevention (CDC) and more than 25 scientific teams, each of which produces a model to forecast the spread of COVID-19. Ray *et al.* (2020) report that, for each of the models, uncertainty increases and accuracy decreases in prediction horizons longer than just four weeks and concludes that to achieve accurate long-term forecasts, models must incorporate realistic human behavior.

Despite the number of social and behavioral studies that examine how individuals respond to pandemics and outbreaks of diseases (Gao *et al.*, 2020; Laato *et al.*, 2020; Rizzo *et al.*, 2013; Van Bavel *et al.*, 2020), ABMs of infectious disease spread do little to incorporate these findings (Funk *et al.*, 2015). Traditionally, ABMs ignore or oversimplify representations of human health behaviors. For example, it is common to exogenously impose behaviors upon agents. The modeler may compare disease outcomes in scenarios where 50% or 75% of agents are randomly selected as vaccinated. These approaches assume that human behavior is temporally stationary, meaning that the change in agent behavior over time is ignored, even as the risk of infection increases or new policies are implemented. Furthermore, many ABMs assume spatial stationarity, ignoring key spatial heterogeneities in human behavior produced by local influences related to social norms or culture (Bauch *et al.*, 2013; Funk *et al.*, 2010) Although work has been done to improve representations of human health behavior in ABMs (Agusto *et al.*, 2020; Barrett *et al.*, 2011; Del Valle *et al.*, 2013), the encoded behavioral response is based on broad conceptual theories such as game theory and economic objective functions rather than empirical observations. Furthermore, there has been limited effort to acknowledge the spatial variation of human behavior due to the individual's sociodemographic profile and their political, social, environmental, historical, and cultural contexts. This is particularly important in locations where many individuals may be less likely to comply with mitigation strategies, including stay-at-home orders, which has a significant impact on disease spread trajectories.

In the case of the COVID-19 pandemic, the *spatial variation* of the behavioral response has been empirically observed (Gao *et al.*, 2020; Weill *et al.*, 2020) using large sets of human mobility that have been made available during the COVID-19 pandemic that capture billions of tweets (Qazi *et al.*, 2020), foot traffic (Gao *et al.*, 2020), and other datasets (Mokbel *et al.*, 2022).

Figure 15.1 shows the variation in stay-at-home behavior in the US at the census block group level. Here, we can directly observe regional differences in the fraction of users who stay-at-home on any given day (Figure 15.1a). Figure 15.1b shows a common distinct spatial pattern of stay-at-home behavior where the fraction of users who stay home is lower in urban areas than in rural areas. Figure 15.1c and Figure 15.1d compare two cities with opposite patterns of stay-at-home behavior. Newark, California has a very low fraction of users who stay home, while Lake Charles, Louisiana has a very high fraction.

Although we can observe such spatial differences in behavior, it remains a challenge to understand and explain them sufficiently. Why do people in some places follow social distancing, mask use, and stay-at-home guidelines more strictly than in other places? Why do some support vaccination and not others? What are the driving factors of such behavior? Moreover, what can we do to increase participation in these regions in preventative behavior?

Recent studies have observed a covariate relationship between behavior and the socio-economic and political profile of a region (Gollwitzer *et al.*, 2020; Painter and Qiu, 2020; Weill *et al.*, 2020). Covariation is a measure of the correlated variation between two variables such that as one variable in a region increases (or decreases),

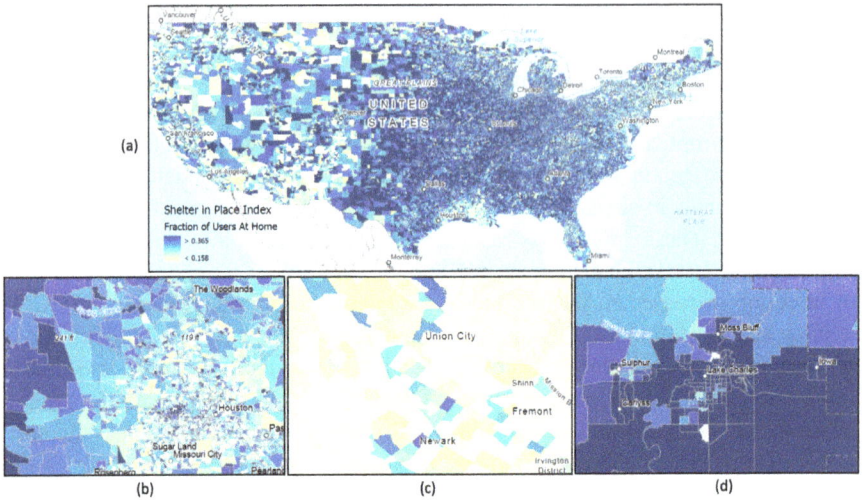

Figure 15.1 Geographic heterogeneity of stay-at-home behavior. (a) Fraction of stay-at-home users for each census block group across the US; (b) Urban and rural pattern of stay-at-home behavior near Houston, TX; (c) Region having a low fraction of stay-at-home users in Newark, CA; (d) Region having a high fraction of stay-at-home users in Lake Charles, LA.

the other also increases (or decreases). Covariate relationships often exhibit spatial trends since individuals who share sociodemographic and thus behavioral common-alities "cluster" together in terms of location. Examining how human behavior varies in geographic space is an emerging area of work in psychology (Hehman *et al.*, 2019, 2020).

Covariate relationships between the regional behavioral response to COVID-19 and the corresponding characteristics of those same regions have been explored at various spatial granularities. For example, strong correlations have been observed at the state level between social distancing behavior and income (Elarde *et al.*, 2021; Weill *et al.*, 2020), where wealthier states engage in stay-at-home behavior signifi-cantly more than lower-income states. Covariation between behavioral response and political tendencies has been observed at the county level, where individuals in Re-publican counties are less likely to stay home than residents in Democratic coun-ties (Elarde *et al.*, 2021; Gollwitzer *et al.*, 2020; Painter and Qiu, 2020). Covari-ation between the number of cases of COVID and behavior has been reported by an early Pew Research Center survey, where states with a higher number of cases have a more significant effect on behavior than states with a lower number of cases (Pew Research Center, 2000). Other variables, including age, race, gender, education, family structure, social connection, religion, and environment, are also likely to be important, but the spatial relationship between these variables and the behavioral re-sponse has not yet been quantified. For example, studies have found that young adults are more likely than adults over 30 to attend a party, restaurant, or small gathering.

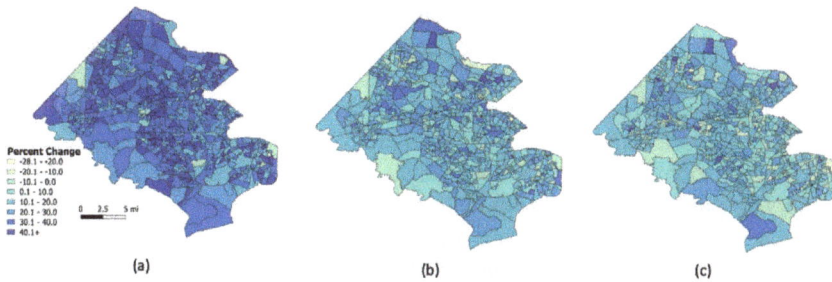

Figure 15.2 Temporal heterogeneity of stay-at-home behavior. (a) Percent change in stay-at-home behavior, February–April 2020; (b) Percent change in stay-at-home behavior, February–July 2020; (c) Percent change in stay-at-home behavior, February–October 2020.

Environmental factors such as temperature and weather are also key in shaping human behavior. Religious gatherings often result in close contact between large numbers of individuals.

In addition to spatial heterogeneity, the behavioral response to a disease outbreak can change over time. Early during the COVID-19 pandemic, individuals reported being uncomfortable with the idea of attending parties, eating at restaurants, voting, and even going to the grocery store, resulting in increased stay-at-home behavior and decreased mobility (Pew Research Center, 2000). As the disease persists, some individuals have maintained stay-at-home behavior and social distancing, while others' physical, financial or social needs may outweigh the risk of the virus. This behavior change can be directly observed by mapping mobility over time using *Safe-Graph's* foot traffic data (Figure 15.2). In the early stages of the pandemic, that is April 2020 (Figure 15.2a), the percentage of people who stay home increases compared to February, where February represents "normal mobility". As the pandemic progresses, mobility increases (Figures 15.2b and 15.2c) and, in some counties, even returns to normal, where the percent change is close to 0. It is unclear whether this is a function of the individuals' social and financial needs that begin to outweigh the risk of infection, the warmer temperatures that Fairfax County experienced in July and October, the changes in local and state-wide policies over time, or a combination of all of the above.

15.4 AI FOR SIMULATION OF HEALTH-RELATED HUMAN BEHAVIOR IN AGENT-BASED MODELS

More recently, especially in light of the COVID-19 pandemic, the modeling community has worked to improve representations of human behavior in models of disease spread by leveraging AI. As summarized by Brearcliffe and Crooks (2021), AI in combination with ABM is used primarily as follows: (1) *to derive behavioral parameter values and patterns* for an ABM to inform dynamics such as human mobility (Kavak *et al.*, 2018; Lamperti *et al.*, 2018; Pesavento *et al.*, 2020) and (2) *as an internal agent decision framework* ranging from simple reactions to complex learning-based decision-making (Ramchandani *et al.*, 2017). AI for agent behavior has been integrated in broader applications of ABM and also in applications to ABMs that are used to simulate the spread of diseases.

ABMs of disease spread typically incorporate agents that decide where they want to go (agent mobility) and how they might adjust their behavior, including their mobility, to prevent disease transmission. Health behaviors commonly modeled include vaccination, social distancing, staying at home, and general preventative behavior (usually framed as a combination of several protective behaviors).

As described in the book "Artificial Intelligence" by Russell and Norvig (2005), there are five types of intelligent agents, increasing in complexity, as follows: reflex agents, model-based reflex agent, goal-based agent, utility-based agent, and learning agent. With each type of agent, behavior and decision-making is self-initiated, meaning that behavior is endogeneously based on the agent's perception of current, past, and/or future state of the world which is either complete or limited. This differs from the vast majority of infectious disease models that impose behaviors on agents. Reflex agents, the least complex, have limited intelligence. Agent's simply receive an input, and, using some condition-action rule, will select an action to follow. For example, if someone in the agent's household is sick, then there is some probability that the agent will stay home (Lukens *et al.*, 2014; Oruc *et al.*, 2021).

Model-based reflex agents perceive the current state of the world and understand how the world evolves, including how the agent's own actions affect the world. Based on this perception, agents use some condition-action to select an action. For example, in Mao (2014), agents perceive information about how to protect themselves from disease, the adoption of protective behaviors among their social networks, and the prevalence of local and global disease prevalence. Based on both the fraction of agents in their social network that adopt protective behavior and the fraction that is infected with the disease, agents will decide whether or not to also adopt.

The agent's perception of the world is not always enough to make decisions. For example, a goal-based agent with the goal of moving through a room while maintaining social distancing may have several options in routes to their target (Sajjadi *et al.*, 2021). The evaluation of possible routes is binary, meaning that the route either does or does not satisfy the agent's goal. For example, some routes may have points that are too crowded to maintain social distancing, and thus taking these routes to get to the destination will not satisfy the agent's goal. Where goals provide a binary distinction between satisfied and unsatisfied states, economists and computer scientists use utility or "the quality of being useful" as a way to measure the degree of satisfaction that each new potential location would give. The agent will then choose the action that maximizes it's utility or payoff. For example, Nardin *et al.* (2016) use an SPIR (susceptible, prophylactic, infectious, recovered) and a rational choice model to represent the choice to adopt prophylactic behavior (handwashing or wearing a face mask) or not. Payoff, rather than utility as a performance measure, is traditionally used to play the "vaccination game" (Fu *et al.*, 2011), where agents make decisions based on game theory. In this game, vaccines provide immunity against disease. Each individual who gets infected pays the cost of infection, and each individual who decides to get vaccinated pays the cost of vaccination. Thus, the cost paid by a "free-rider" who does not get vaccinated or get infected, is zero. The agent weighs costs with risk of infection based on the number of social connections they have (who might infect them) and how many of those connections have decided to opt for vaccination.

Learning agents are the most complex and adjust their actions over time. For example, Tanaka and Tanimoto (2020) use a strategy updating function to incorporate learning into the vaccination game, where the agent learns from their local or global neighbors about which strategy used in previous years might have the best reward in the current year. Where most of these types of agents implement rule-based intelligence, learning agents sometimes leverage machine learning algorithms such as neural networks, Bayesian networks, reinforcement learning, and genetic algorithms to drive agent behaviors (Abdulkareem *et al.*, 2019). Outside of disease ABMs, machine learning has been used to simulate agent residential migration decisions (Sert *et al.*, 2020), spatial optimization of land use allocation (Vallejo *et al.*, 2013), land market decisions (Abdulkareem, 2019) and conflict management (Brearcliffe and Crooks, 2021). However, the use of these algorithms for health-related behavior in ABMs of disease spread is somewhat limited. In one example, Abdulkareem *et al.* (2019) uses survey data and Bayesian networks to train and guide agent behavior to simulate risk perception and response to the cholera outbreak. Fuzzy cognitive maps (FCM) model decision-making as a system by means of concepts (perceptions of individual health state, local and global disease prevalence, memory, etc.) connected by cause and effect relationships. The FCM can be developed manually or can be learned from data. One such learning method is based on a linear Hebbian learning (NHL) method. Fuzzy cognitive maps have been developed to simulate the protective behaviors of individuals in disease scenarios (López *et al.*, 2020; Mei *et al.*, 2013).

15.5   A USE-CASE OF USING AGENT-BASED MODELING TO PREDICT THE SPREAD OF AN INFECTIOUS DISEASE IN FAIRFAX, VA, USA

This section describes an example agent-based simulation for the spread of an infectious disease using a synthetic population of 1.1 million agents in the county of Fairfax, VA, USA. The following sections describe the datasets used to inform the simulation, how data was abstracted to be digested by the simulation, and the simulation results. Details of this simulation can be found in (Pesavento *et al.*, 2020).

15.5.1   DATA SETS

The goal of this use case is to develop an ABM of disease spread based on real-world human mobility patterns rather than relying on simplified and often unrealistic assumptions.

Data from *SafeGraph Inc.*, which aggregates anonymized location data from numerous applications to provide insights into physical presence in places. To enhance privacy, *SafeGraph* aggregates home locations to the census block group level and excludes locations if fewer than five devices visited a POI in a month from a given census block group. *SafeGraph* provides unique and valuable insight into foot-traffic patterns of large-scale businesses and consumer POIs. This work uses *SafeGraph's* "Weekly Patterns" data, which register GPS-identified visits to POIs (primarily businesses) with an exact location in the United States. For each visit by an individual to a POI, the home census block group (derived from nighttime GPS location) is recorded (For detailed information, see `https://docs.safegraph.com/docs/weekly-patterns`). Additionally, *SafeGraph* provides a taxonomy of POIs types in a "Core Places" schema, allowing our simulation to test the closure of specific business categories (e.g., restaurants). *SafeGraph* also includes information on the proportion of residents who stay home or leave the house on any given day for each CBG in a separate "Social Distancing Metrics" dataset, allowing us to establish the probabilities of agents leaving their home at the CBG level.

Due to the sheer size of the datasets, all of the data we used was filtered to include only POI and CBG data from Fairfax County, Virginia. We chose to use data from the week that spanned October 28 to November 3, 2019, as a representative sample of typical movement patterns before the onset of COVID-19. We filter POIs only to include those with a large enough sample of aggregate visitors (30 or more) throughout the week-long period. The filtered dataset resulted in 4,130 unique POIs in Fairfax County and 689,731 recorded visits to these POIs.

We also used United States Census data (`https://www.census.gov`) to map the CBGs to their correct geographic locations. This data also facilitates the initialization of agents and agent households.

15.5.2 POPULATION INITIALIZATION

To initialize our simulation, we first generate households according to CBG-level data provided by the US Census, filling each household with its corresponding number of agents. Between one and seven agents are assigned to each household, with "7-or-more person households" being treated as size seven for simplicity. By default, we simulate approximately 10% of the total population of Fairfax County by only generating 10% of households of each size in each CBG, resulting in a simulation of 106,978 agents. In addition to infection status, agents and households are not assigned any other attributes such as age, income, or race. A small percentage (25%) of agents from a single randomly selected CBG are initially infected, resulting in a default of 26 initially infected agents. For consistency, this CBG's *SafeGraph* ID is *510594804023* for all of our trials. We used the integer *1* as the seed for the pseudorandom number generator in all of our trials for reproducibility.

15.5.3 REPRESENTATION OF DISEASE DYNAMICS

Once the agent population initialization is complete, the simulation begins at midnight and runs until the agents are no longer exposed or infected. Each tick in the model represents fifteen minutes, according to the CDC definition of close contact between individuals (CDC, 2020). The probability that an agent will leave their home location to visit a POI on any given day is based on the *SafeGraph's* "Social

Distancing" dataset. We divide the total daily number of people who did not stay home by the total daily number of people in the CBG between October 28 and November 3, 2019. We calculate that the average daily POI visit probability across each CBG in Fairfax County is 74.8%. To roughly approximate the likelihood that an agent would leave the house at each tick, we divide this probability by the number of ticks in a day (96 by default), resulting in a 0.780% average probability. This finding is also consistent with other travel surveys and mobile mobility studies based on cell phones (Schneider *et al.*, 2013). We consider this probability, calculated based on foot traffic data acquired before the onset of COVID-19 and thus not influenced by the pandemic, as a default parameter of a 100% propensity to leave.

At each 15 minute tick, infectious agents may come into contact with a maximum of five other agents, by default, who are located at the same POI that is not their household. If a susceptible agent comes into contact with an infectious agent, they have a 5% chance of being exposed and subsequently infectious by default (Klompas *et al.*, 2020).

Infectious agents also have the opportunity to spread the virus to susceptible agents in their home. Research indicates that approximately 20.4% of people living in small households (size six or less) will contract the virus if they share a residence with someone infected (Jing *et al.*, 2020). This percentage decreases to 9.1% in large households (size seven or larger). Using these numbers and the median infectious period of the virus according to the Gamma distributions that we use, as described below, we approximated that susceptible agents have a chance of 4.44% and 1.98% of contracting the virus from an infected household member each day in small and large households, respectively. For simplicity, household infection occurs at midnight each day, even if a household member is visiting a POI.

We represent the dynamics of COVID-19 using a generalized SEIR model (Aron and Schwartz, 1984) that is modified to include subclinical, preclinical and clinical subclasses of the infectious stage. Agents undergo the following stages:

- **Susceptible**: An agent who has never been infected or exposed to the virus but has the potential to become exposed.

- **Exposed**: An agent who has caught the virus and will become contagious (infectious) after an incubation period.

- **Infectious**: An agent that can infect others and is contagious. We define three subclasses of infectious agents:

  Subclinical: An asymptomatic infected agent. It is estimated that 40% of infections are subclinical. As these agents will never show symptoms, they are estimated have a 75% relative infectiousness compared to clinical agents.

  Preclinical: An infected agent who is presymptomatic (not currently symptomatic), but will enter the clinical stage and become symptomatic in the future. All agents entering the clinical stage first pass through the preclinical stage. As preclinical agents do not show symptoms, they are also estimated to have a 75% relative infectiousness compared to clinical agents.

Clinical: An infected agent that shows symptoms of the virus and is fully infectious. It is estimated that the remaining 60% infections progress to the clinical stage.

- **Recovered**: A previously infected agent that is non-contagious and immune to the virus. An agent is classified as recovered as long as they cannot actively spread the virus, even if they have lasting complications or symptoms.

The duration of each stage in the SEIR model in days is determined by drawing from the following gamma distributions (Davies *et al.*, 2020):

- Exposed stage duration: $gamma(\mu = 3.0, k = 4)$
- Subclinical stage duration: $gamma(\mu = 5, k = 4)$
- Preclinical stage duration: $gamma(\mu = 2.1, k = 4)$
- Clinical stage duration: $gamma(\mu = 2.9, k = 4)$

15.5.4   FOOT TRAFFIC TOPIC EXTRACTION

15.5.4.1   Latent Topic Modeling

Given the CBG foot traffic data for each POI as obtained from the SafeGraph data, we apply topic modeling using LDA (Blei *et al.*, 2003) – a generative probabilistic model. Although traditionally used to find K latent topics among a corpus of M text documents containing N words per document, we use LDA in our simulation to find K latent topics among a subset of M CBGs each containing N distinct POI visits. This modeling of CBGs as documents and POIs as words allows us to efficiently generate realistic new POI visits of individuals at a CGB. In using the LDA approach, each CBG's POI visits are a mixture of underlying latent topics, and each topic has a latent distribution of more and less likely POIs. In that respect, LDA provides two distributions: (1) a topic probability distribution for each visitor home CBG and (2) a POI probability distribution for each topic. These two distributions allow an ABM to be constructed such that (1) agents are generated and assigned a specific LDA topic according to the topic probability distribution of their home CBG and that (2) agents visit POIs based on the POI probability distribution of their assigned topic.

We are now able to generate agents with specific attributes based on their home CBG, providing the foundation for our ABM. First, for each home CBG in Fairfax County that *SafeGraph* provides data for, agents are generated according to the real population of the CBG, each being assigned a topic according to the first distribution. Agents' topics are static and cannot change throughout the simulation.

After agents are assigned topics, we may randomly sample from the second distribution to determine which POI agents will visit if they decide to leave their house. Due to LDA's nature, it is unlikely that two unrelated POIs, such as a nightclub and a library, will have relatively equal weights in this probability distribution, resulting in a vast improvement from the uniform probability distribution used to select POIs in traditional ABMs. However, this probability distribution does not provide information on whether or not an agent will decide to visit a POI in the first place. Details of the LDA model can be found in (Pesavento *et al.*, 2020).

15.5.4.2   Modeling Hourly Visit Patterns

So far, LDA has generated distributions from data that contain the number of visitors from each CBG to each POI over an entire day. However, this approach is slightly flawed because, in reality, the number of visits is dependent on the time of day. For example, a restaurant would likely have higher concentrations of visits at noon and in the evening and lower concentrations during the mid-afternoon. Similarly, visits to a school POI during the evening would be less likely. To remedy this issue, we use additional *SafeGraph* data that provides individual POI visits for each hour over the entire week-long timeframe and create 24 distinct POI probability distributions for every topic, one for each hour of the day. We do this by reweighting the topic's base POI probability distribution 24 times according to each POI's proportion of visits during the given hour. For each topic, a weighted distribution of visits that take place each hour is constructed according to the topic's POI distribution and the number of visits to those POIs that take place during the given hour. This distribution provides the weighted percentage of visits to the topic that takes place in the given hour compared to the entire day. For example, the midnight hour may have a probability of 0.01, while the noon hour may have a probability of 0.07. By modifying the likelihood, we allow simulating agent POI visits more accurately. Given our simulation's default parameters for the hour of noon example, an agent's chance to visit a POI between 12:15 and 12:30 would be approximately $0.748 * 0.07/4$, or 1.31%, markedly higher than the generic average probability of leaving each tick of 0.780%.

15.5.5   DWELL TIME DISTRIBUTIONS FOR POIs

The *SafeGraph* data allows us to use a data-driven approach to modeling dwell time by fitting a suitable probability distribution to a POI's bucketed dwell time data (provided by *SafeGraph*) and random sampling the said distribution for every agent that "visits" the POI.

15.5.5.1   SafeGraph Bucketed Dwell Time Data

SafeGraph provides a "bucketed" version of each POI's dwell times, where only the number of visits within a range of dwell time is quantified, i.e., "<5 minutes": 266, "5–20 minutes": 4184, "21–60 minutes": 3597, "61–240 minutes": 2492, ">240 minutes": 892. While providing an initial perspective on the potential probability of dwell times for an individual agent's visit to a POI, the raw *SafeGraph* data is not adequate for direct usage due to their nonspecificity.

15.5.5.2   Fitting Probability Distributions

To compensate for the bucketed format of the data ("5–20 minutes":4184), we first impute the bucket ranges of each POI's dwell data by random uniform sampling: for a range of 5–20 minutes with 266 visits, we fill the bucket with 266 random uniform

samples ranging from 5–20. With a full range of dwell data for each bucket, we move
on to methods of sampling. We determine that employing probability distributions
allows for the most optimal method of a random sample due to their "smoothing"
of minor irregularities that may occur from the random uniform imputation of each
POI's dwell time buckets. From these we approximate using the parametric function
with the best fit for each POI dwell time distribution. For example, restaurants might
have a more normal distribution around a mean stay time of 1 hour. Compared to
malls, where a large proportion might drop off or drive by with a large proportion of
visits under 5 minutes, so it might be better represented using an exponential curve.
For each POI, we test the fit of 10 of SciPy's most common probability distributions
for a continuous random variable – normal, generalized extreme value, exponential,
gamma, Pareto, lognormal, double Weibull, beta, Student's t, uniform—and select
the most optimal based on the goodness-of-fit test. See (Taskesen, 2019) for more
information on the probability distributions used and the package used. We initialize
and cache the fitted distribution for the POI in question; the distribution is randomly
sampled five times for each agent visit to said POI. The median is returned, representing an estimated *SafeGraph* data-based dwell time. If the median dwell time is
less than one tick, then one tick is returned. Alternatively, if the median dwell time is
greater than 16 hours, then 16 hours is returned. The median dwell time is rounded
to the nearest tick in all other cases.

15.5.6 SIMULATION RESULTS

Various experiments are implemented to test the impact of various public health interventions on the spread of COVID-19 and the subsequent effect on the epidemic
curve.

15.5.6.1 Generic Quarantine

This intervention requires any agent that is infectious and aware of symptoms (in the
subclinical stage or beyond) to stay home for a specific number of days. Figure 15.3a
shows the infection curves for zero, four, six and ten days of quarantine. We observe
that a quarantine of four days is sufficient to drastically flatten the number of infections per day from 2000. However, since 40% of the agents are subclinical and
unaware of their infection, these agents are not quarantined and continue to spread
the disease.

15.5.6.2 Household Quarantine

This intervention additionally requires that all agents that share the same home as
the quarantined agent remain at home. This intervention is significantly more effective, further flattening the curve to about 1500 infections per day (Figure 15.3b). This
result is intuitive, as agents living in the same household are at highest risk of becoming infected. We also observe that a longer quarantine duration is significantly
more effective. Specifically, agents are quarantined before they show symptoms, thus
dropping the infections per day to 1,000 using a ten-day quarantine.

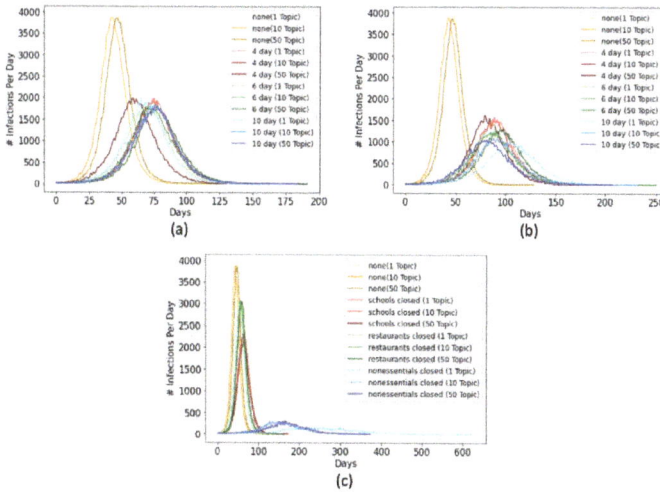

Figure 15.3   Disease spread simulation results after prescribing interventions: (a) generic quarantine; (b) household quarantine; (c) closure of POIs.

15.5.6.3   Closure of POIs

We tested three interventions, each aimed at closing a specific type of POI including schools, restaurants, and non-essential places. Figure 15.3c presents the effect of POI closure on the epidemic curve. Instead of visiting the closed POI, the agent will decide to stay at home. We observe that closing restaurants yields a significant reduction of disease spread from a disease peak of 4000 agents per day down to 3000 agents per day. Interestingly, the closure of schools is far more important, reducing the peak to about 2000. This is likely due to the difference in dwell time between schools and restaurants. As agents typically dwell at school POIs for up to eight hours a day, it becomes extremely likely that during any 15-minute tick they are successfully exposed to the virus by an infectious agent. In contrast, dwell times at restaurants are usually less than an hour (this also includes fast food restaurants), drastically reducing the probability of becoming exposed by a collocated infectious agent. We also see that if we choose to close all POIs that are classified as non-essentials (using the POI classification provided by SafeGraph), we observe that the disease is nearly eradicated.

15.6   FUTURE DIRECTION FOR GEOAI FOR PUBLIC HEALTH: PRESCRIPTIVE ANALYTICS

Existing research in GeoAI excels at predictive analytics, such as predicting road traffic (Pan *et al.*, 2019; Zhao *et al.*, 2019), rental bike flow (Chai *et al.*, 2018; Li *et al.*, 2015), and foot traffic (Islam *et al.*, 2021; Wang *et al.*, 2017). Such solutions excel at finding spatiotemporal patterns to optimize predictions, but lack an understanding of causality that allows one to investigate "what-if" scenarios.

For GeoAI to become useful in public health scenarios, we need to go beyond prediction analytics toward prescriptive analytics (Frazzetto *et al.*, 2019; Kim *et al.*, 2020b). Rather than predicting a variable of interest (such as the number of cases of an infectious disease), the goal of prescriptive analytics is to prescribe optimal actions and policies to optimize the variable of interest (such as minimizing the number of cases). Examples of such actions and policies are social distancing measures and the closure of businesses. Prescriptive analytics use machine learning approaches to find the optimal combination of actions and policies that achieve a desired outcome and have been successfully applied in management science and business analytics (Bertsimas and Kallus, 2020; Lepenioti *et al.*, 2020). A good example of prescriptive analytics for public health was presented at the Prescriptive Analytics for the Physical World (PAPW 2020) workshop held in conjunction with ACM KDD 2020 (`https://prescriptive-analytics.github.io`). The workshop featured a programming challenge in which teams would prescribe policies to mitigate the spread of an infectious disease in a simulated scenario. Teams would be able to observe infectious (but not exposed) agents in the simulation and prescribe interventions such as isolation, quarantine, and hospitalization of agents. Teams were then evaluated using a score weighted by the resulting number of infections but also weighted by the severity of interventions. The top-ranked solutions used diverse machine learning solutions to find optimal policies to minimize infections while also minimizing the severity of interventions (Dong *et al.*, 2020; Kim *et al.*, 2020a; Rizzo, 2020). Unfortunately, these machine learning solutions are not applicable to real-world disease spread, as the simulation used in this challenge was overly simple and assumed that all simulated agents behaved the same way and visited the same locations with the same probability.

To use prescriptive analytics for infectious disease mitigation in the real-world, we require a realistic digital twin of a city, county, or even the entire world. Then we could use this simulation to investigate optimal prescriptions (actions or interventions) to mitigate disease outbreaks. Work toward such a realistic simulation has to be interdisciplinary. Social scientists ensure realistic human behavior in a simulation, epidemiologists ensure realistic infectious disease spread for emerging pathogens, and computer scientists leverage efficient algorithms to allow the simulation to scale to large populations despite high social and epidemiologic complexity.

Such prescriptions may include specific strategies aimed at isolation, quarantine, vaccination of individuals, closure of sites, working from home, and usage of masks. The union of all prescriptions injected into a single simulation is called a policy and each policy injected into a simulation creates a "what-if" simulation outcome called a "possible world". Such policies may be constrained. For example, at a specific time, only 1000 vaccines may be produced per day, such that the goal is to find the optimal set of people to vaccinate, or the goal might minimize disease spread by closing sites but constrained to an economic loss of no more than $100M$ for a study region.

Once the space of injectable actions and the constraints to define viable policies are defined, GeoAI solutions may be leveraged to search for optimal policies.

This can be done by running a massive number of simulations and mining patterns, such as disease hot-spots, from the simulations. Such data mining approaches can identify the different locations that may benefit from the same optimal policies. This approach would allow us to explore the causality between different variables and behavior further. For example, we may find that similar policies are effective in places with similar political or religious beliefs. We can classification models to build supervised models that allow us to predict, for a given place and given mobility and behavior data, what the most effective sets of policies will be. This will give us a broader understanding of how different policies exhibit different degrees of effectiveness among different populations and places. Having experts in epidemiology and policy makers (such as local health departments) in the loop, this optimization may help find optimal policies for expert-defined applications such as the closure of businesses, lock-down, or vaccination rollout.

BIBLIOGRAPHY

Abdulkareem, S.A., 2019. *Enhancing Agent-Based Models with Artificial Intelligence for Complex Decision Making*. Thesis (PhD). University of Twente.

Abdulkareem, S.A., *et al.*, 2019. Bayesian networks for spatial learning: a workflow on using limited survey data for intelligent learning in spatial agent-based models. *Geoinformatica*, 23 (2), 243–268.

Aboubakr, H.A. and Magdy, A., 2020. On improving toll accuracy for COVID-like epidemics in underserved communities using user-generated data. *In*: *Proceedings of the 1st ACM SIGSPATIAL International Workshop on Modeling and Understanding the Spread of COVID-19*. 32–35.

Adhikari, B., *et al.*, 2021. The 4th international workshop on epidemiology meets data mining and knowledge discovery (epidamik 4.0@ kdd2021). *In*: *Proceedings of the 27th ACM SIGKDD Conference on Knowledge Discovery & Data Mining*. 4104–4105.

Adhikari, B., *et al.*, 2019. Epideep: Exploiting embeddings for epidemic forecasting. *In*: *Proceedings of the 25th ACM SIGKDD International Conference on Knowledge Discovery & Data Mining*. 577–586.

Adhikari, B., *et al.*, 2022. epidamik 5.0: The 5th international workshop on epidemiology meets data mining and knowledge discovery. *In*: *Proceedings of the 28th ACM SIGKDD Conference on Knowledge Discovery and Data Mining*. 4850–4851.

Adiga, A., *et al.*, 2020. Mathematical models for COVID-19 pandemic: a comparative analysis. *Journal of the Indian Institute of Science*, 100 (4), 793–807.

Agarwal, R. and Banerjee, A., 2020. Infection risk score: Identifying the risk of infection propagation based on human contact. *In*: *Proceedings of the 1st ACM SIGSPATIAL International Workshop on Modeling and Understanding the Spread of COVID-19*. 1–10.

Agusto, F.B., *et al.*, 2020. To isolate or not to isolate: The impact of changing behavior on COVID-19 transmission. *medRxiv*.

Ajayakumar, J., Curtis, A., and Curtis, J., 2021. A clustering environment for real-time tracking and analysis of COVID-19 case clusters. *In*: *Proceedings of the 2nd ACM SIGSPATIAL International Workshop on Spatial Computing for Epidemiology (SpatialEpi 2021)*. 1–9.

Alam, M., Tanaka, M., and Tanimoto, J., 2019. A game theoretic approach to discuss the positive secondary effect of vaccination scheme in an infinite and well-mixed population. *Chaos, Solitons & Fractals*, 125, 201–213.

Anderson, T. and Dragićević, S., 2020. Neat approach for testing and validation of geospatial network agent-based model processes: case study of influenza spread. *International Journal of Geographical Information Science*, 1–30.

Anderson, T., et al., 2022. *Proceedings of the 3rd ACM SIGSPATIAL International Workshop on Spatial Computing for Epidemiology (SpatialEpi 2022)*. ACM.

Anderson, T., et al., 2021a. *Proceedings of the 2nd ACM SIGSPATIAL International Workshop on Spatial Computing for Epidemiology (SpatialEpi 2021)*. ACM.

Anderson, T., Yu, J., and Züfle, A., 2021b. The 1st acm sigspatial international workshop on modeling and understanding the spread of COVID-19. *SIGSPATIAL Special*, 12 (3), 35–40.

Aron, J.L. and Schwartz, I.B., 1984. Seasonality and period-doubling bifurcations in an epidemic model. *Journal of Theoretical Biology*, 110 (4), 665–679.

Azad, F.T., et al., 2022. Sirtem: Spatially informed rapid testing for epidemic modeling and response to COVID-19. *ACM Transactions on Spatial Algorithms and Systems*, 8 (4), 1–43.

Baldassare, M., 1978. Human spatial behavior. *Annual Review of Sociology*, 4 (1), 29–56.

Barrett, C., et al., 2011. Economic and social impact of influenza mitigation strategies by demographic class. *Epidemics*, 3 (1), 19–31.

Bauch, C., d'Onofrio, A., and Manfredi, P., 2013. Behavioral epidemiology of infectious diseases: an overview. *In: Modeling the Interplay Between Human Behavior and the Spread of Infectious Diseases*. Springer, 1–19.

Behera, S., Dogra, D.P., and Satpathy, M., 2022. Effect of migrant labourer inflow on the early spread of COVID-19 in odisha: A case study. *ACM Transactions on Spatial Algorithms and Systems*, 8 (4), 1–18.

Bertsimas, D. and Kallus, N., 2020. From predictive to prescriptive analytics. *Management Science*, 66 (3), 1025–1044.

Bharti, N., 2021. Linking human behaviors and infectious diseases. *Proceedings of the National Academy of Sciences*, 118 (11).

Bian, L., 2004. A conceptual framework for an individual-based spatially explicit epidemiological model. *Environment and Planning B: Planning and Design*, 31 (3), 381–395.

Blei, D.M., Ng, A.Y., and Jordan, M.I., 2003. Latent dirichlet allocation. *Journal of Machine Learning Research*, 3 (Jan), 993–1022.

Bobashev, G., et al., 2020. Geospatial forecasting of COVID-19 spread and risk of reaching hospital capacity. *SIGSPATIAL Special*, 12 (2), 25–32.

Bossert, A., et al., 2020. Limited containment options of COVID-19 outbreak revealed by regional agent-based simulations for south africa. *arXiv preprint arXiv:2004.05513*.

Brearcliffe, D.K. and Crooks, A., 2021. Creating intelligent agents: Combining agent-based modeling with machine learning. *In: Proceedings of the 2020 Conference of The Computational Social Science Society of the Americas*. Springer, 31–58.

Brooks, L.C., *et al.*, 2018. Nonmechanistic forecasts of seasonal influenza with iterative one-week-ahead distributions. *PLoS Computational Biology*, 14 (6), e1006134.

Burke, D.S., *et al.*, 2006. Individual-based computational modeling of smallpox epidemic control strategies. *Academic Emergency Medicine*, 13 (11), 1142–1149.

Burtner, S. and Murray, A.T., 2022. Covid-19 and minimizing micro-spatial interactions. *ACM Transactions on Spatial Algorithms and Systems*, 8 (3), 1–17.

Cabana, E., *et al.*, 2022. Using mobile network data to color epidemic risk maps. *In: Proceedings of the 3rd ACM SIGSPATIAL International Workshop on Spatial Computing for Epidemiology.* 35–44.

Cardoso, M., *et al.*, 2021. Modeling the geospatial evolution of COVID-19 using spatio-temporal convolutional sequence-to-sequence neural networks. *ACM Transactions on Spatial Systems and Algorithms.*

CDC, 2020. Contact tracing for COVID-19. *Centers for Disease Control and Prevention.*

Chai, D., Wang, L., and Yang, Q., 2018. Bike flow prediction with multi-graph convolutional networks. *In: Proceedings of the 26th ACM SIGSPATIAL International Conference on Advances in Geographic Information Systems.* 397–400.

Chao, D.L., *et al.*, 2010. Flute, a publicly available stochastic influenza epidemic simulation model. *PLoS Comput Biol*, 6 (1), e1000656.

Chen, E. and McKenzie, G., 2021. Mobility response to COVID-19-related restrictions in new york city. *In: Proceedings of the 2nd ACM SIGSPATIAL International Workshop on Spatial Computing for Epidemiology (SpatialEpi 2021).* 10–13.

Chen, L.C., *et al.*, 2006. Model alignment of anthrax attack simulations. *Decision Support Systems*, 41 (3), 654–668.

Coro, G. and Bove, P., 2022. A high-resolution global-scale model for COVID-19 infection rate. *ACM Transactions on Spatial Algorithms and Systems*, 8 (3), 1–24.

Cramer, E.Y., *et al.*, 2022. Evaluation of individual and ensemble probabilistic forecasts of COVID-19 mortality in the united states. *Proceedings of the National Academy of Sciences*, 119 (15), e2113561119.

Davies, N.G., *et al.*, 2020. Age-dependent effects in the transmission and control of COVID-19 epidemics. *MedRxiv.*

Del Valle, S.Y., Mniszewski, S.M., and Hyman, J.M., 2013. Modeling the impact of behavior changes on the spread of pandemic influenza. *In: Modeling the Interplay Between Human Behavior and the Spread of Infectious Diseases.* Springer, 59–77.

Deng, S., *et al.*, 2020. Cola-gnn: Cross-location attention based graph neural networks for long-term ili prediction. *In: Proceedings of the 29th ACM International Conference on Information & Knowledge Management.* 245–254.

Dignum, F., *et al.*, 2020. Analysing the combined health, social and economic impacts of the corovanvirus pandemic using agent-based social simulation. *arXiv preprint arXiv:2004.12809.*

Dong, Y., Yu, C., and Xia, L., 2020. Hierarchical reinforcement learning for epidemics intervention. *In: First International KDD Workshop for Prescriptive Analytics for the Physical World (PAPW 2020).*

D'Orazio, M., Bernardini, G., and Quagliarini, E., 2020. How to restart? an agent-based simulation model towards the definition of strategies for COVID-19" second phase" in public buildings. *arXiv preprint arXiv:2004.12927*.

Elarde, J., *et al.*, 2021. Change of human mobility during COVID-19: A united states case study. *PloS One*, 16 (11), e0259031.

Elsaka, T., *et al.*, 2021. Correlation analysis of spatio-temporal arabic COVID-19 tweets. *In: Proceedings of the 2nd ACM SIGSPATIAL International Workshop on Spatial Computing for Epidemiology (SpatialEpi 2021)*. 14–17.

Fan, Z., *et al.*, 2020. Human mobility based individual-level epidemic simulation platform. *SIGSPATIAL Special*, 12 (1), 34–40.

Fan, Z., *et al.*, 2022. Human mobility-based individual-level epidemic simulation platform. *ACM Transactions on Spatial Algorithms and Systems*, 8 (3), 1–16.

Fanticelli, H.C., *et al.*, 2022. Data-driven mobility analysis and modeling: Typical and confined life of a metropolitan population. *ACM Transactions on Spatial Systems and Algorithms*.

Frazzetto, D., *et al.*, 2019. Prescriptive analytics: a survey of emerging trends and technologies. *The VLDB Journal*, 28 (4), 575–595.

Frias-Martinez, E., Williamson, G., and Frias-Martinez, V., 2011. An agent-based model of epidemic spread using human mobility and social network information. *In: 2011 IEEE Third International Conference on Social Computing*. IEEE, 57–64.

Fu, F., *et al.*, 2011. Imitation dynamics of vaccination behaviour on social networks. *Proceedings of the Royal Society B: Biological Sciences*, 278 (1702), 42–49.

Funk, S., *et al.*, 2015. Nine challenges in incorporating the dynamics of behaviour in infectious diseases models. *Epidemics*, 10, 21–25.

Funk, S., Salathé, M., and Jansen, V.A., 2010. Modelling the influence of human behaviour on the spread of infectious diseases: a review. *Journal of the Royal Society Interface*, 7 (50), 1247–1256.

Gao, S., *et al.*, 2020. Mapping county-level mobility pattern changes in the united states in response to COVID-19. *SIGSPATIAL Special*, 12 (1), 16–26.

Ginsberg, J., *et al.*, 2009. Detecting influenza epidemics using search engine query data. *Nature*, 457 (7232), 1012–1014.

Golledge, R.G., 1997. *Spatial Behavior: A Geographic Perspective*. Guilford Press.

Gollwitzer, A., *et al.*, 2020. Partisan differences in physical distancing are linked to health outcomes during the COVID-19 pandemic. *Nature Human Behaviour*, 1–12.

Halder, N., Kelso, J.K., and Milne, G.J., 2010. Developing guidelines for school closure interventions to be used during a future influenza pandemic. *BMC Infectious Diseases*, 10 (1), 221.

Hehman, E., *et al.*, 2019. Establishing construct validity evidence for regional measures of explicit and implicit racial bias. *Journal of Experimental Psychology: General*, 148 (6), 1022.

Hehman, E., Ofosu, E.K., and Calanchini, J., 2020. Using environmental features to maximize prediction of regional intergroup bias. *Social Psychological and Personality Science*, 1948550620909775.

Hoertel, N., *et al.*, 2020. A stochastic agent-based model of the sars-cov-2 epidemic in france. *Nature Medicine*, 1–5.

Holmdahl, I. and Buckee, C., 2020. Wrong but useful—what COVID-19 epidemiologic models can and cannot tell us. *New England Journal of Medicine*.

Imran, M., Qazi, U., and Ofli, F., 2022. Tbcov: two billion multilingual COVID-19 tweets with sentiment, entity, geo, and gender labels. *Data*, 7 (1), 8.

Islam, S., *et al.*, 2021. Spatiotemporal prediction of foot traffic. *In: Proceedings of the 5th ACM SIGSPATIAL International Workshop on Location-Based Recommendations, Geosocial Networks and Geoadvertising*. 1–8.

Jing, Q.L., *et al.*, 2020. Household secondary attack rate of COVID-19 and associated determinants in Guangzhou, China: a retrospective cohort study. *The Lancet Infectious Diseases*.

Kang, Y., *et al.*, 2020. Multiscale dynamic human mobility flow dataset in the us during the COVID-19 epidemic. *Scientific Data*, 7 (1), 1–13.

Karl, S., *et al.*, 2014. A spatial simulation model for dengue virus infection in urban areas. *BMC Infectious Diseases*, 14 (1), 1–17.

Kavak, H., *et al.*, 2018. Big data, agents, and machine learning: towards a data-driven agent-based modeling approach. *In: Proceedings of the Annual Simulation Symposium*. 1–12.

Kiamari, M., *et al.*, 2020. Covid-19 risk estimation using a time-varying sir-model. *In: Proceedings of the 1st ACM SIGSPATIAL International Workshop on Modeling and Understanding the Spread of COVID-19*. 36–42.

Kim, J.S., Jin, H., and Züfle, A., 2020a. Expert-in-the-loop prescriptive analytics using mobility intervention for epidemics. *In: First International KDD Workshop for Prescriptive Analytics for the Physical World (PAPW 2020)*.

Kim, J.S., *et al.*, 2019. Advancing simulation experimentation capabilities with runtime interventions. *In: SpringSim 2019*. IEEE, 1–11.

Kim, J.S., *et al.*, 2020b. Location-based social simulation for prescriptive analytics of disease spread. *SIGSPATIAL Special*, 12 (1), 53–61.

Kim, J.S., *et al.*, 2020c. Covid-19 ensemble models using representative clustering. *SIGSPATIAL Special*, 12 (2), 33–41.

Klompas, M., Baker, M.A., and Rhee, C., 2020. Airborne transmission of sars-cov-2: theoretical considerations and available evidence. *JAMA*.

Laato, S., *et al.*, 2020. Unusual purchasing behavior during the early stages of the COVID-19 pandemic: The stimulus-organism-response approach. *Journal of Retailing and Consumer Services*, 57, 102224.

Lamperti, F., Roventini, A., and Sani, A., 2018. Agent-based model calibration using machine learning surrogates. *Journal of Economic Dynamics and Control*, 90, 366–389.

Lee, B.Y., *et al.*, 2010. A computer simulation of vaccine prioritization, allocation, and rationing during the 2009 h1n1 influenza pandemic. *Vaccine*, 28 (31), 4875–4879.

Lepenioti, K., *et al.*, 2020. Prescriptive analytics: Literature review and research challenges. *International Journal of Information Management*, 50, 57–70.

Li, Y., *et al.*, 2015. Traffic prediction in a bike-sharing system. *In: Proceedings of the 23rd SIGSPATIAL International Conference on Advances in Geographic Information Systems*. 1–10.

Lopes, G.R., *et al.*, 2022. Multimaps: a tool for decision-making support in the analyzes of multiple epidemics. *In: Proceedings of the 3rd ACM SIGSPATIAL International Workshop on Spatial Computing for Epidemiology*. 22–25.

López, L., Fernández, M., and Giovanini, L., 2020. Influenza epidemic model using dynamic social networks of individuals with cognition maps. *MethodsX*, 7, 101030.

Lorch, L., *et al.*, 2022. Quantifying the effects of contact tracing, testing, and containment measures in the presence of infection hotspots. *ACM Transactions on Spatial Algorithms and Systems*, 8 (4), 1–28.

Lukens, S., *et al.*, 2014. A large-scale immuno-epidemiological simulation of influenza a epidemics. *BMC Public Health*, 14 (1), 1–15.

Manfredi, P. and D'Onofrio, A., 2013. *Modeling the Interplay Between Human Behavior and the Spread of Infectious Diseases*. Springer Science & Business Media.

Mao, L., 2014. Modeling triple-diffusions of infectious diseases, information, and preventive behaviors through a metropolitan social network—an agent-based simulation. *Applied Geography*, 50, 31–39.

Marathe, M. and Vullikanti, A.K.S., 2013. Computational epidemiology. *Communications of the ACM*, 56 (7), 88–96.

Mehrab, Z., *et al.*, 2021. Evaluating the utility of high-resolution proximity metrics in predicting the spread of COVID-19. *ACM Transactions on Spatial Systems and Algorithms*.

Mei, S., *et al.*, 2013. Individual decision making can drive epidemics: a fuzzy cognitive map study. *IEEE Transactions on Fuzzy Systems*, 22 (2), 264–273.

Merler, S., *et al.*, 2015. Spatiotemporal spread of the 2014 outbreak of ebola virus disease in liberia and the effectiveness of non-pharmaceutical interventions: a computational modelling analysis. *The Lancet Infectious Diseases*, 15 (2), 204–211.

Mokbel, M., Abbar, S., and Stanojevic, R., 2020. Contact tracing: Beyond the apps. *SIGSPATIAL Special*, 12 (2), 15–24.

Mokbel, M., *et al.*, 2022. Mobility data science (dagstuhl seminar 22021). *In: Dagstuhl reports*. Schloss Dagstuhl-Leibniz-Zentrum für Informatik.

Nardin, L.G., *et al.*, 2016. Planning horizon affects prophylactic decision-making and epidemic dynamics. *PeerJ*, 4, e2678.

Oruc, B.E., *et al.*, 2021. Homebound by covid19: the benefits and consequences of non-pharmaceutical intervention strategies. *BMC Public Health*, 21 (1), 1–8.

Painter, M. and Qiu, T., 2020. Political beliefs affect compliance with COVID-19 social distancing orders. *Available at SSRN 3569098*.

Pan, Z., *et al.*, 2019. Urban traffic prediction from spatio-temporal data using deep meta learning. *In: Proceedings of the 25th ACM SIGKDD International Conference on Knowledge Discovery & Data Mining*. 1720–1730.

Pechlivanoglou, T., *et al.*, 2022. Microscopic modeling of spatiotemporal epidemic dynamics. *In: Proceedings of the 3rd ACM SIGSPATIAL International Workshop on Spatial Computing for Epidemiology*. 11–21.

Pejó, B. and Biczók, G., 2020. Corona games: Masks, social distancing and mechanism design. *In: Proceedings of the 1st ACM SIGSPATIAL International Workshop on Modeling and Understanding the Spread of COVID-19*. 24–31.

Pesavento, J., *et al.*, 2020. Data-driven mobility models for COVID-19 simulation. *In: Proceedings of the 3rd ACM SIGSPATIAL Workshop on Advances in Resilient and Intelligent Cities, Seattle, WA, USA*.

Pew Research Center, 2000. Most Americans Say Coronavirus Outbreak Has Impacted Their Lives, March 2020.

Qazi, U., Imran, M., and Ofli, F., 2020. Geocov19: a dataset of hundreds of millions of multilingual COVID-19 tweets with location information. *SIGSPATIAL Special*, 12 (1), 6–15.

Ramchandani, P., Paich, M., and Rao, A., 2017. Incorporating learning into decision making in agent based models. *In: EPIA Conference on Artificial Intelligence*. Springer, 789–800.

Ray, E.L., *et al.*, 2020. Ensemble forecasts of coronavirus disease 2019 (COVID-19) in the us. *medRxiv*.

Reich, N.G., *et al.*, 2019. A collaborative multiyear, multimodel assessment of seasonal influenza forecasting in the united states. *Proceedings of the National Academy of Sciences*, 116 (8), 3146–3154.

Rizzo, C., *et al.*, 2013. Survey on the likely behavioural changes of the general public in four european countries during the 2009/2010 pandemic. *In: Modeling the Interplay Between Human Behavior and the Spread of Infectious Diseases*. Springer, 23–41.

Rizzo, S.G., 2020. Balancing precision and recall for cost-effective epidemic containment. *In: First International KDD Workshop for Prescriptive Analytics for the Physical World (PAPW 2020)*.

Rodríguez, A., *et al.*, 2022. Data-centric epidemic forecasting: A survey. *arXiv preprint arXiv:2207.09370*.

Russell, S. and Norvig, P., 2005. Ai a modern approach. *Learning*, 2 (3), 4.

Sajjadi, S., Hashemi, A., and Ghanbarnejad, F., 2021. Social distancing in pedestrian dynamics and its effect on disease spreading. *Physical Review E*, 104 (1), 014313.

Samet, H., *et al.*, 2020. Using animation to visualize spatio-temporal varying COVID-19 data. *In: Proceedings of the 1st ACM SIGSPATIAL International Workshop on Modeling and Understanding the Spread of COVID-19*. 53–62.

Schneider, C.M., *et al.*, 2013. Unravelling daily human mobility motifs. *Journal of The Royal Society Interface*, 10 (84), 20130246.

Sert, E., Bar-Yam, Y., and Morales, A.J., 2020. Segregation dynamics with reinforcement learning and agent based modeling. *Scientific Reports*, 10 (1), 1–12.

Siettos, C., *et al.*, 2015. Modeling the 2014 ebola virus epidemic–agent-based simulations, temporal analysis and future predictions for liberia and sierra leone. *PLoS Currents*, 7.

Silva, P.C., *et al.*, 2020. Covid-abs: An agent-based model of COVID-19 epidemic to simulate health and economic effects of social distancing interventions. *Chaos, Solitons & Fractals*, 139, 110088.

Susarla, A., *et al.*, 2022. Spatiotemporal disease case prediction using contrastive predictive coding. *In: Proceedings of the 3rd ACM SIGSPATIAL International Workshop on Spatial Computing for Epidemiology*. 26–34.

Sydora, C., *et al.*, 2022. Building occupancy simulation and analysis under virus scenarios. *ACM Transactions on Spatial Algorithms and Systems*, 8 (3), 1–20.

Tanaka, M. and Tanimoto, J., 2020. Is subsidizing vaccination with hub agent priority policy really meaningful to suppress disease spreading? *Journal of Theoretical Biology*, 486, 110059.

Taskesen, E., 2019. distfit. https://github.com/erdogant/distfit.

Thakur, G., *et al.*, 2020. Covid-19 joint pandemic modeling and analysis platform. In: *Proceedings of the 1st ACM SIGSPATIAL International Workshop on Modeling and Understanding the Spread of COVID-19*. 43–52.

Vallejo, M., Corne, D.W., and Rieser, V., 2013. Evolving urbanisation policies-using a statistical model to accelerate optimisation over agent-based simulations. In: *ICAART (2)*. Citeseer, 171–181.

Van Bavel, J.J., *et al.*, 2020. Using social and behavioural science to support COVID-19 pandemic response. *Nature Human Behaviour*, 1–12.

Venkatramanan, S., *et al.*, 2018. Using data-driven agent-based models for forecasting emerging infectious diseases. *Epidemics*, 22, 43–49.

Volkova, S., *et al.*, 2017. Forecasting influenza-like illness dynamics for military populations using neural networks and social media. *PloS One*, 12 (12), e0188941.

Wang, X., *et al.*, 2017. Predicting the city foot traffic with pedestrian sensor data. In: *Proceedings of the 14th EAI International Conference on Mobile and Ubiquitous Systems: Computing, Networking and Services*. 1–10.

Wang, Z. and Cruz, I.F., 2020. Analysis of the impact of COVID-19 on education based on geotagged twitter. In: *Proceedings of the 1st ACM SIGSPATIAL International Workshop on Modeling and Understanding the Spread of COVID-19*. 15–23.

Weill, J.A., *et al.*, 2020. Social distancing responses to COVID-19 emergency declarations strongly differentiated by income. *Proceedings of the National Academy of Sciences*, 117 (33), 19658–19660.

Williams, A.D., *et al.*, 2011. An individual-based simulation of pneumonic plague transmission following an outbreak and the significance of intervention compliance. *Epidemics*, 3 (2), 95–102.

Wu, Y., *et al.*, 2018. Deep learning for epidemiological predictions. In: *The 41st International ACM SIGIR Conference on Research & Development in Information Retrieval*. 1085–1088.

Xiong, L., *et al.*, 2020. React: Real-time contact tracing and risk monitoring using privacy-enhanced mobile tracking. *SIGSPATIAL Special*, 12 (2), 3–14.

Ye, W. and Gao, S., 2022. Understanding the spatiotemporal heterogeneities in the associations between COVID-19 infections and both human mobility and close contacts in the united states. In: *Proceedings of the 3rd ACM SIGSPATIAL International Workshop on Spatial Computing for Epidemiology*. 1–9.

Zakaria, C., *et al.*, 2022. Analyzing the impact of COVID-19 control policies on campus occupancy and mobility via wifi sensing. *ACM Transactions on Spatial Algorithms and Systems*, 8 (3), 1–26.

Zhao, L., *et al.*, 2019. T-gcn: A temporal graph convolutional network for traffic prediction. *IEEE Transactions on Intelligent Transportation Systems*, 21 (9), 3848–3858.

Züfle, A. and Anderson, T., 2020. Introduction to this special issue: Modeling and understanding the spread of COVID-19: (part i). *SIGSPATIAL Special*, 12 (1), 1–2.

Züfle, A., Anderson, T., and Gao, S., 2022a. Introduction to the special issue on understanding the spread of COVID-19, part 1.

Züfle, A., Gao, S., and Anderson, T., 2022b. Introduction to the special issue on understanding the spread of COVID-19, part 2.

16 GeoAI for Agriculture

Chishan Zhang, Chunyuan Diao, and Tianci Guo
Department of Geography and Geographic Information Science,
University of Illinois at Urbana-Champaign

CONTENTS

16.1 Introduction 330
16.2 Conceptual Framework of Crop Yield Estimation 332
    16.2.1 Preparation of Geospatial Modeling Inputs.....................................332
    16.2.2 GeoAI-based Yield Estimation Model..334
    16.2.3 Feature Importance Analysis 336
    16.2.4 Uncertainty Analysis 337
16.3 Case Study on Soybean Yield Estimation 338
    16.3.1 Study Site...338
    16.3.2 Data..339
    16.3.3 Method..340
    16.3.4 Results and Discussion 343
16.4 Conclusion..345
    Bibliography..346

16.1 INTRODUCTION

The rising demand for food caused by the planet's population boom, climate change, depletion of natural resources, changes in dietary preferences, as well as safety and health concerns brings global challenges to modern agriculture (Benos *et al.*, 2020; Conrad *et al.*, 2018; Nassani *et al.*, 2019; Thayer *et al.*, 2020). Due to environmental, agronomic, and economic concerns, more sustainable agricultural practices are urgently needed to increase agricultural production while minimizing the environmental footprint (García Pereira *et al.*, 2020). The integration of artificial intelligence (AI) and geographic information system (GIS) has led to the emergence of Geospatial Artificial Intelligence (GeoAI), which is being increasingly applied in smart agriculture for building more sustainable agricultural management systems (VoPham *et al.*, 2018).

In the last decade, the advances in remote sensing have largely facilitated the growth of GeoAI in agriculture. Earth has been increasingly monitored by sensors aboard satellites, manned aircraft, unmanned aerial vehicles (UAVs), and ground systems in unprecedented ways, with rich geospatial data acquired at various spatial,

DOI: 10.1201/9781003308423-16

temporal, and spectral resolutions. For example, the two Sentinel-2 satellites provide 10-60 m imagery of global coverage with 13 spectral bands around every five days (Drusch *et al.*, 2012), while the Terra & Aqua Moderate Resolution Imaging Spectroradiometer (MODIS) provides 250-1000 m global monitoring imagery of 36 spectral bands with an average revisit period of 2 days (Barnes *et al.*, 2003). With the increasing volumes and types of geospatial big data, GeoAI has been employed in a number of agricultural management applications, including crop management, water management, and soil management. GeoAI-based yield prediction (Khaki and Wang, 2019), disease (Anagnostis *et al.*, 2021; Zhang *et al.*, 2021) and weed detection (Islam *et al.*, 2021), as well as crop recognition (Zhang *et al.*, 2020), can provide a wide range of crop biological, chemical, physical and environmental information for smart crop management. Furthermore, GeoAI-based estimation of evapotranspiration (Feng *et al.*, 2017; Patil and Deka, 2016) and daily dew point temperature (Mohammadi *et al.*, 2015) facilitate more efficient water management and sustainable crop production. Accurate estimation of soil properties such as soil dryness (Coopersmith *et al.*, 2014), condition (Morellos *et al.*, 2016), temperature (Nahvi *et al.*, 2016), and moisture (Johann *et al.*, 2016) have also benefited from the development of GeoAI.

Among all these agricultural applications, yield estimation is one of the most important and challenging topics in modern agriculture (Benos *et al.*, 2020), because of the growing concern about food security. Early crop yield estimation plays an important role in reducing famine by estimating the food availability for the growing world population (Archontoulis *et al.*, 2020; Ziliani *et al.*, 2022). Policymakers rely on accurate estimation to make timely import and export decisions to enhance national food security. Seed companies need to predict how new hybrids perform in various environments to breed better hybrids. Farmers can also benefit from yield estimation to make informed management and financial decisions. Current mainstream crop yield estimation models include process-based crop methods and GeoAI-based data-driven models. Crop models simulate the complex soil-crop-atmosphere processes to model the crop yield formation. A large amount of field observations regarding soil characteristics, weather conditions, and management practices (e.g., sowing date, fertilization rate, and irrigation scheme) are usually required to calibrate the underlying physical processes. However, these observations might not always be accessible or be of adequate quality in practice. By contrast, GeoAI-based data-driven models leverage geospatial data (e.g., satellite imagery) alongside AI techniques to build the direct empirical relationships between crop yields and satellite-based as well as environmental factors (e.g., satellite-based vegetation index, meteorological conditions, and soil conditions), which may overcome the calibration issues. Given the complexity and non-linearity of the empirical relationships, AI models, particularly machine learning (ML) and deep learning (DL) models, have been increasingly utilized in empirical crop yield estimation. Some widely used GeoAI-based yield estimation models include Random Forest (RF), Support Vector Regression (SVR), Deep Neural Network (DNN), Convolutional Neural Network (CNN), and Long-Short Term Memory (LSTM) (Barbosa *et al.*, 2020; Kattenborn *et al.*, 2021; Murugananth *et al.*, 2022; Van Klompenburg *et al.*, 2020). The GIS techniques are further employed during the estimation process to develop spatial models, analyze yield spatial patterns and anomalies, and visualize yield estimation maps.

In the following sections, we will introduce the conceptual framework of crop yield estimation using GeoAI (section 16.2). The framework mainly comprises preparation of geospatial modeling inputs, GeoAI-based yield estimation models, and feature importance and uncertainty analysis. We then introduce a case study on crop yield estimation and uncertainty analysis using GeoAI in section 16.3.

16.2 CONCEPTUAL FRAMEWORK OF CROP YIELD ESTIMATION

16.2.1 PREPARATION OF GEOSPATIAL MODELING INPUTS

The preparation of geospatial inputs of yield predictors is the first step to build the empirical relationships for yield estimation using GeoAI-based models. Considering the diverse characteristics of yield predictors, the geospatial input data for crop yield modeling could usually be classified as satellite data, meteorological data, soil data, and others (Figure 16.1).

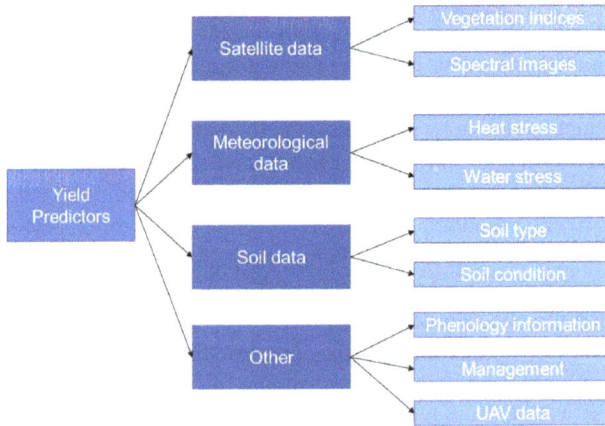

Figure 16.1   Geospatial modeling input diagram for crop yield estimation.

Vegetation Indices (VIs) derived from satellite observations usually have significant correlations with vegetation growth and biomass accumulation, and have been increasingly utilized for estimating crop yield. Normalized Difference Vegetation Index (NDVI) and Enhanced Vegetation Index (EVI) are two commonly used vegetation indices, calculated as Eqs. 16.1 and 16.2:

$$NDVI = \frac{NIR - Red}{NIR + Red} \tag{16.1}$$

$$EVI = 2.5 \times \frac{NIR - Red}{NIR + c_1 \times Red - c_2 \times Blue + L} \tag{16.2}$$

where NIR, Red, and $Blue$ denote the reflectance in near-infrared, red, and blue bands, respectively; L denotes the soil background adjustment factor; and are the coefficients to correct atmospheric interference. These two vegetation indices usually exhibit a high degree of correlation, and EVI is considered to be less sensitive to atmospheric interference and saturation effects than $NDVI$. Besides, other vegetation indices, such as Green Chlorophyll Index (GCI), Normalized Difference Water Index (NDWI), and Normalized Difference Red-Edge Index (NDRE), can also be used as the input to characterize crop canopy greenness, vegetation moisture content, and photosynthetic capacity for estimating crop yield (Kang *et al.*, 2020; Ma *et al.*, 2021; Muruganantham *et al.*, 2022; Van Klompenburg *et al.*, 2020).

In comparison, meteorological data, such as temperature and precipitation, can be used to characterize the environmental conditions that affect the growth and development of crop. Throughout the growing season, environmental stress (i.e., water and heat stress) may impair the plant photosynthetic activities and inhibit the crop growth, leading to reduction in the crop yields. In crop yield estimation, meteorological predictors can generally be grouped into heat or water stress-related predictors (Kang *et al.*, 2020; Ma *et al.*, 2021; Muruganantham *et al.*, 2022; Van Klompenburg *et al.*, 2020). Usually, precipitation (e.g., cumulative precipitation), total evapotranspiration (ET), total potential evapotranspiration (PET), and soil moisture can be used as water stress indicators to represent the crop water supply and atmospheric water cycle conditions. For heat stress, daily air temperature and land surface temperature are commonly used predictors. Growing Degree Days (GDD) and Killing Degree Days (KDD) can also be good heat stress indicators, as they are the estimates of effective thermal units and extreme heat events, respectively. They can be calculated using the temperature thresholds (Jiang *et al.*, 2020a) as Eqs. 16.3 and 16.4:

$$\textbf{GDD} = \begin{cases} 21 & T_{max} > T_{min} \geqslant 29 \\ \frac{min(T_{max},29)+max(T_{min},8)}{2} - 8 & others \\ 0 & 8 \geqslant T_{max} > T_{min} \end{cases} \tag{16.3}$$

$$KDD = max(T_{max}, 29) - 29 \tag{16.4}$$

where T_{max} and T_{min} are the maximum and minimum temperature (◦C), respectively. Vapor pressure deficit (VPD) is also closely related to crop water and heat stress and can be a potential predictor in crop yield estimation.

Apart from meteorological data, soil data (e.g., soil type, clay content, silt content, sand content, bulk density, and organic carbon content) have also been used to characterize environmental conditions for crop yield estimation. The soil-related predictors represent soil water holding capacity and air capacity that have a substantial effect on plant water and oxygen supply and soil nutrient availability (Pourmohammadali *et al.*, 2019; Shirani *et al.*, 2015). Besides, some other predictor inputs, such as phenology information, management practices (e.g., crop density, irrigation, and fertilization), and UAV data, could also be used to improve crop yield estimation. Crop phenology closely corresponds to crop physiological development progress and is a good indicator of crop growth response to climate change (Diao, 2019, 2020; Diao and Li, 2022; Diao *et al.*, 2021; Jiang *et al.*, 2020a). The management practices significantly change the environmental conditions that affect crop yield formation. As for the UAV data, the UAV-based RGB and multispectral imagery can be employed for characterizing the structural and geometric properties of plant canopies (e.g., texture), physiological properties, biochemical concentration (e.g., nitrogen), etc.; the UAV-based Light Detection and Ranging (LiDAR) system can provide the canopy coverage and canopy height information that is directly related to crop biomass.

16.2.2 GEOAI-BASED YIELD ESTIMATION MODEL

The GeoAI-based yield estimation models typically include data-driven ML/DL models and hybrid ML/DL-crop models.

16.2.2.1 ML and DL Models

Several widely used data-driven ML/DL models for crop yield estimation are briefly described as follows:

- *Random Forest (RF)*: RF is an ensemble model as an extension of the bagging classification tree to improve the accuracy of yield estimation. Each bagging tree is trained independently by bootstrap samples of the whole training dataset. RF integrates the results from all the trees to make the final decision. Thus, it can reduce the overfitting of data and improve yield estimation accuracy. The critical hyper-parameters of RF include tree number and maximum node (Jeong *et al.*, 2016).

- *Support Vector Regression (SVR)*: As a widely used regression method, SVR uses the kernel functions to reproject the raw inputs into high-dimensional feature space, in which a linear hyperplane defined by the support vectors can be optimized to fit the training data with minimized errors. Because the non-linear relationships are accommodated by relatively simple kernel functions, SVR is effective in dealing with a large amount of yield predictors and has good generalization capability (Crane-Droesch, 2018; Sishodia *et al.*, 2020). The main hyper-parameters of SVR include kernel function, kernel coefficient, and regularizer weighting parameter.

- *Deep Neural Network (DNN)*: Artificial neural network (ANN) usually comprises one input layer, fully connected hidden layers, and one output layer. As an ANN, DNN typically contains multiple hidden layers to learn complicated linear and non-linear relationships between the output and input yield predictors. DNN also allows the multi-stream modeling structure to synthesize different information effectively, which can achieve good performance in crop yield estimation (Maimaitijiang *et al.*, 2020; Yoosefzadeh-Najafabadi *et al.*, 2021). Before training, the number of layers and cells in each layer should be tuned in DNN.

- *Convolutional Neural Network (CNN)*: CNN is one of the most widely used deep learning architectures in remote sensing applications because of its capability to extract the image spatial and contextual information. In general, CNN includes several convolutional blocks, and each block is composed of several convolutional layers and pooling layers. Convolutional layers use a series of convolutional filters to extract features in the images, and the convolutional filters can be of one to three dimensions (Barbosa *et al.*, 2020; Russello, 2018). Pooling layers are used to change the dimensions of the feature maps, which can help convolutional layers learn different levels of feature representations. Therefore, CNN could learn complex and abstract spatial textures and patterns in the remote sensing images for yield estimations (Hamida *et al.*, 2018; You *et al.*, 2017; Zhong *et al.*, 2019). Compared to DNN, CNN has more parameters (e.g., weights and biases) to be learned and usually requires a large training dataset.

- *Long Short-Term Memory (LSTM)*: Different from CNN, LSTM is a special recurrent neural network (RNN) architecture, which is designed to learn the complex temporally evolving patterns of time series data (Jiang *et al.*, 2020a; Sun *et al.*, 2019; You *et al.*, 2017). The inputs of LSTM are usually sequential vectors; in general, the raw remote sensing images need to be transformed into vectors to train the LSTM for crop yield estimation. During the training process, three gates (i.e., forget gate, input gate, and output gate) are used in LSTM to control the cell state and memorizing process. These gates control how much information can pass through the memory cells and how much information can be reserved. LSTM can overcome the vanishing gradient problems in conventional RNNs and achieve robust yield estimation results in several studies (Lin *et al.*, 2020; Schwalbert *et al.*, 2020; Tian *et al.*, 2021). Before training, the numbers of LSTM layers and hidden units are the important hyper-parameters that need tuning.

16.2.2.2   Hybrid ML/DL-Crop Model

In recent years, the data-driven ML/DL models have also been hybridized with process-based crop models to accommodate the complex mechanisms underlying soil-crop-atmospheric interactive processes for more robust yield estimation. To date there are typically three categories of hybrid ML/DL-crop models: (1) Using crop growth models to create new synthetic data under different environmental conditions and broaden the set of predictors for ML/DL models. A typical example is the scalable crop yield mapping (SCYM) model (Lobell *et al.*, 2015) and its improved version (Jeffries *et al.*, 2020), where the authors use crop model simulations to train a statistical data-driven model for different combinations of possible image acquisition dates. Li *et al.* (2021) uses a process-based crop model to simulate wheat phenology, leaf area index (LAI), and yield under drought scenarios, and calculate wheat growth indicators. The relationship between growth indicators and yield is then obtained through ML to estimate wheat yield under different drought scenarios.

Shahhosseini *et al.* (2021) uses the output variables of a process-based crop model as the input of the ML model to explore the ability of the hybridized model for crop yield prediction at the county level. (2) Using ML/DL models to replace uncertain processes or estimate unknown parameters of process-based crop models. Saha *et al.* (2021) discusses the utilization of the ML models to replace the subroutines of crop models of limited mechanism understanding or equation representation, such as the simulation of soil N_2O emissions. (3) Building metamodels (i.e., model emulators) of crop growth models using ML/DL models to reduce computation costs. Folberth *et al.* (2019) applies the metamodel trained on the low spatial resolution output data of the Global Gridded Crop Model (GGCM) to high-resolution data for spatio-temporal downscaling of GGCM yield estimates, which overcomes the limitation of high computational demands of direct crop model simulations at higher resolutions.

However, the integration of crop models into data-driven ML/DL models has risks and caveats. The crop growth models need accurately simulate the variables employed in the data-driven models as a prerequisite for integration. The incorrectly simulated values of variables would potentially decrease the ML/DL model performance (Maestrini *et al.*, 2022).

16.2.3 FEATURE IMPORTANCE ANALYSIS

Understanding the contribution of predictors to crop yield formation is one of the objectives of building GeoAI-based yield estimation models. Yet the black-box nature of ML/DL models makes the interpretation of those predictors difficult. The approaches to estimating the relative importance of each input predictor in ML/DL models are briefly introduced as below.

- *Gini importance*: Gini importance is a feature importance analysis method used in decision-tree-based models, such as RF and extreme gradient boosting (XGBoost) (Menze *et al.*, 2009; Nembrini *et al.*, 2018). It measures the relative importance of each feature by calculating the total reduction of the impurity index (e.g., Gini impurity) that is achieved by splitting on that feature over all the trees in the model. The resulting Gini importance represents the percentage of the total reduction in impurity that can be attributed to that feature and helps identify the key drivers of the model's predictions. However, the Gini importance may not be well suited when dealing with highly correlated features as it assumes independence between features (Menze *et al.*, 2009).

- *Model-free variable importance*: Some model-free variable importance analysis techniques commonly used for non-linear models are permutation feature importance test and Leave-One-Covariate-Out (LOCO) (Archontoulis *et al.*, 2020; Breiman, 2001; Wang *et al.*, 2020). With the permutation or dropping of a feature from the feature set, the model's performance will degrade. The feature importance is thus measured via the decrease of the model performance when a specific feature is randomly shuffled or dropped. Taking the permutation feature importance test as an example, after the model first fits the original data (X_i, Y_i), the permutation importance test will then calculate the model error based on the new dataset $(X_i(j'), Y_i)$ to investigate the importance of the jth input variable, where $X_i(j')$ represents the jth variable being randomly shuffled. The importance of the feature can be measured via the difference between the yield estimation error using the permutated dataset $(X_i(j'), Y_i)$ and that using the original dataset (X_i, Y_i).

- *Attention mechanism-based feature importance*: For *DL* models, the attention mechanism can also be employed to quantify the feature importance (Lin *et al.*, 2020; Wang *et al.*, 2021; Woo *et al.*, 2018). The attention mechanism is inspired by the cognitive function of human beings, which selectively focuses on parts of an image or a sentence instead of processing a whole image or sentence at once to better understand the information. In DL models, spatial, temporal, and channel-wise attention can be designed to help the models focus on prominent features across spatial, temporal, and channel dimensional spaces. As the attention values of different features weigh their contributions to the results, the distribution of attention values can reflect the importance of features. The higher attention values indicate that the features in the corresponding part have higher impacts on the estimated yield, and the attention mechanism thus provides insights to help us understand how different features contribute to the estimation of crop yield.

16.2.4 UNCERTAINTY ANALYSIS

As ML/DL models are usually constructed using the given training datasets for yield estimation, these models are not always able to make accurate predictions over extended areas or years especially when applied to different environmental settings. Thus a trustworthy representation of uncertainty (i.e., degree of confidence of ML/DL modeling prediction) is desirable, particularly in yield-based decision-making processes that are of agricultural, environmental, and economic significance, including agricultural market planning and government agricultural policy-making (Lencucha *et al.*, 2020; Müller *et al.*, 2021). In this section, two different uncertainty modeling approaches for DL are introduced.

16.2.4.1 Probability Modeling

To incorporate the uncertainty into DL models, the final output of DL can be expanded into a Gaussian distribution $[y, \sigma]$ to account for the observation noise or randomness stemming from different yield predictor inputs, with the mean y denoting the average yield estimate and the standard deviation σ representing the uncertainty/noise in the input observations (Kendall and Gal, 2017; Mobiny *et al.*, 2021; Wang *et al.*, 2020). As the output becomes a distribution, a simplified negative log-likelihood loss function (as Eq. 16.5 shows) rather than commonly used regression loss functions (e.g., mean square error loss function) should be used to optimize the parameters and to simultaneously estimate the two outputs (Kendall and Gal, 2017):

$$L(W) = \sum_{i=1}^{N} (\frac{1}{2\hat{\sigma}_i^2} \|\hat{y}_i - y_i\|^2 + \frac{1}{2} \log \hat{\sigma}_i^2) \tag{16.5}$$

where $\hat{y}_i$ and y_i denote the estimated mean yield and observed yield for sample i, respectively. N is the number of samples in the training data. $\hat{\sigma}_i^2$ denotes the estimated variance for sample i. With the negative log-likelihood loss function, the probability modeling can be used to quantify the uncertainty for DL-based yield estimations.

16.2.4.2   Bayesian Modeling

Besides probability modeling, the Bayesian theory can also be used to model the predictive uncertainty in DL models. Unlike the traditional DL models, all the weights W in the Bayesian DL models are represented by probability distributions, and those probability distributions can be used to characterize the variance of the model weights. Due to the intricacies of modeling weights as distributions, Bayesian DL models are typically approximated using the Monte Carlo (MC) dropout approach. By conducting random dropout of neurons during both training and testing stages, T sets of network weights $\{W_1, \ldots, W_T\}$ can be generated from Bernoulli distributions, resulting in different predicted values for the same sample each time. Thus the predictive uncertainty could be calculated as the standard deviation of the MC dropout-based yield estimations (Gal and Ghahramani, 2016; Shridhar et al., 2019) as Eq. (16.6):

$$\sigma(\mathbf{y}) = \sqrt{\frac{1}{T} \sum_{t=1}^{T} \left(\hat{\mathbf{y}}_t - \frac{1}{T} \sum_{t=1}^{T} \hat{\mathbf{y}}_t \right)^2} \tag{16.6}$$

where T represents the number of MC samplings; $\hat{\mathbf{y}}_t$ is the estimated yield value at the tth simulation.

16.3   CASE STUDY ON SOYBEAN YIELD ESTIMATION

A case study on county-level soybean yield estimation and uncertainty analysis in the U.S. Corn Belt using GeoAI-based models is introduced in this section. The GeoAI-based yield estimation models leverage three AI algorithms: SVR, RF, and DNN. The performance of these three models in soybean yield estimations is compared. With the DNN model as an example, we further quantify the yield estimation uncertainty using probability modeling, as well as evaluating the importance of yield predictors using the permutation feature importance test.

16.3.1   STUDY SITE

The study site of this case study is the US Corn Belt, which covers 12 states of the Midwestern US, including Illinois (IL), Indiana (IN), Iowa (IA), Kansas (KS), Michigan (MI), Minnesota (MN), Missouri (MO), Nebraska (NE), North Dakota

(ND), Ohio (OH), South Dakota (SD), and Wisconsin (WI) (shown in Figure 16.2). As the primary agricultural area in the US, Corn Belt contributed more than 81% of US soybean and corn production in 2020 (Naeve and Miller-Garvin 2019), because of the humid continental climate together with the relatively flat land with fertile soil. The western states (e.g., Nebraska, Missouri, South Dakota, North Dakota, and Kansas), which are close to the arid Great Plains, have more diverse crop types (e.g., wheat and barley), more irrigation area, and relatively more severe heat and water stress during the summer.

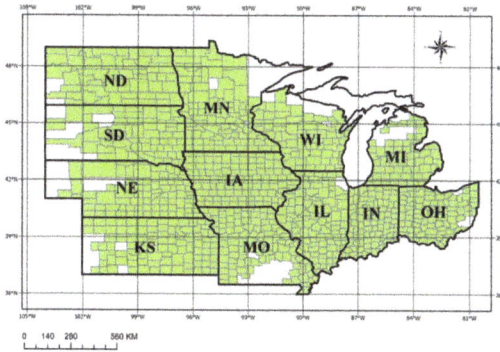

Figure 16.2   The U.S. Corn Belt states with 958 counties selected for analysis.

16.3.2   DATA

In this study, the Cropland Data Layer (CDL) is used to retrieve information related to soybean field locations. The CDL is a 30-m resolution, crop-specific land-cover data layer created annually for the US using satellite imagery and extensive ground measurements (Jiang *et al.*, 2020b). The county-level yield records are retrieved from the United States Department of Agriculture (USDA) National Agricultural Statistics Service (NASS) Database (https://quickstats.nass.usda.gov/). The soybean yield records from 2008-2016 for a total of 958 counties in the US Corn Belt are collected in this study.

The satellite-based yield predictor, two-band Enhanced Vegetation Index (EVI2), is considered in this study as a direct indicator of soybean biomass. EVI2 diminishes the influence of atmospheric and soil background and has enhanced sensitivity in high biomass regions. It is calculated as Eq. (16.7):

$$\text{EVI2} = 2.5 \times \frac{NIR - Red}{NIR + 2.4 \times Red + 1} \tag{16.7}$$

where Red and NIR represent the reflectance in red and near-infrared bands, respectively. EVI2 is extracted from the MODIS MCD43A4 V6 Nadir Bidirectional Reflectance Distribution Function Adjusted Reflectance (NBAR) product (Sharma,

Hara, and Hirayama 2018), which provides 500-meter nadir surface reflectance data on a daily basis. The meteorological predictors, including daily maximum 2-meter air temperature (T_{max}), daily minimum 2-meter air temperature (T_{mix}), and daily total precipitation (Precipitation), are downloaded from the Daymet V4 (Thornton et al. 2021). Daymet V4 provides gridded daily estimates of weather variables for North America. VPDmax is acquired from the daily 5km Parameter-elevation Regressions on Independent Slopes Model (PRISM) dataset. Evapotranspiration (ET) and potential ET are acquired from the 3-hour 0.25 degree Global Land Data Assimilation System (GLDAS) dataset. Six soil-related predictors that are characteristic of soil physical and chemical properties at six depths (0, 10, 30, 60, 100, and 200 cm) are collected from OpenLandMap (Hengl *et al.*, 2017). The soil predictors include clay content mass fraction, sand content mass fraction, water content at 33kPa suction, pH in H2O, bulk density, and organic carbon content. To better understand how crop yields respond to the sub-seasonal changes of environmental factors throughout the growing season, vegetation phenology information from the MODIS MCD12Q2 Land Cover Dynamics product is also considered in our study.

16.3.3 METHOD

16.3.3.1 Data Pre-processing

To model responses of the crop yield to the sub-seasonal changes of environmental factors throughout the growing season, the phenology-based temporal aggregation is conducted for each predictor (shown in Figure 16.3) with the phenological information. First, the whole crop growth season for each pixel is divided into six phenological stages by seven vegetation phenology transition dates of the MCD12Q2, including the onset of greenness (greenup), greenup midpoint (midgreenup), maturity, peak greenness (peak), senescence, greendown midpoint (midgreendown), and dormancy. All the environmental and satellite-based predictors for each pixel are temporally aggregated into the six soybean phenological stages based on the corresponding start and end dates. Specifically, for the precipitation predictor, the cumulative aggregation is conducted for each phenological stage to analyze the accumulation effect of precipitation on crop growth and productivity in each stage (Meng *et al.*, 2017). For all the other predictors, the mean aggregation is applied for each phenological stage to represent its average conditions within each stage. Because of the irregular boundary of each county, the spatial averaged aggregation is further conducted for each county to accommodate the county averaged information. After the temporal and spatial aggregations, 78 variables in total (7 satellite and meteorological predictors of 6 phenological stages as well as 6 soil predictors at 6 depths) are generated as the model input for the soybean yield modeling (as seen in Table 16.1).

16.3.3.2 GeoAI-based Yield Estimation Model

In this case study, we build three GeoAI models, including SVR, RF, and DNN, for county-level soybean yield estimations. We further quantify the yield estimation uncertainty for the DNN model. As discussed above, there are two ways to

Table 16.1

Description of the data.

Category	Variable	Spatial resolution	Temporal resolution	Source
Crop	Yield Record (bu/ac)	County-level	Annual	USDA Quick Statistic Database
	Cropland Map	30 m	Annual	NASS CDL Program
Phenology	Satellite Phenology	500 m	Annual	MODIS MCD12Q2
Vegetation index	EVI2	500 m	Daily	MODIS MCD43A4
Heat-related	Maximum Air Temperature(∘C) Minimum Air Temperature(∘C)	1 km	Daily	Daymet V4
Water-related	Precipitation(mm)	1 km	Daily	Daymet V4
	VPDmax (hPa)	5 km	Daily	PRISM
	Evapotranspiration (ET) (kg/m2/s) Potential ET (W/m2)	0.25	3-Hour	GLDAS
Soil	Clay Content (%) Sand Content (%) pH in H_2O Bulk Density (kg/m^3) Organic Carbon Content (g/kg) Water Content at 33kPa(%)	250 m	Static	OpenLandMap

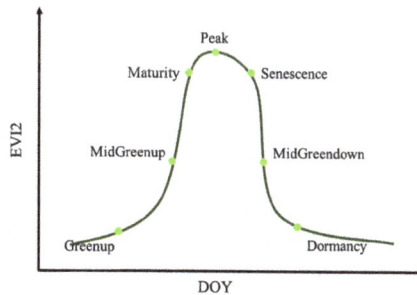

Figure 16.3 Six phenological stages based on crop phenology transition dates.

incorporate uncertainty estimation into the DNN model, namely probability modeling and Bayesian modeling. Considering the MC dropout implementation of Bayesian modeling is computationally intensive and time-consuming (Ma *et al.*, 2021), here we design a probability modeling-based DNN architecture with two outputs (i.e., y and σ) to estimate both the yield and the corresponding predictive uncertainty (shown in Figure 16.4). As the probability modeling distribution of the predicted yield follows the Gaussian distribution, the outputs y and σ represent the mean and standard deviation of the distribution, respectively. The standard deviation can be used to quantify the uncertainty of the predicted value. Through experimental analysis for hyper-parameter tuning, the developed DNN model is tuned with four fully connected hidden layers of 256, 128, 64, and 32 neurons, respectively, to extract high-level features from the yield predictor inputs. Two outputs in the final layer of DNN are utilized for estimating predictive yield distribution. The importance of yield predictors in DNN is then evaluated using the permutation feature importance test. The DNN model is constructed using Keras 2.8 (Python version 3.7), and the SVR and RF models are implemented with scikit-learn library (Python version 3.7). For a fair comparison, all models use the same predictors as the inputs, and their hyper-parameters are tuned based on the ten-fold cross-validation accordingly.

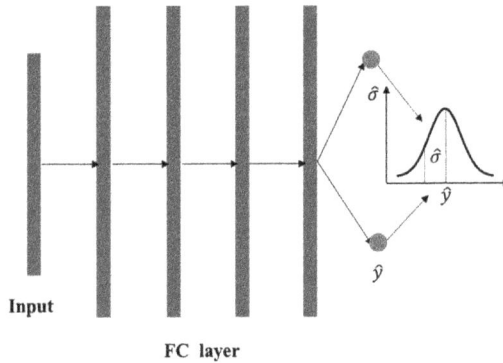

Figure 16.4   The designed DNN architecture with two outputs for estimating predictive yield distribution.

16.3.3.3   Model Training and Evaluation

All the models are trained using the county-level data from 2008 to 2015, and are then tested on the target year 2016. The whole dataset includes 6611 training data samples and 788 testing data samples. During the training, the initial learning rate is set to be 0.0001 and the Adam optimizer is applied to minimize the loss. Early stopping based on the validation loss is used in the learning process to prevent the network from overfitting when the validation loss stops decreasing. The root mean square error (RMSE), coefficient of determination (R^2), and bias are selected as the metrics to evaluate the performances of different models. The three metrics are calculated as Eqs. (16.8), (16.9), and (16.10):

$$\mathbf{RMSE} = \sqrt{\frac{1}{N}\sum_{i=1}^{N}(\hat{y}_i - y_i)^2} \qquad (16.8)$$

$$R_2 = 1 - \frac{\sum_{i=1}^{N}(\hat{y}_i - y_i)^2}{\sum_{i=1}^{N}(\bar{y}_i - y_i)^2} \tag{16.9}$$

$$\mathbf{Bias} = \sum_{i=1}^{N} \frac{(y_i - \hat{y}_i)}{N} \tag{16.10}$$

where N denotes the number of samples; y_i is the observed yield; $\hat{y}_i$ is the estimated yield; $\bar{y}$ is the average yield observation. For DNN, the standard deviation of the predictive yield distribution is used as the measure of the predictive uncertainty.

16.3.4 RESULTS AND DISCUSSION

16.3.4.1 Model Performance and Predictive Uncertainty

The scatter plots and the accuracies of three models, including DNN, SVR, and RF, in the testing year 2016 are shown in Figure 16.5. The results show that all of the models have relatively good performance, with RMSEs ranging from 5.913 to 6.091 bu/ac, R squares ranging from 0.353 to 0.469, and biases ranging from -4.493 to -4.113 bu/ac. In general, the three models have comparable performance. The RF model outperforms other models in terms of RMSE, with this metric value being 5.913 bu/ac. By comparison, the DNN model achieves a higher R^2 of 0.469 and shows better agreement between observed and predicted yields in the scatter plots. More samples are distributed around the 1:1 line in the DNN plot. We also find that all the models tend to underestimate the soybean yield in 2016 as shown in the bias and scatter plots. In particular, the underestimation of yield by RF is the most severe. Few testing samples are distributed over the 1:1 line and the bias of RF is significantly larger than that of the other two models. This general underestimation may partly be caused by the difference between the training dataset and the testing dataset, leading to challenges in modeling the yield variations.

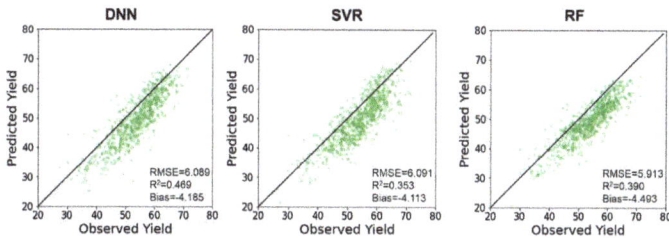

Figure 16.5 The scatter plot and accuracy of DNN, SVR, and RF in estimating the soybean yields in 2016.

With the DNN model as an example, we further map the spatial distributions of predicted yields and errors across counties for the testing year 2016 (Figure 16.6). The spatial pattern of predicted yields aligns with that of observed yields for most counties. Similar to the scatter plot, the predicted yields tend to be underestimated in the study site. The central and eastern regions, especially Illinois and Iowa, have relatively high soybean yield records. Nebraska also has a relatively high soybean yield record possibly due to the extensive irrigation in this area (Grassini *et al.*, 2015). The low yield records are mainly distributed in the Great Plain area (i.e., southwestern states). According to the RMSE map, the errors in most counties are less than 6 bu/ac, and larger errors are mainly observed in South Dakota, Kansas, Missouri, and Ohio, where the soybean yields are relatively low.

Besides the estimated yields, the predictive uncertainty map derived from the DNN model is shown in Figure 16.6(d). Compared to the distribution of RMSE, the spatial distribution of uncertainty is more homogeneous over the study area. The predictive uncertainty of most counties is less than 2.0 bu/ac, and the regions with relatively larger uncertainty are mostly distributed in the west of the study site (e.g., North Dakota, South Dakota, Kansas, and Nebraska), which is possibly caused by the unique climate and soil characteristics (e.g., more severe heat and water stress during the summer) in the Great Plain (Challinor *et al.*, 2014; Ma *et al.*, 2021).

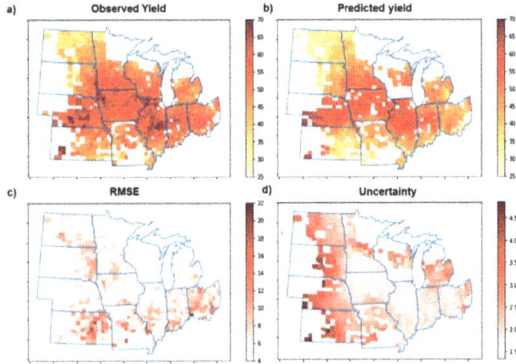

Figure 16.6 Spatial distributions of (a) observed yield, (b) predicted yield, (c) RMSE, and (d) uncertainty using DNN over the study site for the year 2016.

16.3.4.2 Feature Importance Analysis

To evaluate the contributions of each yield predictor in DNN, the permutation importance test is employed in this case study (shown in Figure 16.7). Overall, the EVI2 predictor has significantly higher contributions to the final soybean yield estimation. The importance of this vegetation index is attributed to its direct reflection of crop biomass as well as yield implications. The relatively good performance of vegetation index has also been reported in previous studies (Dadsetan *et al.*, 2021; Raun *et al.*, 2002; Yao *et al.*, 2012). For the meteorological predictors, VPDmax, precipitation, and PET have relatively higher importance, indicating both water stress and heat stress in crop growth may cause a reduction in the crop yield (Jumrani and Bhatia, 2018). It's worth noting that the soil predictors, including soil clay content mass fraction (SC), soil bulk density (SB), and soil pH in H_2O (SP), are more important than the meteorological predictors, since they can be used to characterize the water, oxygen, and nutrient uptake of soybean. Specifically, clay content mainly affects the water availability of root and has an influence on plant growth; Bulk density affects water holding capacity and air capacity for water and oxygen supply; Soil pH can largely affect the soil nutrient availability (Pourmohammadali *et al.*, 2019; Shirani *et al.*, 2015). Therefore, these soil predictors play an important role in estimating the soybean yield.

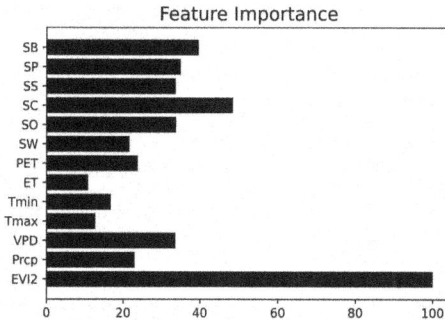

Figure 16.7   Permutation feature importance of DNN in soybean yield estimation (SW: soil water content at 33kPa suction, SO: soil organic carbon content, SC: soil clay content mass fraction, SS: soil sand content mass fraction, SP: soil pH in H_2O, SB: soil bulk density).

16.4   CONCLUSION

To tackle the global food security challenge, it is critical to utilize GeoAI to optimize farm management practices and evaluate agricultural decision-making. In this chapter, we introduce the applications of GeoAI in agriculture, particularly the GeoAI-based framework for crop yield and uncertainty analysis. In the case study, we introduce three GeoAI models (with SVR, RF, and DNN incorporated respectively) for county-level soybean yield estimation and uncertainty quantification in the US Corn Belt. By aggregating satellite, meteorological, and soil predictors across soybean phenological stages, the models achieve the RMSE from 5.913 to 6.091 bu/ac, R squares from 0.353 to 0.469, and biases from -4.493 to -4.113 bu/ac in 2016. According to the feature importance analysis, EVI2 and soil predictors are more critical in soybean yield estimation. The DNN-based uncertainty analysis further shows that larger yield uncertainties tend to be distributed in the regions with unique environmental conditions (e.g., climate and soil conditions). Overall, GeoAI-based models enhance the ability to understand crop yield response to various environmental conditions as well as quantifying the predictive uncertainty, which can further help optimize farm management strategies and agricultural decision-making for sustainable agricultural development.

ACKNOWLEDGMENTS

This research is supported in part by the National Science Foundation (No.2048068), in part by the National Aeronautics and Space Administration (No.80NSSC21K0946), and in part by the United States Department of Agriculture (No.2021-67021-33446).

BIBLIOGRAPHY

Anagnostis, A., *et al.*, 2021. A deep learning approach for anthracnose infected trees classification in walnut orchards. *Computers and Electronics in Agriculture*, 182, 105998.

Archontoulis, S.V., *et al.*, 2020. Predicting crop yields and soil-plant nitrogen dynamics in the us corn belt. *Crop Science*, 60 (2), 721–738.

Barbosa, A., *et al.*, 2020. Modeling yield response to crop management using convolutional neural networks. *Computers and Electronics in Agriculture*, 170, 105197.

Barnes, W.L., Xiong, X., and Salomonson, V.V., 2003. Status of terra modis and aqua modis. *Advances in Space Research*, 32 (11), 2099–2106.

Benos, L., Bechar, A., and Bochtis, D., 2020. Safety and ergonomics in human-robot interactive agricultural operations. *Biosystems Engineering*, 200, 55–72.

Breiman, L., 2001. Random forests. *Machine Learning*, 45, 5–32.

Challinor, A.J., *et al.*, 2014. A meta-analysis of crop yield under climate change and adaptation. *Nature Climate Change*, 4 (4), 287–291.

Chlingaryan, A., Sukkarieh, S., and Whelan, B., 2018. Machine learning approaches for crop yield prediction and nitrogen status estimation in precision agriculture: A review. *Computers and Electronics in Agriculture*, 151, 61–69.

Conrad, Z., *et al.*, 2018. Relationship between food waste, diet quality, and environmental sustainability. *PloS One*, 13 (4), e0195405.

Coopersmith, E.J., *et al.*, 2014. Machine learning assessments of soil drying for agricultural planning. *Computers and Electronics in Agriculture*, 104, 93–104.

Crane-Droesch, A., 2018. Machine learning methods for crop yield prediction and climate change impact assessment in agriculture. *Environmental Research Letters*, 13 (11), 114003.

Dadsetan, S., *et al.*, 2021. Detection and prediction of nutrient deficiency stress using longitudinal aerial imagery. *In*: *Proceedings of the AAAI Conference on Artificial Intelligence*. vol. 35, 14729–14738.

Diao, C., 2019. Innovative pheno-network model in estimating crop phenological stages with satellite time series. *ISPRS Journal of Photogrammetry and Remote Sensing*, 153, 96–109.

Diao, C., 2020. Remote sensing phenological monitoring framework to characterize corn and soybean physiological growing stages. *Remote Sensing of Environment*, 248, 111960.

Diao, C. and Li, G., 2022. Near-surface and high-resolution satellite time series for detecting crop phenology. *Remote Sensing*, 14 (9), 1957.

Diao, C., *et al.*, 2021. Hybrid phenology matching model for robust crop phenological retrieval. *ISPRS Journal of Photogrammetry and Remote Sensing*, 181, 308–326.

Drusch, M., *et al.*, 2012. Sentinel-2: Esa's optical high-resolution mission for gmes operational services. *Remote Sensing of Environment*, 120, 25–36.

Feng, Y., *et al.*, 2017. Modeling reference evapotranspiration using extreme learning machine and generalized regression neural network only with temperature data. *Computers and Electronics in Agriculture*, 136, 71–78.

Folberth, C., *et al.*, 2019. Spatio-temporal downscaling of gridded crop model yield estimates based on machine learning. *Agricultural and Forest Meteorology*, 264, 1–15.

Gal, Y. and Ghahramani, Z., 2015. Bayesian convolutional neural networks with bernoulli approximate variational inference. *arXiv preprint arXiv:1506.02158*.

Gal, Y. and Ghahramani, Z., 2016. Dropout as a bayesian approximation: Representing model uncertainty in deep learning. *In: Proceedings of PMLR*, 1050–1059.

García Pereira, A., *et al.*, 2020. Data acquisition and processing for geoai models to support sustainable agricultural practices.

Grassini, P., *et al.*, 2015. High-yield maize–soybean cropping systems in the us corn belt. *In: Crop Physiology*. Elsevier, 17–41.

Hamida, A.B., *et al.*, 2018. 3-d deep learning approach for remote sensing image classification. *IEEE Transactions on Geoscience and Remote Sensing*, 56 (8), 4420–4434.

Hengl, T., *et al.*, 2017. Soilgrids250m: Global gridded soil information based on machine learning. *PLoS One*, 12 (2), e0169748.

Islam, N., *et al.*, 2021. Early weed detection using image processing and machine learning techniques in an australian chilli farm. *Agriculture*, 11 (5), 387.

Jeffries, G.R., *et al.*, 2020. Mapping sub-field maize yields in Nebraska, USA by combining remote sensing imagery, crop simulation models, and machine learning. *Precision Agriculture*, 21, 678–694.

Jeong, J.H., *et al.*, 2016. Random forests for global and regional crop yield predictions. *PloS One*, 11 (6), e0156571.

Jiang, H., *et al.*, 2020a. A deep learning approach to conflating heterogeneous geospatial data for corn yield estimation: A case study of the us corn belt at the county level. *Global Change Biology*, 26 (3), 1754–1766.

Jiang, Z., *et al.*, 2020b. Predicting county-scale maize yields with publicly available data. *Scientific Reports*, 10 (1), 1–12.

Johann, A.L., *et al.*, 2016. Soil moisture modeling based on stochastic behavior of forces on a no-till chisel opener. *Computers and Electronics in Agriculture*, 121, 420–428.

Jumrani, K. and Bhatia, V.S., 2018. Impact of combined stress of high temperature and water deficit on growth and seed yield of soybean. *Physiology and Molecular Biology of Plants*, 24 (1), 37–50.

Kang, Y., *et al.*, 2020. Comparative assessment of environmental variables and machine learning algorithms for maize yield prediction in the us midwest. *Environmental Research Letters*, 15 (6), 064005.

Kattenborn, T., *et al.*, 2021. Review on convolutional neural networks (cnn) in vegetation remote sensing. *ISPRS Journal of Photogrammetry and Remote Sensing*, 173, 24–49.

Kendall, A. and Gal, Y., 2017. What uncertainties do we need in bayesian deep learning for computer vision? *Advances in Neural Information Processing Systems*, 30.

Khaki, S. and Wang, L., 2019. Crop yield prediction using deep neural networks. *Frontiers in Plant Science*, 10, 621.

Lencucha, R., *et al.*, 2020. Government policy and agricultural production: a scoping review to inform research and policy on healthy agricultural commodities. *Globalization and Health*, 16, 1–15.

Li, Z., Zhang, Z., and Zhang, L., 2021. Improving regional wheat drought risk assessment for insurance application by integrating scenario-driven crop model, machine learning, and satellite data. *Agricultural Systems*, 191, 103141.

Lin, T., *et al.*, 2020. Deepcropnet: a deep spatial-temporal learning framework for county-level corn yield estimation. *Environmental Research Letters*, 15 (3), 034016.

Lobell, D.B., *et al.*, 2015. A scalable satellite-based crop yield mapper. *Remote Sensing of Environment*, 164, 324–333.

Ma, Y., *et al.*, 2021. Corn yield prediction and uncertainty analysis based on remotely sensed variables using a bayesian neural network approach. *Remote Sensing of Environment*, 259, 112408.

Maestrini, B., *et al.*, 2022. Mixing process-based and data-driven approaches in yield prediction. *European Journal of Agronomy*, 139, 126569.

Maimaitijiang, M., *et al.*, 2020. Soybean yield prediction from uav using multimodal data fusion and deep learning. *Remote Sensing of Environment*, 237, 111599.

Meng, T., *et al.*, 2017. Analyzing temperature and precipitation influences on yield distributions of canola and spring wheat in saskatchewan. *Journal of Applied Meteorology and Climatology*, 56 (4), 897–913.

Menze, B.H., *et al.*, 2009. A comparison of random forest and its gini importance with standard chemometric methods for the feature selection and classification of spectral data. *BMC Bioinformatics*, 10, 1–16.

Mobiny, A., *et al.*, 2021. Dropconnect is effective in modeling uncertainty of bayesian deep networks. *Scientific Reports*, 11 (1), 1–14.

Mohammadi, K., *et al.*, 2015. Extreme learning machine based prediction of daily dew point temperature. *Computers and Electronics in Agriculture*, 117, 214–225.

Morellos, A., *et al.*, 2016. Machine learning based prediction of soil total nitrogen, organic carbon and moisture content by using vis-nir spectroscopy. *Biosystems Engineering*, 152, 104–116.

Müller, C., *et al.*, 2021. Exploring uncertainties in global crop yield projections in a large ensemble of crop models and cmip5 and cmip6 climate scenarios. *Environmental Research Letters*, 16 (3), 034040.

Muruganantham, P., *et al.*, 2022. A systematic literature review on crop yield prediction with deep learning and remote sensing. *Remote Sensing*, 14 (9), 1990.

Nahvi, B., *et al.*, 2016. Using self-adaptive evolutionary algorithm to improve the performance of an extreme learning machine for estimating soil temperature. *Computers and Electronics in Agriculture*, 124, 150–160.

Nassani, A.A., *et al.*, 2019. Management of natural resources and material pricing: Global evidence. *Resources Policy*, 64, 101500.

Nembrini, S., König, I.R., and Wright, M.N., 2018. The revival of the gini importance? *Bioinformatics*, 34 (21), 3711–3718.

Nguyen, G., *et al.*, 2019. Machine learning and deep learning frameworks and libraries for large-scale data mining: a survey. *Artificial Intelligence Review*, 52, 77–124.

Patil, A.P. and Deka, P.C., 2016. An extreme learning machine approach for modeling evapotranspiration using extrinsic inputs. *Computers and Electronics in Agriculture*, 121, 385–392.

Pourmohammadali, B., *et al.*, 2019. Effects of soil properties, water quality and management practices on pistachio yield in rafsanjan region, southeast of iran. *Agricultural Water Management*, 213, 894–902.

Raun, W.R., *et al.*, 2002. Improving nitrogen use efficiency in cereal grain production with optical sensing and variable rate application. *Agronomy Journal*, 94 (4), 815–820.

Russello, H., 2018. Convolutional neural networks for crop yield prediction using satellite images. *IBM Center for Advanced Studies*.

Saha, D., Basso, B., and Robertson, G.P., 2021. Machine learning improves predictions of agricultural nitrous oxide (n2o) emissions from intensively managed cropping systems. *Environmental Research Letters*, 16 (2), 024004.

Schwalbert, R., *et al.*, 2020. Mid-season county-level corn yield forecast for us corn belt integrating satellite imagery and weather variables. *Crop Science*, 60 (2), 739–750.

Shahhosseini, M., Hu, G., and Archontoulis, S.V., 2020. Forecasting corn yield with machine learning ensembles. *Frontiers in Plant Science*, 11, 1120.

Shahhosseini, M., *et al.*, 2021. Coupling machine learning and crop modeling improves crop yield prediction in the us corn belt. *Scientific Reports*, 11 (1), 1–15.

Shirani, H., *et al.*, 2015. Determining the features influencing physical quality of calcareous soils in a semiarid region of iran using a hybrid pso-dt algorithm. *Geoderma*, 259, 1–11.

Shridhar, K., Laumann, F., and Liwicki, M., 2019. A comprehensive guide to bayesian convolutional neural network with variational inference. *arXiv preprint arXiv:1901.02731*.

Sishodia, R.P., Ray, R.L., and Singh, S.K., 2020. Applications of remote sensing in precision agriculture: A review. *Remote Sensing*, 12 (19), 3136.

Sun, J., *et al.*, 2019. County-level soybean yield prediction using deep cnn-lstm model. *Sensors*, 19 (20), 4363.

Terliksiz, A.S. and Altỳlar, D.T., 2019. Use of deep neural networks for crop yield prediction: A case study of soybean yield in lauderdale county, Alabama, USA. *In: 2019 8th International Conference on Agro-Geoinformatics (Agro-Geoinformatics)*. IEEE, 1–4.

Thayer, A.W., *et al.*, 2020. Integrating agriculture and ecosystems to find suitable adaptations to climate change. *Climate*, 8 (1), 10.

Tian, H., *et al.*, 2021. An lstm neural network for improving wheat yield estimates by integrating remote sensing data and meteorological data in the Guanzhong plain, PR China. *Agricultural and Forest Meteorology*, 310, 108629.

Van Klompenburg, T., Kassahun, A., and Catal, C., 2020. Crop yield prediction using machine learning: A systematic literature review. *Computers and Electronics in Agriculture*, 177, 105709.

VoPham, T., *et al.*, 2018. Emerging trends in geospatial artificial intelligence (geoai): potential applications for environmental epidemiology. *Environmental Health*, 17 (1), 1–6.

Wang, X., *et al.*, 2020. Winter wheat yield prediction at county level and uncertainty analysis in main wheat-producing regions of China with deep learning approaches. *Remote Sensing*, 12 (11), 1744.

Wang, Y., *et al.*, 2021. A new attention-based cnn approach for crop mapping using time series sentinel-2 images. *Computers and Electronics in Agriculture*, 184, 106090.

Woo, S., *et al.*, 2018. Cbam: Convolutional block attention module. *In*: *Proceedings of the European Conference on Computer Vision (ECCV)*. 3–19.

Yao, Y., *et al.*, 2012. Active canopy sensor-based precision n management strategy for rice. *Agronomy for Sustainable Development*, 32, 925–933.

Yoosefzadeh-Najafabadi, M., Tulpan, D., and Eskandari, M., 2021. Using hybrid artificial intelligence and evolutionary optimization algorithms for estimating soybean yield and fresh biomass using hyperspectral vegetation indices. *Remote Sensing*, 13 (13), 2555.

You, J., *et al.*, 2017. Deep gaussian process for crop yield prediction based on remote sensing data. *In*: *Proceedings of the AAAI Conference on Artificial Intelligence*. vol. 31.

Zhang, J., *et al.*, 2021. Identification of cucumber leaf diseases using deep learning and small sample size for agricultural internet of things. *International Journal of Distributed Sensor Networks*, 17 (4), 15501477211007407.

Zhang, S., *et al.*, 2020. Plant species recognition methods using leaf image: Overview. *Neurocomputing*, 408, 246–272.

Zhong, L., *et al.*, 2019. Deep learning based winter wheat mapping using statistical data as ground references in Kansas and Northern Texas, US. *Remote Sensing of Environment*, 233, 111411.

Ziliani, M.G., *et al.*, 2022. Early season prediction of within-field crop yield variability by assimilating cubesat data into a crop model. *Agricultural and Forest Meteorology*, 313, 108736.

17 GeoAI for Urban Sensing

Filip Biljecki
National University of Singapore

CONTENTS

17.1 Introduction 351
17.2 Recent Examples of GeoAI for Urban Sensing – Case Studies in Singapore352
    17.2.1 Sensing rooftops from high-resolution satellite images 352
    17.2.2 Sensing urban soundscapes from street-level imagery.....................353
    17.2.3 Sensing and understanding human comfort using smartwatches
        and other sensors..355
    17.2.4 Sensing the perception and quality of underinvestigated urban
        spaces 356
17.3 Challenges and Opportunities...357
17.4 Conclusion...360
    Bibliography 361

17.1 INTRODUCTION

Urban sensing can be defined as a collection of methods and techniques to sense and obtain information about the built environment and human activities in cities (Shi, 2021). It is a major pillar of urban analytics, and it involves the collection and management of both static (e.g., buildings, road infrastructure) and dynamic (e.g., traffic, social media, noise) phenomena in urban areas. It has permeated through numerous domains and tasks pertaining to cities, from transportation, tourism and social networks to disaster management, air quality and foodscapes (Abirami and Chitra, 2022; Andris and Lee, 2021; Bai *et al.*, 2022; Calabrese *et al.*, 2013; Liang and Andris, 2021; Shin *et al.*, 2015; Xu *et al.*, 2022a,b; Yang *et al.*, 2023).

Following decades of developments, urban sensing remains a vital topic in spatial information sciences and urban management. Many technologies, such as LiDAR and satellite-based remote sensing, have been developed and employed in urban sensing. These approaches allow processing of data to extract knowledge, leading to meaningful and actionable insights. It has been attracting growing interest thanks to several continuously developing factors. In particular, in the past few years, these factors have multiplied: the increased volume of existing data (e.g., coverage, longitudinal acquisition), proliferation of sensors and supporting platforms (e.g., sensor-equipped vehicles), increasing quality of existing data (e.g., resolution of satellite imagery, accuracy of positioning), emergence of new sources and types of data (e.g.,

DOI: 10.1201/9781003308423-17

social media data, street-level imagery), the rise of citizen science and crowdsourcing (e.g., the success of OpenStreetMap), and greater computing power to process
large amounts of data (Biljecki and Ito, 2021; deSouza *et al.*, 2020; Duarte and Ratti,
2021; Gao *et al.*, 2021; Hu *et al.*, 2015; Lai, 2021; O'Keeffe *et al.*, 2019; Psyllidis
et al., 2022; Tu *et al.*, 2021; Yan *et al.*, 2020). As a result, it is not surprising that
GeoAI techniques, powered by developments in computer science, have flourished
in urban sensing. GeoAI aims to make sense of the vast and diverse data, and it helps
enhancing our understanding of urban environments. It has become a dominant topic
in journals and conferences, giving impetus to new opportunities, use cases, and
sensing insights in cities (Das *et al.*, 2022; Gao *et al.*, 2023; Janowicz *et al.*, 2019;
Kang *et al.*, 2019; Li, 2020; Liu and Biljecki, 2022).

   This chapter gives a brief overview of some developments of GeoAI in urban
sensing, together with an overarching overview beyond specific AI techniques to
give a broader understanding such as use cases, data, and challenges and opportunities. Considering the large volume of papers published on this topic, covering all
aspects would be beyond the scope of a single book chapter. Further, specific GeoAI
techniques, datasets, and use cases have been subject of many review papers (Biljecki and Ito, 2021; Chen *et al.*, 2023; Ghahramani *et al.*, 2020; Hsu and Li, 2023;
Ibrahim *et al.*, 2020; Li and Hsu, 2022; Liu *et al.*, 2015; Lu *et al.*, 2023; Martí *et al.*,
2019; Richter and Scheider, 2023; Shahtahmassebi *et al.*, 2021; Shi, 2021; Song
et al., 2023; Yan *et al.*, 2020). Thus, this chapter focuses on a selected set of insights
that capture the general trends and landscape of the GeoAI developments supporting
urban sensing in the ever-growing complexity and scope of urban environments. It
focuses on examples of research conducted at my research group, which reflect the
developments described above, and it gives a diversity of high-level examples of various GeoAI techniques used, input data, and solutions to challenges across multiple
domains.

17.2   RECENT EXAMPLES OF GEOAI FOR URBAN SENSING – CASE STUDIES IN SINGAPORE

This section reviews a few use cases developed at the research group of Urban Analytics Lab at the National University of Singapore to give an overview of recent
applications of GeoAI for urban sensing. These use cases are based on a variety of
types of input data and techniques to process them, and they span diverse application
domains that benefit from novel urban sensing approaches.

17.2.1   SENSING ROOFTOPS FROM HIGH-RESOLUTION SATELLITE IMAGES

Roofpedia is a project described in Wu and Biljecki (2021), focused on mapping the
content of rooftops of one million buildings around the world, in the context of sustainable development. It has been developed against the backdrop of the increasing
volume of studies focused on understanding the potential of rooftops in cities for
the installation of solar panels (Bódis *et al.*, 2019; Han *et al.*, 2022). It identifies a
gap that – while a large volume of papers has been published with the purpose of

measuring the space provided by rooftops and assessing their potential – not many studies have established the current utilization of rooftops, e.g., understanding how many rooftops have already been used for such purpose. The same goes for rooftops that have greenery, i.e., green roofs.

The developed workflow uses satellite imagery and image segmentation to automatically identify rooftops that are vegetated and/or have photovoltaic system installations, and measure their extent. In the preprocessing stage, the developed workflow uses building footprints sourced from OpenStreetMap to delineate portions of the imagery that represent buildings, in order to detect only photovoltaic panels and green roofs that are located on buildings, and exclude such features that are not on rooftops.

It is an example of research that engages primarily existing data (optical satellite imagery) but introduces a novel use case, thanks to its increasing quality (resolution, coverage) and accessibility, and developments in artificial intelligence. At the same time, it leverages crowdsourced data, an emerging data source, to support the mapping of rooftops.

Since the study is conducted at the city-scale for dozens of cities worldwide, it can be used for urban scale analyses (Figure 17.1) and understanding the penetration of sustainable rooftops. It also includes an index that quantifies the rooftop utilization rate in cities, which can be considered a proxy to gauge how successful cities are when it comes to unlocking the space provided by rooftops.

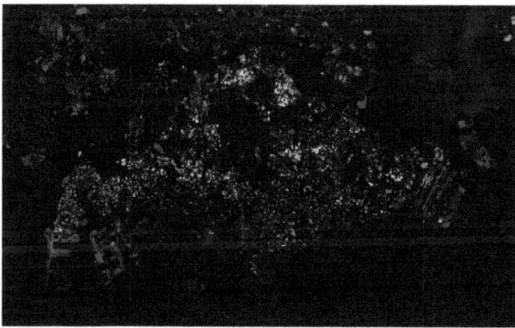

Figure 17.1   Locations of buildings in Singapore that have been detected to have photovoltaic installations on rooftops, using GeoAI techniques. (*Source:* Wu and Biljecki, 2021.)

Such research is an example of urban sensing that has potential to be linked to urban policies and lead to actions, e.g., understand whether policies to install solar panels in residential buildings have had an effect, and whether there is a particular area in a city that did not experience an uptake of rooftops for such purposes.

17.2.2   SENSING URBAN SOUNDSCAPES FROM STREET-LEVEL IMAGERY

Street-level imagery has been used extensively in the past few years in combination with GeoAI techniques (Biljecki and Ito, 2021; Kang *et al.*, 2020b; Zhang *et al.*,

2023a). It is a quintessential example of an emerging data source for urban sensing and it has greatly benefited from the advancements in GeoAI. For example, it has been used for understanding human perception (Kruse *et al.*, 2021), mapping street-level greenery (Yang *et al.*, 2021), understanding factors driving house prices (Kang *et al.*, 2021), generation of 3D city models (Pang and Biljecki, 2022), inferring urban density (Garrido-Valenzuela *et al.*, 2023), and discovering inconspicuous but interesting places in cities (Zhang *et al.*, 2020).

The work developed by Zhao *et al.* (2023) is a new application of GeoAI for urban sensing – predicting the intensity and nature of sounds in urban areas at a large-scale from street view imagery. The key motivation for this work is the increasing importance of soundscapes in the development and management of smart cities. However, techniques to sense the acoustic environment have been limited, especially at a high resolution. Thus, this work establishes a hypothesis that visual representations of streets may suggest the nature and intensity of noise, which can be leveraged by the increasing availability of street view imagery and advancements in GeoAI.

The developed approach uses street view imagery and computer vision to extract relevant visual features, doing so at multiple levels: pixel-level, object-level, semantic-level, and scene-level visual features. For example, the semantic-level feature is the proportion of 19 semantic features in an image, such as sky and vegetation, which was derived thanks to the DeepLabV3+ model trained on the Cityscape dataset (Chen *et al.*, 2018; Cordts *et al.*, 2016). After the visual features have been determined, the work establishes relationships using Gradient Boosted Regression Trees to infer the soundscape at the city-scale at a high resolution (Figure 17.2). The results have been validated using field audio measurements.

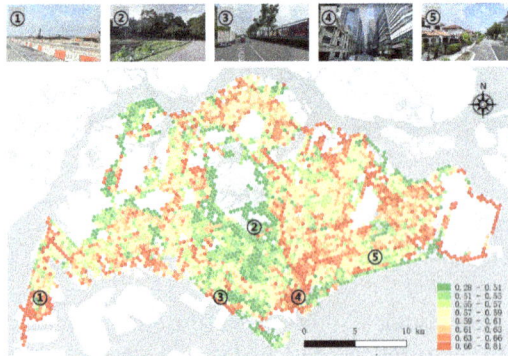

Figure 17.2   Predicted sound intensity from street view imagery across Singapore at high resolution. (Adapted from Zhao *et al.*, 2023.)

This work demonstrates how an emerging dataset in conjunction with GeoAI techniques can uncover a new use case and be leveraged to fill data gaps and derive new insights for multiple domains, i.e., enable inferring urban soundscapes for purposes such as measuring liveability, understanding impact on house prices, and support urban planning.

17.2.3 SENSING AND UNDERSTANDING HUMAN COMFORT USING SMARTWATCHES AND OTHER SENSORS

Walkability and outdoor comfort have been topical subjects in urban planning, and GeoAI techniques have been used to assess them (He and He, 2023). A recent paper by Liu *et al.* (2023a) is an example of GeoAI engaged for sensing and understanding human comfort – it introduces GraphSAGE-LSTM (GraphSAGE (Hamilton *et al.*, 2017) and Long short-term memory network (LSTM) (Hochreiter and Schmidhuber, 1997)) on crowdsourced data and computer vision to predict human comfort on the sidewalks.

A particularity of the work is that it combines multiple datasets (Figure 17.3): it collects data from a wearable device (smartwatch) and a developed app to gather physiological data and human comfort (Jayathissa *et al.*, 2019; Tartarini *et al.*, 2022), and couples it with a variety of environmental data such as noise, solar irradiation, and street-level imagery (which was used to establish what pedestrians see at a particular location).

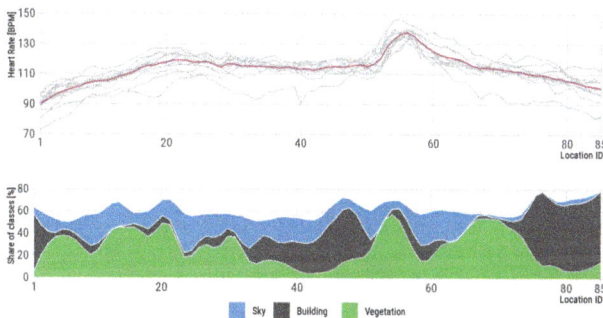

Figure 17.3 Collecting a variety of sensory experiences of dozens of people during walking through a specific path: their heart rate (top plot) and what they see (bottom plot), which was collected using different sensors. GeoAI techniques have been used to understand the dynamic interactions between the environment and individual factors in the context of walkability and outdoor comfort. The plots have been generated using the data collected in the experiment conducted by Liu *et al.* (2023a).

The developed GeoAI model is spatio-temporal-explicit because it captures the interactive nature of humans and surrounding built and unbuilt environments to predict human comfort through sequential movements. It is an example of a study that demonstrates the superiority of GeoAI approaches for urban sensing as it delivers substantially better accuracy than traditional machine learning models and some state-of-art deep learning frameworks.

The work also sets the scene for the consideration of dynamic data and GeoAI in urban digital twins, as it may help establishing the integration of GeoAI-empowered models in digital twins to achieve location-attentive predictions and human-centric simulations.

17.2.4   SENSING THE PERCEPTION AND QUALITY OF UNDERINVESTIGATED URBAN SPACES

Street-level imagery, an emerging dataset overviewed in Section 17.2.2, has been largely confined to roads due to being acquired from cameras mounted on cars, affecting downstream analyses that have almost always considered only driveable streets instead of the entire urban environment.

Luo *et al.* (2022b) noticed that in platforms offering street-level imagery (e.g., Google Street View and Mapillary), in some cities, imagery has been collected also from boats (e.g., on water bodies such as urban rivers). The paper dubs such data as "water view imagery", and positions it in the broader context of sensing urban waterscapes, an increasingly relevant topic in urban planning.

The paper presents a comprehensive perception study of multiple dimensions of waterfronts using GeoAI. It performed semantic segmentation of water level scenarios using SegFormer (Xie *et al.*, 2021), and developed a comprehensive set of indexes for urban waterscape evaluation. These are then linked to the perception of these scenes, which were collected through a survey. Thanks to the large-scale availability of data and AI techniques employed, for the first time, a global study was conducted. Its findings are several – the analysis reveals the heterogeneity of riverscapes around the world, and based on the analyses of the relationship between the developed indexes and the subjective visual perception, it finds the drivers of appealing urban waterscape design.

The work of Chen and Biljecki (2023) has focused on "off-road" imagery, a growing subset of street-level imagery that covers public open spaces such as parks, largely enabled by the emergence of crowdsourced street-level imagery, in which contributors use heterogeneous approaches, equipment, and platforms to collect imagery in cities (Figure 17.4). This example doubles as one that reflects another trend mentioned in the Introduction – it is an image that has been obtained from a crowdsourced platform (Mapillary), which has been seeing a surge in volume and coverage (Ding *et al.*, 2021; Hou and Biljecki, 2022; Juhász and Hochmair, 2016; Ma *et al.*, 2019), and it is increasingly used in urban studies (Yap *et al.*, 2022).

Thanks to such data, a method, which relies on computer vision techniques, was developed to establish an automated approach to evaluate open spaces and understand their quality from the human perspective. In a case study conducted across two cities, 800 public open spaces have been evaluated using traditional geospatial and remote sensing data, with the addition of street-level imagery. Thanks to GeoAI, the work shows that this emerging dataset can be used to automatically sense the rarely considered off-road areas, and it provides a convincing advancement over traditionally used data, potentially contributing to policy-making related to open spaces.

Figure 17.4 An unconventional street-level image, taken on a footpath by a pedestrian in the Botanic Gardens in Singapore, as opposed to the typical imagery collected from cars on driveable roads. New forms of existing types of data open new opportunities for urban sensing, which need to be accompanied by the development of new GeoAI approaches to extract reliable insights from such unorthodox platforms. (*Source:* Mapillary.)

17.3 CHALLENGES AND OPPORTUNITIES

Quality of data

The reliability of GeoAI techniques in urban sensing much relies on the quality of data. However, quality is not a topic that is often mentioned in papers, thus, there is an impression that it is usually taken for granted. This matter is especially important considering the increasing use of data that has a crowdsourced provenance, such as OpenStreetMap. Further, there is a lack of data quality standards and assessment procedures tailored for emerging urban datasets (but also some traditional data), which hinders understanding and quantifying the level of quality of data. Such may be relevant for establishing data requirements for certain GeoAI approaches.

On the flip side, in general, we are witnessing an increasing quality of urban data. Various data sources have been increasing in completeness. Further, thanks to the advancements in sensors, remotely sensed data has been increasing in resolution. Such developments will benefit the application of GeoAI, and may potentially lead to the introduction of new use cases (such as the one described in Section 17.2.1) and increased reliability.

In terms of quality, some datasets are also increasing in timeliness, i.e., the temporal resolution, reflecting urbanization and the dynamic nature of urban environments. These may present an opportunity for further use cases. For example, as some cities have been imaged multiple times in the past decade, street-level imagery is often available across multiple epochs. However, very few studies take advantage of the availability of historical data (Byun and Kim, 2022; Li *et al.*, 2022; Zhang *et al.*, 2023b). Moving forward, certain GeoAI techniques may need to be improved to cater to temporal studies, especially as some may require an increased level of reliability for detecting subtle changes between two periods in time.

Emerging data streams

A notable development in urban sensing, and one that has implications for GeoAI, is the multiplication of the types and sources of urban data in the last several years. Besides examples of datasets described earlier in the chapter, emerging urban data streams such as ground-based infrared thermography sensing the cityscape (Arjunan *et al.*, 2021; Dobler *et al.*, 2021; Martin *et al.*, 2022), crowdsourcing indoor images (e.g., from Airbnb listings) to understand local culture (Liu *et al.*, 2019), using take-out data to extract dietary patterns across different population groups in a city and link dietary habits to the sense of place (Xu *et al.*, 2022b), text-mining hotel reviews to learn environmental quality complaints at a large-scale (Ma *et al.*, 2023) and scraping real estate listings (text and photos) to collect data on buildings (Chen and Biljecki, 2022), are just examples among many instances. These datasets provide new perspectives and insights that contribute to understanding urban environments and more reliable decision-making, but at the same time, provide also certain challenges. Further, new variants of existing data, such as the one presented in Figure 17.4 and Section 17.2.4, present particular challenges and opportunities as well.

Some of the challenges include the integration of such data with traditional urban data, entailing the development of data fusion techniques, and the lack of understanding of quality and bias, and data requirements for the development of GeoAI techniques. The two case studies described in Section 17.2.4 demonstrate that there are large and distinct unexplored subsets of data, which may necessitate the adaptation of GeoAI workflows to suit them. While focusing on "off-road imagery", the work of Chen and Biljecki (2023) uses well established approaches developed to extract insights from standard street-level imagery collected on driveable roads, but it reveals limitations as they are not perfectly adequate for such data and use case, e.g., many facilities in public open space, such as amenities for children in parks are difficult to be detected by existing models developed for streetscapes. It calls attention to the development of methods specifically developed for such type of imagery. Next, some emerging data sources should not be taken for granted. Researchers focusing on social sensing have benefited greatly from the availability of social media about a decade ago, but since then, such platforms have experienced curbs and restricted access, depriving an entire research line of suitable data. Finally, because of the novelty of such data streams, privacy and bias that GeoAI models might introduce may not be fully understood yet.

However, opportunities are several. As exemplified earlier in the chapter and with dozens of references, emerging datasets, coupled with GeoAI, enable penetrating into new application areas and developing new use cases previously not possible or bringing considerable improvement to existing ones. The novelty of some of these developments also offers an opportunity to establish data standards and benchmarks, which have been common in traditionally used datasets and have been used to support the development of AI techniques (Rottensteiner *et al.*, 2014). The development of standardised data formats and data structures, and accompanying elements such as metadata, may foster data interoperability, which might be relevant for reproducibility and collaboration. Finally, some of the emerging datasets can provide a better insight into the citizenry: their perception and behaviour. GeoAI may amplify such benefit by extracting more human-centric insights, which has been exemplified in Section 17.2.3 with an application on sensing outdoor human comfort thanks to the development of a novel GeoAI technique and crowdsourced data together with a few other emerging data sources such as wearables.

Reproducible workflows and open data

Reproducibility is an increasingly important topic in GIScience (Wilson *et al.*, 2020), and it is described later in the book in a chapter specifically dedicated to this topic.

Reviewing the publications presenting applications of GeoAI for urban sensing, there is an impression that the research community has room for improvement regarding reproducibility. For example, a recent review of 250 papers using street-level imagery for urban sensing (Biljecki and Ito, 2021), a prominent and emerging data source in this domain (e.g., see Section 17.2.2), revealed that only a fraction of the studies has shared data and/or code openly, and there is often little information for reproducing the methods.

While recently there has been momentum in releasing open data related to urban sensing obtained with GeoAI techniques or developed to support them (Ju *et al.*, 2022; Kang *et al.*, 2020a; Luo *et al.*, 2022a,b; Piadyk *et al.*, 2023; Wu *et al.*, 2023; Zhang *et al.*, 2022), such efforts are still scarce, and an increasing number of studies employing GeoAI techniques for urban sensing calls attention to open data (He and He, 2023). Such matter is hindered by the fact that much of the developments rely on proprietary data, which limits their sharing, but also technical challenges – some workflows may simply be too complex and too computationally intensive to be reproduced by many others, especially those with limited resources, which may also impact equity in GeoAI research.

This aspect offers lots of opportunities for advancements. Researchers may adopt reproducible research practices and help ensure that the GeoAI workflows can be reproduced.

Ethical concerns and bias

Ethical concerns and bias are being increasingly discussed in the GeoAI community (Janowicz, 2023), with many concerns applicable in the domain of urban sensing. This is in particular important in research that includes human-related aspects, such as perception, as GeoAI can perpetuate biases present in the data. Recent work commenced investigations in this line, e.g., Kang *et al.* (2023) worked on understanding the local variations of human perception of safety (i.e., comparing safety perception scores of people living in the study area versus using a global model), tackling potential ethical issues in GeoAI such as population bias.

With the increasing deployment of sensors and increasing data quality, privacy is another concern. GeoAI researchers need to ensure to employ techniques that guarantee the protection of personal data. Further, as a large number of studies involves human participants (e.g., to collect data on perception), researchers need to ensure that they have obtained informed consent and the approval from institutional review boards.

Many of potentially sensitive datasets are collected and managed by corporations, which is a point raised also by Duarte and Ratti (2021), and which may present an opportunity for the development of legal and regulatory acts to safeguard privacy and safety.

Capacity building and domain knowledge

GeoAI for urban sensing requires an intricate set of skills. These are not only related to artificial intelligence and geospatial data but include also domain knowledge, which may often not be available. As such combination is rare and unique, entry barriers remain high. Further, urban sensing research involving GeoAI benefits from understanding the entire geospatial process, which does not include only processing and analysing data but also considering its provenance and quality. An opportunity is the development of educational initiatives to support the development of holistic skills.

17.4   CONCLUSION

This chapter has discussed recent trends and developments of applications of GeoAI for urban sensing across multiple types of data, urban challenges, and application domains. While a single book chapter cannot give justice to all the GeoAI-powered research related to urban sensing, it gives a high-level overview of a few recent efforts that are representative of the use of GeoAI for advancing urban sensing under the umbrella of urban informatics.

The proliferation of urban sensing and GeoAI techniques, while providing unprecedented opportunities and advancements, entails some challenges and issues, such as privacy and ethics concerns. The aspects described in the previous section are by no means complete. There are many further topics that require attention and may represent viable research directions.

The lack of interpretability and the "black-box" nature of existing GeoAI approaches has spurred discussions on the need for explainable GeoAI (GeoXAI) (Hsu and Li, 2023; Xing and Sieber, 2023). This topic is certainly relevant for urban sensing as well. Some of the recent urban sensing work involving GeoAI started tackling this matter (Liu *et al.*, 2023b).

Much of emerging datasets are textual, which may be propelled with the recent popularity of large language models, that is, their improved text analysis capabilities may enable deriving more reliable and meaningful insights from such data, leading to the improvement of applications or introduction of new use cases.

Generalization of GeoAI models is another important topic. Much of the approaches have been developed in a single or a few cities, and may not be fully generalised elsewhere. Further, some GeoAI approaches that are ostensibly designed to be global may have a variable performance across cities, and potentially introduce bias toward certain geographies and types of urban areas.

Finally, while GeoAI and urban sensing have made great strides in the past few years, research rarely translates into actionable policies and adoption by policy makers. At the moment, it appears that the gulf between the two is too wide to be bridged in many cases, and it can be attributed to issues such as limited capacity of policymakers to implement GeoAI-based solutions, regulatory barriers, lack of trust of reliability of the approaches, and limited interpretability of GeoAI models.

In conclusion, GeoAI, largely thanks to developments in computer science, proliferation of emerging datasets and increasing computational resources, continues to revolutionise urban sensing, by introducing more efficient, more comprehensive, and more accurate insights and analyses of urban environments. It holds immense potential for supporting the development of sustainable and liveable cities with continued methodological advancements, inclusion of stakeholders from multiple domains, growing efforts of integrating data sources, and increasing consideration of ethical considerations.

ACKNOWLEDGMENTS

Discussions with the members of the NUS Urban Analytics Lab are gratefully acknowledged. This research is part of the projects: Large-scale 3D Geospatial Data for Urban Analytics, which is supported by the National University of Singapore under the Start Up Grant R-295-000-171-133; and Multi-scale Digital Twins for the Urban Environment: From Heartbeats to Cities, which is supported by the Singapore Ministry of Education Academic Research Fund Tier 1.

BIBLIOGRAPHY

Abirami, S. and Chitra, P., 2022. Probabilistic air quality forecasting using deep learning spatial–temporal neural network. *GeoInformatica*, 27 (2), 199–235.

Andris, C. and Lee, S., 2021. Romantic relationships and the built environment: a case study of a U.S. college town. *Journal of Urbanism: International Research on Placemaking and Urban Sustainability*, 1–22.

Arjunan, P., *et al.*, 2021. Operational characteristics of residential air conditioners with temporally granular remote thermographic imaging. *In: BuildSys '21: Proceedings of the 8th ACM International Conference on Systems for Energy-Efficient Buildings, Cities, and Transportation.* 184–187.

Bai, Y., *et al.*, 2022. Knowledge distillation based lightweight building damage assessment using satellite imagery of natural disasters. *GeoInformatica*, 27 (2), 237–261.

Biljecki, F. and Ito, K., 2021. Street view imagery in urban analytics and GIS: A review. *Landscape and Urban Planning*, 215, 104217.

Bódis, K., *et al.*, 2019. A high-resolution geospatial assessment of the rooftop solar photovoltaic potential in the European Union. *Renewable and Sustainable Energy Reviews*, 114, 109309.

Byun, G. and Kim, Y., 2022. A street-view-based method to detect urban growth and decline: A case study of Midtown in Detroit, Michigan, USA. *PLOS ONE*, 17 (2), e0263775.

Calabrese, F., *et al.*, 2013. Understanding individual mobility patterns from urban sensing data: A mobile phone trace example. *Transportation Research Part C: Emerging Technologies*, 26, 301–313.

Chen, L.C., *et al.*, 2018. Encoder-decoder with atrous separable convolution for semantic image segmentation. *In*: *Proceedings of the European Conference on Computer Vision (ECCV)*. 801–818.

Chen, M., *et al.*, 2023. Artificial intelligence and visual analytics in geographical space and cyberspace: Research opportunities and challenges. *Earth-Science Reviews*, 104438.

Chen, S. and Biljecki, F., 2023. Automatic assessment of public open spaces using street view imagery. *Cities*, 137, 104329.

Chen, X. and Biljecki, F., 2022. Mining real estate ads and property transactions for building and amenity data acquisition. *Urban Informatics*, 1 (1), 12.

Cordts, M., *et al.*, 2016. The cityscapes dataset for semantic urban scene understanding. *In*: *Proceedings of the IEEE Conference on Computer Vision and Pattern Recognition*. 3213–3223.

Das, S., Sun, Q.C., and Zhou, H., 2022. GeoAI to implement an individual tree inventory: Framework and application of heat mitigation. *Urban Forestry & Urban Greening*, 74, 127634.

deSouza, P., *et al.*, 2020. Air quality monitoring using mobile low-cost sensors mounted on trash-trucks: Methods development and lessons learned. *Sustainable Cities and Society*, 60, 102239.

Ding, X., Fan, H., and Gong, J., 2021. Towards generating network of bikeways from Mapillary data. *Computers, Environment and Urban Systems*, 88, 101632.

Dobler, G., *et al.*, 2021. The urban observatory: A multi-modal imaging platform for the study of dynamics in complex urban systems. *Remote Sensing*, 13 (8), 1426.

Duarte, F. and Ratti, C., 2021. What urban cameras reveal about the city: The work of the senseable city lab. *In*: *Urban Informatics*. Springer Singapore, 491–502.

Gao, S., *et al.*, 2023. Special issue on geospatial artificial intelligence. *GeoInformatica*, 27 (2), 133–136.

Gao, S., *et al.*, 2021. User-generated content: A promising data source for urban informatics. *In*: *Urban Informatics*. Springer Singapore, 503–522.

Garrido-Valenzuela, F., Cats, O., and van Cranenburgh, S., 2023. Where are the people? Counting people in millions of street-level images to explore associations between people's urban density and urban characteristics. *Computers, Environment and Urban Systems*, 102, 101971.

Ghahramani, M., Zhou, M., and Wang, G., 2020. Urban sensing based on mobile phone data: approaches, applications, and challenges. *IEEE/CAA Journal of Automatica Sinica*, 7 (3), 627–637.

Hamilton, W., Ying, Z., and Leskovec, J., 2017. Inductive representation learning on large graphs. *Advances in Neural Information Processing Systems*, 30.

Han, J.Y., Chen, Y.C., and Li, S.Y., 2022. Utilising high-fidelity 3D building model for analysing the rooftop solar photovoltaic potential in urban areas. *Solar Energy*, 235, 187–199.

He, X. and He, S.Y., 2023. Using open data and deep learning to explore walkability in Shenzhen, China. *Transportation Research Part D: Transport and Environment*, 118, 103696.

Hochreiter, S. and Schmidhuber, J., 1997. Long short-term memory. *Neural Computation*, 9 (8), 1735–1780.

Hou, Y. and Biljecki, F., 2022. A comprehensive framework for evaluating the quality of street view imagery. *International Journal of Applied Earth Observation and Geoinformation*, 115, 103094.

Hsu, C.Y. and Li, W., 2023. Explainable GeoAI: can saliency maps help interpret artificial intelligence's learning process? an empirical study on natural feature detection. *International Journal of Geographical Information Science*, 37 (5), 963–987.

Hu, Y., *et al.*, 2015. Extracting and understanding urban areas of interest using geotagged photos. *Computers, Environment and Urban Systems*, 54, 240–254.

Ibrahim, M.R., Haworth, J., and Cheng, T., 2020. Understanding cities with machine eyes: A review of deep computer vision in urban analytics. *Cities*, 96, 102481.

Janowicz, K., 2023. Philosophical foundations of geoai: Exploring sustainability, diversity, and bias in geoai and spatial data science.

Janowicz, K., *et al.*, 2019. GeoAI: spatially explicit artificial intelligence techniques for geographic knowledge discovery and beyond. *International Journal of Geographical Information Science*, 34 (4), 625–636.

Jayathissa, P., *et al.*, 2019. Is your clock-face cozie? A smartwatch methodology for the in-situ collection of occupant comfort data. *Journal of Physics: Conference Series*, 1343 (1), 012145.

Ju, Y., Dronova, I., and Delclòs-Alió, X., 2022. A 10 m resolution urban green space map for major Latin American cities from Sentinel-2 remote sensing images and OpenStreetMap. *Scientific Data*, 9 (1).

Juhász, L. and Hochmair, H.H., 2016. User contribution patterns and completeness evaluation of Mapillary, a crowdsourced street level photo service. *Transactions in GIS*, 20 (6), 925–947.

Kang, Y., *et al.*, 2023. Assessing differences in safety perceptions using GeoAI and survey across neighbourhoods in Stockholm, Sweden. *Landscape and Urban Planning*, 236, 104768.

Kang, Y., *et al.*, 2020a. Multiscale dynamic human mobility flow dataset in the U.S. during the COVID-19 epidemic. *Scientific Data*, 7 (1).

Kang, Y., Gao, S., and Roth, R.E., 2019. Transferring multiscale map styles using generative adversarial networks. *International Journal of Cartography*, 5 (2-3), 115–141.

Kang, Y., *et al.*, 2020b. A review of urban physical environment sensing using street view imagery in public health studies. *Annals of GIS*, 26 (3), 261–275.

Kang, Y., *et al.*, 2021. Human settlement value assessment from a place perspective: Considering human dynamics and perceptions in house price modeling. *Cities*, 118, 103333.

Kruse, J., *et al.*, 2021. Places for play: Understanding human perception of playability in cities using street view images and deep learning. *Computers, Environment and Urban Systems*, 90, 101693.

Lai, W.W.L., 2021. Underground utilities imaging and diagnosis. *In: Urban Informatics*. Springer Singapore, 415–438.

Li, M., *et al.*, 2022. Marked crosswalks in US transit-oriented station areas, 2007–2020: A computer vision approach using street view imagery. *Environment and Planning B: Urban Analytics and City Science*, 50 (2), 350–369.

Li, W., 2020. GeoAI: Where machine learning and big data converge in GIScience. *Journal of Spatial Information Science*, (20).

Li, W. and Hsu, C.Y., 2022. GeoAI for Large-Scale Image Analysis and Machine Vision: Recent Progress of Artificial Intelligence in Geography. *ISPRS International Journal of Geo-Information*, 11 (7), 385.

Liang, X. and Andris, C., 2021. Measuring McCities: Landscapes of chain and independent restaurants in the United States. *Environment and Planning B: Urban Analytics and City Science*, 49 (2), 585–602.

Liu, P. and Biljecki, F., 2022. A review of spatially-explicit GeoAI applications in Urban Geography. *International Journal of Applied Earth Observation and Geoinformation*, 112, 102936.

Liu, P., *et al.*, 2023a. Towards human-centric digital twins: Leveraging computer vision and graph models to predict outdoor comfort. *Sustainable Cities and Society*, 93, 104480.

Liu, X., *et al.*, 2019. Inside 50,000 living rooms: an assessment of global residential ornamentation using transfer learning. *EPJ Data Science*, 8 (1).

Liu, Y., *et al.*, 2015. Social sensing: A new approach to understanding our socioeconomic environments. *Annals of the Association of American Geographers*, 105 (3), 512–530.

Liu, Y., *et al.*, 2023b. An interpretable machine learning framework for measuring urban perceptions from panoramic street view images. *iScience*, 26 (3), 106132.

Lu, Y., *et al.*, 2023. Assessing urban greenery by harvesting street view data: A review. *Urban Forestry & Urban Greening*, 83, 127917.

Luo, J., *et al.*, 2022a. Semantic Riverscapes: Perception and evaluation of linear landscapes from oblique imagery using computer vision. *Landscape and Urban Planning*, 228, 104569.

Luo, J., *et al.*, 2022b. Water View Imagery: Perception and evaluation of urban waterscapes worldwide. *Ecological Indicators*, 145, 109615.

Ma, D., *et al.*, 2019. The State of Mapillary: An Exploratory Analysis. *ISPRS International Journal of Geo-Information*, 9 (1), 10.

Ma, N., *et al.*, 2023. Learning building occupants' indoor environmental quality complaints and dissatisfaction from text-mining Booking.com reviews in the United States. *Building and Environment*, 237, 110319.

Martí, P., Serrano-Estrada, L., and Nolasco-Cirugeda, A., 2019. Social media data: Challenges, opportunities and limitations in urban studies. *Computers, Environment and Urban Systems*, 74, 161–174.

Martin, M., *et al.*, 2022. Infrared thermography in the built environment: A multi-scale review. *Renewable and Sustainable Energy Reviews*, 165, 112540.

O'Keeffe, K.P., *et al.*, 2019. Quantifying the sensing power of vehicle fleets. *Proceedings of the National Academy of Sciences*, 116 (26), 12752–12757.

Pang, H.E. and Biljecki, F., 2022. 3D building reconstruction from single street view images using deep learning. *International Journal of Applied Earth Observation and Geoinformation*, 112, 102859.

Piadyk, Y., *et al.*, 2023. StreetAware: A high-resolution synchronized multimodal urban scene dataset. *Sensors*, 23 (7), 3710.

Psyllidis, A., *et al.*, 2022. Points of interest (POI): a commentary on the state of the art, challenges, and prospects for the future. *Computational Urban Science*, 2 (1).

Richter, K.F. and Scheider, S., 2023. Current topics and challenges in geoAI. *KI - Künstliche Intelligenz*.

Rottensteiner, F., *et al.*, 2014. Results of the ISPRS benchmark on urban object detection and 3D building reconstruction. *ISPRS Journal of Photogrammetry and Remote Sensing*, 93, 256–271.

Shahtahmassebi, A.R., *et al.*, 2021. Remote sensing of urban green spaces: A review. *Urban Forestry & Urban Greening*, 57, 126946.

Shi, W., 2021. Introduction to urban sensing. *In*: *Urban Informatics*. Springer Singapore, 311–314.

Shin, D., *et al.*, 2015. Urban sensing: Using smartphones for transportation mode classification. *Computers, Environment and Urban Systems*, 53, 76–86.

Song, Y., *et al.*, 2023. Advances in geocomputation and geospatial artificial intelligence (GeoAI) for mapping. *International Journal of Applied Earth Observation and Geoinformation*, 103300.

Tartarini, F., Miller, C., and Schiavon, S., 2022. Cozie Apple: An iOS mobile and smartwatch application for environmental quality satisfaction and physiological data collection.

Tu, W., *et al.*, 2021. User-generated content and its applications in urban studies. *In*: *Urban Informatics*. Springer Singapore, 523–539.

Wilson, J.P., *et al.*, 2020. A Five-Star Guide for Achieving Replicability and Reproducibility When Working with GIS Software and Algorithms. *Annals of the American Association of Geographers*, 111 (5), 1311–1317.

Wu, A.N. and Biljecki, F., 2021. Roofpedia: Automatic mapping of green and solar roofs for an open roofscape registry and evaluation of urban sustainability. *Landscape and Urban Planning*, 214, 104167.

Wu, W.B., *et al.*, 2023. A first Chinese building height estimate at 10 m resolution (CNBH-10 m) using multi-source earth observations and machine learning. *Remote Sensing of Environment*, 291, 113578.

Xie, E., *et al.*, 2021. SegFormer: Simple and efficient design for semantic segmentation with transformers. *Advances in Neural Information Processing Systems*, 34.

Xing, J. and Sieber, R., 2023. The challenges of integrating explainable artificial intelligence into GeoAI. *Transactions in GIS*.

Xu, S., *et al.*, 2022a. Detecting spatiotemporal traffic events using geosocial media data. *Computers, Environment and Urban Systems*, 94, 101797.

Xu, Y., *et al.*, 2022b. Perception of urban population characteristics through dietary taste patterns based on takeout data. *Cities*, 131, 103910.

Yan, Y., *et al.*, 2020. Volunteered geographic information research in the first decade: a narrative review of selected journal articles in GIScience. *International Journal of Geographical Information Science*, 34 (9), 1–27.

Yang, J., *et al.*, 2021. The financial impact of street-level greenery on New York commercial buildings. *Landscape and Urban Planning*, 214, 104162.

Yang, Y., *et al.*, 2023. Embracing geospatial analytical technologies in tourism studies. *Information Technology & Tourism*.

Yap, W., Chang, J.H., and Biljecki, F., 2022. Incorporating networks in semantic understanding of streetscapes: Contextualising active mobility decisions. *Environment and Planning B: Urban Analytics and City Science*, 239980832211388.

Zhang, F., *et al.*, 2023a. Street-level imagery analytics and applications. *ISPRS Journal of Photogrammetry and Remote Sensing*, 199, 195–196.

Zhang, F., *et al.*, 2020. Uncovering inconspicuous places using social media check-ins and street view images. *Computers, Environment and Urban Systems*, 81, 101478.

Zhang, Y., Liu, P., and Biljecki, F., 2023b. Knowledge and topology: A two layer spatially dependent graph neural networks to identify urban functions with time-series street view image. *ISPRS Journal of Photogrammetry and Remote Sensing*, 198, 153–168.

Zhang, Z., *et al.*, 2022. Vectorized rooftop area data for 90 cities in China. *Scientific Data*, 9 (1).

Zhao, T., *et al.*, 2023. Sensing urban soundscapes from street view imagery. *Computers, Environment and Urban Systems*, 99, 101915.

Section IV

Perspectives for the Future of GeoAI

18 Reproducibility and Replicability in GeoAI

Peter Kedron, Tyler D. Hoffman, and Sarah Bardin
School of Geographical Sciences and Urban Planning, Arizona
State University

CONTENTS

18.1 Introduction 369
18.2 The Scientific Investigation of GeoAI...370
    18.2.1 GeoAI Methods and Applications 370
    18.2.2 The Scientific Investigation of GeoAI...371
    18.2.3 Reproducing and Replicating GeoAI Research...............................372
18.3 Facilitating the Reproducibility of GeoAI Research by Sharing Prove-
    nance Information...373
    18.3.1 Study Design..374
    18.3.2 Spatial Data, Data Processing, and Metadata 375
    18.3.3 Analytical Procedures and Code..377
    18.3.4 Computational Environment...378
18.4 A Research Agenda to Improve the Reproducibility and Replicability of
    GeoAI 379
    18.4.1 Research Directions to Enhance the Reproducibility of GeoAI.......379
    18.4.2 Research Directions Examining the Replicability of GeoAI............382
18.5 Concluding Remarks 383
    Bibliography...384

18.1 INTRODUCTION

As this handbook demonstrates, the use of artificial intelligence (AI) to study geographic problems has quickly moved to the forefront of the geographic research
agenda (see Smith 1984; Couclelis 1986; S. Openshaw and C. Openshaw 1997;
Janowicz et al. 2020). In recent years, researchers have applied AI to geographically-
referenced data to address a wide array of research problems. From these efforts, a
healthy literature has emerged centered on adapting and extending traditional AI
techniques for spatial applications (see Wang and Li 2021; Li and Hsu 2022; Liu

DOI: 10.1201/9781003308423-18

and Biljecki 2022; Zhu et al. 2021). Collectively, these studies and techniques have become known as GeoAI.

While GeoAI research has rapidly progressed, particularly with respect to the development of new GeoAI methods and applications, less attention has been given to how GeoAI fits within the broader traditions and function of scientific inquiry within the field of Geography. While many GeoAI articles broadly frame methodological developments and applications within the context of a Fourth Paradigm for scientific exploration (Hey 2009), the authors of those articles rarely directly discuss how their work might relate to, or benefit from, core principles of scientific discovery. We highlight two such principles in this chapter, reproducibility and replicability, and discuss how each relates and contributes to the development and use of GeoAI. Across paradigms, researchers reproduce prior studies by repeating procedures using the same data to understand and assess the methods used and the claims made in that work. Researchers replicate prior studies by gathering new data and using similar procedures to assess whether existing methods and results hold in new settings. Through repeated reproductions and replications, the research community as a whole can accumulate evidence about the reliability and generalizability of prior work and establish an understanding of how GeoAI methods function across different contexts.

The remainder of this chapter emphasizes the roles reproductions and replications can play in the advancement of GeoAI and examines how GeoAI researchers might go about making their work more reproducible and replicable. To frame the discussion, we use the next section to identify the target objects of GeoAI research and outline how those objects can be subjected to scientific investigation. We highlight the roles reproduction and replication play in that process. Building on the existing literature, we also define reproducibility and replicability (R&R) in the context of GeoAI and discuss how distinguishing between degrees of R&R is useful for amassing knowledge about GeoAI. To help readers make their own work more reproducible, we use the third section to identify information that when documented and shared make it easier for an independent research team to reproduce a prior GeoAI study. We give particular attention in this section to some of the unique challenges presented when studying spatial phenomena using spatial data and GeoAI. Building on that discussion, we use the final section of the chapter to look ahead. We present several lines of reproduction and replication related inquiry that researchers could pursue to quicken the development of GeoAI.

18.2 THE SCIENTIFIC INVESTIGATION OF GEOAI

18.2.1 GEOAI METHODS AND APPLICATIONS

GeoAI research consists of two distinct objects of study: (1) methods, which include the mathematical formulas, models, and techniques that are used to analyze spatial data, and (2) phenomena, which are studied through the application of GeoAI methods. In this chapter, we focus first on the study of GeoAI methods, how research examining GeoAI methods can be made more reproducible, and the potential benefits of attempting reproductions of such studies. We do so because GeoAI methods are the critical analytical engines of GeoAI studies of geographic phenomena, which we return to in the final section.

GeoAI methods comprise a wide range of techniques from geographic knowledge graphs (Dsouza et al. 2021; Li 2022) to terrain feature detection (Zhou et al. 2021; Li and Hsu 2022) and image geolocalization (Lin, Belongie, and Hays 2013; Workman, Souvenir, and Jacobs 2015), among others and have been applied to study research problems across the spectrum of geographic subdisciplines, including urban geography (Liu and Biljecki 2022), transportation planning (Yao et al. 2020), and public health (Brdar et al. 2016). Although specific GeoAI methods vary in their purpose and implementation, this group of methods is often differentiated from conventional AI by the encoding of location-based information, which is subsequently accessible and used in learning tasks (Mai et al. 2022). It is this location encoding which differentiates GeoAI from traditional AI methods that rely solely on aspatial attribute information (Janowicz et al. 2020; Li 2020).

18.2.2   THE SCIENTIFIC INVESTIGATION OF GEOAI

Investigations of GeoAI methods follow a pattern that aligns with other forms of scientific inquiry. First, a research team formulates a hypothesis regarding how a GeoAI method is expected to operate under certain conditions. This hypothesis is typically based on the team's prior knowledge of similar methods, data structures, spatial relationships between objects, and other factors that may affect performance. The research team then devises an experiment with which to test the hypothesis. In the context of GeoAI, an experiment will likely involve the use of either synthetic data, where the spatial structure of the data is controlled by the research team, or ground truth data for which the true outcome is known to the research team. After testing the hypothesis, the research team analyzes the results to determine whether they support the hypothesis or not, and the team may adjust or update their hypothesis in light of the new information provided from the experiment.

In practice, changes to a GeoAI method, the computational environment in which it is implemented, or the spatial data to which it is applied can all affect the performance of the method. As such, in order to design, execute, and independently evaluate GeoAI methods, it is necessary to control for as many of these conditions as possible. For example, a researcher interested in understanding the effect different approaches to location encoding (e.g., encoding features of only the location of interest, encoding features from points in a defined neighborhood around the location of interest) have on the performance of a GeoAI method would need to hold constant other aspects of the programmatic implementation of the method, the computational environment, and application to isolate the variation in outcomes caused by changing the location encoding procedure.

Since the results from even a well-executed test of a GeoAI method provide only a single piece of evidence regarding the method's performance, repeated testing of the method is needed to establish its reliability. A central tenet of the scientific method is that when shared this process of repeated analysis will allow the research community as a whole to root out incorrect beliefs about the GeoAI method and will ultimately lead to methodological improvements such as increased efficiency or wider applicability. By progressively refining our methods to match the information gained from repeated analyses, researchers should be able to incrementally create GeoAI methods that can be reliably used to study geographic phenomena.

18.2.3   REPRODUCING AND REPLICATING GEOAI RESEARCH

The scientific investigation of GeoAI plays out across numerous studies and relies on independent researchers conducting reproduction and replication studies to check the credibility and reliability of prior work. Although reproductions and replications are distinct activities that serve different epistemological purposes[1], they each provide a means for identifying and evaluating the decisions made during the research process. The two activities are commonly differentiated by whether the study uses (1) the same data and (2) the same or similar procedures as the research it is evaluating, which helps distinguish the type of diagnostic evidence the activity will provide.

Following the Committee on Reproducibility and Replicability in Science (National Academies of Sciences, Engineering, and Medicine and others, 2019), and focusing on computational research most relevant to GeoAI, reproductions use the same data, computational steps, methods, code, and conditions of analysis to verify prior research. In contrast, replications use similar methods to answer the same scientific question, but gather new data to assess the consistency of a result. More succinctly, research is reproducible when the same data and same methods and code can be used to create the same result. Research is replicable when consistent results to the same question are obtained by examining new data with similar methods.

The terms "similar", "consistent", and "results" within the definitions of the Committee on Reproducibility and Replicability in Science leave space to debate what constitutes a reproduction or replication study and whether there are different types. Exploring these nuances, researchers have proposed "degrees of reproducibility" to differentiate studies intended to serve different epistemological purposes. For example, Goodman, Fanelli, and Ioannidis (2016) differentiate reproducibility by focusing on whether the reproducing author is attempting to recreate the methods, results, or inferences of the prior study. In contrast, the Turing Community[2] makes a distinction between reproducible, replicable, and robust research, where reproducible research holds constant the data and procedures, robust research subjects the same data to different analytical procedures, and replicable research subjects new data to the same analytical procedures. Gundersen and Kjensmo (2018) propose a similar three part distinction for the study of AI, but define each degree around the separation of AI algorithms from their programmatic implementation.

Blending and applying these approaches to GeoAI, we define three degrees of reproducibility[3]:

---

[1]Disciplinary literatures often invert the use of the terms reproduction and replication leading to much confusion (Barba 2018; Plesser 2018).

[2]Guide to Reproducible Research. Available at: `https://the-turing-way.netlify.app/reproducible-research/overview/overview-definitions.html`

[3]The reproducibility definitions assume the second study is conducted based on documentation created by the original research team.

R1: Experiment Reproducible: The results of a study are experiment reproducible when an independent research team can produce the same results using the
same implementation of a GeoAI method executed on the same data.

R2: Method Reproducible: The results of a study are method reproducible when
an independent research team can produce the same results using an alternative implementation of a GeoAI method executed on the same data.

R3: Replicable: The results of a study are replicable when an independent research team can produce the same results using an alternative but similar implementation of a GeoAI method executed on different data.

Both the definitions for experiment reproducibility (R1) and replicability (R3)
are consistent with the Committee on Reproducibility and Replicability in Science's
more general definitions offered at the top of this section, however, method reproducibility (R2) provides an important nuance for evaluating how the specific implementation of a GeoAI method may affect the results of a data analysis. For example, a
method reproduction could use different programs and computational environments
to implement a geographically neural network weighted regression (GNNWR). Variations in results across implementations could illustrate important differences in how
the methods is coded across packages or performs across environments.

By repeatedly performing independent experiment and method reproductions,
the research community can progressively accumulate evidence regarding how underlying algorithms and programmatic implementations affect the results of a GeoAI
analysis for a given dataset. Moreover, reproductions can help support replication
studies by ensuring that the procedures, code, and data used in the original analysis are sufficiently documented and accessible, which makes it easier for other researchers to reapply or extend these methods to new geographic contexts. When
a replication produces results consistent with prior reproductions, there is reason to
believe that the GeoAI method functions across datasets and implementations. In this
way reproductions, and reproducible research more broadly, facilitate the expansion
of scientific knowledge and the application of GeoAI methods. Similarly, through
performing independent replications, it is possible to progressively assess the generalizability of GeoAI methods to new data that may be drawn from new geographic
contexts or from a different time period.

18.3 FACILITATING THE REPRODUCIBILITY OF GEOAI RESEARCH BY SHARING PROVENANCE INFORMATION

When researchers in a field are unable to repeat studies because key information is
not available, or unwilling to repeat studies because the incentives to do so do not
exist, an error correcting mechanism of the scientific method cannot function. The
nascent literatures on the reproducibility and replicability of empirical AI research
and geographic research suggest that neither field has, as of yet, developed the documentation practices or incentive systems needed to support this form of continuous
error correction (Gundersen, Gil, and Aha 2018; Hutson 2018; Konkol, Kray, and
Pfeiffer 2019; Haibe-Kains et al. 2020; Kedron et al. 2021b). Only a small fraction
of AI researchers share the code or test data used to create their results (Gundersen

and Kjensmo 2018), despite the numerous platforms that exist to make AI research more transparent (Haibe-Kains et al. 2020). Geographers, similarly, have only recently begun to examine the reproducibility of work in their field (Brunsdon 2017; Nüst et al. 2018; Kedron et al. 2021a) and develop systems to facilitate the reporting of study characteristics necessary for reproducing and replicating geographic research (Leipzig et al. 2021; Nüst and Pebesma 2021; Wilson et al. 2021; Holler and Kedron 2022; Kedron et al. 2022).

A foundational challenge to the scientific investigation of GeoAI research, therefore, is the simple lack of sharing study information at a level of detail that is sufficient for independent researchers to interpret, recreate, and extend prior work (Janowicz et al. 2020; Goodchild et al. 2021). To help GeoAI researchers overcome this issue, we identify the information from a GeoAI study that needs to be documented and shared to facilitate experiment and method reproductions of that study. We focus on the information and research artifacts necessary for reproducibility, as this same information is also needed to design and implement replication studies.

Reproductions and replications of any type are possible when the original research team documents and shares information relevant to the creation of their results in sufficient detail that others can understand and repeat their work. This record of how a study is designed and executed is referred to as the provenance of the study and includes both the data-focused descriptive metadata and the process-focused contextual metadata linked to a result (Davidson and Freire 2008; Tullis and Kar 2021). The need to record the provenance of results is commonly translated into a call to share a study's data and code. However, closer consideration of the scientific function of reproductions suggests that more than just data and code is needed to interpret an existing study, undertake an experiment or method reproduction of that work, and evaluate and extend that work. Gundersen and Kjensmo (2018) suggest there is a sliding scale to the level of provenance documentation required to execute data and experiment reproductions. Experiment reproductions require information about study design, spatial data, GeoAI method, and computational environment. It is possible to conduct method reproductions with more limited information about the study design because the focus of the investigation is often the relative performance of different implementations. The authors claim that replications depend on still less documentation because the same data is not used. However, we argue that more complete provenance documentation not only makes it easier to design any type of reproduction study but also to interpret and compare the results of such a study to the original work.

To facilitate the evaluation and extension of a study through reproduction, the research team responsible for the original study should provide information about the following study elements: study design; spatial data, data processing, and metadata; analytical procedures and code; and computational environment.

18.3.1 STUDY DESIGN

Provenance begins with documenting and sharing the overarching design and setup of a GeoAI study. Documentation should explain the purpose of the study and the hypotheses and predictions being tested. While it may appear obvious, clearly stated, falsifiable hypotheses about the phenomena being studied or how a method is expected to perform are not included in many AI studies (Gundersen and Kjensmo 2018). Reproducing or replicating a study disconnected from its scientific purpose does little to advance understanding.

It is similarly important to include a description of how geographic space has been conceptualized within the GeoAI analysis, particularly with respect to decisions regarding spatial data representation and model selection. Communicating this information is important for several reasons. First, adopting a field-based or object-based representation of space can constrain the methods available to researchers (Miller and Wentz 2003). For example, returning to the issue of location encoding, adopting a raster data model may lend itself to discretization-based encoding, but foreclose the possibility of kernel-based encoders. Second, spatial data model selection can have cascading effects on the performance and interpretation of a GeoAI analysis. For instance, suppose a researcher is interested in using point of interest (POI) data for a GeoAI analysis. They may choose to study the data as points or to aggregate them to existing areal units (e.g., counties). In this scenario, point-based GeoAI methods may be less computationally efficient (requiring all the POIs) but may permit prediction of new POI locations, while using a GeoAI method for the areal data may be more computationally efficient but not allow location prediction. Finally, these decisions have important implications for method reproductions because reproductions rely on existing data.

18.3.2   SPATIAL DATA, DATA PROCESSING, AND METADATA

Attempting to reproduce the results of a study requires access to the data used in the original study but also a full description of how that data was processed. Ideally, GeoAI researchers will make the original unprocessed data available along with a metadata file that describes its origin and structure. If the data was processed prior to analysis, those changes should be documented and the resulting data derived from that process should be preserved alongside the original datafile. The FAIR principles (Wilkinson et al. 2016) provide useful guidelines for data management and stewardship.

Although access to the original data is critical for ensuring the reproducibility of GeoAI research, it may not always be possible, or appropriate, to disseminate the original data outside of the research team. For instance, the research team may have obtained restricted-use data that cannot be distributed publicly or the data may be too large to make available in a data repository. In these circumstances, it is important to include information regarding from whom the data was obtained, the version numbers or access dates of the information, and other details related to the data request, as many datasets change continuously. Indeed, the dynamic nature of many of the datasets used in GeoAI research is an identified concern in the existing literature (Janowicz et al. 2020).

In particular, GeoAI researchers must also consider key ethical issues related to data privacy, confidentiality, and security prior to making their data assets publicly

available. Because GeoAI methods typically rely on large amounts of data that may
include sensitive information, such as residential addresses of individuals, which
can be used alone or in combination with other data to identify individuals even
in anonymized data files (Crigger and Khoury 2019; Parasidis, Pike, and McGraw
2019), it is important to consider how data will be shared, stored, and subsequently
used once released. Balancing the needs of reproducibility and replicability with
safeguarding sensitive information is an ongoing area of concern for geography and
AI (Hartter et al. 2013; Janowicz, Sieber, and Crampton 2022; Kang, Gao, and Roth
2022).

Whenever spatial data is shared, however, GeoAI researchers should also be sure
to include metadata that describes the location, spatial extent, and spatial support
of their data. If the data used in a study is not available, a researcher attempting
to reproduce a study will not know when and where the prior work was conducted
which will not allow them to match the data used in the prior study. This information
is critical because spatial heterogeneities and spatial dependencies can make GeoAI
analyses sensitive to changes in location and spatial scale (Openshaw 1984). For
example, training a GeoAI method on point of interest (POI) data within a city versus
POI data for an entire state or country will yield different trained models reflecting
the dynamics of each region. As a result, each model may give different predictions
on the same testing region simply due to having been trained at different scales.

In GeoAI research, spatial metadata should also include information about how
the data was segmented into training, validation, and test sets to mitigate the potential
for data leakage. Data leakage occurs whenever a spurious relationship between the
predictor and response variables of an analysis exists due to the data collection, sampling, or processing decisions of the research team (Kaufman et al. 2012). Kapoor
and Narayanan (2022) identify data leakages of several types across train-test splits
as a principle challenge to the reproducibility and advancement of machine learning
research. Spatial data may be particularly susceptible to unexpected forms of leakage
due to the autocorrelated nature of many spatial datasets.

Of the leakage types identified by Kapoor and Narayanan (2022), two are particularly relevant to the provenance requirements for GeoAI research. First, a lack of
separation between training and test datasets can allow a GeoAI model to learn relationships between predictors and a response that would not be available if new data
were drawn randomly. Separating training and test data can be particularly challenging in GeoAI research because of the structure of spatial data. Even when data appear
to be cleanly separated, spatial dependencies can share information across the data
split and lead to improved model performance. This information sharing can become
a problem during method reproductions if the implementation of the GeoAI method
changes how information is shared across the split during analysis. Similar issues
can occur during replications. For example, if the new analysis shifted the scale at
which the training data was analyzed, it may well change the degree to which that
data is similar (share information) to data in the test set, leading to a reduction in
performance due solely to the researcher's choice of scale.

The second type of data leakage that should be addressed in the design of GeoAI
research and the record of its provenance is whether the data used to assess model

performance matches the distribution of scientific interest. This question is particularly challenging for geographic research where spatial heterogeneity calls into question whether a GeoAI model trained and tested in one location should be expected to perform the same in another location. In evaluating this question, Goodchild and Li (2021) argue that replication across space and time must be weak, meaning some but not all aspects of a model should hold across locations. As a result, an inability to replicate the findings of a study may be due to a number of factors, including (1) sampling bias, (2) variation in the strength of relationships across space and time, or (3) a truly spurious conclusion in the original analysis. Parsing out the cause of a lack of replicability remains an open area of research that we return to in the next section.

Addressing data leakages is difficult whenever a researcher has incomplete knowledge of the data generating process and the resulting heterogeneity and dependency structure of the spatial data. Nonetheless this is often the situation that a GeoAI researcher or someone reproducing or replicating their work faces. To facilitate reproducibility, a researcher should provide as much information as possible about the origins of their data and how they created training-test splits. Whenever possible, a researcher should also articulate in the reporting of their work how decisions to split spatial data are connected to their conceptualization of the data generating process. For example, a researcher studying species distributions on a mountainside may believe that changes in altitude create spatial dependencies in the species data that follow the slope of the environment. The researcher may translate this belief into their training-test split in different ways. If the researcher believes these dependencies create homogenous groups within different elevation bands, they may choose to segment the species data along elevation bands and create training-test splits within each band. If the researcher instead believes these dependencies reflect a linear improvement in fitness along the elevation gradient, they may design their train-test splits to sample equally along this gradient. A clear explanation of which data processing decision was made and how it relates to the motivation and conceptualization of the study allows an independent researcher reproducing the study to evaluate the execution of a study in relation to its reasoning. In cases where no reasoning is provided, the researcher pursuing a reproduction or replication should test the available data and any training-test splits for spatial dependencies and consider how the existence of spatial structure within the data may impact the performance of a GeoAI method.

18.3.3 ANALYTICAL PROCEDURES AND CODE

As presented above, how data are processed before being analyzed in a GeoAI study should be documented and shared. This same rule applies to how data are analyzed to generate results. Researchers should share the code used to implement a GeoAI analysis. Experiment reproductions depend on an exact match of data and GeoAI methods across studies. Sharing code helps researchers conducting experimental reproductions match and control the implementation of GeoAI methods. Accessible and interpretable code is also essential to designing and implementing method reproductions in which the implementation of a GeoAI method is changed. Having a coded implementation of a method at the start of a method reproduction minimizes the chances that an independent researcher unintentionally changes aspects of its implementation.

Current best practice is to document the implementation of the GeoAI method in the form of a digital notebook using literate programming. The code in these notebooks should also include information about the study's variables and hyperparameters and should be shared in a findable repository with a persistent link. However, it is important to preserve and present more than code. Including high-level descriptions of why the GeoAI method was selected, what it is intended to do, and description of how it operates are all important for making a clear, interpretable link between the conceptualization, implementation, and results. Sharing the motivation for method selection and implementation is particularly important because it makes a clear link to the hypotheses and predictions under investigation in the study. These connections allow researchers reproducing or replicating a study to evaluate the reasoning motivating a study as well as the consistency of results across studies.

18.3.4 COMPUTATIONAL ENVIRONMENT

Possessing detailed code and a clear understanding of how a method was selected and implemented does not control for the possibility that a result is the product of the computational environment of a study. As such, study documentation should include information about the computing environment, including the hardware and software used in a study. The hardware used in a GeoAI study typically centers on the computers and servers used but can also include devices used in data collection. Documentation should include information about the operating system, programming environment, programming libraries, and related software dependencies used. Nüst and Hinz (2019) provide a recipe for creating container files for geospatial research that capture a study's software stack. Container files package code and all related dependencies, so programs can run identically across computing environments. The preservation and sharing of such containers is an area of open research (Emsley and De Roure 2018; Watada et al. 2019). How study components are preserved and shared are connected to the computing environment, but so too are things like version control and code commenting practices. While reporting these may not be necessary for a reproduction study, they can bear greatly on the interpretability of study documentation and by extension the chances of a reproduction succeeding.

Nüst and Pebesma (2021) advocate compiling all of the above components into an executable research compendium that connects a manuscript describing the study results with the container file and workflow script that can be used to generate them. Assuming the manuscript adequately describes the experimental set-up and the container and script are executable, such a package can itself achieve experiment reproducibility. Resources such as Whole Tale (Brinckman et al. 2019; Chard et al. 2020) and BinderHub (Jupyter et al. 2018) provide systems and platforms to produce and share research compendia.

While resources are available, there are presently few computational environments specifically designed for GeoAI research (Stewart et al. 2022). Existing AI software platforms, such as PyTorch, Keras and TensorFlow (Abadi et al. 2015; Paszke et al. 2019), are building geographically specific libraries (see Stewart et al. 2022), however, there is yet to be widespread adoption of these coding environments. As a result, many GeoAI analyses are conducted using custom code. Without a common platform, recording, sharing, and subsequently understanding the logical component of the computing environment used in a GeoAI study can be difficult. As with other information about the computational environment, customizations to facilitate GeoAI studies should be recorded and shared to convey the nonstandard or novel steps the researcher took to carry out an analysis. If possible, packaging custom implementations into software libraries conforming to external geospatial data and programming standards can help facilitate reproducibility.

18.4 A RESEARCH AGENDA TO IMPROVE THE REPRODUCIBILITY AND REPLICABILITY OF GEOAI

Sharing the provenance information and research artifacts associated with GeoAI studies will make it easier for independent researchers to evaluate prior studies and continue to build GeoAI methods through reproduction and replication. However, identifying beneficial practices is not the same as implementing them. Both practical and conceptual barriers remain to the widespread pursuit of reproduction and replication studies of GeoAI research. In this section, we present several lines of reproduction and replication related inquiries that researchers could pursue to address such barriers and quicken the development of GeoAI.

18.4.1 RESEARCH DIRECTIONS TO ENHANCE THE REPRODUCIBILITY OF GEOAI

Perhaps the simplest way for researchers to improve the reproducibility of GeoAI research is to attempt and share more rigorously documented reproductions of work in the field. While this chapter, and several entries into the literature (Janowicz et al. 2020; Goodchild and Li 2021; Kedron et al. 2021b), offer a perspective on how to go about making GeoAI research more reproducible, actually attempting and sharing reproductions will create the evidence needed to assess our existing GeoAI methods and to identify areas where further work is needed.

To date, reproductions of AI and geographic research have primarily been pursued by individual researchers or small teams focused on specific subsets of studies (see Nüst et al. 2018). It may be possible to accelerate progress in GeoAI by organizing multiple researchers to conduct coordinated sets of reproductions targeted at specific challenges in the field. For example, a group of researchers reproducing the cellular automata and neural network based studies reviewed by Liu and Biljecki (2022) could shed light on which methods perform best for which applications. A coordinated series of method reproductions could also provide information about how to best implement those methods, potentially leading to some degree of convergence as to which GeoAI methods are most promising for further development in this application area.

GeoAI researchers can facilitate coordinated reproduction efforts by developing and openly sharing a well-documented collection of benchmark datasets designed to support spatial methods testing. Benchmarking typically relies on a single dataset. Use of a common benchmark dataset by independent researchers testing alternative methods allows those researchers to assess the relative performance of competing methods free from the influence of variation across datasets. To understand GeoAI performance in the presence of spatial heterogeneity benchmarking needs to be scaled to include data from multiple times and locations, which should be developed and curated for specific GeoAI problems. For example, a GeoAI researcher could develop a collection of population flow datasets from several cities known to vary in urban form and national context and use that data to assess the relative performance of the different convolutional and graph neural network methods for urban population flow prediction identified by Liu and Biljecki (2022). By documenting and sharing that analysis, and the data that underlies it, other researchers could test future methodological and programmatic advancements against these known techniques using the same collection of data.

A key challenge when developing benchmark data will be identifying which changes in location are likely to impact the performance of GeoAI methods when studying a particular phenomenon. This task amounts to identifying which locations share data generating processes and the contextual factors that mediate and moderate those processes. In empirical settings, researchers almost never possess this knowledge. If they did, GeoAI methods could be directly assessed in light of processes. Absent knowledge about processes and confounders, researchers can group locations using the similarity of specific variables. While there are many techniques to assign similarities across locations on particular dimensions (Hastie, Tibshirani, and Friedman 2009; Duque, Anselin, and Rey 2012), there is limited literature on how to relate these assessments to predictions of replicability when studying specific phenomena. Understanding, or at least having an initial set of hypotheses about, which similarities matter for replicability and how they are distributed in space is necessary for developing benchmark data collections that capture variation along those key dimensions. An approach based on first geographic principles might suggest that locations nearer in space can be expected to be more similar, or that spatial anisotropies may direct similarity in space, such as moving outward from city centers or down elevation gradients. However, researchers have yet to explore these connections or institute them in the creation of benchmark dataset collections.

Another barrier to making reproductions more common is the lack of incentives to pursue this type of work. While awareness of reproducibility has risen in recent years, documenting work, preparing and sharing data, and carefully communicating all aspects of a GeoAI project all require resource commitments, which can make reproducing GeoAI research burdensome. Policy changes on the demand side of the research equation, such as data and code sharing requirements by individual journal publications and badging systems that assess and share the reproducibility of research are first steps toward incentivizing and reducing the barriers to entry for reproducible research. Indeed, the International Journal of Geographic Information Sciences now require that authors share the data and code that support their work matching similar requirements in other prominent journals. However, both practices seem to lack the level of enforcement needed to change behavior. The recent push by funding agencies, such as the U.S. National Science Foundation (NSF), to encourage research on how to improve reproducibility may offer a way forward, but linking project funding to monitorable and enforceable reproducibility requirements would likely create more rapid change.

Similar investments by funding agencies into the computational infrastructure needed to change the supply side of the GeoAI research equation also present opportunities to improve the reproducibility of GeoAI studies and increase the number of reproductions in the field. The complexity of methods and size of the data used in GeoAI research can limit reproducibility by constraining the set of researchers capable of reproducing a study to those with access to specialized hardware or powerful supercomputing clusters (Yin et al. 2017). Research focused on the development of scalable training methods and models can improve reproducibility in GeoAI by relaxing these hardware constraints. An alternative approach is to centralize the computational resources needed and create accessible ecosystems in which researchers can create reproducible GeoAI studies. Perhaps the best example of this approach is the NSF-supported infrastructure available through the CyberGIS Center for Advanced Digital and Spatial Studies[4]. This center provides virtual access to high performance computing resources and supports computationally reproducible GeoAI workflow development using Jupyter Notebooks and the CyberGIS-Compute geospatial middleware framework. Using these resources, a researcher can document the provenance of their GeoAI study and share their research artifacts on a platform that another independent researcher could access and experimentally reproduce.

Linking these emerging computational ecosystems, and GeoAI studies more broadly, to a common provenance model and reporting standard would increase the legibility of provenance information and interoperability across studies. Provenance models standardize reporting about the entities (e.g., data, code, reports) made during the research process, the researchers that were involved in creating those things (agents), and the actions those researchers took to make them (activities). One option is to adopt the W3C PROV model (Missier, Belhajjame, and Cheney 2013), which has a specific syntax and ontology for recording entity, agent, and activity information. The W3C PROV model can be used to track these relationships between entities and agents and ascribe specific actions to individuals and times, which makes the development history of a study easier to understand. Similarly, if sections of the code used in a GeoAI program are borrowed from previous studies those relationships can also be tracked thereby recording the intellectual legacy of the new implementation.

As resources like CyberGIS become more accessible, the question of who is capable and can be expected to reproduce GeoAI research arises. GeoAI research requires highly specialized skills and training. However, a well-documented, well coded, and containerized GeoAI study hosted on an accessible virtual environment

---

[4]https://cybergis.illinois.edu

can be experimentally reproduced with the press of a few buttons. While few GeoAI studies are so well prepared, this possibility means that a wide range of people could potentially reproduce a study. Nonetheless, experimentally reproducing a result does not mean a person has the capacity to understand the methodology, coding, or implications of that result. Research into what capabilities and training is needed to not only reproduce but interpret well-documented and accessible GeoAI studies can help answer these questions. Inviting researchers of varied backgrounds to reproduce studies and documenting their experiences, as has been done in the Geospatial Software Institutes Fellowship programs[5], is one way to gather this information. Expanding these opportunities with supporting training programs to the non-research community would also offer further evidence about how to engage a larger audience in GeoAI research.

18.4.2   RESEARCH DIRECTIONS EXAMINING THE REPLICABILITY OF GEOAI

Goodchild and Li (2021) suggest that it may be possible to advance both AI research and geographic research by focusing on the replicability of GeoAI methods within and across heterogeneous locations. The authors suggest that due to the spatially heterogeneous nature of phenomena we should expect replication across locations to be weak, meaning we should expect some aspects of a GeoAI study to hold across locations but not all. For example, it may be reasonable for a researcher to expect the functional form and implementation of a GeoAI method to operate across locations, but that same researcher might not expect the method to produce the same parameter estimates or identify the same relationship structures in different places. The fundamental challenge identified by Goodchild and Li (2021) is that replications across locations are a form of out-of-sample prediction in which the researcher conducting the replication does not know if a change in location corresponds to a change in the population or a change in the data generating process. As a result, the error incurred from using a GeoAI model trained with data from a known training region to study a new region is not well-defined.

If GeoAI researchers believe that replications conducted in different locations must be weak, an important secondary consequence of that position is the need to move away from binary conceptualizations of the evidence offered by replications as simply supporting or refuting a prior study. Instead, a replication may provide evidence that supports some claims of a prior study but not others. For example, a replication that uses a machine learning model trained to detect landforms in a desert to predict landforms in a jungle may perform poorly, but that does not mean that the same machine learning method could not be trained in the jungle and then be used to predict in other jungles. Moving away from a binary understanding of replication, it will be useful for GeoAI researchers to identify which components of a study they are attempting to replicate before beginning their replication attempt. Replications across locations could be minimally differentiated by whether they are being conducted to test the performance of GeoAI methods or the inferences researchers made about a phenomenon in a prior study. That division could be further refined as each

---

[5]https://gsi.cigi.illinois.edu/geospatial-fellows-members

grouping itself has many components. For example, a researcher could identify if it is the location encoding technique, train-test split, or loss function used in a GeoAI method they wish to test across locations. Communicating the connection between the diagnostic evidence of a replication and the claims of a prior study will allow other researchers to build on the new evidence each replication provides.

As replications of GeoAI research are conducted across locations it will be essential to develop evidence accumulation systems, so the information they collectively provide can be leveraged to improve GeoAI methods and their application to the study of phenomena. One way forward is to use meta-analysis to assess the evidence generated from a series of GeoAI replication studies. Meta-analysis provides a systematic means for aggregating results from multiple studies to generate a distribution of plausible results. This distribution allows for a comparison of the magnitude, direction, and heterogeneity of effects based on all available evidence as opposed to the findings from a single study, which increases the power and generalizability of the evidence generated. Additionally, incorporating information regarding where individual replication studies were performed within the parameters of the meta-analysis can provide insights into how effects may vary across different locations and shed light onto the extent to which study findings are expected to be weakly replicable across space.

18.5 CONCLUDING REMARKS

GeoAI researchers have yet to develop an extensive literature on the ethics of the development and application of their methods and programs. As Janowicz, Sieber, and Crampton (2022) note, the ethics of GeoAI should rest on the social and environmental responsibilities involved in designing and applying GeoAI methods. Related to reproducibility and replicability this means a researcher should consider if and when it is appropriate to make their methods available for others to use and build upon or to reproduce or replicate prior work. Throughout this chapter, we have placed this question in the background and instead have highlighted the important role reproduction and replication play in assessing existing claims and methods. However, it is not clear how that potential benefit should be weighed against the potential downside of accelerating the development of techniques that could be repurposed to identify natural locations for development or marginalize selected groups.

GeoAI researchers also have yet to extensively discuss which ethical frameworks should be used as the foundation for these decisions. For example, if we recognize that data purchased or accessed for use in GeoAI research may have been gathered under ethical conditions that would fail to meet our standards, should we then use the data in our research? This question may produce contradictory answers under alternative approaches to normative ethics. For example, a deontological approach that emphasizes duties to scientific principles may reasonably suggest that we should not use these data to uphold those duties. Alternatively, a consequentialist approach focused on maximizing "the good" may reasonably suggest that data should be used to maximize total or distributed benefit to society. Both approaches face further challenges when placed in the context of replication. If a dataset similar to that used in a prior study cannot be gathered or created due to ethical concerns, then it may become difficult or impossible to assess that work through replication. From the consequentialist perspective, the inability to assess the work through replication may reduce the credibility of that work and therefore diminish the benefit it might provide. From the deontological perspective, validating the finding through replication would violate another set of duties compounding our unethical behavior. Advancing this type of discussion and making similar discussions a core part of GeoAI education will help not only improve the reproducibility and replicability of research but guide future researchers and the field as a whole toward better actions.

In this chapter, we examine the nexus of the AI, geography, and R&R literatures and attempt to point those working with GeoAI toward what we believe will be productive streams of research. As GeoAI studies are reproduced and replicated, this provisional agenda is sure to change. New evidence will emerge as researchers test and retest existing methods, new systems will be created to manage that process, and new standards of practice will emerge. It is therefore an exciting time to not only develop GeoAI methods and apply them to varied geographic phenomena, but to participate in the potential development of a new research infrastructure based on principles from the past and innovations from the present.

BIBLIOGRAPHY

Abadi, M., *et al.*, 2015. Tensorflow: Large-scale machine learning on heterogeneous systems. *arXiv preprint arXiv:1603.04467*.

Arnold, B., *et al.*, 2019. The turing way: a handbook for reproducible data science. *Zenodo*.

Barba, L.A., 2018. Terminologies for reproducible research. *arXiv preprint arXiv:1802.03311*.

Brdar, S., *et al.*, 2016. Unveiling spatial epidemiology of hiv with mobile phone data. *Scientific Reports*, 6 (1), 19342.

Brinckman, A., *et al.*, 2019. Computing environments for reproducibility: Capturing the "whole tale". *Future Generation Computer Systems*, 94, 854–867.

Brunsdon, C., 2017. Quantitative methods ii: Issues of inference in quantitative human geography. *Progress in Human Geography*, 41 (4), 512–523.

Chard, K., *et al.*, 2020. Toward enabling reproducibility for data-intensive research using the whole tale platform. *In: Parallel Computing: Technology Trends*. IOS Press, 766–778.

Couclelis, H., 1986. Artificial intelligence in geography: Conjectures on the shape of things to come. *The Professional Geographer*, 38 (1), 1–11.

Crigger, E. and Khoury, C., 2019. Making policy on augmented intelligence in health care. *AMA Journal of Ethics*, 21 (2), 188–191.

Davidson, S.B. and Freire, J., 2008. Provenance and scientific workflows: challenges and opportunities. *In: Proceedings of the 2008 ACM SIGMOD International Conference on Management of Data*. 1345–1350.

Dsouza, A., *et al.*, 2021. Worldkg: A world-scale geographic knowledge graph. *In: Proceedings of the 30th ACM International Conference on Information & Knowledge Management*. 4475–4484.

Duque, J.C., Anselin, L., and Rey, S.J., 2012. The max-p-regions problem. *Journal of Regional Science*, 52 (3), 397–419.

Emsley, I. and De Roure, D., 2017. A framework for the preservation of a docker container. *International Journal of Digital Curation*, 12 (2), 125–135.

Goodchild, M.F., *et al.*, 2021. Introduction: Forum on reproducibility and replicability in geography. *Annals of the American Association of Geographers*, 111 (5), 1271–1274.

Goodchild, M.F. and Li, W., 2021. Replication across space and time must be weak in the social and environmental sciences. *Proceedings of the National Academy of Sciences*, 118 (35), e2015759118.

Goodman, S.N., Fanelli, D., and Ioannidis, J.P., 2016. What does research reproducibility mean? *Science Translational Medicine*, 8 (341), 341ps12–341ps12.

Gundersen, O.E., Gil, Y., and Aha, D.W., 2018. On reproducible ai: Towards reproducible research, open science, and digital scholarship in ai publications. *AI Magazine*, 39 (3), 56–68.

Gundersen, O.E. and Kjensmo, S., 2018. State of the art: Reproducibility in artificial intelligence. *In: The AAAI Conference on Artificial Intelligence*. vol. 32.

Haibe-Kains, B., *et al.*, 2020. Transparency and reproducibility in artificial intelligence. *Nature*, 586 (7829), E14–E16.

Hartter, J., *et al.*, 2013. Spatially explicit data: stewardship and ethical challenges in science. *PLoS Biology*, 11 (9), e1001634.

Hastie, T., *et al.*, 2009. *The Elements of Statistical Learning: Data Mining, Inference, and Prediction*. vol. 2. Springer.

Hey, A.J., *et al.*, 2009. *The Fourth Paradigm: Data-Intensive Scientific Discovery*. vol. 1. Microsoft Research.

Holler, J. and Kedron, P., 2022. Mainstreaming metadata into research workflows to advance reproducibility and open geographic information science. *The Archives of Photogrammetry, Remote Sensing and Spatial Information Sciences*, 48, 201–208.

Hutson, M., 2018. Artificial intelligence faces reproducibility crisis. *Science*, 359 (6377), 725–726.

Janowicz, K., *et al.*, 2020. Geoai: spatially explicit artificial intelligence techniques for geographic knowledge discovery and beyond.

Janowicz, K., Sieber, R., and Crampton, J., 2022. Geoai, counter-ai, and human geography: A conversation. *Dialogues in Human Geography*, 12 (3), 446–458.

Jupyter, P., *et al.*, 2018. Binder 2.0-reproducible, interactive, sharable environments for science at scale. *In: Proceedings of the 17th Python in Science Conference*. 113–120.

Kang, Y., Gao, S., and Roth, R., 2022. A review and synthesis of recent geoai research for cartography: Methods, applications, and ethics. *In: Proceedings of AutoCarto*. 2–4.

Kapoor, S. and Narayanan, A., 2022. Leakage and the reproducibility crisis in ml-based science. *arXiv preprint arXiv:2207.07048*.

Kaufman, S., *et al.*, 2012. Leakage in data mining: Formulation, detection, and avoidance. *ACM TKDD*, 6 (4), 1–21.

Kedron, P., *et al.*, 2021a. Reproducibility and replicability in geographical analysis. *Geographical Analysis*, 53 (1), 135–147.

Kedron, P., *et al.*, 2022. Reproducibility, replicability, and open science practices in the geographical sciences. *OSF, Center for Open Science*, 1–10.

Kedron, P., *et al.*, 2021b. Reproducibility and replicability: opportunities and challenges for geospatial research. *International Journal of Geographical Information Science*, 35 (3), 427–445.

Konkol, M., Kray, C., and Pfeiffer, M., 2019. Computational reproducibility in geoscientific papers: Insights from a series of studies with geoscientists and a reproduction study. *International Journal of Geographical Information Science*, 33 (2), 408–429.

Leipzig, J., *et al.*, 2021. The role of metadata in reproducible computational research. *Patterns*, 2 (9), 100322.

Li, W., 2020. Geoai: Where machine learning and big data converge in GIScience. *Journal of Spatial Information Science*, (20), 71–77.

Li, W., 2022. Geoai in social science. *Handbook of Spatial Analysis in the Social Sciences*, 291–304.

Li, W. and Hsu, C.Y., 2022. Geoai for large-scale image analysis and machine vision: Recent progress of artificial intelligence in geography. *ISPRS International Journal of Geo-Information*, 11 (7), 385.

Lin, T.Y., Belongie, S., and Hays, J., 2013. Cross-view image geolocalization. *In: Proceedings of the IEEE Conference on Computer Vision and Pattern Recognition*. 891–898.

Liu, P. and Biljecki, F., 2022. A review of spatially-explicit geoai applications in urban geography. *International Journal of Applied Earth Observation and Geoinformation*, 112, 102936.

Mai, G., *et al.*, 2022. A review of location encoding for geoai: methods and applications. *International Journal of Geographical Information Science*, 36 (4), 639–673.

Miller, H.J. and Wentz, E.A., 2003. Representation and spatial analysis in geographic information systems. *Annals of the Association of American Geographers*, 93 (3), 574–594.

Missier, P., Belhajjame, K., and Cheney, J., 2013. The w3c prov family of specifications for modelling provenance metadata. *In: Proceedings of the 16th International Conference on Extending Database Technology*. 773–776.

National Academies of Sciences, Engineering, and Medicine and others, 2019. *Reproducibility and Replicability in Science*. National Academies Press.

Nüst, D., *et al.*, 2018. Reproducible research and giscience: an evaluation using agile conference papers. *PeerJ*, 6, e5072.

Nüst, D. and Hinz, M., 2019. Containerit: Generating dockerfiles for reproducible research with r. *Journal of Open Source Software*, 4 (40), 1603.

Nüst, D. and Pebesma, E., 2021. Practical reproducibility in geography and geosciences. *Annals of the American Association of Geographers*, 111 (5), 1300–1310.

Openshaw, S., 1984. The modifiable areal unit problem. *Concepts and Techniques in Modern Geography*, 60–69.

Openshaw, S. and Openshaw, C., 1997. *Artificial Intelligence in Geography*. John Wiley & Sons, Inc.

Parasidis, E., Pike, E., and McGraw, D., 2019. A belmont report for health data. *The New England Journal of Medicine*, 380 (16), 1493–1495.

Paszke, A., *et al.*, 2019. Pytorch: An imperative style, high-performance deep learning library. *Advances in Neural Information Processing Systems*, 32.

Plesser, H.E., 2018. Reproducibility vs. replicability: a brief history of a confused terminology. *Frontiers in Neuroinformatics*, 11, 76.

Rey, S.J. and Franklin, R.S., 2022. *Handbook of Spatial Analysis in the Social Sciences*. Edward Elgar Publishing.

Smith, T.R., 1984. Artificial intelligence and its applicability to geographical problem solving. *The Professional Geographer*, 36 (2), 147–158.

Stewart, A.J., *et al.*, 2022. Torchgeo: deep learning with geospatial data. *In: Proceedings of the 30th International Conference on Advances in Geographic Information Systems*. 1–12.

Tullis, J.A. and Kar, B., 2021. Where is the provenance? ethical replicability and reproducibility in GIScience and its critical applications. *Annals of the American Association of Geographers*, 111 (5), 1318–1328.

Wang, S. and Li, W., 2021. Geoai in terrain analysis: Enabling multi-source deep learning and data fusion for natural feature detection. *Computers, Environment and Urban Systems*, 90, 101715.

Watada, J., *et al.*, 2019. Emerging trends, techniques and open issues of containerization: a review. *IEEE Access*, 7, 152443–152472.

Wilkinson, M.D., *et al.*, 2016. The fair guiding principles for scientific data management and stewardship. *Scientific Data*, 3 (1), 1–9.

Wilson, J.P., *et al.*, 2021. A five-star guide for achieving replicability and reproducibility when working with gis software and algorithms. *Annals of the American Association of Geographers*, 111 (5), 1311–1317.

Workman, S., Souvenir, R., and Jacobs, N., 2015. Wide-area image geolocalization with aerial reference imagery. *In: Proceedings of the IEEE International Conference on Computer Vision*. 3961–3969.

Yao, X., *et al.*, 2020. Spatial origin-destination flow imputation using graph convolutional networks. *IEEE Transactions on Intelligent Transportation Systems*, 22 (12), 7474–7484.

Yin, D., *et al.*, 2017. A cybergis-jupyter framework for geospatial analytics at scale. *In: Proceedings of the Practice and Experience in Advanced Research Computing on Sustainability, Success and Impact*. 1–8.

Zhou, F., *et al.*, 2021. Land deformation prediction via slope-aware graph neural networks. *In: AAAI Conference on Artificial Intelligence*. vol. 35, 15033–15040.

Zhu, D., *et al.*, 2021. Spatial regression graph convolutional neural networks: A deep learning paradigm for spatial multivariate distributions. *GeoInformatica*, 26, 645–676.

19 Privacy and Ethics in GeoAI

Grant McKenzie
Platial Analysis Lab, Department of Geography, McGill
University, Montréal, Canada

Hongyu Zhang
Platial Analysis Lab, Department of Geography, McGill
University, Montréal, Canada

Sébastien Gambs
Département d'informatique, Université du Québec à Montréal,
Canada

CONTENTS

19.1 Introduction 388
    19.1.1 Data Privacy & AI 389
    19.1.2 Geoprivacy & GeoAI..390
19.2 Data Privacy Methods in GeoAI...392
    19.2.1 Obfuscation & anonymization 392
    19.2.2 Synthetic data generation...393
    19.2.3 Cryptography 394
    19.2.4 Re-identification methods & privacy attacks............................394
19.3 Application Areas..395
    19.3.1 Advertising 395
    19.3.2 Health care 396
    19.3.3 Security & surveillance..397
19.4 The Future of Privacy in GeoAI 398
    19.4.1 Suggested areas for improvement..398
    19.4.2 Emerging privacy topics in GeoAI 399
19.5 Summary..400
    Bibliography..400

19.1 INTRODUCTION

The number of companies, agencies, and institutions using artificial intelligence (AI) techniques has grown substantially over the past few years. Their goals are diverse and span application areas ranging from cashier-less grocery stores to breast cancer screening. As with any technology, these advancements have lead to important

DOI: 10.1201/9781003308423-19

discussions related to ethics. In particular, ethical concerns associated with such technologies range from the collection and storage of personal data to biases in model development and implementation. These concerns also encompass questions on how best to explain their predictions. While ethics is its own domain of research, the rapid development and adoption of AI techniques in many sectors of society has given rise to the field of ethical artificial intelligence (Mittelstadt, 2019). Researchers of ethics in AI aim to identify and investigate issues facing society that can specifically be attributed to the introduction and application of AI and related methods. Approaches to the topic most often include exploration and analysis of one or more themes such as privacy, surveillance, bias and/or discrimination (Naik *et al.*, 2022; Stahl and Wright, 2018).

Like many other aspects of AI, ethical concerns are also shifting. The field is changing so rapidly that legal experts, policy makers, and researchers are forced to continually revise their assessments of bias, transparency, social manipulation, and privacy in AI. Through increased public pressure, many leading technology companies have hired experts to help them navigate these waters and develop policies related to the ethical use of AI. Many private companies and government agencies regularly publish technical reports outlining AI guidelines and principles. A recent scoping review of 84 existing guidelines on ethical AI by Jobin *et al.* (2019) identified a set of ethical principles commonly included in these reports. The top five include *transparency, justice & fairness, non-maleficence, responsibility* and *privacy*. Each of these principles is worthy of its own book chapter, with numerous books having already been published on these topics (see Dubber *et al.* (2020), for instance).

In this chapter, we choose to focus our discussion on the ethical principle of privacy. To understand why, we must examine ethics as it relates to the topic of this book, namely *geographic* artificial intelligence (GeoAI). We argue that the same common set of AI ethical principles identified by Jobin *et al.* also apply to GeoAI, but that the relative importance, or ranking, of these principles has been modified. AI techniques that leverage the relationships of objects, events, and people in geographic space make GeoAI a unique subset of artificial intelligence. We argue here that ethical issues related to privacy are fundamentally different when viewed through a geographic lens. Thus, while a discussion on ethics and all of its themes are essential to the future of GeoAI research, the unique aspects of location privacy will be the focus of this chapter.

19.1.1 DATA PRIVACY & AI

In today's technocratic society, the privacy of one's personal information is of the utmost importance. Given big tech's penchant for collecting data for AI training purposes, people have become increasingly concerned about how their data are being used and how much control they retain over their data. Historically, the broader concept of privacy has been difficult to grasp, with definitions differing substantially depending on the domain considered. The word *private* is derived from the Latin *privatus*, which means to set apart from what is public, personal and belonging to oneself, and not to the state. Various efforts have been made to categorize privacy into different dimensions (Finn *et al.*, 2013; Pedersen, 1979) but many of them come to the conclusion that privacy is the right of an individual or group to control how information about them is accessed, shared, and used, thus being related to the concept of self-information determination. This is a data-centric definition of privacy, which is arguably the most applicable to the GeoAI context.

When the terms privacy, data, and AI are combined, most readers' minds go to a futuristic surveillance state reminiscent of George Orwell's Big Brother. While such a scenario is worthy of further discussion, there are a number of less Orwellian representations of privacy, or privacy violations, that should also be acknowledged. Many of these are less dramatic, but should be no less concerning to those that use AI technologies. As many have noted, the heart of most AI techniques is the data on which the models are trained – sometimes referred to as the *petrol* of AI. The provenance of these data, and details on the individuals from which these data are collected, continue to be a topic of much discussion among privacy researchers. In this era of Big Data we have also seen the emergence of data brokers purchasing and selling data for a variety of uses. Ethics related to data handling, and the confidentiality, anonymity, and privacy of the data all then become topics for further investigation. As the commercial appetite for data grows, we have seen a societal shift from people trading commodities to the information of those people now *being* the commodities. This has led to a significant change in our perception of privacy and the steps we take to ensure it (Zhang and McKenzie, 2022).

With respect to AI, a lot of what is being discussed is not about individual privacy from a philosophical position, but rather *data privacy*, or the rights of the individual to control what information is being collected, accessed, shared and analyzed. More precisely, privacy has the potential to be viewed as a value to uphold or a right to be protected. This latter definition is less about the "right to be left alone" and more about the right to control one's own information. There is a separate philosophical discussion to be had about privacy and AI but in this work we focus on the ethical concerns over data privacy in AI, and specifically GeoAI.

19.1.2   GEOPRIVACY & GEOAI

It has been two decades since Dobson and Fisher (2003) published their paper *Geoslavery*, an evocative call to action showcasing how geographic information systems, global navigation satellite systems, and location-based services can be used to control individuals. While technology trends have deviated from those mentioned in the paper, the idea that location is a unique attribute capable of exposing highly sensitive information remains. Location is inherently tied to identity. Indeed, a plethora of research has demonstrated that socio-economic and demographic characteristics such as race, income, education, and many others correlate with location (Riederer *et al.*, 2016; Zhong *et al.*, 2015). The places that we visit (*e.g.*, restaurants, bars, parks, etc.) and times we visit them are also closely tied to our demographics characteristics (Liu *et al.*, 2017; McKenzie and Mwenda, 2021). The mobility behaviour of an individual uniquely characterizes them and can be used for re-identification even from so called "anonymous data" (Gambs *et al.*, 2014). Thus, publicly sharing the places that one visits, without their knowledge, can be a major violation of their privacy. For instance, exposing the bar one patrons on a Saturday evening may be of little concern for a cisgender male in a North American city, but it may be of appreciable concern to a non-binary gender individual living in a nation in which it is illegal to identify as such. The link between location and identity make such data particularly sensitive and valuable. For developers of AI methods and tools, these data are an extraordinary resource on which to train models for applications areas such as human behavior and crime prediction, local business recommendations, or determining health insurance rates.

Geographers and demographers understand that access to an individual's location data is only the tip of the proverbial "privacy exposure iceberg". Paraphrasing the first law of geography, we know that things that are closer together in geographic space tend to me more similar (Tobler, 1970). From a data privacy ethics perspective, this means that gaining access to socio-demographic information about my neighbor (*e.g.*, income, race and age) means that one can infer my socio-demographic characteristics with a high degree of accuracy. This presents the uncomfortable reality that the privacy of an individual's personal information depends on the privacy of information of those in close proximity. The dilemma here is that, while I do not have control over the personal information that my neighbor chooses to share, I am impacted by the disclosure of such content. In the era of social media, user-generated content, and other sources of geo-tagged data, this means that it is possible to infer information about me purely based on my location and the contributions of people around me (Pensa *et al.*, 2019). This is often referred to as *co-location privacy*. AI technologies have amplified this allowing for data from multiple sources to be combined, multiplying probabilities by probabilities to infer details about people with shocking levels of accuracy. This leads to an entire new set of ethical considerations as we now see that sharing individual location information impacts collective or group privacy.

Despite the fact that location information is so important to our identity, it is surprisingly easy to capture. As outlined by Keßler and McKenzie (2018) in their *Geoprivacy Manifesto*, "ubiquitous positioning devices and easy-to-use APIs make information about an individual's location much easier to capture than other kinds of personally identifiable information". There are so many accessible data out there that the privacy of individual's locations has become a domain of research in and of itself. For instance, research has identified that the location of individuals can be inferred purely based on the text that people share online (Adams and Janowicz, 2012), the photos they post (Hasan *et al.*, 2020) or the time of day that they share information (McKenzie *et al.*, 2016). Armstrong *et al.* (2018) provide an excellent overview of the domain of *geoprivacy* including examples of some of the leading issues in location privacy research. Additional work has specifically reviewed the state of location privacy issues in mobile applications (Liu *et al.*, 2018) and cartographic publications (Kounadi and Leitner, 2014). Like many research domains, those working in geographic information science have renewed calls to investigate ethics as it relates to location privacy and many other themes (Nelson *et al.*, 2022). While not always purposeful, we are increasingly seeing GeoAI techniques used to de-anonymize location data, identify individuals, and violate individual privacy (Wang *et al.*, 2016). As we witness the emergence of GeoAI built on massive amounts of personal, location-tagged content and geospatial data, scientists are reminded of Dobson and Fisher's warning from the early 2000s. If GPS and GIS were perceived to be the harbingers of a geotechnology-enabled surveillance state, what then is GeoAI?

It is not all doom and gloom. The emergence of GeoAI has substantially impacted our society in a number of positive ways (many of which are showcased throughout this book). From a data privacy perspective, advances in GeoAI and affiliated machine learning models have made major contributions to privacy *preservation*. Numerous research teams have contributed to the emergence of new methods, techniques, and tools for obfuscating, anonymizing, encrypting, and protecting location information (Jiang *et al.*, 2021). Public-sharing location applications such as *Koi* (Guha *et al.*, 2012) or *PrivyTo* (McKenzie *et al.*, 2022) are being created that use many of these location obfuscation and data encryption techniques to put users back in control of their personal location information.

19.2 DATA PRIVACY METHODS IN GEOAI

A wide range of artificial intelligence and machine learning techniques exist that touch on privacy as it relates to geospatial data. These can be split between one group that focuses on protection mechanisms such as privacy-preservation, anonymization, and obfuscation, and a second group dedicated to privacy attacks such as re-identification, de-anonymization, and privacy exposure.

19.2.1 OBFUSCATION & ANONYMIZATION

A standard approach for preserving the privacy of a dataset involves obfuscating the dataset, or its properties, in some way. Typical approaches include adding noise either randomly or following some structured probability distribution. These approaches are not unique for location data, but location-specific noise-based obfuscation techniques have been developed. For instance, geomasking or spatial-temporal cloaking, refer to a broad set of methods used for obfuscating location data (Armstrong *et al.*, 1999). Methods for obfuscating point coordinates include reporting a broader geometric region (*e.g.*, circle or annulus) in which the point exists, displacing the point by some distance and direction or reporting the political or social boundary in which the point is contained (Seidl *et al.*, 2016). A variety of tools, such as *MaskMy.XYZ* (Swanlund *et al.*, 2020) have been developed to help the average privacy-conscious user geomask their location content.

Anonymization is another way of preserving individual privacy, which aims to keep one's identity private but not necessarily one's actions. In contrast to obfuscation techniques, the objective is not necessarily to hide sensitive information through the addition of noise but rather to reduce the accuracy of the information disclosed in order to limit the possibility of re-identifying a particular mobility profile. Various approaches have been developed to guarantee some degree of geospatial data anonymity. For instance in *k*-anonymity, the objective is to hide the particular mobility behaviour of a user among other users sharing similar patterns. More precisely, a dataset is said to be *k*-anonymized if a record within the set cannot be differentiated from *k*-1 other records. While the seminal work on this topic (Sweeney, 2002) did not specifically focus on location data, subsequent efforts have highlighted the ways in which one can *k*-anonymize spatial datasets (Ghinita *et al.*, 2010). Geographic obfuscation methods such as Adaptive Areal Elimination (Charleux and Schofield, 2020; Kounadi and Leitner, 2016) leverage this *k*-anonymity property of the data to identify regions that offer a measurable level of privacy.

Differential privacy is often heralded as one of the field's most significant advances, offering strong and formal privacy guarantees (Dwork, 2006). The objective of differential privacy is to extract and publish global usable patterns from a set of data while maintaining the privacy of the individual records in the set. This approach involves adding noise to a dataset such that exposure of one, or a set of attributes, will not expose the identity of a record or individual. Since 2015, differential privacy has been used by leading technology companies to monitor how products are used along with purchasing and mobility trends. Within the geographic domain, variations on differential privacy have been introduced, such as geo-indistinguishability (Andrés *et al.*, 2013), that acknowledge the unique properties of geographic data and obfuscate location details through tailored geomasking techniques (Kim *et al.*, 2021).

With the growth in GeoAI, a variety of new obfuscation and anonymity methods have emerged that leverage network graphs (Jiang *et al.*, 2019), discrete global grids (Hojati *et al.*, 2021), and decentralized collaborative machine learning (Rao *et al.*, 2021), to name a few. In addition, the continued growth of contextually-aware devices has led to advances in obfuscation techniques for mobile device users (Jiang *et al.*, 2021).

19.2.2   SYNTHETIC DATA GENERATION

An alternative to obfuscating or anonymizing real location data is to instead generate *synthetic* data. Sometimes referred to as fake or dummy data, the privacy of a dataset can be maintained by not reporting any piece of the original data at all. Instead, a new set of data are generated that exhibit similar properties of the original dataset. Such an approach can be tailored to specific use cases by only selecting the properties of interest from the original dataset. Methods of synthesizing data are often devised to protect the privacy and confidentiality of particular parts of a dataset, or the data as a whole. The generation of synthetic data through generative models is a hot topic in machine learning and numerous data synthesis methods have been developed and are actively in use in a variety of domains (Nikolenko, 2021). With respect to geospatial data, synthetic population data has a long history in demography (Beckman *et al.*, 1996) with governmental programs, such as the census, often generate synthetic data for regions with small or susceptible populations. In such cases, a population may be so small that even reporting aggregate values may expose unique individuals in a region. Synthetic data can be generated based on properties of the original data, but be adjusted such that the privacy of individuals can be maintained. With respect to location privacy, synthetic data have been used to understand crowd dynamics (Wang *et al.*, 2019), analyze mobility trajectories (Rao *et al.*, 2020) and more generally address a wide array of pressing geographic problems (Cunningham *et al.*, 2021).

19.2.3 CRYPTOGRAPHY

The previously mentioned techniques aim to preserve privacy either through distortion of the original data or generating dummy data. An alternative to these approaches is to simply hide the data using cryptographic techniques. Encryption is a widely used technique for storing and sharing information when the content needs to remain private. The limitation of such an approach is that once encrypted, the utility of the data is basically non-existent for someone that does not have the associated decryption key. Whereas geographic coordinates obfuscated to a neighborhood may still provide utility for location-based services, encrypted data are useless to anyone but those with the ability to decrypt them.

Researchers working with geographic data have proposed a variety of ways to encrypt geospatial data but still maintain some degree of utility. For instance, some approaches rely on partial encryption of the data meaning that some properties are exposed while others remain hidden (Jiang *et al.*, 2021; Sun *et al.*, 2019). Similar to some of the methods mentioned in the previous section, this means that identifiable and confidential information will be encrypted while spatial properties of a dataset (*e.g.*, degree of clustering), may be published. Geospatial communication platforms such as *Drift* (Brunila *et al.*, 2022), are being developed that encrypt geospatial data but maintain utility.

On the advanced cryptographic primitives side, we have seen the recent adoption of homomorphic encryption in a variety of applications (Acar *et al.*, 2018). Homomorphic encryption is an encryption method that allows one to analyze encrypted data without first decrypting it. Such analysis can result in the extraction of patterns and insight without having access to the original unencrypted private information. This technique is actively being used in health research and demography (Munjal and Bhatia, 2022). There are limits to homomorphic encryption, not least of which are the types of analyses that can be performed and the computational costs of such analyses. The unique types of analyses that are conducted on geospatial data offer challenges for homomorphic encryption techniques (Alanwar *et al.*, 2017) but advances in this area are sure to be made in the coming years.

19.2.4 RE-IDENTIFICATION METHODS & PRIVACY ATTACKS

While the methods described in the previous sections aim at preserving privacy and anonymity, another set of methods relevant for privacy researchers are those used for de-anonymizing data and conducting other privacy attacks. While there is not a single leading approach to focus on, we instead highlight a few examples of how this is being done with location data.

De-anonymization approaches often involve the inclusion of an external dataset reflecting the knowledge of a potential adversary during analysis (Harmanci and

Gerstein, 2016). One possible approach to de-anonymization is through a linkage attack that leverage relationships between the external dataset and the anonymized one, reducing the anonymity of individual records in the process (Narayanan and Shmatikov, 2008). Unique properties of location data such as the habitual movement patterns of people can also be leveraged to de-anonymize a dataset. For example, Gambs *et al.* (2014) trained a Mobility Markov Chain model on a set of known mobility trajectories and used this model to identify individuals in an anonymized set of trajectories. When the data represents the location of individuals, *co-location analysis* can be used to reduce the privacy of seemingly obfuscated or anonymized data. For instance, geosocial media users frequently report their co-locations with other users through tags or photographs. Internet protocol (IP) addresses are also a means of co-location identification. Knowing the relationships in a social network can be leveraged to identify an individual (Olteanu *et al.*, 2016). This is part of a broader discussion on *interdependent privacy* in which the privacy of one individual is impacted by the privacy decisions and data sharing of others (Liu *et al.*, 2018). As mentioned in the introduction, if my neighbor chooses to share personal information and an adversary knows that we live in close proximity, they could infer a lot of information (*e.g.*, race, income, education) about me.

With the increase in computational power and access to massive amounts of data, GeoAI techniques are able to re-identify records (*e.g.*, people) in datasets through inference and probabilistic modeling. For instance, large language models use AI techniques to process large volumes of textual data, much of which include geographic elements. Trained on such data, these models can be used to infer mobility patterns, identify individuals, and re-identified seemingly anonymized datasets based on the massive amount of additional (contextual) data on which they are trained. Such models are concerning to privacy advocates as public facing tools built from these models (*e.g.*, chat bots) give immense power to average citizens, power that can be used to reduce the privacy of individuals (Pan *et al.*, 2020).

19.3 APPLICATION AREAS

While privacy is a pervasive concern through arguably all application areas of GeoAI, we thought it useful to highlight a subset of sectors in which privacy is at the forefront of the discussion.

19.3.1 ADVERTISING

Location-based advertising involves targeting advertisements to groups and individuals based on their geographic location. A study of user attitudes toward targeted advertising found that targeted ads were generally preferred to non-target ones but targeted ads were seen as a privacy concern (Zhang *et al.*, 2010). While not new, the adoption of context-aware devices and advanced in predictive analytics have changed the landscape of location-based advertising. An analysis of mobile device ad libraries found that a large number of them track a user's location (Stevens *et al.*, 2012), even if the location is not needed for the functionality provided by a particular application.

Location data, along with a variety of other attributes are used by AI companies for tailored advertising and to target particular users and groups (Boerman *et al.*, 2017). In addition, the knowledge of someone's location can be combined with other factors such as the time of day or mode of transportation to further refine targeted ads and track users across devices and platforms. Studies have shown that location-based tracking works (Dhar and Varshney, 2011) and given the importance of training data for advertising models, significant efforts are underway to collect and sell such data. As these data are transferred between data providers, brokers, and agencies, maintaining the privacy of the individual records often falls by the wayside. For instance, in 2019 the New York Times was provided access to detailed information, including locations, for 12 million mobile devices (Thompson and Warzel, 2019). The source of the data was apparently unauthorized to share such content, yet the full records were shared without any attempt to preserve the privacy of the individuals in the data. Though not an advertising example, this does highlight the market for private data. While location-based advertising is unlikely to disappear in the near future, advances in GeoAI will enable advertisers and advertisees to strike a balance between privacy preservation and advertising utility.

19.3.2   HEALTH CARE

A large percentage of the research on location privacy preservation and spatial anonymization was originally done for the purposes of maintaining data confidentiality in health. Understandably, medical researchers and practitioners are highly incentivized to maintain the confidentiality and privacy of patient data yet it is necessary to share data to access the collaborative expertise of those in the medical field. While geomasking and other obfuscation techniques are used to preserve data privacy as well as maintaining utility, newer methods are being developed that guarantee privacy while still permitting a level of analysis. As discussed in Section 19.2.3, cryptographic techniques such as homomorphic encryption are on the verge of dramatically changing how medical health records are stored and analyzed.

AI techniques are also being actively used in disease prevention and epidemiological research with impressive results (Munir *et al.*, 2019). GeoAI too is having a significant impact with methods having been designed to model unique conditions such as spatial non-stationarity, variation in scale, and data sparsity. These are relevant to fields such as environmental epidemiology (VoPham *et al.*, 2018), precision medicine, and healthy cities (Kamel Boulos *et al.*, 2019). All of these fields have a strong privacy and confidentiality component and many of the models being developed today are designed with privacy in mind. These are often referred to as *privacy-aware* or *privacy-enhancing* technologies. As mentioned previously, models that deal with location data are particularly vulnerable to privacy inference attacks as knowledge of one's location allows for the inference of different characteristics. Not surprisingly, this has impacted the other side of the medical industry, namely health insurance. While some of us are aware that AI techniques are being used to analyze our driving records (Arumugam and Bhargavi, 2019), we should also be conscious that they are being used to estimate risk and set health insurance rates (Naylor, 2018).

The COVID-19 pandemic gave rise to a new era of health-related privacy concerns with many agencies and industry partners using AI for contact tracing (Grekousis and Liu, 2021) and predicting outbreaks (Vaishya *et al.*, 2020). During the Covid-19 pandemic, many of the privacy mechanisms that went into securing public and private health care data were reduced or removed to support contact tracing and epidemiological modeling efforts. Ribeiro-Navarrete *et al.* (2021) provide an overview of Covid-19 related privacy discussions and surveillance technologies.

19.3.3 SECURITY & SURVEILLANCE

The quintessential domain that one thinks of when discussing privacy in GeoAI is surveillance. Concern over AI technologies used to monitor citizens has received quite a bit of attention in the news media in recent years. This is not unwarranted but the relationship between AI and surveillance is more complex than it is often made out to be. There are plenty of examples in the literature of machine learning methods and tools that are used to track the locations of objects (*e.g.*, people, vehicles). Tracking technologies range from collecting locations of people through GNSS, Wi-Fi, or cellular trilateration, to license plate identification on traffic cameras. Other surveillance efforts monitor animal movement through image recognition for habitat delineation, conservation, and poaching prevention (Kumar and Jakhar, 2022). Tracking or surveilling an object, by definition, involves the collection of information about that object and while the act itself is not a privacy violation, in certain circumstances, it can be. Aside from the actual data collection, AI has contributed to advances in how such tracking data are analyzed. Improvements in image classification and high performance computing mean that people can be monitored across different regions through CCTV surveillance cameras (Fontes *et al.*, 2022). Tracking and surveillance can be less explicit as well. Existing research has demonstrated that humans are creatures of habit and are highly predictable in their activity behavior. Through the analysis of user-contributed and crowd-sourced data, *social sensing* techniques can be used to identify when and where someone may visit a place (Janowicz *et al.*, 2019).

Tools and methods for crime prediction and counter-terrorism are often seen as being at odds with privacy preservation. The role of AI in crime forecasting specifically has received considerable interest in recent years (Dakalbab *et al.*, 2022; Kounadi *et al.*, 2020). Many of the techniques used in these fields are design for de-anonymization and re-identification in the name of safety and security. Most of the discussion related to privacy stems from surveillance being viewed as an infringement on individual rights. Given that criminal activity clusters geographically, one must be concerned about the privacy of one's data and, when the data are exposed, how that data is being used. A large body of research has investigated mass surveillance for security purposes and few results have indicated that AI models built on such data are more accurate at predicting crimes (Verhelst *et al.*, 2020) or identifying repeat offenders (Dressel and Farid, 2018). Work by Mayson (2019) demonstrated that the personal data used as input to such prediction models have dire consequences on the resulting actions taken by law enforcement. Predictive AI modeling has been shown to incorrectly identify individuals as criminals (Crawford and Schultz, 2014) and that some AI predictive recidivism tools demonstrate concerning bias in their recommendations either as a result of the input data or the model designs.

19.4 THE FUTURE OF PRIVACY IN GEOAI

In this section, we look to how privacy within GeoAI is changing and identify some of the leading concerns that should be addressed by the community. Specifically, we outline three ways in which privacy within GeoAI can be improved and highlight three emerging topics related to location privacy.

19.4.1 SUGGESTED AREAS FOR IMPROVEMENT

While there are multiple ways that privacy can be further addressed within GeoAI, we provide the following three suggestions as starting points.

- **Privacy by design**. Despite the significant body of work on privacy from legal experts, policy makers, and ethical AI researchers, privacy concerns are still typically a secondary factor in the advancement of artificial intelligence. This is not only true for GeoAI, but for the broader field of AI and related technologies. Rather than being considered as an after thought, future directions of GeoAI research should integrate data privacy principles from the outset. Furthermore, data privacy should be considered at all stages of development from conception through delivery. Those with expertise in privacy and ethics should be consulted in the development and assessment of new algorithms that will impact the privacy of individuals or certain demographic groups. Privacy impact assessments (Clarke, 2009) or audits, similar to ethics-based audits (Mökander and Floridi, 2021), may be one such solution.

- **Spatial privacy is special**. Building off *Spatial is Special*, the alliterative phrase commonly uttered by geographic information scientists, there continues to be a need for wider acknowledgment within the artificial intelligence community that geographic data are unique due to the relationship between entity similarity and spatiotemporal proximity. This is particularly true when the privacy of an individual is at stake. Ignoring spatial properties of a dataset can substantially impact one's privacy (Griffith, 2018). Working with geographic data requires an understanding of basic geographic concepts such as spatial heterogeneity, autocorrelation, and inference, and how they can be leveraged to either preserve or divulge private details.

- **Enhancing regulations**. Since data are the foundation on which virtually all AI technologies are built, access to such data for AI development should be scrutinized. Currently there is very little oversight or transparency on what types of data are collected, how they are collected, and how they are being used. We need independent assessment and inter-governmental regulations pertaining to data collection, storage, and its use. The European Union's General Data

Protection Regulation (GDPR) is a good, but flawed first step. For instance, each European country is responsible for investigating the companies that are registered within it. This means that a country like Ireland is responsible for regulating a massive percentage of big tech. The actual number of penalties placed on violators as a result of the GDPR are much lower than predicted five years ago (Burgess, 2022). Additional efforts must be made to ensure that users of digital platforms have the right to control how their data are collected, stored, and analyzed. The need for such transparency is paramount.

19.4.2 EMERGING PRIVACY TOPICS IN GEOAI

Aside from these recommendations there are a number of new challenges and emerging opportunities within GeoAI privacy research (Richardson *et al.*, 2015). Some of these are actively being investigated while other are merely proposal for future research directions within this domain. Below we identify three directions that we feel are of particular interest to the GeoAI community.

- **Fake geospatial data**. The methods introduced earlier in this chapter highlight techniques for preserving the privacy of real people sharing real data. Synthetic data generation is one such approach, but new disinformation campaigns are focused on generating *fake* location data. Similar to how *deep fake* algorithms have emerged as practical tools for communicating disinformation visually, we are beginning to see similar approaches used to generating fake, but geospatially probable data. We are already seeing the emergence of a new subdomain of deep fake geography (Zhao *et al.*, 2021). The reasons for generating fake location data include identity theft, political or social disruption, or bypassing security protocols. Note that fake data generation, while similar to synthetic data generation, is substantially different in its design and motivation. As our security tools increasingly relying on location information for verification (e.g., known IP address for banking), a new focus on detecting fake location information is required and the GeoAI community is well situated to address this challenge.

- **Publicly accessible and integrated tools**. We have only just scratched the surface in developing techniques for privacy preservation. As AI development and data availability grow, so too will the need for privacy preservation tools. Similar to how efforts are under way to detect text generated by large language model chat bots, we need publicly accessible tools to help users detect privacy violations and help users take control of their data. While many of the techniques and tools mentioned in this chapter are realized through theoretical models published by academics, real-world applications of these approaches have been slow to emerge. This is doubly true for methods generated by GeoAI developers. Future research will involve (1) the further integration of privacy preservation methods into existing location data sharing platforms and (2) more investment in the development of publicly accessible location privacy tools. Finally, educational efforts from geographers and computational scientists will focus on investigating the ways in which these tools educate and inform the public as to what is possible with personal location information.

- **Policy development**. From a social, political, and ethical perspective, future research will undoutably focus on developing policies in partnership with commercial entities and government agencies. Historically, government regulation and laws follow technological advances – often years behind. As highlighted in our suggestions above, regulatory bodies need to rise to the occasion, but these regulations need to be driven by evidence produced by ethical AI researchers and domain experts. As GeoAI emerges as it's own subdomain from within AI and geography, we have an opportunity to include the study of ethical and privacy implications within our research principles. The inclusion and reporting of such research will help inform regulators and policy makers when considering the impact of GeoAI on local communities and the global population.

19.5 SUMMARY

In this chapter we presented an overview of data privacy as it related to geographic artificial intelligence. Geographic data are a unique type of information in that knowledge of one person's location reveals highly sensitive information about nearby individuals or groups. The growth of AI and associated techniques has forced researchers, companies, governments, and the public to think seriously about the privacy implications of sharing, collecting, and analyzing such data. Within GeoAI, particular attention needs to be made to how personal location and movement data are being analyzed and what can be inferred through geospatial analysis. A growing body of AI methods and tools are focused on privacy preservation with respect to geographic data within a wide range of domains. We encourage continued discussion on ethics and privacy as advances in GeoAI continue to shape the world around us.

BIBLIOGRAPHY

Acar, A., *et al.*, 2018. A survey on homomorphic encryption schemes: Theory and implementation. *ACM Computing Surveys*, 51 (4), 1–35.

Adams, B. and Janowicz, K., 2012. On the geo-indicativeness of non-georeferenced text. *In*: *Proceedings of the International AAAI Conference on Web and Social Media*. vol. 6, 375–378.

Alanwar, A., *et al.*, 2017. Proloc: Resilient localization with private observers using partial homomorphic encryption. *In*: *Proceedings of the 16th ACM/IEEE International Conference on Information Processing in Sensor Networks*. 41–52.

Andrés, M.E., *et al.*, 2013. Geo-indistinguishability: Differential privacy for location-based systems. *In*: *Proceedings of the 2013 ACM SIGSAC Conference on Computer & Communications Security*. 901–914.

Armstrong, M.P., Rushton, G., and Zimmerman, D.L., 1999. Geographically masking health data to preserve confidentiality. *Statistics in Medicine*, 18 (5), 497–525.

Armstrong, M.P., Tsou, M.H., and Seidl, D.E., 2018. 1.28-geoprivacy. *In: Comprehensive Geographic Information Systems*. Elsevier Inc, 415–430.

Arumugam, S. and Bhargavi, R., 2019. A survey on driving behavior analysis in usage based insurance using big data. *Journal of Big Data*, 6, 1–21.

Beckman, R.J., Baggerly, K.A., and McKay, M.D., 1996. Creating synthetic baseline populations. *Transportation Research Part A: Policy and Practice*, 30 (6), 415–429.

Boerman, S.C., Kruikemeier, S., and Zuiderveen Borgesius, F.J., 2017. Online behavioral advertising: A literature review and research agenda. *Journal of Advertising*, 46 (3), 363–376.

Brunila, M., *et al.*, 2022. Drift: E2ee spatial feature sharing & instant messaging. *In: Proceedings of the 6th ACM SIGSPATIAL International Workshop on Location-Based Recommendations, Geosocial Networks and Geoadvertising*. 1–11.

Burgess, M., 2022. How gdpr is failing. *Wired Magazine*.

Charleux, L. and Schofield, K., 2020. True spatial k-anonymity: Adaptive areal elimination vs. adaptive areal masking. *Cartography and Geographic Information Science*, 47 (6), 537–549.

Clarke, R., 2009. Privacy impact assessment: Its origins and development. *Computer Law & Security Review*, 25 (2), 123–135.

Crawford, K. and Schultz, J., 2014. Big data and due process: Toward a framework to redress predictive privacy harms. *BCL Rev.*, 55, 93.

Cunningham, T., Cormode, G., and Ferhatosmanoglu, H., 2021. Privacy-preserving synthetic location data in the real world. *In: 17th International Symposium on Spatial and Temporal Databases*. 23–33.

Dakalbab, F., *et al.*, 2022. Artificial intelligence & crime prediction: A systematic literature review. *Social Sciences & Humanities Open*, 6 (1), 100342.

Dhar, S. and Varshney, U., 2011. Challenges and business models for mobile location-based services and advertising. *Communications of the ACM*, 54 (5), 121–128.

Dobson, J.E. and Fisher, P.F., 2003. Geoslavery. *IEEE Technology and Society Magazine*, 22 (1), 47–52.

Dressel, J. and Farid, H., 2018. The accuracy, fairness, and limits of predicting recidivism. *Science Advances*, 4 (1), eaao5580.

Dubber, M.D., Pasquale, F., and Das, S., 2020. *The Oxford Handbook of Ethics of AI*. Oxford Handbooks.

Dwork, C., 2006. Differential privacy. *In: Automata, Languages and Programming: 33rd International Colloquium, ICALP 2006, Venice, Italy, July 10-14, 2006, Proceedings, Part II 33*. Springer, 1–12.

Finn, R.L., Wright, D., and Friedewald, M., 2013. Seven types of privacy. *European Data Protection: Coming of Age*, 3–32.

Fontes, C., *et al.*, 2022. Ai-powered public surveillance systems: why we (might) need them and how we want them. *Technology in Society*, 71, 102137.

Gambs, S., Killijian, M.O., and del Prado Cortez, M.N., 2014. De-anonymization attack on geolocated data. *Journal of Computer and System Sciences*, 80 (8), 1597–1614.

Ghinita, G., *et al.*, 2010. A reciprocal framework for spatial k-anonymity. *Information Systems*, 35 (3), 299–314.

Grekousis, G. and Liu, Y., 2021. Digital contact tracing, community uptake, and proximity awareness technology to fight covid-19: a systematic review. *Sustainable Cities and Society*, 71, 102995.

Griffith, D.A., 2018. Uncertainty and context in geography and giscience: reflections on spatial autocorrelation, spatial sampling, and health data. *Annals of the American Association of Geographers*, 108 (6), 1499–1505.

Guha, S., *et al.*, 2012. Koi: A location-privacy platform for smartphone apps. *In: NSDI*. vol. 12, 14.

Harmanci, A. and Gerstein, M., 2016. Quantification of private information leakage from phenotype-genotype data: linking attacks. *Nature Methods*, 13 (3), 251–256.

Hasan, R., *et al.*, 2020. Automatically detecting bystanders in photos to reduce privacy risks. *In: 2020 IEEE Symposium on Security and Privacy (SP)*. IEEE, 318–335.

Hojati, M., *et al.*, 2021. Decentralized geoprivacy: leveraging social trust on the distributed web. *International Journal of Geographical Information Science*, 35 (12), 2540–2566.

Janowicz, K., *et al.*, 2019. Using semantic signatures for social sensing in urban environments. *In: Mobility Patterns, Big Data and Transport Analytics*. Elsevier, 31–54.

Jiang, H., *et al.*, 2021. Location privacy-preserving mechanisms in location-based services: A comprehensive survey. *ACM Computing Surveys*, 54 (1), 1–36.

Jiang, J., *et al.*, 2019. A survey on location privacy protection in wireless sensor networks. *Journal of Network and Computer Applications*, 125, 93–114.

Jobin, A., Ienca, M., and Vayena, E., 2019. The global landscape of ai ethics guidelines. *Nature Machine Intelligence*, 1 (9), 389–399.

Kamel Boulos, M.N., Peng, G., and VoPham, T., 2019. An overview of geoai applications in health and healthcare. *International Journal of Health Geographics*, 18, 1–9.

Keßler, C. and McKenzie, G., 2018. A geoprivacy manifesto. *Transactions in GIS*, 22 (1), 3–19.

Kim, J.W., *et al.*, 2021. A survey of differential privacy-based techniques and their applicability to location-based services. *Computers & Security*, 111, 102464.

Kounadi, O. and Leitner, M., 2014. Why does geoprivacy matter? the scientific publication of confidential data presented on maps. *Journal of Empirical Research on Human Research Ethics*, 9 (4), 34–45.

Kounadi, O. and Leitner, M., 2016. Adaptive areal elimination (aae): A transparent way of disclosing protected spatial datasets. *Computers, Environment and Urban Systems*, 57, 59–67.

Kounadi, O., *et al.*, 2020. A systematic review on spatial crime forecasting. *Crime Science*, 9 (1), 1–22.

Kumar, D. and Jakhar, S.D., 2022. Artificial intelligence in animal surveillance and conservation. *Impact of Artificial Intelligence on Organizational Transformation*, 73–85.

Liu, B., *et al.*, 2018. Location privacy and its applications: A systematic study. *IEEE Access*, 6, 17606–17624.

Liu, Y., *et al.*, 2017. Point-of-interest demand modeling with human mobility patterns. *In: Proceedings of the 23rd ACM SIGKDD International Conference on Knowledge Discovery and Data Mining*. 947–955.

Mayson, S.G., 2019. Bias in, bias out. *The Yale Law Journal*, 128 (8), 2218–2300.

McKenzie, G., Janowicz, K., and Seidl, D., 2016. Geo-privacy beyond coordinates. *In: Geospatial Data in a Changing World: Selected Papers of the 19th AGILE Conference on Geographic Information Science*. Springer, 157–175.

McKenzie, G. and Mwenda, K., 2021. Identifying regional variation in place visit behavior during a global pandemic. *Journal of Spatial Information Science*, 1 (23), 95–124.

McKenzie, G., *et al.*, 2022. Privyto: A privacy-preserving location-sharing platform. *Transactions in GIS*.

Mittelstadt, B., 2019. Principles alone cannot guarantee ethical ai. *Nature Machine Intelligence*, 1 (11), 501–507.

Mökander, J. and Floridi, L., 2021. Ethics-based auditing to develop trustworthy ai. *Minds and Machines*, 31 (2), 323–327.

Munir, K., *et al.*, 2019. Cancer diagnosis using deep learning: a bibliographic review. *Cancers*, 11 (9), 1235.

Munjal, K. and Bhatia, R., 2022. A systematic review of homomorphic encryption and its contributions in healthcare industry. *Complex & Intelligent Systems*, 1–28.

Naik, N., *et al.*, 2022. Legal and ethical consideration in artificial intelligence in healthcare: who takes responsibility? *Frontiers in Surgery*, 266.

Narayanan, A. and Shmatikov, V., 2008. Robust de-anonymization of large sparse datasets. *In: 2008 IEEE Symposium on Security and Privacy (sp 2008)*. IEEE, 111–125.

Naylor, C.D., 2018. On the prospects for a (deep) learning health care system. *JAMA*, 320 (11), 1099–1100.

Nelson, T., Goodchild, M., and Wright, D., 2022. Accelerating ethics, empathy, and equity in geographic information science. *Proceedings of the National Academy of Sciences*, 119 (19), e2119967119.

Nikolenko, S.I., 2021. *Synthetic Data for Deep Learning*. vol. 174. Springer.

Olteanu, A.M., *et al.*, 2016. Quantifying interdependent privacy risks with location data. *IEEE Transactions on Mobile Computing*, 16 (3), 829–842.

Pan, X., *et al.*, 2020. Privacy risks of general-purpose language models. *In: 2020 IEEE Symposium on Security and Privacy (SP)*. IEEE, 1314–1331.

Pedersen, D.M., 1979. Dimensions of privacy. *Perceptual and Motor Skills*, 48 (3˙suppl), 1291–1297.

Pensa, R.G., Di Blasi, G., and Bioglio, L., 2019. Network-aware privacy risk estimation in online social networks. *Social Network Analysis and Mining*, 9, 1–15.

Rao, J., *et al.*, 2020. LSTM-TrajGAN: A Deep Learning Approach to Trajectory Privacy Protection. *In: K. Janowicz and J.A. Verstegen, eds. 11th International Conference on Geographic Information Science (GIScience 2021) - Part I*, Dagstuhl, Germany. Schloss Dagstuhl–Leibniz-Zentrum für Informatik, Leibniz International Proceedings in Informatics (LIPIcs), vol. 177, 12:1–12:17. Available from: `https://drops.dagstuhl.de/opus/volltexte/2020/13047`.

Rao, J., *et al.*, 2021. A privacy-preserving framework for location recommendation using decentralized collaborative machine learning. *Transactions in GIS*, 25 (3), 1153–1175.

Ribeiro-Navarrete, S., Saura, J.R., and Palacios-Marqués, D., 2021. Towards a new era of mass data collection: Assessing pandemic surveillance technologies to preserve user privacy. *Technological Forecasting and Social Change*, 167, 120681.

Richardson, D.B., *et al.*, 2015. Replication of scientific research: addressing geoprivacy, confidentiality, and data sharing challenges in geospatial research. *Annals of GIS*, 21 (2), 101–110.

Riederer, C., *et al.*, 2016. Linking users across domains with location data: Theory and validation. *In*: *Proceedings of the 25th International Conference on World Wide Web*. 707–719.

Seidl, D.E., Jankowski, P., and Tsou, M.H., 2016. Privacy and spatial pattern preservation in masked gps trajectory data. *International Journal of Geographical Information Science*, 30 (4), 785–800.

Stahl, B.C. and Wright, D., 2018. Ethics and privacy in ai and big data: Implementing responsible research and innovation. *IEEE Security & Privacy*, 16 (3), 26–33.

Stevens, R., *et al.*, 2012. Investigating user privacy in android ad libraries. *In*: *Workshop on Mobile Security Technologies (MoST)*. vol. 10, 195–197.

Sun, G., *et al.*, 2019. Location privacy preservation for mobile users in location-based services. *IEEE Access*, 7, 87425–87438.

Swanlund, D., Schuurman, N., and Brussoni, M., 2020. Maskmy. xyz: An easy-to-use tool for protecting geoprivacy using geographic masks. *Transactions in GIS*, 24 (2), 390–401.

Sweeney, L., 2002. k-anonymity: A model for protecting privacy. *International Journal of Uncertainty, Fuzziness and Knowledge-Based Systems*, 10 (05), 557–570.

Thompson, S.A. and Warzel, C., 2019. Twelve million phones, one dataset, zero privacy. *In*: *Ethics of Data and Analytics*. Auerbach Publications, 161–169.

Tobler, W.R., 1970. A computer movie simulating urban growth in the detroit region. *Economic Geography*, 46 (sup1), 234–240.

Vaishya, R., *et al.*, 2020. Artificial intelligence (ai) applications for covid-19 pandemic. *Diabetes & Metabolic Syndrome: Clinical Research & Reviews*, 14 (4), 337–339.

Verhelst, H.M., Stannat, A., and Mecacci, G., 2020. Machine learning against terrorism: how big data collection and analysis influences the privacy-security dilemma. *Science and Engineering Ethics*, 26, 2975–2984.

VoPham, T., *et al.*, 2018. Emerging trends in geospatial artificial intelligence (geoai): potential applications for environmental epidemiology. *Environmental Health*, 17 (1), 1–6.

Wang, Q., *et al.*, 2019. Learning from synthetic data for crowd counting in the wild. *In*: *Proceedings of the IEEE/CVF Conference on Computer Vision and Pattern Recognition*. 8198–8207.

Wang, R., *et al.*, 2016. A de-anonymization attack on geo-located data considering spatio-temporal influences. *In*: *Information and Communications Security: 17th International Conference, ICICS 2015, Beijing, China, December 9–11, 2015, Revised Selected Papers 17*. Springer, 478–484.

Zhang, H., *et al.*, 2010. Privacy issues and user attitudes towards targeted advertising: A focus group study. *In*: *Proceedings of the Human Factors and Ergonomics Society Annual Meeting*. SAGE Publications Sage CA: Los Angeles, CA, vol. 54, 1416–1420.

Zhang, H. and McKenzie, G., 2022. Rehumanize geoprivacy: from disclosure control to human perception. *GeoJournal*, 1–20.

Zhao, B., *et al.*, 2021. Deep fake geography? when geospatial data encounter artificial intelligence. *Cartography and Geographic Information Science*, 48 (4), 338–352.

Zhong, Y., *et al.*, 2015. You are where you go: Inferring demographic attributes from location check-ins. *In*: *Proceedings of the Eighth ACM International Conference on Web Search and Data Mining*. 295–304.

20 A Humanistic Future of GeoAI

Bo Zhao and Jiaxin Feng
Department of Geography, University of Washington, Seattle

CONTENTS

20.1 Introduction 406
20.2 Envision a Humanistic Future of GeoAI 407
    20.2.1 Examining GeoAI practices and their social implications 407
    20.2.2 Incorporating humanistic values into the design of GeoAI 408
    20.2.3 Promoting human-GeoAI cooperation 408
    20.2.4 Fostering an inclusive and diverse GeoAI community 409
20.3 Conclusion..409
    Bibliography..409

20.1 INTRODUCTION

Humanism is a philosophical stance that emphasizes the value of human beings and encourages critical thinking and evidence-based decision-making, rather than the acceptance of dogma or superstition. This perspective has been incorporated into humanistic GIS, which prioritizes human experiences, values, and perspectives in the study of space and place (Zhao 2022). GeoAI, an emerging topic in GIScience, has demonstrated its potential in various fields. For instance, GeoAI has been used to optimize land use in Singapore's urban planning (Huang 2021), improve traffic management in Los Angeles (Netzley 2023), enhance crop yields in precision agriculture projects (Michaux 2019), monitor deforestation in the Amazon rainforest (Torres et al. 2021), and predict flood impacts in Bangladesh's emergency response scenarios (Uddin et al. 2019). Despite these advancements, developing and applying GeoAI without considering humanistic values can lead to unintended consequences and adverse effects on society and the environment. For example, Waze, a traffic navigation app, has been criticized for rerouting traffic through residential areas, causing congestion and safety concerns (Foderaro 2017). Likewise, the use of facial recognition technology by law enforcement agencies has raised privacy and civil liberties concerns, leading to human rights violations. The absence of humanistic values in AI can also perpetuate existing power structures and inequalities. For instance, the Gender Shades project revealed biases in commercial facial recognition systems, which

DOI: 10.1201/9781003308423-20

struggled to accurately recognize individuals with darker skin tones and women, leading to discrimination (Buolamwini and Gebru 2018). Such outcomes can worsen existing racial and gender disparities and reinforce systemic inequality. Therefore, it is crucial to incorporate humanistic values into the design and development of GeoAI. By doing so, we can ensure that this technology benefits society and the environment while upholding essential ethical principles, such as social justice, equity, and sustainability.

In this commentary, we advocate for a humanistic rewire of GeoAI, emphasizing its ethical, inclusive, and human-guided development. A humanistic rewire refers to an approach that seeks to integrate humanistic principles, values, and perspectives into a field or technology that has traditionally been dominated by technical and scientific perspectives. The goal of a humanistic rewire is to consider the needs and impact of GeoAI on people, places, and other sentient beings, while also acknowledging the diversity of lived experiences and other types of creative forces. In the rest of this commentary, we will provide some suggestions to inspire people to envision a humanistic future of GeoAI.

20.2 ENVISION A HUMANISTIC FUTURE OF GEOAI

20.2.1 EXAMINING GEOAI PRACTICES AND THEIR SOCIAL IMPLICATIONS

GeoAI is a rapidly developing technology that can have significant and far-reaching impacts on society, particularly on marginalized communities and nonhuman entities. As with any new technology, GeoAI has the potential to reinforce existing power structures and exacerbate social inequalities. For example, predictive policing algorithms, which utilize GeoAI to anticipate crime hotspots, have been criticized for perpetuating racial bias and leading to discriminatory outcomes for certain groups of people (Heaven 2020). Similarly, biased data used in housing algorithms could reinforce patterns of discrimination and social exclusion, exacerbating inequalities in access to housing and resources (Mcilwain 2020). Additionally, the use of GeoAI may have unintended consequences on nonhuman entities and the natural environment, such as disrupting ecosystems or infringing on the rights of nonhuman animals. For instance, drone-based wildlife monitoring systems, while providing valuable conservation data, can inadvertently disturb nesting birds or disrupt the behavior of other sensitive species (Weston et al. 2020). The rapid expansion of data centers required to support GeoAI systems can also contribute to significant energy consumption and carbon emissions, potentially exacerbating climate change and its impacts on ecosystems and biodiversity (Labbe 2021). Therefore, by analyzing the impact of GeoAI on different stakeholders, including marginalized communities and nonhuman entities, we can identify potential ethical and social issues and work to address them proactively.

20.2.2   INCORPORATING HUMANISTIC VALUES INTO THE DESIGN OF GEOAI

Incorporating humanistic values into the design of GeoAI can foster ethical, inclusive, and socially responsible applications. By prioritizing principles such as fairness, transparency, and accountability, we can address potential biases and promote equitable outcomes. Achieving this requires developers and stakeholders to actively engage with diverse perspectives and experiences, conduct user research, and create intuitive, accessible user interfaces. GeoAI can thus be developed not only as a technically advanced tool but also as one that respects and prioritizes human experiences, values, and perspectives. For instance, emotions about a place, one of important human spatial experiences, can be learned from brainwaves captured by an EEG headset as a human subject walk along a route. Those emotions, combined with the subject's personal account, have great potential to be integrated into route planning and urban development. Mapillary, a platform for crowdsourcing street-level imagery, adopted a user-centered approach by working closely with volunteers to understand their needs and preferences and develop features to meet those needs (Joseph 2018). However, potential limitations may arise from the complexities of reconciling competing human values and interests, as well as the challenges of translating humanistic principles into concrete design guidelines. Despite these challenges, embracing humanistic values in GeoAI design and development remains vital for creating responsible solutions that respect human dignity and well-being while fostering sustainable and equitable advancements.

20.2.3   PROMOTING HUMAN-GEOAI COOPERATION

Promoting human-GeoAI cooperation is essential, as GeoAI technologies are intended to assist humans in making more informed and effective decisions, rather than replacing human decision-making. By incorporating GeoAI technologies into decision-making processes, humans can benefit from the improved accuracy, speed, and scale that AI can provide. A notable example of human-GeoAI cooperation is the Global Forest Watch platform, which uses satellite imagery analysis and machine learning algorithms to monitor deforestation in the Amazon rainforest and other regions worldwide (Anika Berger et al. 2022). While GeoAI algorithms can rapidly process vast amounts of satellite data to identify potential areas of deforestation, human experts can verify these findings and provide valuable context for a more accurate understanding of the situation on the ground. No less importantly, human-GeoAI cooperation can lead to more equitable outcomes by empowering humans to play an active role in the decision-making process and reducing the risk of bias or discrimination. For instance, CrisisMappers, a network of volunteers, combines GeoAI with human expertise to support disaster response efforts by analyzing and verifying data from social media, satellite imagery, and other sources (Zastrow 2014). By working collaboratively with GeoAI systems, humans can gain a better understanding of the strengths and limitations of these systems and develop new insights and approaches to complex problems.

20.2.4   FOSTERING AN INCLUSIVE AND DIVERSE GEOAI COMMUNITY

Fostering an inclusive and diverse GeoAI community is increasingly urgent as GeoAI becomes more integrated into our daily lives and has the potential to impact different groups of people in various ways. An inclusive and diverse community can bring a wider range of perspectives and experiences to the development and application of GeoAI. For example, the AI for Good initiative by Microsoft aims to address global challenges by engaging diverse teams to develop AI solutions that are inclusive and socially responsible (Microsoft 2023). This can help identify and address potential biases and ethical issues that may not have been considered otherwise. Additionally, encouraging participation from underrepresented groups, such as women and minorities, can promote equity and social justice in the development and use of GeoAI. For instance, the Women+ in Geospatial initiative (2023) strives to empower women and other underrepresented groups within the geospatial community. This helps ensure that the benefits of GeoAI are shared among different communities and not limited to a privileged few. Furthermore, promoting diversity can inspire more young people from underrepresented communities to pursue careers in GeoAI, contributing to a more diverse and equitable workforce in the future. Programs like ESRI's GeoMentors and the USGIF's K-12 educational outreach efforts support underrepresented students in learning about geospatial technologies and careers (Esri-AAG 2023; US-GIF 2023). Overall, an inclusive and diverse GeoAI community can help ensure that GeoAI is developed and applied in a way that benefits all members of society.

20.3   CONCLUSION

In conclusion, envisioning a humanistic future for GeoAI is not only desirable but imperative. By weaving humanistic values into the fabric of GeoAI development and application, we can harness the transformative power of GeoAI to create a more just, equitable, and sustainable world. Embracing the principles of fairness, transparency, accountability, and inclusivity will help ensure that GeoAI serves as a force for good, empowering people and communities, while mitigating potential adverse effects on society and the environment. By fostering an inclusive and diverse GeoAI community, we can bring together a rich tapestry of perspectives and experiences that drive innovation and enable us to collectively address the complex challenges facing our world. As we embark on this journey toward a humanistic future of GeoAI, we must remain vigilant and steadfast in our commitment to uphold the ethical principles that form the bedrock of humanism. It is our shared responsibility to ensure that GeoAI not only propels us forward into a technologically advanced future but also safeguards the well-being, dignity, and rights of all sentient beings and the planet we call home.

BIBLIOGRAPHY

Anika Berger, Teresa Schofield, Amy Pickens, Johannes Reiche, and Y. Gou, 2022. Looking for the quickest signal of deforestation? Turn to GFW's integrated alerts. *Global Forest Watch*.

Buolamwini, J., and T. Gebru. 2018. Gender shades: Intersectional accuracy disparities in commercial gender classification. In *Proceedings of the 1st Conference on Fairness, Accountability and Transparency*, eds. A. F. Sorelle and W. Christo, 77–91. Proceedings of Machine Learning Research: PMLR.

Esri-AAG, 2023. GeoMentors - supporting K-12 geography and GIS education.

Foderaro, L. W., 2017. Navigation apps are turning quiet neighborhoods into traffic nightmares. *The New York Times*.

Heaven, W. D., 2020. Predictive policing algorithms are racist. They need to be dismantled. *MIT Technology Review*.

Huang, Z., 2021. AI in urban planning: 3 ways it will strengthen how we plan for the future. *Urban Redevelopment Authority*.

Joseph, D., 2018. How Red Cross uses data during global disasters. *Mapillary*.

Labbe, M., 2021. Energy consumption of AI poses environmental problems. *TechTarget*.

Mcilwain, C., 2020. AI has exacerbated racial bias in housing. Could it help eliminate it instead? *MIT Technology Review*.

Michaux, S., 2019. Farming for the future: UGA leads the way in precision agriculture. *University of Georgia Research*.

Microsoft, 2023. Using AI for good with Microsoft AI.

Netzley, L., 2023. Artificial intelligence used to reduce traffic congestion. *Downtown Los Angeles News*.

Torres, D. L., J. N. Turnes, P. J. Soto Vega, R. Q. Feitosa, D. E. Silva, J. Marcato Junior, and C. Almeida, 2021. Deforestation detection with fully convolutional networks in the Amazon forest from Landsat-8 and Sentinel-2 images. *Remote Sensing*, 13 (24), 5084.

Uddin, K., M. A. Matin, and F. J. Meyer, 2019. Operational flood mapping using multitemporal Sentinel-1 SAR images: A case study from Bangladesh. *Remote Sensing*, 11 (13), 1581.

USGIF, 2023. USGIF K-12 educational outreach. Accessed on March 31 2023, available from `https://usgif.org/k-12`.

Weston, M. A., C. O'brien, K. N. Kostoglou, and M. R. E. Symonds, 2020. Escape responses of terrestrial and aquatic birds to drones: Towards a code of practice to minimize disturbance. *Journal of Applied Ecology*, 57 (4), 777-785.

Women+ in Geospatial, 2023. Accessed on March 31 2023, available from `https://womeningeospatial.org`.

Zastrow, M., 2014. Crisis mappers turn to citizen scientists. *Nature*, 515 (7527), 321-321.

Zhao, B., 2022. Humanistic GIS: Toward a research agenda. *Annals of the American Association of Geographers*, 112 (6), 1576-1592.

21 Fast Forward from Data to Insight: (Geographic) Knowledge Graphs and Their Applications

Krzysztof Janowicz
University of Vienna, Austria University of California, Santa Barbara

Kitty Currier
University of California, Santa Barbara

Cogan Shimizu
Wright State University

Rui Zhu
University of Bristol

Meilin Shi
University of Vienna, Austria

Colby K. Fisher
Hydronos Labs

Dean Rehberger
Michigan State University

Pascal Hitzler
Kansas State University

Zilong Liu
University of Vienna, Austria

Shirly Stephen
University of California, Santa Barbara

DOI: 10.1201/9781003308423-21

CONTENTS

21.1 What are Knowledge Graphs?..412
21.2 The Value Proposition of Knowledge Graphs 415
21.3 Application Areas for Knowledge Graphs 417
  21.3.1 COVID-19 Data Management 417
  21.3.2 The Enslaved.org Hub Ontology 417
  21.3.3 Disaster Response..418
  21.3.4 Food Supply Chain 418
21.4 KnowWhereGraph...419
21.5 Summary..422
  Bibliography..423

21.1 WHAT ARE KNOWLEDGE GRAPHS?

Knowledge Graphs (KGs) are a combination of technologies, specifications, and data cultures for representing, publishing, accessing, integrating, and interlinking individual data points by modeling them as nodes in a densely connected network of places, people, events, and other entities, thereby giving more attention to the relationship between them instead of merely focusing on their properties (Hogan *et al.*, 2021; Janowicz *et al.*, 2022). More formally, a knowledge graph is a node- and edge-labeled directed multigraph where each node-edge-node statement is called a <subject-predicate-object> triple.

It is often useful to distinguish between two functions of edges, called *predicates*, *properties*, or *relations*: as a link that joins a subject node to (1) an object node, which, itself, can act as the subject node in another triple; and (2) a literal node, e.g., a string or number. Contains, for instance, is an object property in a statement such as California contains SantaBarbara, whereas hasPopulation may be a datatype property in a statement such as Vienna hasPopulation 1.897 million. It is those object properties that, together with the usage of globally unique node identifiers called Internationalized Resource Identifiers (IRIs), create a dense network of interconnected statements. While most of these connections are within a graph, they can also link nodes across graphs, thereby forming an ecosystem of KGs such as the Linked Data Cloud (Bizer *et al.*, 2011; Hart and Dolbear, 2013; Janowicz *et al.*, 2012; Lehmann *et al.*, 2015; Shbita *et al.*, 2020; Stadler *et al.*, 2012) or the Open Knowledge Network (OKN) currently envisioned in the US.

While most KGs of substantial size already represent data from a multitude of highly heterogeneous data sources, these ecosystems and the federated queries, i.e., cross-walks, that can be run against them enable users to rapidly gain a holistic view of the requested entities. For instance, information about Cyclone Idai, which devastated Mozambique, Malawi, and Zimbabwe in 2019, might be found in several graphs: one about natural disasters and their damage, which puts the impact of Idai into perspective; another that contains information about all impacted places; and a third, containing medical knowledge, that may provide contextual data about cholera outbreaks. This ability to run open-ended exploratory graph queries across domains, themes, regions, and so on, is part of the value proposition of KGs for industry, government agencies, nonprofit organizations, and universities.

Ecosystems such as the Linked Data Cloud consist of hundreds of graphs jointly encompassing more than 50 billion graph statements. Technically speaking, joining them requires establishing identity links, also called sameAs links, between nodes (their IRIs) across graphs that represent the same real-world entities. For instance, Mozambique in one graph needs to be matched with Q1029–the Wikidata graph (Vrandečić *et al.*, 2023) identifier for the country of Mozambique, but not Q12126300, the identifier for a 1956 movie called "Mozambique". As KGs are usually constructed from multiple sources, methods for establishing links within and between different graphs in order to identify whether multiple IRIs point to the same real-world entity have received a lot of attention and remain an active research topic. Depending on the research field, this task is known as *deduplication* or *co-reference resolution* (Amini *et al.*, 2020). For geospatial KGs, this includes determining whether places and events, as well as their types, within or across graphs are the same (Zhu *et al.*, 2016). For example, a wildfire may be reported multiple times by different agencies, recorded multiple times if it spans administrative borders, e.g., states, or the same fire may be present in separate point- and area-feature datasets.

In addition, nodes and edges can also describe abstract concepts and their relationships, e.g., to state that a TropicalStorm is a specific kind (a subclass) of TropicalCyclone, which in turn is a subclass of Cyclone. Similarly, other triples could specify that storms have a (geographic) trajectory and that they can cause damage when making landfall. These (taxonomic) statements can then be used within other statements to make data self-descriptive, e.g., by stating that Idai is an instance of a cyclone (ex:Idai rdf:type ex:Cyclone). This set of statements (here axioms), which defines the terminology to be used by a KG, is also called a KG *schema* or, more commonly, an *ontology*. Simply put, ontologies are *shared specifications of conceptualizations* that define the meaning (semantics) of a domain's vocabulary using a formal knowledge representation (KR) language (Frank, 1997; Kuhn, 2005; Smith and Mark, 2001). The expressiveness of such a language determines the types of formal statements that can be made and their model-theoretic interpretation. For instance, a KR language such as the Web Ontology Language (OWL)[1] supports cardinality restrictions, thereby enabling knowledge engineers to express that every cyclone has exactly one trajectory (Allemang and Hendler, 2011). Reusing these ontologies, e.g., by realizing that trajectories have common elements across storms, human mobility, animal tracking, and so on, is key to their successful application for data integration and reuse (Hu *et al.*, 2013). How best to represent geographic entities, geographic feature types, their relation, and the spatial and temporal scope of statements remain an active area of study (Kuhn *et al.*, 2021; Wiafe-Kwakye *et al.*, 2022; Zacharopoulou *et al.*, 2022). Today, the most interesting work at the intersection between KG and GeoAI happens at the intersection between classical top-down knowledge engineering and bottom-up approaches from data mining and machine learning (Mai *et al.*, 2022).

---

[1] https://www.w3.org/TR/owl-features

Ontologies are not only useful for capturing explicit statements. They can generate new, implicit statements, thereby growing the knowledge encoded in a knowledge graph. For instance, we can automatically infer that since Idai is a tropical cyclone, it is also a cyclone, and that as each cyclone has (exactly) one trajectory, Idai had a trajectory (but, whether known or not is another issue). In the domain of AI and KR, the term *TBox* is used for the collection of all terminological statements, while the *ABox* is the set of all assertions (statements) about the world. Hence, a KG is the union of TBox and ABox. A very popular KG language is the Resource Description Framework (RDF)[2], which is part of the broader Semantic Web and Linked Data technology stacks, including the schema language OWL, the constraint language SHACL [3], the query language (Geo)SPARQL[4], and so on. Other (partially compatible) technology stacks for KR include so-called property graphs.

A KG entails a (not explicitly stated) triple α if it follows from explicit triples: $KG \models \alpha$. Put differently, α is a logical consequence of triples in the KG. The knowledge representation and reasoning (KRR) literature (Hitzler *et al.*, 2009) distinguishes between several types of reasoning services, most of which can be (re)phrased as a *satisfiability* problem. Two common examples are (concept) subsumption $KG \models C \sqsubseteq D$, e.g., $\{TropicalStorm \sqsubseteq TropicalCyclone, TropicalCyclone \sqsubseteq Cyclone\} \models \{TropicalStorm \sqsubseteq Cyclone\}$ and instance checking $KG \models C(a)$, e.g., if we would define cyclone categories based on wind speed we could check whether Idai was an instance of a Category 3 vs. a Category 4 storm. Ontologies and reasoning services can play multiple roles during knowledge engineering, including being used for inferential reasoning, informing humans, modeling constraints, supporting semantic interoperability (Scheider and Kuhn, 2015), and so forth (Hitzler and Krisnadhi, 2016).

Finally, querying one or multiple KGs can be done via query languages such as SPARQL. This requires publicly available query endpoints on top of graph databases (called *triplestores*) and differs from API-based access in the sense that SPARQL endpoints support arbitrary queries over the entire graph and return fragments of this graph in multiple output types, e.g., RDF, JSON-LD, or CSV. It is worth noting that each of the entities described in a KG and represented via a globally unique IRI should be Web-accessible and *dereferenceable*. Simply put, a client can look up an IRI via HTTP and will receive a description of the resource, e.g., a city or cyclone, in a form specified by said client, e.g., RDF, or as a human-readable webpage. This part is also known as *content negotiation*. Taken together, KGs are human- and machine-readable and reason-able (in the sense that machines can use the semantics encoded in the ontology to reason over the KG).

Additionally, a growing literature complements the top-down engineering of ontologies with bottom-up approaches from data mining and machine learning to learn representations, better handle uncertainty and noise, and provide softer reasoning services such as similarity and analogy-based reasoning in addition to classical deduction. The most well-known of these representation learning methods are so-called

---

[2]https://www.w3.org/RDF/
[3]https://www.w3.org/TR/shacl/
[4]https://www.w3.org/TR/rdf-sparql-query/

KG embeddings (Bordes *et al.*, 2013; Wang *et al.*, 2017), such as TransE, where entities (nodes) and relations (edges) are represented in an n-dimensional vector space $\mathbb{R}^d$. Subject-predicate-object triples are called head-relation-tail (**h-r-t**) triples in the graph embedding literature, and the relationship between h and t is interpreted as a translation vector such that adding the head to the relation would approximate the tail, i.e., $h + r \approx t$. To give a first impression, learning vectors for relations and entities in a KG could be achieved by a loss function that tries to minimize the distance between the $(h + r)$ vector and the t vector for each triple. The resulting embeddings can then be used to measure similarity using well-established (Geographic) Information Retrieval (IR) (Purves *et al.*, 2018) measures, such as cosine similarity. Common downstream tasks include recommending missing tails (or relations, respectively) – a task called link prediction. Pre-trained place embeddings (Yan *et al.*, 2017) from KGs, social media, or other sources can be used in downstream tasks such as POI recommender systems, question-answering systems (Li *et al.*, 2019; Mai *et al.*, 2020; Scheider *et al.*, 2020), trajectory analysis, neighborhood summarization, and so on.

But how to scale the semi-automatic construction of such graphs? This research question is known as Knowledge Graph Construction (KGC) and describes the challenge of automatically creating graphs from unstructured and semi-structured data. In addition to classical Extract, Transform, Load (ETL) workflows, KG construction from sources such as texts requires advanced Named Entity Recognition (NER) and relation-extraction methods. NER aims to find people, places, events, and other entities in text and assign the correct KG identifier to them. In the case of location identifiers, this requires two distinct steps, namely toponym recognition and toponym resolution, e.g., using a gazetteer (Hu, 2018). Relation extraction (and classification) detects relations in text and then assigns the correct relationship from the ontology used. This also involves extracting causal relationships, e.g., a storm led to a landslide.

With the rapid uptake of KGs in web search engines, question-answering systems, foundation models, health research (e.g., on COVID vaccines), and so forth, ethical considerations are moving to the forefront of KG research and practice. An example is the challenge of debiasing graphs and their embeddings (Fisher *et al.*, 2020; Janowicz *et al.*, 2018).

21.2 THE VALUE PROPOSITION OF KNOWLEDGE GRAPHS

Kuhn *et al.* (2014) identified several dimensions along which (Geo)KGs provide new perspectives on key problems of Web-scale data management and knowledge representation. All data across scales, modes, and domains are represented in the form of statements (triples), providing a simple yet powerful data structure. These statements can (locally) contradict without posing problems for the entire KG or even the ecosystem of graphs. This is a central feature of KGs. Trying to enforce a single perspective is neither desired nor possible, especially as systems such as the Linked Data Cloud aim to create a decentralized ecosystem (at least in theory (Polleres *et al.*, 2020)). Inconsistencies can occur on the assertion, the terminology, and the query level. To give a few concrete examples, KGs remain usable even if data are partially missing. For instance, even if a TBox axiom requires earthquakes to have an associated location and magnitude, both may be missing from historical records where only the broadly affected area has been documented. While such cases may be flagged during reasoning or constraint checking, e.g., using SHACL, the larger graph remains operational. ABox statements across and even within graphs can contradict; e.g., a federated graph can return different population counts for Vienna from multiple graphs. Finally, even TBox statements can be inconsistent, something that should be avoided but that does not necessarily impact the entire graph or ecosystem.

Another major value proposition of knowledge graphs is their handling of metadata. Simply put, KGs remove this distinction by making data self-descriptive. The underlying idea is to *create smart data, not smart applications* (Janowicz *et al.*, 2015). While smart applications are not only difficult to develop, changing technology stacks make their maintenance even more challenging. This is evident in the many generations of geospatial *one-stop portals*, all promising to be the ultimate discovery and access point for geo-data, only to be rendered obsolete by outdated technologies (e.g., Flash or Silverlight), changes in user interaction, and so forth. Not only do these portals take considerable resources to develop and maintain, they also do not leave anything of value behind, as new portals must be developed almost from scratch. Smart data, i.e., triples defined using (expressive) ontologies, make off-the-shelf software smarter by improving the retrieval, integration, and reuse of data. Hence, it is not surprising that KGs are among the leading technical solutions for implementing FAIR principles-based (Wilkinson *et al.*, 2016) research data management systems such as Pheidra[5]. More generally, the ability to uniquely identify, describe (mark up), and access a resource at a granular level – such as an observation or webpage fragment – instead of at the data*set* level, has revolutionized data management and Web search, alike, as the success of `Schema.org` has shown.

Finally, KGs also act as a key resource for contextual information that can be used to enrich data on the fly or serve as the basis for question-answering systems and digital assistants. Today, many web searches are not answered by users clicking on links but by search engines embedding results directly in their main page (e.g., as knowledge panels). For geographic and environmental data, for instance, the KnowWhereGraph, discussed in more detail in Section 21.4, offers billions of graph statements about physical and human geography for any place in the US (and some worldwide) within seconds and in a form that can easily be integrated into bespoke user interfaces, e.g., for supply chain management or humanitarian relief.

Summing up, knowledge graphs ease data retrieval, integration, and reuse, provide contextual information and help to surface hidden insights, scale well across very large and highly heterogeneous data sources, can encode different perspectives, and make developing end-user applications less resource intensive.

---

[5]https://phaidra.univie.ac.at/

21.3 APPLICATION AREAS FOR KNOWLEDGE GRAPHS

Knowledge graphs have been used in numerous application areas in industry, academia, and government. These applications can be classified by the tasks the KGs have been used for, such as semantic search, data integration, conflation, and research data management, as well as by the application domains, such as healthcare, intelligence, and the Internet of Things (IoT). Here we list selected use cases by example where KGs have been used in different roles, in different domains, and by different actors.

21.3.1 COVID-19 DATA MANAGEMENT

During the COVID-19 pandemic, teams across the world attempted to predict the trend of COVID-19's impact on multiple aspects of society, including health and economics. For instance, according to the COVID-19 Forecast Hub[6], about 94 teams around the world built approximately 117 models to predict key variables such as the daily number of deaths, the daily number of new cases, and the cumulative number of hospitalizations. These predicted variables were key to depicting the situation in a timely way for decision-makers and the general public. However, each team adopted quite different assumptions, applied distinct models, and trained them on different types of datasets. Contextualizing these forecasts in an accessible way for non-experts was a challenge. By leveraging knowledge graphs, Zhu et al. (2022) introduced a COVID-Forecast-Graph, where the different forecasts were formally represented using international standards (e.g., SOSA[7] and OWL Time[8]). In contrast to traditional data portals, the graph format made forecasting results more easily accessible to the general public by presenting them with contextual information about their respective models to support the interpretation of the results in a single framework. Moreover, the COVID-19 pandemic was a dynamic and complex challenge that no single domain could fully explain and tackle. A holistically integrated approach was, therefore, needed. For instance, the aforementioned health-related variables (e.g., the daily number of new cases) often varied proportionately with local and global economic indicators (e.g., credit card spending on entertainment). To integrate these diversely themed data from different sectors, the COVID-Forecast-Graph proposed using place and time, two geographic attributes common to all of the observations and events it documented, as the nexus of the graph. Finally, AstraZeneca, for instance, used knowledge graphs for drug discovery and also in the COVID research.

21.3.2 THE ENSLAVED.ORG HUB ONTOLOGY

Within the digital humanities, and more broadly, there has been a significant shift in perceptions about what can be known about enslaved Africans and how we can make sense of records of enslavement. As a result, a growing number of knowledge bases have been made accessible for public consumption. This proliferation of silos

---

[6]https://covid19forecasthub.org/
[7]https://www.w3.org/TR/vocab-ssn/
[8]https://www.w3.org/TR/owl-time/

presents a number of challenges, which the use of KGs can address: (a) individual projects tend to be isolated, preventing federated and cross-project analysis; (b) disambiguating (or merging) individuals across multiple datasets is nearly impossible given their current silo-ed nature; and (c) there are no best practices for (digital) data creation agreed upon by the (digital) humanities community. To address these challenges, the Enslaved.Org project pioneered a new model for humanities scholarship by following state of art to establish an RDF-based KG equipped with an OWL ontology, called the Enslaved.Org Hub Ontology; the modeling approach and its core concepts are detailed in (Shimizu *et al.*, 2022, 2023, 2020). In contrast to other use cases, the Enslaved.Org Hub Ontology is not actually used for inference but as a guide for integrating the heterogeneous datasets and understanding the KG as a whole (Hitzler and Krisnadhi, 2016). Further documentation can be found online[9].

21.3.3 DISASTER RESPONSE

In the face of a humanitarian disaster, a KG can support situational awareness that helps responders act efficiently and effectively. By providing seamless access to a variety of environmental and social data within the same framework, a KG can help to answer questions like *Who is most vulnerable, given the nature of the disaster, historical patterns of loss, socioeconomic inequalities, and critical health risks particular to local populations?* and *Who has expertise relevant to this situation and place?* The KnowWhereGraph (described in detail in Section 21.4) is one such KG that provides this kind of information to analysts at Direct Relief[10], a private humanitarian nonprofit organization based in Santa Barbara, California, whose core mission is to improve the health and lives of people worldwide affected by poverty or emergencies (Zhu *et al.*, 2021). Similarly, the Urban Flooding Open Knowledge Network (UFOKN) (Johnson *et al.*, 2022))[11] incorporates KGs to relate information about urban infrastructure to real-time hydrologic forecasts, helping responders and planners understand short- and long-term impacts of urban flooding.

21.3.4 FOOD SUPPLY CHAIN

It is critically important to understand and improve the robustness and adaptability of our food supply chains, making them more resilient to disturbances in food supply and demand networks. There is inherently a risk of network fracturing and delayed recovery during extreme weather events, wildfires, floods, and other natural hazards. In the face of uncertain natural hazards of increasing frequency and severity, it is vital that the implications of these disruptions are evaluated for the source nodes of our supply chains, such that resiliency in the whole supply chain can be promoted. Ensuring this requires bridging across topics and domains, but data are silo-ed, distributed, variously described and have regional variations that create different needs and solutions. To solve this challenge, the Food Industry Association (FMI), which

---

[9]https://enslaved.org/ and https://docs.enslaved.org/
[10]https://www.directrelief.org/
[11]https://ufokn.com

has identified food safety and food quality issues arising from environmental disasters or disturbances as a high-priority industry concern, tried to leverage a KG via a joint pilot project together with KnowWhereGraph (described in detail following). The ideas was to combine a KG with enrichment services to cross-walk between data silos in multiple domains with rich public data in order to address the following pilot use cases: (1) ashfall on leafy greens and vegetables during wildfires; (2) smoke damage to crops during wildfires, such as smoke-tainted grapes and their impact on wine; and (3) transportation routes and/or retail operations severely impacted by closures and rerouting during wildfires or other crises.

21.4   KNOWWHEREGRAPH

The KnowWhereGraph (KWG) (Janowicz *et al.*, 2022) is a densely connected, highly heterogeneous and geographically enabled KG consisting of more than 15 billion graph statements across 30 different data layers (Figure 21.1), including environmental and demographic themes. These data layers are divided into two groups, namely region identifiers and thematic data. The first group comprises millions of named geospatial units such as administrative areas, climate divisions, drought zones, metropolitan areas, and many other geographic feature types. In addition, KnowWhereGraph implements the S2 hierarchical discrete global grid system at level 13, which means that any area (in the US) of about 1 square kilometer can be addressed individually or as a combination of cells to form arbitrary areal features. For each of these named or S2-based regions, the graph contains data about previous events such as natural disasters, critical infrastructure, soil types, crop cover, the social vulnerability of the population, air quality, and so on from the group of thematic layers. As these layers are all integrated, KnowWhereGraph can provide environmental intelligence for any region within seconds to answer questions such as:

- What is here?

- What happened here before?

- Who knows more?

- How does this region or event compare to other regions or previous events?

By doing so, KnowWhereGraph assists decision-makers and data scientists in enriching their data with billions of connected, up-to-date graph statements at the interface between humans and their environment to rapidly gain the situational awareness required for successful decision-making. Put differently, KnowWhereGraph can be considered a gazetteer of gazetteers, backed with rich data. In addition, KnowWhereGraph also links to other KGs such as Wikidata (Vrandečić and Krötzsch, 2014) and collaborates with graphs such as SPOKE (Baranzini *et al.*, 2022) and UFOKN (Johnson *et al.*, 2022) to enable cross-walks between graphs with the overall mission of creating an Open Knowledge Network for (governmental) data in the US and beyond.

Thematic Datasets					Place-Centric Datasets		
Dataset Name/ Theme	Source Agency	Key Attributes	Spatial Coverage	Temporal Coverage	Place-Centric Dataset	Defining Authority	Spatial Coverage
Soil Properties	USDA	soil type, farmland class	Targeted regions in US	Current	S2 Cells	Google	Lvl 9 (Global), Lvl 13 (US),
Wildfires	USGS, USDA, USFS, NIFC	wildfire type, burn severity, num. acres burned, contained date	US	1984–current	Global Administrative Regions	University of Berkeley, Museum of Vertebrate Zoology and the International Rice Research Institute	Global
Earthquakes	USGS	magnitude, length, width, geometry	Global (mag. over 4.5)	2011-01-01 to 2022-01-18			
Climate Hazards	NOAA	injuries, deaths, property damages	US	1950–2022			
Expert - Covid-19 Mobility	Direct Relief (DR)	name, affiliation, expertise	Global	2021	US Federal Judicial District	DoJ, ESRI	US
Expert - General	KWG, UC System, DR, Semantic Scholar	name, affiliation, expertise with spatiotemporal scopes	Global	unlimited	National Weather Zones	NOAA	US
Cropland Types	USDA	crop types (raster data)	US	2008-2021	FIPS Codes	NRCS	US
Air Qual. Obs.	U.S. EPA	AQI value, CO concentration	US	1980–2022	Designated Market Area	Nielen	US
Smoke Plumes	NOAA	daily smoke plumes extent	US	2010-2022	ZIP	ZCTA	US
Climate Observations	NOAA	temperature, precipitation, PDSI, PHSI	US	1950 - 2022	Climate Division	NOAA	US
Disaster Declaration	FEMA	designated area, program, amount approved, program designated date	US	1953 - 2022	Census Metropolitan Area	US Census	US
Smoke Plume Extents	NOAA	Smoke extent	US	2017 - 2022	Drought Zone	NDMC, USDA,NOAA	US
BlueSky Forecasts	Bluesky	PM10, PM5	US	2022-03-07	Geographic Name Information System	USGS	US
Transportation (highway network)	DOT	road type, road length, road sign	US	2014			
Public Health Observations	CDC, US Census, University of Wisconsin Population Health Institute	below poverty level, diabetes, obesity, mental health provider rate, annual mammogram, vaccinated, injury death	US	2017, 2021			
Public Health Infrastructure	HIFLD	pharmacies, hospitals, dialysis centers, public health departments	US	2017 - 2022			
Social Vulnerability	CDC/ATSDR	social vulnerability index	US	2018			
Hurricane Tracks	NOAA	max wind speed, min pressure	US	1851-2020			

Figure 21.1   Data layers in KnowWhereGraph as of March 2023.

To foster integration, interoperability, and querying, KnowWhereGraph puts all statements in a homogeneous framework by representing most of them in the form of observations using the Sensors, Observations, Samples, Actuators (SOSA) ontology (Janowicz *et al.*, 2019). Every property, such as the magnitude of fires, storms, and earthquakes, crop-cover and soil types, population, and so on, is modeled as being an observed property (sosa:ObservableProperty) with a timestamp (sosa:resultTime), feature of interest (sosa:FeatureOfInterest; e.g., an S2 cell, named geographic region, or event), a result (sosa:hasSimpleResult), and so forth. For instance, the Thomas Fire that burned in Santa Barbara county in 2017 is represented as a feature of interest of a certain type, here a kwg:Wildfire, with one of its observed properties being the area burned, namely 114,078 ha. Features such as events or regions are then represented using GeoSPARQL (Battle and Kolas, 2011) to encode their geometries and the topological relations among them, while temporal aspects are represented using OWL-Time (Cox *et al.*, 2017). Put differently, most of KnowWhereGraph relies on a small set of well-established and internationally standardized ontologies. On top of this layer sit various so-called Ontology Design Patterns (ODP) (Gangemi and Presutti, 2009) that jointly form the KnowWhere-Graph Schema[12]. This schema also allows for disambiguating terms such as the Wildfire class, whose instances are physical fires, i.e., events, from the Wildfire class in the hazards branch of the KnowWhereGraph that describes types of hazards and human expertise about them. Together the KnowWhereGraph Schema consists of 312 classes and 2980 axioms describing their relationships. This is interesting insofar as the relationship between ABox and TBox size seems to vary greatly among KGs, and in the case of KnowWhereGraph, they are at least six orders of magnitude apart, thereby differing substantially from a graph such as SPOKE. Given that KnowWhereGraph runs on a GraphDB triplestore with RDFS-PLUS entailment enabled, the materialization ratio, i.e., the relation between explicitly stated and inferred triples, is about 1.55.

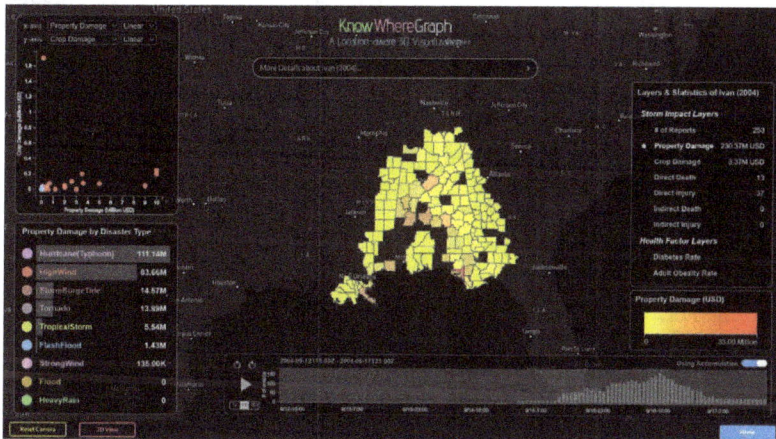

Figure 21.2 KnowWhereGraph accessed via a bespoke end-user interface for Direct Relief.

The (dense) relationship links within KnowWhereGraph and outgoing links to other graphs enable exploratory queries such as starting with a specific event, e.g., a hurricane, then exploring the geographic areas, e.g., counties, it affected, and then comparing these areas based on multiple thematic layers – e.g., previous events, soils, critical infrastructure, demographic variables – to understand why one region along the storm's trajectory was impacted more than its neighboring regions. Figure 21.2 showcases such an example using a custom interface built for the humanitarian relief organization Direct Relief on top of KnowWhereGraph (Li *et al.*, 2023). In addition to gaining situational awareness quickly, e.g., to coordinate their

---

[12]https://stko-kwg.geog.ucsb.edu/lod/ontology

supply chain, Direct Relief is interested in matching experts to emergencies in order to quickly get the right "boots on the ground". Hence, KnowWhereGraph uses a hazard ontology to link its thematic layers, e.g., floods, to experts. Other application areas for KnowWhereGraph include supply chain management and food safety, providing indices for Environmental, Social, and corporate Governance (ESG), and acting as a provider of contextual information for downstream tasks such as location optimization, agent-based models for vector-born diseases, and so on.

This practice is frequently known as *geo-enrichment*. There are plenty of tools that offer this functionality (e.g., Esri's GeoEnrichment service), but they tend to suffer from the same sorts of limitations: (1) they only serve data for a small set of predefined categories, such as demographic data; (2) they are closed data silos that encode just one domain or cultural perspective; (3) because they are centrally maintained, scalability and timely updates become bottlenecks when those services try to incorporate more (diverse) data; and (4) they lack an integration layer that enables follow-up queries over the enriched data. Consequently, a new approach is needed that combines the strength of geo-enrichment services, i.e., seamless access to contextual information for an analyst's areas of concern, with a technology that provides open, densely integrated, cross-domain data across a wide range of perspectives (Janowicz *et al.*, 2022). KnowWhereGraph, on the other hand, is poised to overcome these issues.

The KnowWhereGraph's public SPARQL query endpoint is available at `https://stko-kwg.geog.ucsb.edu/sparql`, while an exploratory search interface with dereferencing is available at `https://stko-kwg.geog.ucsb.edu/`.

21.5 SUMMARY

In this chapter, we introduced knowledge graphs as a paradigm for representing structured and unstructured data in a human- and machine-readable and reason-able way with a strong focus on establishing relations between entities such as places, people, events, and objects. Densely interlinked knowledge graphs that combine datasets from highly heterogeneous sources foster data integration, retrieval, and exploration; provide a uniform means of accessing data; simplify data management by getting rid of the data–metadata distinction, serving self-descriptive, smart data; scale reasonably well; and provide contextual information to almost any application or use case that benefits from situational awareness. In addition, knowledge graphs also power many FAIR-based digital asset repositories, thereby contributing to many other scientific domains indirectly. Finally, many Fortune 500 companies maintain public or private knowledge graphs, and Esri recently released their *ArcGIS Knowledge* enterprise knowledge graph with direct integration into their ecosystem and analytical capabilities.

While knowledge graphs as such do not depend on a specific technology stack, the W3C- and OGC-supported Semantic Web and Linked Data stack, which consists of technologies such as RDF, SHACL, OWL, and (Geo)SPARQL, provides additional benefits. Their well-established and standardized open-source ecosystems offer software solutions for all steps – including data lifting, storage, visualization, etc.,

– and are available off the shelf. This means that smart data created using these technologies will remain accessible and (re)usable for the foreseeable future and that the entrance hurdles, e.g., financial commitment and the availability of documentation, remain low.

At the same time, knowledge graphs are also an active research area with many open and often geo-specific research questions such as graph summarization, debiasing, trust, handling uncertainty and noise, graph fusion, and the learning of knowledge graph embeddings for downstream tasks, including question answering, recommender systems, and so forth. Given their ability to homogeneously represent data across domains and modes, knowledge graphs may also serve as a promising data provider for (geo-)foundation models and/or autonomous GIS agents of the near future.

ACKNOWLEDGMENTS

This work was partially funded by the National Science Foundation under Grant 2033521 A1: KnowWhereGraph: Enriching and Linking Cross-Domain Knowledge Graphs using Spatially-Explicit AI Technologies. Any opinions, findings, conclusions, or recommendations expressed in this material are those of the authors and do not necessarily reflect the views of the National Science Foundation.

BIBLIOGRAPHY

Allemang, D. and Hendler, J., 2011. *Semantic Web for the Working Ontologist: Effective Modeling in RDFs and OWL*. Elsevier.

Amini, R., Zhou, L., and Hitzler, P., 2020. Geolink cruises: A non-synthetic benchmark for co-reference resolution on knowledge graphs. *In*: *Proceedings of the 29th ACM International Conference on Information & Knowledge Management*. 2959–2966.

Baranzini, S., *et al.*, 2022. A biomedical open knowledge network harnesses the power of ai to understand deep human biology. *AI Magazine*, 43 (1), 46–58.

Battle, R. and Kolas, D., 2011. Geosparql: enabling a geospatial semantic web. *Semantic Web Journal*, 3 (4), 355–370.

Bizer, C., Heath, T., and Berners-Lee, T., 2011. Linked data: The story so far. *In*: *Semantic Services, Interoperability and Web Applications: Emerging Concepts*. IGI global, 205–227.

Bordes, A., *et al.*, 2013. Translating embeddings for modeling multi-relational data. *Advances in Neural Information Processing Systems*, 26.

Cox, S., *et al.*, 2017. Time ontology in owl. *W3C Recommendation*, 19.

Fisher, J., *et al.*, 2020. Debiasing knowledge graph embeddings. *In*: *Proceedings of the 2020 Conference on Empirical Methods in Natural Language Processing (EMNLP)*. 7332–7345.

Frank, A.U., 1997. Spatial ontology: A geographical information point of view. *Spatial and Temporal Reasoning*, 135–153.

Gangemi, A. and Presutti, V., 2009. Ontology design patterns. *In*: *Handbook on Ontologies*. Springer, 221–243.

Hart, G. and Dolbear, C., 2013. *Linked Data: A Geographic Perspective*. Taylor & Francis.

Hitzler, P. and Krisnadhi, A., 2016. On the roles of logical axiomatizations for ontologies. *In*: P. Hitzler, A. Gangemi, K. Janowicz, A. Krisnadhi and V. Presutti, eds. *Ontology Engineering with Ontology Design Patterns - Foundations and Applications*. Studies on the Semantic Web, vol. 25. IOS Press, 73–80.

Hitzler, P., Krotzsch, M., and Rudolph, S., 2009. *Foundations of Semantic Web Technologies*. CRC Press.

Hogan, A., *et al.*, 2021. Knowledge graphs. *ACM Computing Surveys (CSUR)*, 54 (4), 1–37.

Hu, Y., 2018. Geo-text data and data-driven geospatial semantics. *Geography Compass*, 12 (11), e12404.

Hu, Y., *et al.*, 2013. A geo-ontology design pattern for semantic trajectories. *In*: *Spatial Information Theory: 11th International Conference, COSIT 2013, Scarborough, UK, September 2-6, 2013. Proceedings 11*. Springer, 438–456.

Janowicz, K., *et al.*, 2019. Sosa: A lightweight ontology for sensors, observations, samples, and actuators. *Journal of Web Semantics*, 56, 1–10.

Janowicz, K., *et al.*, 2022. Know, know where, knowwheregraph: A densely connected, cross-domain knowledge graph and geo-enrichment service stack for applications in environmental intelligence. *AI Magazine*, 43 (1), 30–39.

Janowicz, K., *et al.*, 2012. Geospatial semantics and linked spatiotemporal data–past, present, and future. *Semantic Web*, 3 (4), 321–332.

Janowicz, K., *et al.*, 2015. Why the data train needs semantic rails. *AI Magazine*, 36 (1), 5–14.

Janowicz, K., *et al.*, 2018. Debiasing knowledge graphs: Why female presidents are not like female popes. *In*: *ISWC (P&D/Industry/BlueSky)*.

Johnson, J.M., *et al.*, 2022. Knowledge graphs to support real-time flood impact evaluation. *AI Magazine*, 43 (1), 40–45.

Kuhn, W., 2005. Geospatial semantics: why, of what, and how? *In*: *Journal on Data Semantics III*. Springer, 1–24.

Kuhn, W., *et al.*, 2021. The semantics of place-related questions. *Journal of Spatial Information Science*, (23), 157–168.

Kuhn, W., Kauppinen, T., and Janowicz, K., 2014. Linked data-a paradigm shift for geographic information science. *In*: *Geographic Information Science: 8th International Conference, GIScience 2014, Vienna, Austria, September 24-26, 2014. Proceedings 8*. Springer, 173–186.

Lehmann, J., *et al.*, 2015. Dbpedia–a large-scale, multilingual knowledge base extracted from wikipedia. *Semantic Web*, 6 (2), 167–195.

Li, W., Song, M., and Tian, Y., 2019. An ontology-driven cyberinfrastructure for intelligent spatiotemporal question answering and open knowledge discovery. *ISPRS International Journal of Geo-Information*, 8 (11), 496.

Li, W., *et al.*, 2023. Geographvis: a knowledge graph and geovisualization empowered cyber-infrastructure to support disaster response and humanitarian aid. *ISPRS International Journal of Geo-Information*, 12 (3), 112.

Mai, G., *et al.*, 2022. Symbolic and subsymbolic geoai: Geospatial knowledge graphs and spatially explicit machine learning. *Transactions in GIS*, 26 (8), 3118–3124.

Mai, G., *et al.*, 2020. Se-kge: A location-aware knowledge graph embedding model for geographic question answering and spatial semantic lifting. *Transactions in GIS*, 24 (3), 623–655.

Polleres, A., *et al.*, 2020. A more decentralized vision for linked data. *Semantic Web*, 11 (1), 101–113.

Purves, R.S., *et al.*, 2018. Geographic information retrieval: Progress and challenges in spatial search of text. *Foundations and Trends in Information Retrieval*, 12 (2-3), 164–318.

Scheider, S. and Kuhn, W., 2015. How to talk to each other via computers: Semantic interoperability as conceptual imitation. *Applications of Conceptual Spaces: The Case for Geometric Knowledge Representation*, 97–122.

Scheider, S., *et al.*, 2020. Ontology of core concept data types for answering geo-analytical questions. *Journal of Spatial Information Science*, (20), 167–201.

Shbita, B., *et al.*, 2020. Building linked spatio-temporal data from vectorized historical maps. *In: The Semantic Web: 17th International Conference, ESWC 2020, Heraklion, Crete, Greece, May 31–June 4, 2020, Proceedings 17.* Springer, 409–426.

Shimizu, C., *et al.*, 2022. Ontology design facilitating Wikibase integration – and a worked example for historical data. *CoRR*, abs/2205.14032.

Shimizu, C., Hammar, K., and Hitzler, P., 2023. Modular ontology modeling. *Semantic Web*, 14 (3), 459–489.

Shimizu, C., *et al.*, 2020. The enslaved ontology: Peoples of the historic slave trade. *J. Web Semant.*, 63, 100567.

Smith, B. and Mark, D.M., 2001. Geographical categories: an ontological investigation. *International Journal of Geographical Information Science*, 15 (7), 591–612.

Stadler, C., *et al.*, 2012. Linkedgeodata: A core for a web of spatial open data. *Semantic Web*, 3 (4), 333–354.

Vrandečić, D. and Krötzsch, M., 2014. Wikidata: a free collaborative knowledgebase. *Communications of the ACM*, 57 (10), 78–85.

Vrandečić, D., Pintscher, L., and Krötzsch, M., 2023. Wikidata: The making of. *In: Companion Proceedings of the ACM Web Conference 2023.* 615–624.

Wang, Q., *et al.*, 2017. Knowledge graph embedding: A survey of approaches and applications. *IEEE Transactions on Knowledge and Data Engineering*, 29 (12), 2724–2743.

Wiafe-Kwakye, K., Hahmann, T., and Beard, K., 2022. An ontology design pattern for spatial and temporal aggregate data (stad).

Wilkinson, M.D., *et al.*, 2016. The fair guiding principles for scientific data management and stewardship. *Scientific Data*, 3 (1), 1–9.

Yan, B., *et al.*, 2017. From itdl to place2vec: Reasoning about place type similarity and relatedness by learning embeddings from augmented spatial contexts. *In: Proceedings of the 25th ACM SIGSPATIAL International Conference on Advances in Geographic Information Systems.* 1–10.

Zacharopoulou, D., *et al.*, 2022. A web-based application to support the interaction of spatial and semantic representation of knowledge. *AGILE: GIScience Series*, 3, 70.

Zhu, R., *et al.*, 2021. Providing humanitarian relief support through knowledge graphs. *In: Proceedings of the 11th on Knowledge Capture Conference.* 285–288.

Zhu, R., *et al.*, 2016. Spatial signatures for geographic feature types: Examining gazetteer ontologies using spatial statistics. *Transactions in GIS*, 20 (3), 333–355.

Zhu, R., *et al.*, 2022. Covid-forecast-graph: An open knowledge graph for consolidating covid-19 forecasts and economic indicators via place and time. *AGILE: GIScience Series*, 3, 21.

22 Forward Thinking on GeoAI

Shawn Newsam
Department of Computer Science and Engineering, University of
California, Merced

CONTENTS

22.1 A Simple Knowledge Retrieval Scenario 427
22.2 Continued Interaction between Communities 428
22.3 The Challenges of (Rewarding) Interdisciplinary Research......................... 428
22.4 The Role of Industry.. 429
22.5 Opportunities 429
    22.5.1 Location as a Key for Unsupervised Training 429
    22.5.2 Mutli-Modal Foundational Models... 430
    22.5.3 Spatial Turing Tests for GeoAI Models.. 431
22.6 Interesting Recent Developments in Generative Modeling......................... 431
22.7 Closing Remarks 433
    Bibliography... 433

22.1 A SIMPLE KNOWLEDGE RETRIEVAL SCENARIO

Entering "How long is the coast of Italy" into Google search[1] at the time of writing results in the following answer "Italy has 4,723 miles of coastline, dotted with some of the most beautiful beaches and seaside towns in the entire world". A website[2] is indicated as the source of this information. The answer is followed by links to other websites. Entering the same query into Bing search[3] results in the answer "7,600 km" with links to three sources[4][5][6]. This is also followed by links to other websites. While answering spatially related queries like this is only one service that GeoAI can support, this retrieval scenario is worth further inspection for two reasons. First, because of how reliant people are on online search. It is often how the most recent technological advancements meet the masses. This is unlikely to change. And, second, it is interesting due to some very recent developments in generative models.

---

[1] https://www.google.com/
[2] https://www.cntraveler.com/galleries/2014-08-01/italy-s-most-beautiful-beaches
[3] https://www.bing.com/
[4] https://www.worldatlas.com/articles/countries-in-europe-with-the-longest-coastline.html
[5] https://science.blurtit.com/749871/what-is-the-length-and-width-of-italy
[6] https://www.indexmundi.com/italy/coastline.html

22.2   CONTINUED INTERACTION BETWEEN COMMUNITIES

How you perceive the answers above likely depends on what research community you (primarily) belong to. Researchers on the computer science end of the GeoAI spectrum might accept these answers as reasonable since they seem to provide useful information and, assuming the search engines perform correctly, are likely based on credible sources. Researchers on the geography end of the spectrum might find these answers problematic since, among other things, they ignore the coastline paradox (Richarson, 1961) which states that the length of a fractal-like object, like as a coastline, depends on the scale at which you measure it. Thus, there is no definite answer such as "4,723 miles". While I won't speculate here on the answer that more geography-leaning researchers might find more acceptable (more on this later), hopefully this simple example highlights that GeoAI is (still) a coming together of different research communities and that progress requires interdisciplinary collaboration. Lots of good thinking on this has already been written in various recent GeoAI workshop reports (e.g., Lunga *et al.* (2022)), etc., so I won't elaborate other than to say the various communities involved in GeoAI still have a lot to contribute jointly as we move the frontier forward. I reside on the computer science side of the spectrum and acknowledge this limits my perspective in this chapter on Forward Looking on GeoAI. I'll end with the anecdote that I probably would not be aware of the coastline paradox had I not attended a more geography-leaning meeting many years ago at which GIScience pioneer Michael Goodchild described it in a presentation.

22.3   THE CHALLENGES OF (REWARDING) INTERDISCIPLINARY RESEARCH

There are many challenges to conducting interdisciplinary research such as the effort needed for collaborators to learn domain specific goals and terminology, teams often being in different locations, etc. A fundamental challenge that I believe can and must be overcome is recognizing and rewarding interdisciplinary work similar to intradisciplinary, especially in the academy. Promotion and tenure committees at research universities still retreat to the comfort of reviewing a candidate based solely on their contributions to the field of their home department even when the institution touts itself as being a champion of interdisciplinarity. While we seem to do better with Ph.D. students, often requiring their dissertation committees to contain members from outside their programs and involving the students in interdisciplinary projects (which thankfully have become more prevalent due to funding agencies realizing the importance of interdisciplinary research in solving societal-scale problems), this recognition and reward of interdisciplinarity usually does not carry over to the review of assistant professors, especially in the promotion process. I faced this challenge when I went from being a Ph.D. student in the highly interdisciplinary NSF-funded Alexandria Digital Library research project at the University of California, Santa Barbara, to an assistant professor of computer science and engineering at the University of California, Merced. Even though a fundamental tenet of UC Merced's mission statement is to foster and encourage cross-disciplinary inquiry and discovery, my experience is that this principle is not sufficiently reflected in the faculty review process. The academy must strive to do better at recognizing and rewarding interdisciplinary research in exciting and important frontiers like GeoAI.

22.4   THE ROLE OF INDUSTRY

It is through the products and services provided by industry (companies) that technologies such as AI touch the masses. While these technologies can aid efforts by academia and government labs to solve societally important problems, it is through their commercialization that they have the most impact due to the scale of uptake by the general public. Thus any forward thinking on GeoAI must consider the role that industry has to play, especially with respect to ethical and moral concerns.

Yet, none of the contributors to this Handbook are from, or represent, industry. This observation is not meant as a criticism of the editors but rather a summons that industry should be engaged and involved in shaping the GeoAI frontier.

Take, for example, the well-developed task of location-based search, the ability to retrieve information based on location. The underlying science and technology may have originated in academia and government labs, and, indeed, location-based search has broad application in work being done in these institutions. However, the impact of this work is much less than that of the many location-based services enabled by location-based search but provided by industry such as traffic-aware route planning and ride-hail transportation. This is where location-based search meets the masses. While these services are clearly beneficial, they raise ethical issues especially surrounding privacy. It is not clear how well these issues are being addressed.

The products and services that result from advances in GeoAI promise to be even more impactful than the location-based services enabled by location-based search. Their ethical and moral concerns will be even more formidable. Industry generally does not have the types of checks-and-balances found at universities (e.g, institutional research boards) and government labs. Mechanisms must be established to ensure the ethical use of GeoAI technologies. Engagement with industry will be essential to develop and deploy these safeguards.

22.5   OPPORTUNITIES

This section discusses some near- to medium-term opportunities in GeoAI, specifically, leveraging location as a key for associating data to enable self-supervised training of large AI models; multi-modal foundational models; and spatial Turing tests.

22.5.1   LOCATION AS A KEY FOR UNSUPERVISED TRAINING

Recent large language models (LLM), such as Open AI's generative pre-trained (GPT) models, have shown remarkable success on a number of challenging AI tasks such as the dialog interface ChatGPT. Two developments in particular point to this

success. First, is the size of the models which can have hundreds of billions of parameters. Second, is the unsupervised pre-training on large corpora of text such as
the Common Crawl archive of the Internet. This pre-training does not require manually labeled data since it largely involves simply determining whether the model
correctly predicts the next word in a string of text. Once trained, LLMs can be applied to a number of downstream tasks after limited fine-tuning on labelled data.

There is an almost limitless amount of text digitally available to pre-train LLMs.
By leveraging location to associate different kinds of information, there could similarly be a limitless amount of spatially related data to pre-train large geographic
models (LGMs). Even though the pre-training of LLMs only explicitly involves predicting the next word in a sentence given the previous words, the models appear to
implicitly encapsulate higher-level concepts. An LGM similarly trained only to explicitly predict one piece of held-out information given spatially co-located or nearby
information could potentially implicitly encapsulate higher-level spatial knowledge.
With a limited amount of fine-tuning, the pre-trained LGM could then be applied to
various downstream tasks involving analysis, prediction, etc.

There are great opportunities to pre-train large geographic models in an unsupervised manner by exploiting location as a key to connect varied and large amounts of
spatial information.

22.5.2   MUTLI-MODAL FOUNDATIONAL MODELS

There have been recent breakthroughs in the AI community on multi-modal foundational models (Wang *et al.*, 2022). These models are pre-trained in an unsupervised
fashion through simple data masking on uni- and multi-modal data: predicting a left-
out word in a sentence, predicting a left-out patch in an image, or predicting either
a left-out word or image patch in image-text pairs. (The masking and prediction are
actually done on the tokens that are input to the transformer models–these tokens
are embeddings of the words and patches.) These unsupervised learning tasks allow
the models to learn effective image and text embeddings as well as correlations between modalities. The significant finding is that fine-tuning the pre-trained models
on down-stream tasks can significantly outperform state-of-the-art models specifically designed and trained for a variety of challenging tasks. This includes both
vision-language tasks like visual questioning and answering, visual reasoning, image captioning, and image-text retrieval, as well as, perhaps somewhat surprisingly,
pure vision tasks like object detection and instance segmentation, semantic segmentation, and image classification.

There is opportunity to similarly develop multi-modal foundational models for
GeoAI applications. As described earlier, location can be used as a key to associate
different modalities of geospatial data, and similar data masking strategies could be
used to perform model pre-training. Some of the interesting open research questions
include what kinds of data modalities to include beyond images and text, how to
embed these new data modalities as tokens for input to the transformers (e.g., how
do you embed a map into a latent space common to other modalities?), and what
downstream tasks are the pre-trained models fine-tuned for.

22.5.3 SPATIAL TURING TESTS FOR GEOAI MODELS

Janowicz *et al.* (2020) citing Goodchild (2001) state that a model can be called spatially explicit if it satisfies at least one of the following requirements:

- Invariance test: the results of spatially explicit models are not invariant under relocation of the studied phenomena.

- Representation test: spatially explicit models contain spatial representations of the studied phenomena in their implementations (this can be in the form of coordinates, spatial relations, place names, and so on).

- Formulation test: spatially explicit models make use of spatial concepts in their formulations, e.g., the notion of a neighborhood.

- Outcome test: The spatial structures/forms of inputs and outcomes of the model differ.

They go on to describe the moonshot goal of developing an artificial GIS analyst that passes a domain-specific Turing test by 2030. With the recent progress in large language models paving the way for large geospatial models described in Section 22.5.1, this moonshot might be realized sooner. Further, the foundational models described in Section 22.5.2 might enable a broader set of functionality than just analysis, such as prediction or even the discovery of unknown spatial phenomena. An interesting research question then becomes how to design spatial Turing tests to determine whether GeoAI models are spatially explicit according to criteria such as those listed above.

22.6 INTERESTING RECENT DEVELOPMENTS IN GENERATIVE MODELING

Entering "How long is the coastline of Italy?" into OpenAI's ChatGPT3[7] results in the following response:

> "The length of the coastline of Italy depends on the scale at which it is measured, as the irregular shape of the coast means that the measurement will vary depending on the level of detail captured. According to the CIA World Factbook, Italy has a coastline that stretches for 7,600 km (4,722 miles) when measured at a scale that takes into account all the small bays, inlets, and islands along its shores. However, other estimates may provide slightly different figures depending on the method used to measure the coastline."

This response seems to be a significant improvement over what Google and Bing return since it encapsulates the coastline paradox. But it is worth studying this response further in order to shed light on potential near-term developments in GeoAI.

---

[7] https://openai.com/blog/chatgpt

While the ChatGPT response encapsulates the coastline paradox, it is not clear whether this establishes that the underlying large language model is spatially explicit. It seems to be spatially aware in that it not only provides spatially referenced information–the length of the coastline of Italy when measured at a particular scale– but it also acknowledges the importance of scale in geography. But is this *spatial explicit-ness*? What other kinds of inquiries could be developed through the new paradigm of "prompt engineering" that could serve as spatial Turing tests to establish that the GPT model is spatially explicit?

ChatGPT's response is problematic, however, along a number of dimensions which need addressing as work progresses on large generative models that are pre-trained in an unsupervised fashion. First are the issues of model explainability and data provenance. These large models are based on transformer networks which, like other deep-learning architectures, do not lend themselves to understanding how they produce their outputs. (That OpenAI has so far provided little technical detail about their GPT model further exacerbates this problem and points to the need for meaningful industry engagement, as argued in Section 22.4.) Beyond lacking explainability, these generative pre-trained models typically provide no provenance about their output even when it is seemingly factual. These models simply take an input (such as a text prompt) and provide an output. There is no looking-up or referencing of information in sources that could provide meta-data about the output. There is no search to find relevant documents that themselves might have provenance information. (In this sense, the results from ChatGPT are inferior to that of traditional Internet searches which usually provide links to sources.) The output is determined solely based on the large number of parameters (connection weights) that are learned during the unsupervised pre-training and supervised fine tuning. There is no way to trace back the specific training data that contributed to a particular output. Even if the training data has provenance information, there is no way to associate it with an output. (Even though ChatGPT's response above mentions the CIA World Factbook as a source, this could be completely fictional as is often the case with the chatbot's output.)

Finally, it must be understood that these models are *generative*. As mentioned above, there is no look-up, reference, retrieval, or other direct connection made to existing (read "real") information when producing an output. The output is in a sense thought-up or, to use a term from the generative AI community, hallucinated based solely on the interaction between the input and the model's connections. The models can also be stochastic in that the same input can result in different but generally consistent outputs. In the context of GeoAI, large generative pre-trained models have the potential for *generative geography*, that is, results that might seem plausible but are not real.

I will end with an anecdote concerning the pace at which AI is progressing. When I set out to write this chapter, I originally expected that ChatGPT would produce a definite answer similar to Google and Bing search when asked about the length of the coastline of Italy. It was to my surprise that its response encapsulates awareness of the coastline paradox. I did not expect this so soon from a model which is essentially just a large set of connections with weights learned in an unsupervised fashion from general knowledge corpora such as an archive of the Internet. Deep-learning models (connections with weights) have advanced over the last decade to outperform AI methods that were developed over the previous five decades. It appears that generative pre-trained models may have made similar progress, but in a matter of months and not years.

22.7 CLOSING REMARKS

In this chapter, I have laid out some of the specific challenges facing GeoAI such as making sure interdisciplinary research is appropriately rewarded especially in the academy and the need to engage industry particularly around ethics and privacy. I have also described some near- to medium-term opportunities such as using location as a key for the unsupervised training of deep learning models; developing multi-modal foundational models for GeoAI applications; and the approaching need for spatial Turing tests. Finally, I noted some very recent developments in generative models, specifically how they appear to be more spatially explicit than previous knowledge bases but also pointed out that such generative frameworks are fraught with issues around explainability, provenance, and hallucination. I'll end by saying that while it has been over 25 years since I first conducted GeoAI research as a Ph.D. student on the Alexandria Digital Library project at the University of California, Santa Barbara (although we didn't refer it as GeoAI then), I still find the nexus of fields in GeoAI as intellectually stimulating and enticing as I did then. It remains an exciting frontier.

BIBLIOGRAPHY

Goodchild, M., 2001. Issues in spatially explicit modeling. *In*: D.C. Parker, T. Berger and S.M. Manson, eds. *Agent-Based Models of Land-Use and Land-Cover Change Report and Review of an International Workshop*, October. 12–15.

Janowicz, K., *et al.*, 2020. GeoAI: spatially explicit artificial intelligence techniques for geographic knowledge discovery and beyond. *International Journal of Geographical Information Science*, 34 (4), 625–636.

Lunga, D., *et al.*, 2022. GeoAI at ACM SIGSPATIAL: The New Frontier of Geospatial Artificial Intelligence Research. *SIGSPATIAL Special*, 13 (3), 21–32.

Richarson, L.F., 1961. The problem of contiguity: An appendix to statistics of deadly quarrels. *General Systems Yearbook*, 6, 139–187.

Wang, W., *et al.*, 2022. Image as a Foreign Language: BEiT Pretraining for All Vision and Vision-Language Tasks. arXiv:2208.10442.

Index

A

abstraction 63
accessible 218
accountability 27, 408
ACM SIGSPATIAL 307
acoustic environment 354
activation function 47, 127
activation maximization 185
adaptation 29
adjacency matrix 55
administrative boundaries 271
aerial photography 76
agent population initialization 314
agent-based models 273, 306
agent-based simulation 306
agriculture 345
AI 10
AI guidelines 389
air pollution 124
air quality 351
AlexNet 4, 48, 81, 192
algorithm 30
analogy-based reasoning 414
analytical 124
analytical engines 371
ancient Greek 18
ancillary features 187
animal-related help 294
ANN 249
annotated data 219
anonymity 28
anonymization 29
anonymized data 376
anonymizing 392
anthropogenic 218
anthropogenic disasters 261
application programming interface (API) 224
ArcGIS Knowledge 422
ArcGIS Living Atlas 11
areal units 124
ARIMA 248

artifacts 180
artificial 23
artificial GIS analyst 40
artificial intelligence 4, 25, 26, 46
artificial life 18
artificial neural networks 4
assertion 34
assertions 414
assistance 288
association rule 34
assumption 27
assumptions 130
astronomers 77
atmospheric interference 333
attention 52
attention mechanism 337
attribution map 179
authoritative datasets 263
autoencoder 82
automata 18
automatic number plate recognition 248
automation 30
autonomous 24
autonomous GIS agents 423
autonomous vehicles 103

B

backpropagation 49, 80
BART 231
Bayesian deep learning 142
Bayesian Kriging 142
Bayesian methods 134
Bayesian modeling 338
behavioral response 311
benchmark 33
benchmark datasets 380
benefit 28
BERT 100, 231, 295
between-area prediction 202
bias 27
biased decisions 300

435

biases 47, 389
big data 49, 219
BigEarthNet 11
bilinear interpolation 131
biological 306
black boxes 178
black-box 27, 188
bottom-up approach 413
bottom-up method 266
bounding boxes 232
building block 223
building codes 288
building footprints 11
buildings 351, 352
built environment 269, 351
built-up areas 236

C

canopy 334
carpooling 252
Cartesian 105
cartographic 124
cartography 81
cascading effects 375
catastrophic events 288
categorical 129
causal inference 143
cell 48
cellular automata 379
census blocks 122
charge-coupled device 77
chatbot 432
chatbots 22
ChatGPT 4, 22, 37
check-ins 123
circuit diagram 50
cities 353
city-scale 353
cityscape 354
classification 48
climate change 330
climate change mitigation 28
climate crisis 261
climate migration 271
clinical 316
clustering-based spatial CV 211
CNN 48, 192
co-location 49
coast guards 290

code of conduct 262
cognitive science 26
color 220
color homogeneity 220
communities 288
community 265
complexity 91, 178
compliance 64
comprehensive insights 361
computation paradigm 123
computational environment 374
computational geography 10
computational performance 154
computer 30
computer vision 49, 100, 253
    concept recursive activation
        factorization 179
concept relevance propagation 179
conceptual sensitivity 185
confidence 179
confidentiality 28
conflict 261
congestion 249
    connected autonomous vehicles
        (CAVs) 252
connectivity 126
conservation 29
contact tracing 307
contagious 315
contemporary 227
context-aware devices 395
convolutional 48, 53
convolutional layers 335
convolutional neural network 48
convolutional operation 126
convolutional weights 130
correlation coefficients 122
correlogram 126
counterfactuals 143
covariance 126
covariate relationships 310
coverage 353
COVID-19 pandemic 306
crater detection 90
creative forces 407
Crete 17
critic 61
critical thinking 406
crop production 331

crop yield estimation 331
crop yields 331
cross-validate 179
cross-validation 201
crowd-sourced 263
crowd-sourcing 232
crowd-sourcing efforts 262
crowdsourced 239
crowdsourced data 353
crowdsourcing 352
cryptographic techniques 394
cultural heritage 262
culture 34
curriculum 27
CyberGIS-Compute 381
CycleGAN 58
cyclic connections 50

D

damage assessment 261
dasymetric mapping 268
data acquisition 279
data annotations 219
data centers 407
data culture 34
data generation processes 153
data integration 413
data management 375
data mining 219
data privacy 390
data quality 90, 299, 360
data representation 99
data science 26
data scientist 31
data sparsity 104
data streams 358
data volume 90
data-driven 178
data-driven epidemic forecasting 306
data-driven perspective 124
data-focused descriptive metadata 374
data-intensive 46
DCNN 49
de-anonymization 395
de-anonymizing 394
debias 27
debiasing 423
decentralized ecosystem 415
decision 27

decision-making 29
decoder 110
decomposition-based methods 178
deduction 414
deep fake 399
deep learning 4, 46
deep neural network 47
deep neural networks 4
DeepLabV3+ 354
democratization 281
DenseNet 81
deterministic function 105
deterministic methods 122
diagnostic 122
dietary preferences 330
differential equations 187
differential privacy 393
diffuse horizontal irradiance 131
digital elevation model 59
digital humanities 417
digital image processing 77
digital twins 356
digitization 261
dilemma 35
direct normal irradiance 131
Direct Relief 421
direction 105
disaster damages 290
disaster impacts 288, 289
disaster management 261
disaster management 288
disaster resilience 288
disaster response 288
disasters 288
discoverable 218
discrete grid cells 125
discrete lattice systems 122
discrimination 30
discriminator 57
disease leaves 191
disease spread 306
dissemination 29
distance 55
distance bands 127
distance decay 56, 122
distance-based weights 46
distribution shift 154
diverse community 409
diversity 29

DNN 345
document understanding 219
domain 100
domain knowledge 180
domain vocabulary 34
downscaling 131
downstream tasks 27
dynamic phenomena 351

E

earth 25
earth observation 82
EarthNets 11
earthquakes 271
edge devices 255
edges 54
education 29
electric vehicle 252
elevation 124
embedding 47
embedding space 103
embedding techniques 100
emergency relief 261
emergency rescue 300
emerging data source 353
empirical AI research 373
empirical sciences 26
encoder-decoder 52
encrypting 392
engineering 28
enlightenment 28
ensemble models 306
enthusiasm 25
entities 63
environment 29
environmental conditions 345
environmental disasters 264
environmental epidemiology 396
environmental footprint 330
environmental sensors 280
environments 61
epistemological 26
equation 110
estimation 122
ethical 25, 27, 389, 407
ethical principles 389
ethics 27, 389
ETL workflows 415
Euclidean space 112

evacuation 288, 291
evapotranspiration 187, 331
evidence-based decision-making 406
experiment reproducible 373
expert 180
expert systems 4
explainability 143
explainable AI 178
explainable artificial intelligence 178
explanation assessment 179
explicit statements 414
exposed 315
expressiveness 63
extensibility 63

F

facial recognition systems 406
FAIR 29, 64
FAIR principles 375
fairness 389
fairness-driven learning 154
faithfulness 179
fake geospatial data 399
falsifiable hypotheses 375
feature extraction 221
feature importance 143, 178
features 99
FEMA 290
fertilization 333
field observations 331
findability 29
fine-tune 53
first law of geography 59
first responders 290
flexibility 63
flood 187
food 330
food supply chains 418
foot-traffic 307
forced migration 261
forest 34
formalism 66
formulation 124
foundation models 34, 280, 415
foundations 26, 46
Fourier mapping 106
Fourier transformation 113
Fourth Paradigm 370
frontier 433

fully connected layers   48
functional types   137
fuzz logics   4

G

Gamma distribution   315
GAT   55
gated recurrent unit   51
Gaussian   110
Gaussian distribution   337
GCNN   249, 250
generalizability   152, 281, 370
generalization   142
generative   432
generator   57
genetic programming   4
geo-attribute-based spatial CV 211
geo-enrichment   422
geo-entity linking   225
geo-entity typing   240
geo-foundation models   32, 423
geo-tags   296
GeoAI   3, 26, 178
GEOBIA   79
geoethics   29
geographers   22
geographic coordinates   85
geographic distribution   103
geographic expeditions   76
geographic feature types   413
geographic features   223
geographic information   124
geographic information observatories
                               38
geographic information system 78
geographic knowledge discovery 122
geographic modeling   121
geographic questions   38
geographic space   122
geographic units   56
geographical   46
geographical disparities   297
geographical factors   187
geographical knowledge   195
geographical proximity   52
geographically-referenced data 369
geography   27
geohistorical information   218
GeoImageNet   11

GeoJSON   239
GeoKG   48
geolocation   232
geomasking   300, 392
geometries   224
geomorphic   76
geoparsing   33
geoprivacy   4
geoprivacy manifesto   391
georeferenced place   232
geoscience processes   178
geosciences   27
geosocial media   395
GeoSPARQL   65, 226, 420
geospatial   10
geospatial artificial intelligence 46, 178
geospatial big data   288
geospatial community   409
geospatial contexts   66
geospatial coordinate   152
geospatial entities   223
geospatial knowledge graph   65
geospatial knowledge-informed models
                              142
geospatial uncertainty   142
geostatistics   122
geotagged   49
geovisual   271
GIR   4
GIScience   4, 55
global   121
global explanations   181
GML   65
GNN   53
government agencies   389
governmental organizations   290
governments   265
GPS2Vec   104
GPT   429
gradient   47
gradient boosted regression trees 354
gradient-based methods   178
graph   53
graph attention   55
graph convolution   54
graph convolutional filters   134
graph convolutional neural networks
                              134
graph neural network   53

grayscale   76
grid representations   104
grid-based spatial CV   211
gridded population distribution   265
ground-truth data   227
GRU   51

H

hallucination   433
hardware   46
health insurance   391
heat stress   333
heterogeneity   122
heterogeneous   34, 153
heuristic search   4
hierarchical bi-partitioning process 157
hierarchical process   162
hierarchical representation   157
hierarchical structure   226
hierarchy-network synchronization 159
  high performance computing (HPC)
                    269
high-altitude   77
high-dimensional   48
High-resolution   266
high-resolution   266
historical   33
historical maps   218
homomorphic encryption   394
hotspots   83
house prices   354
household surveys   275
human activity   56
human behavior   28, 306
human comfort   355
human dynamics   262
human experience   408
human feedback   62
human geography   10
human impacts index   271
human intelligence   23
human interaction   29
human mobility   277, 306
human reactions   179
human responses   275
human rights   262
human settlement   236
human-annotated   219
human-centric   262, 359

human-centric simulations   356
human-environment interaction   28
human-guided development   407
human-in-the-loop   280
human-understandable   179
humanism   406
humanistic   407
humanistic GIS   406
humanistic principles   407
humanistic rewire   407
humanistic values   406
humanitarian action   277
humanitarian assistance   261
humanitarian crises   272
humanitarian help requests   294
humanitarian practice   261
humanitarian principles   262
humanitarian relief   27
humanity   262
hurricane   288
hybrid intelligent systems   4
hydrological   188
hyperparameters   47

I

ICICLE   66
IKONOS   78
image geolocalization   371
image processing   75
image segmentation   353
ImageNet   4, 48
immediate actions   288
impartiality   262
implicit neural representations   103
implicit statements   414
incentive systems   373
inclusive   407
independence   262
independent   124
  independent and identical distributed
        (IID)   110
  independent and identical distribution
        (IID)   152
independent research   370
independent variable   138
indoor mapping   103
industry engagement   432
infectious   315
infectious disease spread   306

inferential reasoning 414
information retrieval 63
information-theoretic 27
infrared thermography sensing 358
infrastructure damages 293
infrastructure malfunction 294
infrastructure mapping 261
infrastructure resilience 272
insights 179
instrumental variables 134
integrated gradients 184
integration 123
intelligence 23
intelligent agents 312
intelligent decision systems 252
intelligent spatial analysis 124
intelligent transportation systems 248
interdisciplinarity 428
interdisciplinary collaborations 307
international humanitarian law 262
internationalized resource identifiers
(IRIs) 412
interoperability 29
interpolation 122
interpret 179, 374
interpretability 143
interpretable 188
interpretable model 186
interpretation 28
invariance 114
inverse distance weighting (IDW) 122
irregular neighborhood 127
irrigation 333

J

JSON-LD 414
judgment 31
justice 28, 360

K

k-fold 202
k-nearest neighbors 55
kernel density estimation 104
kernels 48
knowledge bases 219
knowledge discovery 4, 288
knowledge engineering 413
knowledge graph 63
knowledge graph construction 415

knowledge graph embeddings 415
knowledge Graphs 412
knowledge graphs 5, 27, 46
knowledge representation 413, 415
KnowWhereGraph 419
kriging 122
Kriging interpolation 126

L

label 84
label correction algorithm 220
labeling 91
land cover 49
land surface 187
land use 261
land-cover 270
landform 82
Landsat 82, 218
LandScan 268
landscape 218
landscape features 289
lane-changing 252
large geographic models (LGMs) 430
large language model (LLM) 294
large language models 4, 39
large-scale 234
latent topics 316
layer-wise relevance propagation 143, 178
leaf area index (LAI) 335
learnable 130
learning-friendly 107
leave-one-out 202
LeNet-5 48
Levenshtein edit distance 231
LiDAR 334
linear object 236
link geographic features 219
link prediction 54, 415
linked data cloud 413
linked geospatial data 223
linked metadata 219, 232
LinkedGeoData 233
linking map text 219
liveable cities 361
LLMs 429
local 121
local connectivity 124
local explanations 181

local spatial relationships 126
local surrogate 182
localization 179
location 123
location encoders 106
location encoding 102
location privacy 389
location resolution 292
location variance 46
location-attentive predictions 356
location-based advertising 395
location-based information 371
longitudinal acquisition 351
loss function 47
LSTM 51, 249
LSTM-TrajGAN 107

M

machine learning 4, 27
machine translation 100
machine-readable 414
man-made features 142
management practices 333
manipulation 30
manned aircraft 330
map sheet 224
map text 228
Mapillary 356
mapKurator 232
MapProcessor 234
marginalized communities 407
Markov 60
Markov decision process 251
markup language 65
Mask R-CNN 49
masked autoencoders 100
masked entity prediction 242
masked language modeling 242
mathematics 28
MAUP 27
maximum likelihood estimation 134
mean squared error 131
measurements 129
mechanism 130
mechanistic models 306
medical 396
medical diagnoses 181
metadata 63, 219, 226
meteorological data 332

method reproducible 373
micromobility 254
misalignment 220
mobility 49, 66
model development 279
model evaluation 212
model-agnostic 183
model-free approaches 61
MODIS 331
MoveBank 11
multi-agent 252
multi-criteria Decision analysis 271
multi-head 52
multi-modal foundational models 430
multi-scale geographically weighted
regression 271
multigraph 412
multilayer perceptions 58
multimodal 63
multipolygons 112
multiscale 79
multispectral imagery 334
multispectral scanner 77

N

named entity recognition (NER) 415
National Renewable Energy Laboratory
131
natural 142
natural disaster 288
natural disasters 261
natural hazards 299
natural language processing 53, 219,
289
natural resources 330
NDVI 333
NDWI 333
near-infrared 333
neighborhood summarization 415
neighborhoods 24, 54
neighbors 127
Netlogo 276
networks 47
neural nets 80
neural networks 46
neural radiance fields 105
neuron 50
neutral 33
neutrality 27, 262

NLP 53
node classification 54
nodes 54
non-governmental organizations 290
non-maleficence 389
nonhuman animals 407
nonlinear model decisions 178

O

obfuscating 392
object detection and tracking 253
object recognition 104
objects 125
observation 26
observations 122
off-road imagery 358
OGC 112
ontologies 35, 63, 414
ontology 4, 413
Ontology Design Patterns (ODP) 421
Open Geospatial Consortium 112
Open Knowledge Network (OKN) 412
open-source 11
OpenStreetMap 220, 357
optical character recognition (OCR) 231
optimization 251
optimize 62
ordinary least squares 134
Ordnance Survey 218
organization 63
organizations 270
orientation angle 232
OSMNx 234
out-of-domain 228
out-of-sample prediction 382
outdoor comfort 355
outliers 180
overfitting 152
oversight mechanisms 279
OWL 64

P

paradigm 123
parameterization 35
partial derivatives 183
participants 29
partition 153
pedestrians 254
people 63

Perceptron 80
performance-driven learning 154
performance-wise symmetric solutions 155
personal data 389
personal location information 392
personally identifiable information 391
perturbation-based 183
phenology 333
philosophical 26
philosophy 23, 28
photogrammetry 76
photographic 75
physical geography 10
physical processes 77
physiological data 355
pixel-level 181
pixels 220
pixels-of-interest 221
place 28
placially 27
planet 34
plant species 191
platial 46
POI recommender systems 415
point cloud 103
PointNet 104
points 47
points of interest 52
POIs 52, 137
policy guidelines 305
policymakers 331
polygon encoding 102
polygons 47
polyline encoding 102
polylines 47
pooling 48
population boom 330
population displacement 261
population groups 358
population mapping 261
post-OCR 231
PostGIS 224
PostgreSQL 224
poverty 34
practices 29
practitioners 28
pre-trained language model 219
precipitation 187, 333

precision agriculture   27
precision medicine   396
preclinical   315
prediction   59
preparedness   261
prescriptive analytics   320
preventative behaviors   305
prevention   29
principles   29
prior knowledge   90
priori knowledge   178
privacy   358, 389
privacy protection   299
privacy-aware   396
private companies   389
probability modeling   338
probability-based methods   129
process-based   335
  process-focused contextual metadata
                                        374
prompt engineering   432
propagation   184
property graphs   414
provenance   357, 374
provenance documentation   374
public health   306
public sector   279
public transit   248
public-sharing   392

Q

Q-Learning   252
QGIS   229
qualitative spatial relations   65
quality criteria   179
quarantine   318
Queen adjacency   55
question answering   100
question-answering systems   415

R

railroad   223
rainfall-runoff   187
random cross-validation   202
random CV   211
random forest   272
randomization   179
randomly splits   202
raster   47, 125

RDF   64, 226
read comprehension   100
real-time   289
recommender systems   35, 423
recovery   261
recreate   374
recurrent neural networks   50
regional   121
regional science   122
regularization terms   171
reinforcement learning   60, 251
relevance score   182
remote sensing   49, 192
repeated testing   371
replicability   370
replicable   370
replications   370
representation learning   34, 99
representation schema   123
representative   27
representativeness   27
reproducibility   370
reproducible   370
reproductions   370
rescue operations   288
research compendium   378
research ethics   28
research infrastructure   384
research questions   9
residential   122
resilient communities   288
ResNet   49, 81
resolution   32
resource allocation   252
  Resource Description Framework
    (RDF)   414
resources   11
response phase   288
responsibility   28, 389
retrieval   59
return beam vidicon   77
reusability   29
reverse-geocoding   225
reward   60
reward function   61
right for the right reasons   180
risk   28
risk analysis   262
risk assessment   262

risk mitigation 261
riverscapes 356
RNN 50, 249
road infrastructure 351
road vector 236
robustness 30, 179
rock types 187
Roofpedia 352
rooftop 352
rooftop utilization rate 353
Rook adjacency 55
routes 249

S

sampled locations 122
sampling bias 377
satellite imagery 60, 331, 353
scale 78
scanned historical maps 219
schema 34, 413
science 28
scientific discovery 46
scientific exploration 370
scientific inquiry 370
scientific investigation 370
second law of geography 59
self-supervised learning 280
semantic interoperability 414
semantic segmentation 49, 219
semantic web 219
semantics 4, 414
semi-variogram 122
sensitive 391
sensor data 290
sensor-equipped vehicles 351
sensors 103
sentient beings 407
seq2seq 108
sequence 50
sequence model 108
SHAP 183
shared weights 124
shelter 294
short-range 125
short-term events 261
short-term forecasts 249
shortest path 36
sigmoid 51
similarity 56

situational awareness 418
sliced Wasserstein distance 131
sliding windows 48
smart agriculture 330
smart cities 354
social distancing 315
social interactions 306
social media 49, 280
social media data 289
social sensing 358
socially responsible 408
socioeconomic 28, 262
socioeconomic characteristics 289
socioeconomic factors 122
socioeconomic indicators 280
soil 187
soil dryness 331
soil moisture 292
soil-crop-atmosphere processes 331
solar 131
solar radiation 131
SOSA ontology 420
soybean yield estimation 338
SpaBERT 241
space 34
Space as a Distribution of Partitionings
(SPAD) 165
space-aware loss 46
space-partitioning 154
space-time 249
space-time scan statistic 307
SPARQL 64, 414
spatial aggregation 125
spatial and temporal scope 413
spatial anisotropies 380
spatial autocorrelation 78, 133, 152
spatial computation 226
spatial computing 307
spatial concepts 46
spatial constraints 56
spatial contiguity 133
spatial continuity 124
spatial cross-validation 202
spatial CV 202
spatial data 152
spatial data distribution 126
spatial data imputation 59
spatial data science 26, 39
spatial data types 102

spatial decision-making 62
spatial dependence 27, 122
spatial dependencies 377
spatial dependency 53
spatial domain 122
spatial Durbin model 133
spatial econometrics 122, 133
spatial effects 122
spatial embedding 47
spatial estimations 129
spatial fairness 155
spatial granularities 310
spatial heterogeneity 153
spatial hierarchy 158
spatial interaction 56
spatial interpolation 122
spatial leave-one-out CV 211
spatial metadata 376
spatial optimization 62
spatial partition 154
spatial patterns 122
spatial phenomena 370
spatial prediction 122
spatial principles 49
spatial privacy 398
spatial process 122
spatial reasoning 22, 50
spatial regression 122
  spatial regression convolutional neural
        networks 135
spatial relatedness 49
spatial relationships 51, 122
spatial representation 124
spatial representation learning 47, 100
spatial scene 49
spatial structure 46, 123
spatial thinking 56
spatial transformation 162
Spatial Turing Tests 431
spatial variations 122
spatial weights 134
spatial-aware 289
spatial-temporal 53
spatial-temporal analysis 293
spatial-temporal cloaking 392
spatial-temporal modeling 288
spatial-temporal patterns 301
spatially 27
spatially embedded graphs 101

spatially explicit 4, 46
spatially explicit AI 101
spatially extensive 125
spatially intensive 124
spatially-split validation 203
spatio-temporal network 251
spatio-temporal unit 190
spatiotemporal 18, 56
Spatiotemporal AI 249
spatiotemporal behavior 305
spatiotemporal correlation 249
spatiotemporal data 306
spatiotemporal linked data 219
specification 122
spectral bands 331
spectral resolution 331
spectral-based 250
split 201
spotting text content 219
SQL 64
SRGCNNs 135
SRL 100
stakeholders 300
standardization 64
state-of-the-art 122
statements 413
statistical 77
statistical methods 122
statistical principles 123
statistics 122
stay-at-home 311
street view 49
street-level imagery 352
streets 122
subclinical 315
subsymbolic 4
summarization 423
super-resolution 131
supply chains 28
support vector regression (SVR) 334
surveillance 397
susceptible 315
sustainability 27
  sustainable agricultural management
        330
sustainable agriculture 330
sustainable development 262, 352
  sustainable development goals (SDGs)
        261

symbolic	4
symptoms	318
synthetic	57
synthetic data	393
synthetic datasets	228
synthetic map	229
synthetic population dataset	274
system	221
systematic	77
systems of equations	187

T

T-GCN	56
Talos	17
tanh	51
target variable	123
task-specific	178
taxi	190
taxonomy	178
technology	30
temperature	124, 187
temporal dependency	56
temporal pattern	57
temporal resolution	131
temporally	27
text annotations	228
text spotting pipeline	228
thermal units	333
third law of geography	59
time	34
time series	181
time-series	59
timestamps	50
tokens	103
tomographic reconstruction	77
top-down approach	266
top-down engineering	414
topic modeling	316
topographic maps	219
topology	38, 111
toponym recognition	296, 415
toponym resolution	415
tourism	351
traffic conditions	249
traffic data	249
traffic flows	249
traffic forecasting	248, 249
traffic monitoring	253
traffic networks	249

traffic signaling systems	255
traffic trajectories	181
training	34
training-test split	377
trajectories	248
trajectory	52
trajectory analysis	415
trajectory embedding	52
trajectory generation	57
transdisciplinary	273
transductive	134
transfer learning	100
transferability	142
transformative power	409
Transformer	52
transparency	27, 300, 389, 398
transport	187
transport planners	249
transportation	248
transportation networks	262
transportation planning	371
travel behavior	276
travel time	249
TreeExplainer	186
triangular irregular networks	122
triangulated irregular networks	100
triple stores	414
triplet	50
trips	254
trust	423
trustworthiness	300
Turing test	40

U

U-Net	49
U.S. Geological Survey	76
UAVs	330
uncertainty	143
uncertainty analysis	332
unconditional	58
United Nations	262
unmanned aerial vehicle (UAV)	290
unobserved locations	123
unrepresentative	33
unsupervised	131
unsupervised pre-training	430
unsupervised training	100
urban areas	236, 351
urban digital twins	356

urban environment 356
urban environments 249
urban flooding 418
urban geography 371
urban informatics 360
urban management 351
urban planning 28
urban sensing 351
urban soundscapes 354
urban transit systems 251
URI 64
USDA 339
user-centered approach 408
user-generated content 33
USGS 218

V

validation 202
valuable 391
value proposition 416
variable importance analysis 336
variable relationships 122
variables 122
variational autoencoder 109
variogram 126, 129
vector 47
vector data 223
vector-to-raster 220
vegetation indices 332
vehicle dispatch 252
vehicle routing 252
vehicle-to-everything 254
videos 253
vision transformer 100
visual art 22
visual features 48
visualize 178
volumetric representations 104

volunteered geographic information
  (VGI) 269
vulnerability assessments 261
vulnerable 396
vulnerable populations 265

W

W3C 64, 226
walkability 38
Wasserstein GAN 58
water management 331
waterfronts 356
waterscape 356
weak replicability 382
wearable device 355
weather 293
Web Ontology Language (OWL) 413
weights 47
wildlife 407
wind 131
within-area prediction 202
WKT 65, 226
word embeddings 100
Word2Vec 100
WorldPop 268

X

XAI 178

Y

yield estimation 345
yield estimation models 334
YOLOv3 83

Z

zones 252
zoning regulations 288

For Product Safety Concerns and Information please contact our EU
representative GPSR@taylorandfrancis.com
Taylor & Francis Verlag GmbH, Kaufingerstraße 24, 80331 München, Germany

www.ingramcontent.com/pod-product-compliance
Lightning Source LLC
Chambersburg PA
CBHW060742220326
41598CB00022B/2302

* 9 7 8 1 0 3 2 3 1 1 6 7 8 *